# Fracture and Fatigue Control in Structures

# Fracture and Fatigue Control in Structures

## Applications of Fracture Mechanics

SECOND EDITION

### John M. Barsom

*Senior Consultant*
*United States Steel Corporation*

### Stanley T. Rolfe

*Ross H. Forney Professor of Engineering*
*University of Kansas*

Prentice-Hall, Inc., *Englewood Cliffs, New Jersey 07632*

*Library of Congress Cataloging-in-Publication Data*

BARSOM, JOHN M.
    Fracture and fatigue control in structures.

    (Prentice-Hall international series in civil
engineering and engineering mechanics)
    Rev. ed. of: Fracture and fatigue control in struc-
tures / Stanley T. Rolfe, John M. Barsom.  1977.
    Includes bibliographical references and index.
    1. Fracture mechanics.  2. Metals—Fatigue.
I. Rolfe, S. T. (Stanley Theodore) (date).
II. Rolfe, S. T. (Stanley Theodore) (date).  Fracture
and fatigue control in structures.  III. Title.
IV. Series.
TA409.B37  1987        620.1'126        86–16870
ISBN  0–13–329863–9

Editorial/production supervision and
  interior design: *Joan L. Stone*
Cover design: Lundgren Graphics, Ltd.
Manufacturing buyer: *Rhett Conklin*

Prentice-Hall International Series in Civil Engineering
and Engineering Mechanics
William J. Hall, Editor

© 1987, 1977 by Prentice-Hall, Inc.
A Division of Simon & Schuster
Englewood Cliffs, New Jersey 07632

Printed in the United States of America

10  9  8  7  6  5  4  3  2  1

ISBN  0-13-329863-9  025

PRENTICE-HALL INTERNATIONAL (UK) LIMITED, *London*
PRENTICE-HALL OF AUSTRALIA PTY. LIMITED, *Sydney*
PRENTICE-HALL CANADA INC., *Toronto*
PRENTICE-HALL HISPANOAMERICANA, S.A., *Mexico*
PRENTICE-HALL OF INDIA PRIVATE LIMITED, *New Delhi*
PRENTICE-HALL OF JAPAN, INC., *Tokyo*
PRENTICE-HALL OF SOUTHEAST ASIA PTE. LTD., *Singapore*
EDITORA PRENTICE-HALL DO BRASIL, LTDA., *Rio de Janeiro*

*To*
Valentina and Phyllis

# Contents

*chapter 3*

# Experimental Determination of $K_{Ic}$ and Other $K_c$ Values, *64*

## chapter 4

## Effect of Temperature, Loading Rate, and Plate Thickness on Fracture Toughness, 109

## chapter 5

## Correlations Between Fracture Mechanics and Other Fracture-Toughness Test Results, 159

*chapter 17*

# Elastic-Plastic Fracture Mechanics, *542*

## *appendix A*
## Problems, *593*

## Index, *619*

# Foreword

In his well-known text on "Mathematical Theory of Elasticity," Love inserted brief discussions of several topics of engineering importance for which linear elastic treatment appeared inadequate. One of these topics was rupture. Love noted that various safety factors, ranging from 6 to 12 and based upon ultimate tensile strength, were in common use. He commented that "the conditions of rupture are but vaguely understood." The first edition of Love's treatise was published in 1892. Fifty years later, structural materials had been improved with a corresponding decrease in the size of safety factors. Although Love's comment was still applicable in terms of engineering practice in 1946, it is possible to see in retrospect that most of the ideas needed to formulate the mechanics of fracturing on a sound basis were available. The basic content of modern fracture mechanics was developed in the 1946 to 1966 period. Serious fracture problems supplied adequate motivation and the development effort was natural to that time of intensive technological progress.

Mainly what was needed was a simplifying viewpoint, progressive crack extension, along with recognition of the fact that real structures contain discontinuities. Some discontinuities are prior cracks and others develop into cracks with applications of stress. The general idea is as follows. Suppose a structural component breaks after some general plastic yielding. Clearly a failure of this kind could be traced to a design error which caused inadequate section strength or to the application of an overload. The fracture failures which were difficult to understand are those which occur in a rather brittle manner at stress levels no larger than were expected when the structure was designed. Fractures of this second kind, in a special way, are also due to overloads. If one considers the stress redistribution around a pre-existing crack subjected to tension, it is clear that the region adjacent to the perimeter of the crack is overloaded due to the severe stress concentration and that local plastic strains must occur. If the toughness is limited, the plastic strains at the crack border may be accompanied by crack extension. However, from similitude, the crack border overload increases with crack size. Thus progressive crack extension tends to be self stimulating. Given a prior crack, and a material of limited toughness, the possibility for development of rapid fracturing prior to general yielding is therefore evident.

Analytical fracture mechanics provides methods for characterizing the "overload" at the leading edge of a crack. Experimental fracture mechanics collects information of practical importance relative to fracture toughness, fatigue cracking, and corrosion cracking. By centering attention on the active region involved in progressive fracturing, the collected laboratory data are in a form which can be transferred to the leading edge of a crack in a structural component. Use of fracture mechanics analysis and data has explained many service fracture failures with a satisfactory degree of quantitative accuracy. By studying the possibilities for such fractures in advance, effective fracture control plans have been developed.

Currently the most important task is educational. It must be granted that all aspects of fracture control are not yet understood. However, the information now available is basic, widely applicable, and should be integrated into courses of instruction in strength of materials. The special value of this book is the emphasis on practical use of available information. The basic concepts of fracture mechanics are presented in a direct and simple manner. The descriptions of test methods are clear with regard to the essential experimental details and are accompanied by pertinent illustrative data. The discussions of fracture control are well-balanced. Readers will learn that fracture control with real structures is not a simple task. This should be expected and pertains to other aspects of real structures in equal degree. The book provides helpful fracture control suggestions and a sound viewpoint. Beyond this the engineer must deal with actual problems with such resources as are needed. The adage "experience is the best teacher" does not seem to be altered by the publication of books. However, the present book by two highly respected experts in applications of fracture mechanics provides the required background training. Clearly the book serves its intended purpose and will be of lasting value.

*George R. Irwin*

University of Maryland
College Park, Maryland

# Preface

The field of fracture mechanics has become the primary approach to controlling brittle fracture and fatigue failures in structures. This book introduces the field of fracture mechanics from an application viewpoint. Since the first publication of this book in 1977, the field of fracture mechanics has grown significantly. Almost all specifications for fracture and fatigue control now either use fracture mechanics directly or are based on concepts of fracture mechanics. In this book, we emphasize applications of fracture mechanics to the fields of fracture and fatigue control in structures. We believe that the book will serve as an introduction to the field of fracture mechanics for seniors or beginning graduate students, but more importantly, it introduces the practicing engineer to a field that has become increasingly important. In recent years, structural failures and the desire for increased safety and reliability of structures have led to the development of various fracture criteria for many types of structures, including bridges, airplanes, pipelines, and nuclear pressure vessels.

The development of fracture-control plans for new and unusual structures such as offshore drilling rigs, nuclear power plants, space shuttles, etc., has become more widespread. Each of the topics of fracture criteria and fracture control is developed from an engineering viewpoint, including economic and practical considerations. The textbook should assist engineers to become aware of the fundamentals of fracture mechanics, and, in particular, of their responsibility in controlling brittle fracture and fatigue failures in structures.

Chapter 1 serves as an overview of the problem of fracture and fatigue in structures as well as an introduction to the field of fracture mechanics. Chapter 2 provides the theoretical development of stress-intensity factors, $K_I$, and Chapter 3 describes the test methods for obtaining critical stress intensity factors, $K_{Ic}$, $K_{Ic}$ ($t$) for intermediate loading rates, or $K_{Id}$ for impact loading rates. Chapter 4 describes the effect of temperature, loading rate, and plate thickness on the fracture toughness of a wide variety of structural materials, primarily structural steels. Because many structural materials have fracture toughness levels outside the range of linear-elastic fracture mechanics at service temperatures and loading rates, correlations with other more common notch toughness tests are widely used, and these are described in Chapter 5. Chapter 6 describes the relationship between stress, flaw size, and material toughness, with specific design examples. These first six chapters provide the fundamental basis of linear-elastic fracture mechanics and the principles for applications of fracture mechanics.

Chapters 7 through 13 deal with sub-critical crack initiation and growth by fatigue, stress corrosion, or corrosion fatigue. It is in these areas that linear-elastic fracture mechanics has perhaps had its greatest field of application. Chapter 7 introduces the field of fatigue; Chapter 8 describes a method of analyzing fatigue-crack initiation from a blunt-notch using fracture mechanics terminology. Chapter 9 describes one of the most widespread uses of fracture mechanics, namely fatigue-crack propagation under constant-amplitude load fluctuation. Chapter 10 describes fatigue-crack propagation under variable-amplitude load fluctuation and the use of the root-mean-square $\Delta K$ value, $\Delta K_{RMS}$.

Chapters 11, 12, and 13 introduce the effects of environments on subcritical crack initiation and growth, namely stress corrosion cracking and corrosion fatigue initiation and propagation. Chapter 14 describes the fatigue behavior of weldments.

Chapter 15 introduces the various factors affecting fracture criteria and presents examples of existing fracture criteria currently in use. Chapter 16 develops the principles of fracture-control and explains existing fracture-control plans for nuclear pressure vessels and bridges.

Chapter 17 introduces the field of elastic-plastic fracture-mechanics as analyzed by crack-opening displacement (CTOD), $R$-curve, or $J$-integral.

Based on the student's background and course hours, an introductory course might consist of Chapters 1 through 7, with the more advanced topics covered in Chapters 8 through 17. After reading Chapters 1, 2, and 7, the practicing engineer who is generally familiar with behavior of structural materials could move to any chapter of particular interest.

The authors wish to express their appreciation to their respective organizations, U.S. Steel Corporation and The University of Kansas, for their support in preparing this revision. Most importantly, however, we wish to acknowledge the support of our many colleagues, both within and outside our organizations, who have contributed to the development of this book, as well as the continued encouragement and support of our families.

*John Barsom*
*Stan Rolfe*

# Fracture and Fatigue Control
# in Structures

# 1

# Overview of the Problem of Fracture and Fatigue in Structures

## 1.1. Historical Background

Although the total number of structures that have failed by brittle fracture*
is low, brittle fractures have occurred and do occur in structures. The
following limited historical review is not meant to be complete but only to
illustrate the fact that brittle fractures can occur in engineering structures
such as tanks, pressure vessels, ships, bridges, airplanes, and the like.

Shank[1] and Parker[2] have reviewed many structural failures, beginning
in the late 1800s when members of the British Iron and Steel Institute
reported the mysterious cracking of steel in a brittle manner. In 1886, a
250-ft-high standpipe in Gravesend, Long Island, failed by brittle fracture
during its hydrostatic acceptance test. During this same period, other catastrophic brittle failures of riveted structures such as gas holders, water
tanks, and oil tanks were reported even though the materials used in these
structures had met all existing tensile and ductility requirements.

One of the most famous tank failures was that of the Boston molasses
tank, which failed in January 1919 while it contained 2,300,000 gal of molasses.
Twelve persons were drowned in molasses or died of injuries, 40 others
were injured, and several horses were drowned. Houses were damaged,
and a portion of the Boston Elevated Railway structure was knocked over.
An extensive lawsuit followed, and many well-known engineers and scientists
were called to testify. After years of testimony, the court-appointed auditor

---

* Brittle fracture is a type of catastrophic failure in structural materials that usually
occurs without prior plastic deformation and at extremely high speeds (as high as 7000 ft/sec).
The fracture is usually characterized by a flat fracture surface (cleavage) with little or no
shear lips, as shown in Figure 1.1, and at average stress levels below those of general yielding.
Brittle fractures are not so common as fatigue, yielding, or buckling failures, but when they
do occur, they may be more costly in terms of human life and/or property damage.

1

**Figure 1.1**   Photograph of typical brittle-fracture surface.

handed down the decision that the tank failed by overstress. In commenting on the conflicting technical testimony, the auditor stated in his decision, "amid this swirl of polemical scientific waters, it is not strange that the auditor has at times felt that the only rock to which he could safely cling was the obvious fact that at least one half of the scientists must be wrong . . . ." His statement fairly well summarized the state of knowledge among engineers regarding the phenomenon of brittle fracture. At times, it seems that the statement is still true today.

Prior to World War II, several welded vierendeel truss bridges in Europe failed shortly after being put into service. All the bridges were lightly loaded, the temperatures were low, the failures were sudden, and the fractures were brittle. Results of a thorough investigation indicated that most failures were initiated in welds and that many welds were defective (discontinuities were present). The Charpy impact test results showed that most steels were brittle at the service temperature.

However, in spite of these and other brittle failures, it was not until the large number of World War II ship failures that the problem of brittle fracture was fully appreciated by the engineering profession. Of the approximately 5000 merchant ships built during World War II, over 1000 had developed cracks of considerable size by 1946. Between 1942 and 1952, more than 200 ships had sustained fractures classified as serious, and at least 9 T-2 tankers and 7 Liberty ships had broken completely in two as a result of brittle fractures. The majority of fractures in the Liberty ships started at square hatch corners or square cutouts at the top of the sheer strake. Design changes involving rounding and strengthening of the hatch corners, removing square cutouts in the sheer strake, and adding riveted crack arresters in various locations led to immediate reductions in the incidence of failures.[3,4]

Most of the fractures in the T-2 tankers originated in defects in the bottom-shell butt welds. The use of crack arresters and improved work quality reduced the incidence of failures in these vessels.

Studies indicated that in addition to design faults steel quality also was a primary factor that contributed to brittle fracture in welded ship hulls.[5]

Therefore, in 1947, the American Bureau of Shipping introduced restrictions on the chemical composition of steels, and in 1949, Lloyds Register stated that "when the main structure of a ship is intended to be wholly or partially welded, the committee may require parts of primary structural importance to be steel, the properties and process of manufacture of which have been specially approved for this purpose."[6]

In spite of design improvements, the increased use of crack arresters, improvements in quality of work, and restrictions on the chemical composition of ship steels during the late 1940s, brittle fractures still occurred in ships in the early 1950s.[2] Between 1951 and 1953, two comparatively new all-welded cargo ships and a transversely framed welded tanker broke in two. In the winter of 1954, a longitudinally framed welded tanker constructed of improved steel quality using up-to-date concepts of good design and welding quality broke in two.[7]

Since the late 1950s (although the actual number has been low) brittle fractures still have occurred in ships as is indicated by Boyd's description of ten such failures between 1960 and 1965 and a number of unpublished reports of brittle fractures in welded ships since 1965.[8]

The brittle fracture of the 584-ft-long Tank Barge I.O.S. 3301 in 1972,[9] in which the 1-yr-old vessel suddenly broke almost completely in half while in port with calm seas (Figure 1.2), shows that this type of failure continues to be a problem.

In this particular failure, the material had adequate notch toughness as measured by one method of testing (Charpy V-notch) and marginal toughness as measured by another more severe method of testing (dynamic tear). However, the primary cause of failure was established to be an unusually high loading stress caused by improper ballasting. This failure illustrated the fact that human factors can contribute to brittle fractures in structures, including overloads.

In the mid-1950s two Comet aircraft failed catastrophically while at high altitudes.[10] An exhaustive investigation indicated that the failures initiated from very small fatigue cracks originating from rivet holes near openings in the fuselage. Numerous other failures of aircraft landing gear and rocket motor cases have occurred from undetected defects or from subcritical crack growth either by fatigue or stress corrosion. The failures of F-111

**Figure 1.2**   Photograph of I.O.S. 3301 barge failure.

aircraft were attributed to brittle fractures of members with preexisting flaws. Also in the 1950s, several failures of steam turbines and generator rotors occurred, leading to extensive brittle-fracture studies by manufacturers and users of this equipment.

In 1962, the Kings Bridge in Melbourne failed by brittle fracture at a temperature of 40°F. Poor details and fabrication resulted in cracks which were nearly through the flange *prior to* any service loading. Although this bridge failure was studied extensively (and other bridges had failed previously by brittle fracture), the bridge-building industry did not pay particular attention to the possibility of brittle fractures in bridges until the failure of the Point Pleasant Bridge at Point Pleasant, West Virginia. On December 15, 1967, this bridge collapsed without warning, resulting in the loss of 46 lives. Photographs of an identical eyebar suspension bridge before the collapse and of the Point Pleasant Bridge after the collapse are shown in Figures 1.3 and 1.4.

An extensive investigation of the collapse was conducted by the National Transportation Safety Board (NTSB),[11] and its conclusion was "that the cause of the bridge collapse was the cleavage fracture in the lower limb of the eye of eyebar 330 at joint C13N of the north eyebar suspension chain in the Ohio side span." Because the failure was unique in several ways, numerous investigations of the failure were made.

Extensive use of fracture mechanics was made by Bennett and Mindlin[12] in their metallurgical investigation and they concluded that

**Figure 1.3**    Photograph of St. Mary's Bridge similar to the Point Pleasant Bridge.

**Figure 1.4**    Photograph of Point Pleasant Bridge after collapse.

1. "The fracture in the lower limb of the eye of eyebar 330 was caused by the growth of a flaw to a critical size for fracture under normal working stress.

2. The initial flaw was due to stress-corrosion cracking from the surface of the hole in the eye. There is some evidence that hydrogen sulfide was the reagent responsible for the stress-corrosion cracking. The final report indicates that the initial flaw was due to fatigue, stress-corrosion cracking and/or corrosion fatigue.[11]

3. The composition and heat treatment of the eyebar produced a steel with very low fracture toughness at the failure temperature.

4. The fracture resulted from a combination of factors; in the absence of any of these, it probably would not have occurred: (a) the high hardness of the steel which rendered it susceptible to stress-corrosion cracking; (b) the close spacing of the components in the joint which made it impossible to apply paint to the most highly stressed region of the eye, yet provided a crevice in this region where water could collect; (c) the high design load of the eyebar chain, which resulted in a local stress at the inside of the eye greater than the yield strength of the steel; and (d) the low fracture toughness of the steel which permitted the initiation of complete fracture from the slowly propagating stress-corrosion crack when it had reached a depth of only 0.12 in. (3.0 mm) (Figure 1.5)."

**Figure 1.5** Photograph showing origin of failure in Point Pleasant Bridge.

Since the time of the Point Pleasant Bridge failure, other brittle fractures have occurred in steel bridges and other types of structures as a result of unsatisfactory fabrication methods, design details, or material properties.[13,14] Fisher[15] has described numerous fractures in a text on case studies.

These and other brittle fractures led to an increasing concern about the possibility of brittle fractures in steel bridges and resulted in the AASHTO (American Association of State Highway and Transportation Officials) Material Toughness Requirements being adopted for bridge steels. (A complete description of these requirements is presented in Chapter 16.)

Other industries have developed or are developing fracture-control plans for Arctic construction, offshore drilling rigs, and more specific applications such as the space shuttle.

Fracture mechanics has shown that because of the *interrelation* among *materials, design, fabrication,* and *loading,* brittle fractures cannot be eliminated in structures merely by using materials with improved notch toughness. The designer still has fundamental responsibility for the overall safety and reliability of his or her structure. It is the objective of this book to describe the fracture, fatigue, and stress-corrosion behavior of structural materials and to show how fracture mechanics can be used in design to *prevent* brittle fractures and fatigue failures of engineering structures.

As will be described throughout this textbook, the science of *fracture mechanics* can be used to describe *quantitatively* the trade-offs among these three factors (stress, material toughness, and flaw size) so that the designer can determine the relative importance of each of them during *design* rather than during *failure analysis*.

## 1.2. Notch-Toughness Testing

In addition to the traditional mechanical property tests that measure strength, ductility, modulus of elasticity, and so on, many tests measure some form of notch toughness. *Notch toughness* is defined as the ability of a material to absorb energy (usually when loaded dynamically) in the presence of a flaw, whereas *toughness* of a material is defined as the ability of a smooth member (unnotched) to absorb energy, usually when loaded slowly. Notch toughness is measured with a variety of specimens such as the Charpy V-notch impact specimen, dynamic tear test specimen, $K_{Ic}$, crack-tip opening displacement, $R$-curve, and so on, while toughness is usually characterized by the area under a stress-strain curve in a slow tension test. It is the presence of a notch or some other form of stress raiser that makes structural materials susceptible to brittle fracture under certain conditions.

Traditionally, the notch-toughness characteristics of low- and intermediate-strength steels have been described in terms of the transition from brittle to ductile behavior as measured by various types of impact tests.

Most structural steels can fail in either a ductile or brittle manner depending on several conditions such as temperature, loading rate, and constraint. Ductile fractures are generally preceded by large amounts of plastic deformation and usually occur at 45° to the direction of the applied stress. Brittle or cleavage fractures generally occur with little plastic deformation and are usually normal to the direction of the principal stress. The transition from one type of fracture behavior to the other generally occurs with changes in service conditions such as the state of stress, temperature, or strain rate.

This transition in fracture behavior can be related schematically to various fracture states, as shown in Figure 1.6. Plane-strain behavior refers to fracture under elastic stresses with little or no shear-lip development, and it is essentially brittle. Plastic behavior refers to ductile failure under general yielding conditions accompanied usually, but not necessarily, with large shear lips. The transition between these two extremes is the elastic-plastic region, which is also referred to as the mixed-mode region.

For static loading, the transition region occurs at lower temperatures than for impact (or dynamic) loading. Thus, for structures subjected to static loading, the static transition curve should be used to predict the level of performance at the service temperature. For structures subjected to impact or dynamic loading, the impact transition curve should be used to predict the level of performance at the service temperature.

For structures subjected to some intermediate loading rate, an intermediate loading-rate transition curve should be used to predict the level of performance at the service temperature. Because the actual loading rates for many structures are not well defined, the impact loading curve (Figure

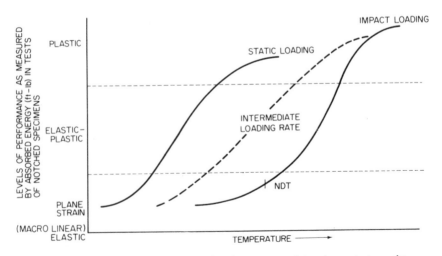

**Figure 1.6**  Schematic showing relation between notch-toughness test results and levels of structural performance for various loading rates.

1.6) often is used to predict the service performance of structures even though the actual loading may be slow or intermediate. This practice is somewhat conservative and helps to explain why many structures that have low notch toughness as measured by impact tests have not failed even though their service temperatures are well below an impact transition temperature. As noted in Figure 1.6, a particular notch-toughness value called the nil-ductility transition (NDT) temperature generally defines the upper limits of plane-strain behavior under conditions of impact loading.

One of the fundamental questions to be resolved regarding the interpretation of any particular toughness test for large structures is as follows: What level of material performance should be required for satisfactory performance in a particular structure? That is, as shown schematically in Figure 1.7 for impact loading, one of the following three general levels of material performance could be established at the service temperature for a structural material:

1. Plane-strain behavior—steel 1.
2. Elastic-plastic behavior—steel 2.
3. Fully plastic behavior—steel 3.

Although fully plastic behavior would be a very desirable level of performance for structural materials, it may not be necessary or even economically feasible for many structures. That is, for a large number of structures, a reasonable level of elastic-plastic behavior (steel 2, Figure 1.7)

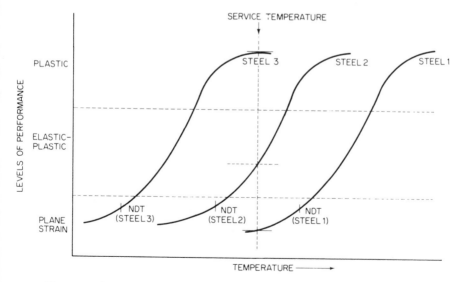

**Figure 1.7**   Schematic showing relation between level of performance and transition temperature for three arbitrary steels.

is often satisfactory to prevent initiation of brittle fractures provided the design and fabrication are satisfactory. A more detailed description of fracture criteria and levels of performance is provided in Chapter 15.

Not all structural materials exhibit a brittle-ductile transition. For example, some of the very-high-strength structural steels or other structural materials such as aluminum or titanium do not undergo the brittle-ductile transition shown in Figure 1.6. For these materials, temperature has a rather small effect on toughness, as shown in Figure 1.8 for a 250-ksi yield-strength steel.

The general purpose of the various kinds of notch-toughness tests is to model the behavior of actual structures so that the laboratory test results can be used to predict service performance. In this sense many different tests have been used to measure the notch toughness of structural materials. These include Charpy V-notch (CVN) impact, drop weight NDT, dynamic tear (DT), wide plate, Battelle drop weight tear test (DWTT), crack-tip opening displacement (CTOD), as well as many others. A complete description of most of these tests can be found in References 1–7. Generally these notch-toughness tests were developed for specific purposes. For example, the CVN test is widely used as a screening test in alloy development as well as a fabrication and quality control test. In addition, because of correlations with service experience, the CVN test is often used in steel specifications for various structural, marine, and pressure-vessel applications. The NDT test is used to establish the minimum service temperature for various naval and marine applications, whereas the Battelle DWTT test was developed to measure the fracture appearance of line pipe steels as a function of temperature.

All these notch-toughness tests generally have one thing in common, however, and that is to produce fracture in steels under carefully controlled laboratory conditions. It is hoped that the results of the test can be correlated with service performance to establish levels of performance, as shown in Figure 1.6, for various materials being considered for specific applications. In fact, the results of the foregoing notch-toughness tests have been extremely useful in many structural applications.

However, even if correlations are developed for existing structures, they do not necessarily hold for certain designs, new operating conditions, or new materials because the results, which are expressed in terms of energy, fracture appearance, or deformation, cannot always be translated into structural design and engineering parameters such as stress and flaw size. Thus, a much better way to measure notch toughness is with principles of fracture mechanics. Fracture mechanics is a method of characterizing the fracture behavior in structural parameters that can be used directly by the engineer, namely, stress and flaw size. Fracture mechanics is based on a stress analysis as described in Chapter 2 and thus does not depend

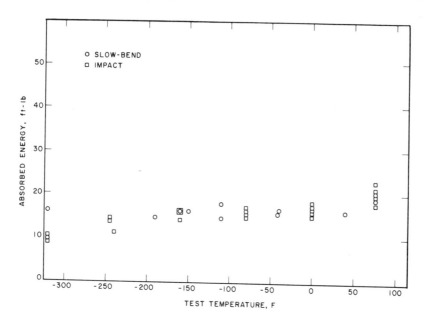

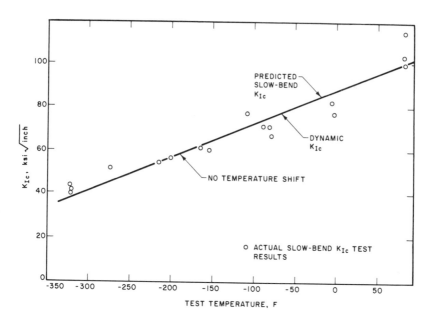

**Figure 1.8**  CVN and $K_{Ic}$ test results for 250 ksi yield strength steel showing no effect of loading rate and a minor effect of temperature.

on the use of extensive service experience to translate laboratory results into practical design information so long as the engineer can determine the material toughness, nominal stress, and flaw size in a particular structural member.

Fracture mechanics can account for the effect of temperature and loading rate on the behavior of structural members that have sharp cracks. It is becoming recognized that most large complex structures have discontinuities of some kind. Thus, the results of a fracture-mechanics analysis for a particular application (specimen size, service temperature, and loading rate) will yield the combinations of stress level and flaw size that would be required to cause fracture. The engineer can then *quantitatively* establish allowable stress levels and inspection requirements so that fractures cannot occur. In addition, fracture mechanics can be used to analyze the growth of small cracks (for example, by fatigue loading or stress corrosion) to critical size. Therefore, fracture mechanics has several very definite advantages compared with traditional notch-toughness tests and finally offers the designer a very definitive method of quantitatively designing to prevent brittle fracture in structures.

This is not to imply that the traditional notch-toughness tests are not still useful. In fact, as discussed extensively in Chapter 5, there are many empirical correlations between fracture-mechanics values and existing toughness test results such as the Charpy V-notch, dynamic tear, NDT, CTOD, and so on that are extremely useful to the engineer. In many cases, because of the current limitations on test requirements for measuring $K_{Ic}$ (Chapter 3), existing notch-toughness tests must be used to help the designer estimate $K_{Ic}$ values for a particular material. These estimates and correlations are described in Chapter 5.

## 1.3. Brittle-Fracture Design Considerations

In addition to the catastrophic failures described in Section 1.1, there have been *numerous* "minor" failures of structures during construction or service that have resulted in delays, repairs, and inconveniences, some of which are very expensive. Nonetheless, compared with the total number of engineering structures that have been built throughout the world, the number of catastrophic brittle fractures has been very small. As a result, the designer seldom concerns himself or herself with the notch toughness of structural materials because the failure rate of most structures is very low. Nonetheless,

1. When designs become more complex,
2. When the use of high-strength thick welded plates becomes more common compared with the use of lower-strength thin riveted plates,
3. When the choice of construction practices becomes more dependent on minimum cost,

4. When the magnitude of loadings increases, and

5. When actual factors of safety decrease because of more precise computer designs,

the possibility of brittle fractures in large complex structures must be considered, and the designer must become more aware of available methods to prevent brittle failures.

The state of the art *is* that fracture mechanics concepts *are* available that can be used in the design of structures to prevent brittle fractures.

Design codes often include this fact, and in the early 1970s, several design and materials specifications *based on concepts of fracture mechanics* were adopted by various engineering professions. These include

1. ASME Boiler and Pressure Vessel Code Section III—Nuclear Power Plant Components, Appendix G—Protection Against Nonductile Failure.[16]

2. American Association of State Highway and Transportation Officials—Notch Toughness Requirements for Bridge Steels.[17]

3. Air Force Aircraft Designed to Fracture Mechanics Criteria in Addition to the Usual Static Load Limits.[18]

A detailed discussion of various specifications or design procedures used in various types of structures is presented in Chapter 16.

The traditional design approach for most structures is generally based on the use of safety factors to limit the maximum calculated stress level to some percentage of either the yield or ultimate stress. It is suggested by Weck[19] that this approach is somewhat outdated.

The factor of safety approach, by itself, does not always give the proper assurance of safety with respect to brittle fracture because large complex structures are not fabricated without some kind of discontinuities. Numerous failure investigations and inspections have shown this to be true. Research by Fisher and Yen[20] has shown that discontinuities exist in practically all structural members, either from manufacture or from the process of fabricating the members by rolling, machining, punching, or welding. The sizes of these discontinuities range from very small microdiscontinuities (<0.01 in.) to several inches long.

In almost every brittle fracture that has occurred in structures, some type of discontinuity was present. These included very small arc strikes in some of the World War II ship failures, very small fatigue cracks (~0.07 in.) in the Comet airplane failures, and stress-corrosion or corrosion-fatigue cracks (0.12 in.) in the critical eyebar of the Point Pleasant Bridge. Other failures, such as the 260-in.-diameter missile motor case that failed during hydrotest or the F-111 aircraft failure, had somewhat larger, but still undetected, cracks.

Bravenec[21] has reviewed various brittle fractures during fabrication and testing and has shown that cracks have originated from torch-cut edges, mechanical gouges, corrosion pits, weld repairs, severe stress concentrations, and the like. Dolan[22] has made the flat statement that *"every structure contains small flaws whose size and distribution are dependent upon the material and its processing. These may range from nonmetallic inclusions and microvoids to weld defects, grinding cracks, quench cracks, surface laps, etc."*

The significant point is that discontinuities or cracks *are* present in many large fabricated structures even though the structure may have been "inspected." Methods of inspection or nondestructive testing are gradually improving, with the result that smaller and smaller discontinuities are becoming detectable. But the fact is that discontinuities are present regardless of whether or not they are discovered. In fact, the problem of establishing acceptable levels of discontinuities in welds is becoming somewhat of an economic problem since techniques that minimize the size and distribution of discontinuities are available if the engineer chooses to use them.

Weck[19] has stated that "Some authorities—in the API-ASME Code for instance—have produced porosity charts and rules for permissible sizes of other defects which, in the complete absence of any factual or experimental basis, must have been the result of divine inspiration of the code makers. What is good enough for the job cannot be established by divine inspiration but only by patient experimental research. Such researches have amply demonstrated that for many jobs far less than absolute perfection is adequate. Whether a given defect is permissible or not depends on the extent to which the defect increases the risk of failure of the structure. It is quite clear that this will vary with the type of structure, its service conditions and the material from which it is constructed."

As described in the following sections and the remainder of this book, fracture mechanics is the best available science that can correctly account *quantitatively* for the factors that influence the true factor of safety or degree of reliability of a structure. Fracture mechanics is not only a quantitative research tool and method of failure analysis, but it has been and should be used in structural design to help determine acceptable stress levels, acceptable discontinuity sizes, and desired material properties for specific service conditions.

## 1.4. Introduction to Fracture Mechanics

An overwhelming amount of research on brittle fracture in structures of all types has shown that numerous factors (e.g., service temperature, material toughness, design, welding, residual stresses, fatigue, constraint, etc.) can contribute to brittle fractures in large welded structures.[23-30] However, the development of fracture mechanics[31-37] has shown that there are three *primary* factors that control the susceptibility of a structure to brittle fracture:

**1. Material toughness ($K_c$, $K_{Ic}$, $K_{Id}$).** Material toughness can be defined as the ability to carry load or deform plastically in the presence of a notch and can be described in terms of the critical stress-intensity factor under conditions of plane stress ($K_c$) or plane strain ($K_{Ic}$) for slow loading and linear-elastic behavior. $K_{Id}$ is a measure of the critical material toughness under conditions of maximum constraint (plane strain) and impact or dynamic loading, also for linear-elastic behavior. (These terms are described more completely in Chapters 2 and 3.) For elastic-plastic behavior (materials with higher levels of notch toughness than linear-elastic behavior), the material toughness is measured in terms of parameters such as $R$-curve resistance, $J_{Ic}$, and CTOD as described in Chapter 17. In addition to metallurgical factors such as composition and heat treatment, the notch toughness of a steel also depends on the application temperature, loading rate, and constraint (state of stress) ahead of the notch, as described in Chapter 4.

**2. Crack size (a).** Brittle fractures initiate from discontinuities of various kinds. These discontinuities can vary from extremely small cracks within a weld arc strike (as was the case in the brittle fracture of a T-2 tanker during World War II) to much larger weld or fatigue cracks. Complex welded structures are not fabricated without discontinuities (porosity, lack of fusion, toe cracks, mismatch, etc.), although good fabrication practice and inspection can minimize the original size and number of these discontinuities. Thus, these discontinuities will be present in many welded structures even after all inspections and weld repairs are finished. Furthermore even though only "small" discontinuities may be present initially, these discontinuities can grow by fatigue or stress corrosion, possibly to a critical size.

**3. Stress level ($\sigma$).** Tensile stresses (nominal, residual, or both) are necessary for brittle fractures to occur. These stresses are determined by conventional stress analysis techniques for particular structures.

These three factors generally are the primary ones that control the susceptibility of a structure to brittle fracture. However, it is possible for brittle fractures to occur without all three factors being present if the other factors are sufficiently severe. The failure of the I.O.S. 3301 Tank Barge (Figure 1.2) is an example of a brittle fracture that occurred at a severe structural detail at a high stress level but with no preexisting crack. Other factors such as temperature, loading rate, stress concentrations, residual stresses, and so on will affect the foregoing three *primary* factors.

Engineers have known these facts for many years and have reduced the susceptibility of structures to brittle fractures by controlling the three factors in their structures *qualitatively*. That is, good design (e.g., adequate sections, minimum stress concentrations) and fabrication practices (decreased discontinuity size because of proper welding control and inspection), as well as the use of materials with good notch-toughness levels (e.g., as measured with a Charpy V-notch impact test), will minimize and have minimized the probability of brittle fractures in structures. However, the engineer has not had specified design guidelines to evaluate the relative performance and economic trade-offs among design, fabrication, and materials in a *quantitative* manner.

The emergence of fracture mechanics as an applied science has shown that all three of the above primary factors can be interrelated to predict (or to design against) the susceptibility of various structures to brittle fracture. Fracture mechanics is a method of characterizing fracture behavior in structural parameters familiar to the engineer, namely, stress and crack size. Linear-elastic fracture-mechanics technology is based on an analytical procedure that relates the stress-field magnitude and distribution in the vicinity of a crack tip to the nominal stress applied to the structure; to the size, shape, and orientation of the crack or crack-like discontinuity; and to the material properties. In Figure 1.9 are the equations that describe the elastic-stress field in the vicinity of a crack tip in a body subjected to tensile stresses normal to the plane of the crack (Mode I deformation, as will be described in Chapter 2). The stress-field equations show that the distribution of the elastic-stress field in the vicinity of the crack tip is invariant in all

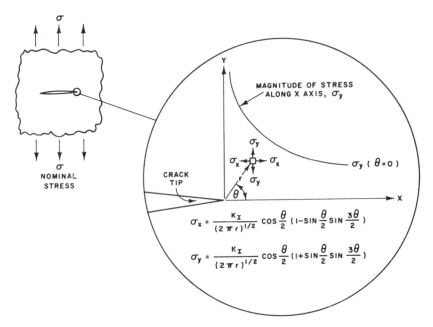

**Figure 1.9**   Elastic-stress-field distribution ahead of a crack.

structural components subjected to this type of deformation and that the magnitude of the elastic-stress field can be described by a single parameter, $K_I$, designated the stress-intensity factor. Consequently, the applied stress, the crack shape, size, and orientation, and the structural configuration associated with structural components subjected to this type of deformation affect the value of the stress-intensity factor but do not alter the stress-field distribution. Thus it is possible to translate laboratory results into practical design information without the use of extensive service experience or correlations. Relationships between the stress-intensity factor and various body configurations, crack sizes, shapes, orientations, and loading conditions are presented in Chapter 2; however, examples of some of the more widely used stress–flaw-size relations are presented in Figure 1.10.

One of the underlying principles of fracture mechanics is that unstable fracture occurs when the stress-intensity factor at the crack tip reaches a critical value, $K_c$. For Mode I deformation and for small crack-tip plastic deformation (plane-strain conditions), the critical-stress-intensity factor for

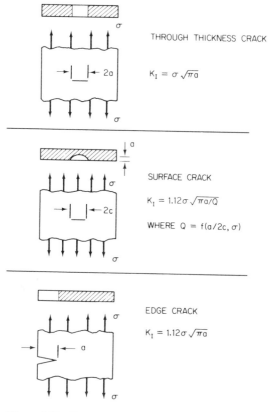

THROUGH THICKNESS CRACK

$$K_I = \sigma \sqrt{\pi a}$$

SURFACE CRACK

$$K_I = 1.12\sigma \sqrt{\pi a/Q}$$

WHERE $Q = f(a/2c, \sigma)$

EDGE CRACK

$$K_I = 1.12\sigma \sqrt{\pi a}$$

**Figure 1.10**   $K_I$ values for various crack geometries.

fracture instability is designated $K_{Ic}$. $K_{Ic}$ represents the inherent ability of a material to withstand a given stress-field intensity at the tip of a crack and to resist progressive tensile crack extension under plane-strain conditions. Thus, $K_{Ic}$ represents the fracture toughness of the material and has units of $ksi\sqrt{in.}$ ($MN/m^{3/2}$). However, this material-toughness property depends on the particular material, loading rate, and constraint as follows:

$K_c$ = critical stress-intensity factor for static loading and plane-stress conditions of variable constraint. Thus, this value depends on specimen thickness and geometry, as well as on crack size.

$K_{Ic}$ = critical stress-intensity factor for static loading and plane-strain conditions of maximum constraint. Thus, this value is a minimum value for thick plates.

$K_{Ic}$ $(t)$ = critical stress-intensity factor for rapid-loading rates (intermediate between slow and impact) and plane-strain condition of maximum constraint. The time to failure, generally about $t$ second, should be indicated in the parentheses.

$K_{Id}$ = critical stress-intensity factor for dynamic (impact) loading and plane-strain conditions of maximum constraint,

where $K_c$, $K_{Ic}$, $K_{Ic}$ $(t)$ or $K_{Id} = C\sigma\sqrt{a}$.

$C$ = constant, function of specimen and crack geometry (Chapter 2).

$\sigma$ = nominal stress, ksi ($MN/m^2$).

$a$ = flaw size, in. (mm).

Each of these values is also a function of temperature, particularly for those structural materials exhibiting a transition from brittle to ductile behavior.

By knowing the critical value of $K_I$ at failure ($K_c$, $K_{Ic}$, or $K_{Id}$) for a given material of a particular thickness and at a specific temperature and loading rate, the designer can determine flaw sizes that can be tolerated in structural members for a given design stress level, as described in Chapter 6. Conversely, the designer can determine the design stress level that can be safely used for an existing crack that may be present in a structure.

This general relationship among material toughness ($K_c$), nominal stress ($\sigma$), and crack size ($a$) is shown schematically in Figure 1.11. If a particular combination of stress and crack size in a structure ($K_I$) reaches the $K_c$ level, fracture can occur. Thus there are *many* combinations of stress and flaw size (e.g., $\sigma_f$ and $a_f$) which may cause fracture in a structure that is fabricated from a steel having a particular value of $K_c$ at a particular service temperature, loading rate, and plate thickness. Conversely, there are *many* combinations of stress and flaw size (e.g., $\sigma_o$ and $a_o$) that will *not* cause failure of a particular structural material.

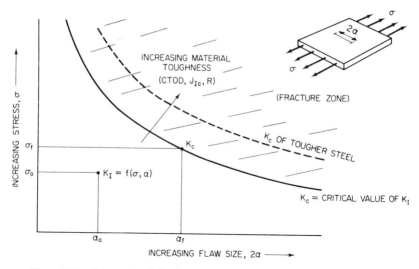

**Figure 1.11** Schematic relation between stress, flaw size, and material toughness.

### 1.4.1. Example

As an introductory numerical example of the design application of fracture mechanics, consider the equation relating $K_I$ to the applied stress and flaw size for a through-thickness crack in a wide plate, that is, $K_I = \sigma\sqrt{\pi a}$ (Figure 1.10). Assume that laboratory test results show that for a particular structural steel with a yield strength of 80 ksi (552 MPa), the $K_c$ is 60 ksi$\sqrt{\text{in.}}$ (66 MPa · m$^{1/2}$) at the service temperature, loading rate, and plate thickness used in service. Also assume that the design stress is 20 ksi (138 MPa). Substituting $K_I = K_c = 60$ ksi$\sqrt{\text{in.}}$ (66 MPa · m$^{1/2}$) into the appropriate equation in Figure 1.12 results in $2a = 5.7$ in. (145 mm). Thus, for these conditions, the tolerable flaw size would be about 5.7 in. (145 mm). For a design stress of 45 ksi (310 MPa), the same material could tolerate a flaw size, $2a$, of only about 1.1 in. (27.9 mm). If residual stresses such as may be due to welding are present so that the total stress in the vicinity of a crack is 80 ksi (552 MPa), the tolerable flaw size is reduced considerably. Note from Figure 1.12 that if a tougher steel is used, for example, one with a $K_c$ of 120 ksi$\sqrt{\text{in.}}$ (132 MPa · m$^{1/2}$), the tolerable flaw sizes at all stress levels are significantly increased. If the fracture toughness of a steel is sufficiently high, brittle fractures will not occur, and failures under tensile loading can occur only by general plastic yielding, similar to the failure of a tension test specimen. Fortunately, most structural steels have a high enough level of fracture toughness at service temperatures and loading rates such that brittle fractures rarely occur.

A useful analogy for the designer is the relation among applied load ($P$), nominal stress ($\sigma$), and yield stress ($\sigma_{ys}$) in an unflawed structural

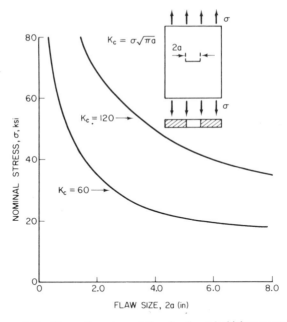

**Figure 1.12** Stress–flaw-size relation for through-thickness crack.

member and among applied load ($P$), stress intensity ($K_I$), and critical stress intensity for fracture ($K_c$, $K_{Ic}$, or $K_{Id}$) in a structural member with a flaw. In an unflawed structural member, as the load is increased the nominal stress increases until an instability (yielding at $\sigma_{ys}$) occurs. As the load is increased in a structural member with a flaw (or as the size of the flaw grows by fatigue or stress corrosion), the stress intensity, $K_I$, increases until an instability (fracture at $K_c$, $K_{Ic}$, $K_{Id}$) occurs. Thus the $K_I$ level in a structure with flaws should always be kept below the appropriate $K_c$ value in the same manner that the nominal design stress ($\sigma$) is kept below the yield strength ($\sigma_{ys}$) in structures without flaws. This is the fundamental approach to fracture control, and it is discussed in detail in Chapter 16.

Another analogy that may be useful in understanding the fundamental aspects of fracture mechanics is the comparison with the Euler column instability (Figure 1.13). The stress level required to cause instability in a column (buckling) decreases as the $L/r$ ratio increases. Similarly, the stress level required to cause instability (fracture) in a flawed tension member decreases as the flaw size ($a$) increases. As the stress level in either case approaches the yield strength, both the Euler analysis and the $K_c$ analysis are invalidated because of yielding. To prevent buckling, the actual stress for a given ($L/r$) value must be below the Euler curve. To prevent fracture, the actual stress for a given flaw size, $a$, must be below the $K_c$ curve. Obviously, using a material with a high level of notch toughness [e.g., a

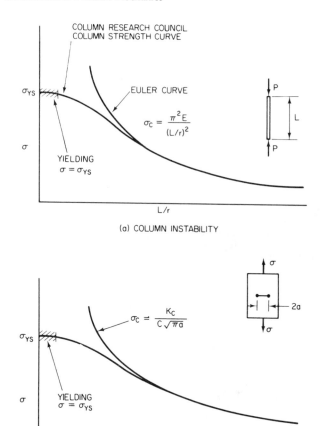

**Figure 1.13**  Column instability and crack instability (after Madison and Irwin, Ref. 37).

$K_c$ level of 120 ksi$\sqrt{\text{in.}}$ (132 MPa · m$^{1/2}$) compared with 60 ksi$\sqrt{\text{in.}}$ (66 MPa · m$^{1/2}$) in Figure 1.12] will increase the possible combinations of design stress and flaw size that a structure can tolerate without fracturing.

The critical-stress-intensity factor, $K_{Ic}$, represents the terminal conditions in the life of a structural component. The total useful life of the component is determined by the time necessary to initiate a crack and to propagate the crack from subcritical dimensions to the critical size, $a_c$. Crack initiation and subcritical crack propagation may be caused by cyclic stresses in the absence of an aggressive environment (Chapters 8, 9, and 10), by an aggressive environment under sustained load (Chapter 11), or by the combined effects of cyclic stresses and an aggressive environment (Chapters 12 and 13). Because all these modes of subcritical crack propagation are localized phenomena that depend on the boundary conditions at the crack tip, it is logical to expect the rate of subcritical crack propagation to depend on the stress-intensity factor, $K_I$, which serves as a single-term parameter representative of the stress conditions in the vicinity of the crack tip. Thus fracture mechanics theory can be used to analyze the behavior of a structure throughout its entire life.

Many low- to medium-strength structural materials in the section sizes of interest for most large structures are of insufficient thickness to maintain plane-strain conditions under slow loading and at normal service temperatures. For these cases, the linear-elastic analysis used to calculate $K_{Ic}$ values is invalidated by elastic-plastic behavior and the formation of large plastic zones. Under these conditions, which occur in the transition range between plane-strain and fully plastic behavior, analyses other than linear-elastic fracture mechanics (LEFM) must be used. The most promising extensions of LEFM into plane-stress as well as elastic-plastic fracture mechanics are the following:

1. *R-curve analysis.* A procedure used to characterize the resistance to fracture of materials during incremental slow-stable crack extension, $K_R$. At instability, $K_R = K_c$, the plane-stress fracture toughness which is dependent upon specimen thickness and geometry, as well as temperature and loading rate.

2. *Crack-tip opening displacement* (CTOD). A measure of the prefracture deformation at the tip of a sharp crack under conditions of inelastic behavior.

3. *J-integral.* Path-independent integral which is an average measure of the elastic-plastic stress-strain field ahead of a crack. For elastic conditions, $J_{Ic} = K_{Ic}^2/E(1 - \mu^2)$.

4. *Tearing modulus.* A material parameter, $T$, that is proportional to the slope of the $J-\Delta a$ curve during stable crack extension by "tearing," $T \simeq dJ/da \cdot E/\sigma_0^2$.

The first three methods, which are standardized, are described in Chapter 17.


## 1.5. Fatigue and Stress-Corrosion Crack Growth

Conventional procedures that are used to design structural components subjected to fluctuating loads provide the engineer with a design fatigue curve which characterizes the basic unnotched fatigue properties of the material and a fatigue-strength-reduction factor. The fatigue-strength-reduction factor incorporates the effects of all the different parameters characteristic of the specific structural component that make it more susceptible to fatigue failure than the unnotched specimen. The design fatigue curves are based on the prediction of cyclic life from data on nominal stress (or strain) versus elapsed cycles to failure (*S–N* curves) as determined from laboratory test specimens. Such data represent both the number of cycles required to initiate a crack in the specimen and the number of cycles required to propagate the crack from a subcritical size to a critical dimension which often varies from laboratory to laboratory. The dimension of the critical crack required to cause "failure" in the fatigue specimen also depends on the magnitude of the applied stress and on the test specimen size. Alternatively, fatigue specimens that incorporate the actual geometry, welding, and so on can be tested to obtain an *S–N* type curve directly.

Figure 1.14 is a schematic *S–N* curve divided into an initiation component and a propagation component. The number of cycles corresponding to the endurance limit represents initiation life primarily, whereas the number of cycles expended in crack initiation at a high value of applied stress is negligible. Consequently, *S–N* type data do not necessarily provide information regarding safe-life predictions in structural components (particularly in components having surface irregularities different from those of the test specimens) and in components containing crack-like discontinuities because

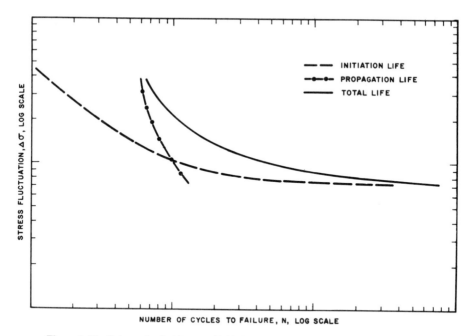

**Figure 1.14**    Schematic *S–N* curve divided into "initiation" and "propagation" components.

the existence of surface irregularities and crack-like discontinuities reduces and may eliminate the crack-initiation portion of the fatigue life of structural components.

Many attempts have been made to characterize the fatigue behavior of metals. The results of some of these attempts have proved invaluable in the evaluation and prediction of the fatigue strength of structural components, and an introduction to fatigue is presented in Chapter 7. However, these fatigue-strength evaluation procedures are subject to many limitations, caused primarily by the failure to distinguish adequately between fatigue-crack initiation and fatigue-crack propagation.

As described in Chapter 8, fracture mechanics can be used to determine the initiation life of fatigue cracks from various stress-concentration factors. In Chapters 9 and 10, we describe the use of fracture mechanics to evaluate the behavior of propagating cracks.

Thus, although $S-N$ curves have been widely used to analyze the fatigue behavior of steels and weldments, closer inspection of the overall fatigue process in complex welded structures indicates that a more rational analysis of fatigue behavior may be possible by using concepts of fracture mechanics. Specifically, small (possibly large) fabrication discontinuities are invariably present in welded structures, even though the structure has been "inspected" and "all injurious flaws removed" according to some specifications. Accordingly, a conservative approach to designing to prevent fatigue failure is to assume the presence of an initial flaw and analyze the fatigue-crack-growth behavior of the structural member. The size of the initial flaw is obviously highly dependent on the quality of fabrication and inspection. However, such an analysis would minimize the need for expensive fatigue testing for many different types of structural details.

A schematic diagram showing the general relation between fatigue-crack initiation and propagation is given in Figure 1.15. The question of when does a crack "initiate" to become a "propagating" crack is somewhat philosophical and depends on the level of observation of a crack, that is, crystal imperfection, dislocation, microcrack, and lack of penetration, for example. The fracture-mechanics approach to fatigue is to assume an initial imperfection on the basis of the quality of fabrication or inspection and then to calculate the number of cycles required to initiate a sharp crack from that imperfection and then to grow that crack to the critical size for brittle fracture. Obviously if the initial imperfection is very sharp, the initiation life can be very short. Using this approach, inspection requirements can be established logically.

In addition to subcritical crack growth by fatigue, small cracks also can grow by stress corrosion during the life of structures. Although crack growth by either fatigue or stress corrosion does not represent catastrophic failure for structures fabricated from materials having reasonable levels of notch toughness, in both mechanisms small cracks can become large enough

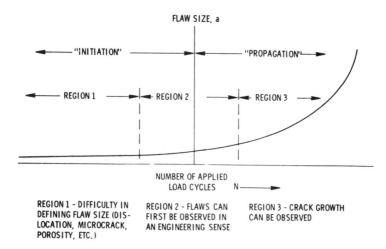

Figure 1.15 Schematic showing relation between "initiation" life and "propagation" life.

to cause a decrease in structural efficiency and can thereby require repairs. Furthermore the possibility of both mechanisms operating at once by corrosion fatigue also exists. Thus, a knowledge of the fatigue, corrosion-fatigue, and stress-corrosion behavior of materials is required to establish an overall fracture-control plan that includes inspection requirements.

By testing precracked specimens under static loads in specific environments (such as salt water) and analyzing the results according to fracture-mechanics concepts, a $K_I$ value can be determined, below which subcritical crack propagation does not occur. This threshold value is called $K_{Iscc}$. The $K_{Iscc}$ value for a particular material and environment is the plane-strain stress-intensity threshold that describes the value below which subcritical cracks (scc) will not propagate and also has units of ksi$\sqrt{\text{in}}$.

For materials that are susceptible to crack growth in a particular environment, the $K_{Iscc}$ value is used as the failure criterion rather than $K_{Ic}$. Thus, in the schematic relation among material toughness, design stress, and flaw size shown in Figure 1.11, $K_{Iscc}$ would replace $K_c$ as the critical value of $K_I$. This procedure is described in Chapter 11. The combined effects of corrosion and fatigue are described in Chapter 12—Corrosion-Fatigue-Crack Initiation—and in Chapter 13—Corrosion-Fatigue-Crack Propagation.

## 1.6. Fracture Criteria

Criteria selection, i.e., the determination of how much toughness is necessary as well as how much the designer is willing to pay to obtain materials with superior toughness, is one of the most important parts of developing an adequate fracture-control plan and should be based on a very careful study of the particular requirements for a structure. Some of the factors involved in the development of a criterion are

1. A knowledge of the service conditions (temperature, loading, loading rate, etc.) to which the structure will be subjected.
2. The desired level of performance in the structure (plane-strain, elastic-plastic, or plastic).
3. The consequences of structural failure.

Examples of criteria that have been developed for structural applications are the AASHTO material-toughness requirements and the PVRC recommendations for ferritic materials for nuclear pressure vessels. Fracture criteria are discussed in Chapter 15.

## 1.7. Fracture-Control Plans

Most engineering structures in existence are performing safely and reliably. The comparatively few service failures in structures indicate that present-day practices governing material properties, design, and fabrication procedures are generally satisfactory. However, the occurrence of infrequent failures indicates that further understanding and possible modifications in present-day practices are needed. The identification of the specific modifications needed requires a thorough study of material properties, design, fabrication, inspection, erection, and service conditions.

Structures are fabricated in various sizes and are subjected to numerous service conditions. Thus, it is very difficult to develop a set of rules that would ensure the safety and reliability of all structures when each structure involves a unique set of operating conditions. The use of data obtained by testing laboratory specimens to predict the behavior of complex structural components may result in approximations or in excessively conservative estimates of the life of such components but does not guarantee a correct prediction of the fracture or fatigue behavior. The safety and reliability of structures and the correct prediction of their overall resistance to failure by fracture or fatigue can be approximated best by using a fracture-control plan. A fracture-control plan is a detailed procedure used

1. To identify all the factors that may contribute to the fracture of a structural detail or to the failure of the entire structure.
2. To assess the contribution of each factor and the synergistic contribution of these factors to the fracture process.
3. To determine the relative efficiency and trade-off of various methods to minimize the probability of fracture.

4. To assign responsibility for each task that must be undertaken to ensure the safety and reliability of the structure.

The development of a fracture-control plan for complex structures is very difficult. Despite the difficulties, attempts to formulate a fracture-control plan for a given application, even if only partly successful, should result in a better understanding of the fracture characteristics of the structure under consideration.

A fracture-control *plan* is a procedure tailored for a given application and cannot be extended indiscriminately to other applications. However, general fracture-control *guidelines* that pertain to classes of structures (such as bridges, ships, pressure vessels, etc.) can be formulated for consideration in the development of a fracture-control plan for a particular structure within any particular class of structures.

The correspondence among fracture-control plans based on crack initiation, crack propagation, and fracture toughness of materials can be readily demonstrated by using fracture-mechanics concepts. The fact that crack initiation, crack propagation, and fracture toughness are functions of the stress-intensity fluctuation, $\Delta K_I$, and of the critical-stress-intensity factor, $K_{Ic}$—which are in turn related to the applied nominal stress (or stress fluctuation)—demonstrates that a fracture-control plan for various structural applications depends on

1. The fracture toughness, $K_{Ic}$ (or $K_c$), of the material at the temperature and loading rate representative of the intended application. The fracture toughness can be modified by changing the material used in the structure.
2. The applied stress, loading rate, stress concentration, and stress fluctuation, which can be altered by design changes, loading changes, and by proper fabrication.
3. The initial size of the discontinuity and the size and shape of the critical crack, which can be controlled by design changes, fabrication, and inspection.

The total useful life of structural components is determined by the time necessary to initiate a crack and to propagate the crack from subcritical dimensions to the critical size. The life of the component can be prolonged by extending the crack-initiation life and the subcritical-crack-propagation life. Consequently, crack initiation, subcritical crack propagation, and fracture characteristics of structural materials are primary considerations in the formulation of fracture-control guidelines for structures.

Fracture-control guidelines and fracture-control plans for various classes of structures are presented in Chapter 16.

# References

1. M. E. Shank, "A Critical Review of Brittle Failure in Carbon Plate Steel Structures Other than Ships," *Ship Structure Committee Report, Serial No. SSC-65*, National Academy of Sciences–National Research Council, Washington, D.C., Dec. 1, 1953 (also reprinted as *Welding Research Council Bulletin*, No. 17).

2. E. R. Parker, *Brittle Behavior of Engineering Structures*, prepared for the Ship Structure Committee under the general direction of the Committee on Ship Steel–National Academy of Sciences–National Research Council, John Wiley, New York, 1957.

3. D. B. Bannerman and R. T. Young, "Some Improvements Resulting from Studies of Welded Ship Failures," *Welding Journal, 25*, No. 3, Mar. 1946.

4. H. G. Acker, "Review of Welded Ship Failures," *Ship Structure Committee Report, Serial No. SSC-63*, National Academy of Sciences–National Research Council, Washington, D.C., Dec. 15, 1953.

5. *Final Report of a Board of Investigation—The Design and Methods of Construction of Welded Steel Merchant Vessels, 15 July, 1946*, GPO, Washington, D.C., 1947.

6. G. M. Boyd and T. W. Bushell, "Hull Structural Steel—The Unification of the Requirements of Seven Classification Societies," *Quarterly Transactions: The Royal Institution of Naval Architects (London), 103*, No. 3, Mar. 1961.

7. J. Turnbull, "Hull Structures," *The Institution of Engineers and Shipbuilders of Scotland, Transactions, 100*, pt. 4, Dec. 1956–1957, pp. 301–316.

8. G. M. Boyd, "Fracture Design Practices for Ship Structures," in *Fracture*, Vol. V: *Fracture Design of Structures*, edited by H. Libowitz, Academic Press, New York, 1969, pp. 383–470.

9. Marine Casualty Report, "Structural Failure of the Tank Barge I.O.S. 3301 Involving the Motor Vessel Martha R. Ingram on 10 January 1972 Without Loss of Life," *Report No. SDCG/NTSB*, Mar. 1974.

10. T. Bishop, "Fatigue and the Comet Disasters," *Metal Progress*, May 1955, p. 79.

11. "Collapse of U.S. 35 Highway Bridge, Point Pleasant, West Virginia," *NTSB Report No. NTSB-HAR-71-1*, Oct. 4, 1968.

12. J. A. Bennett and Harold Mindlin, "Metallurgical Aspects of the Failure of the Pt. Pleasant Bridge," *Journal of Testing and Evaluation*, Mar. 1973, pp. 152–161.

13. "State Cites Defective Steel in Bryte Bend Failure," *Engineering News Record, 185*, No. 8, Aug. 20, 1970.

14. "Joint Redesign on Cracked Box Girder Cuts into Record Tied Arch's Beauty," *Engineering News Record, 188*, No. 13, Mar. 30, 1972.

15. J. W. Fisher, *Fatigue and Fracture in Steel Bridges—Case Studies*, John Wiley, New York, 1984.

16. *ASME Boiler and Pressure Vessel Code*, Section III, Section G, *ASME*, New York, 1972.

17. *American Association of State Highway and Transportation Officials—Material Specifications*, Association General Offices, Washington, D.C., 1978.

18. *MIL-Std—1530*, "Aircraft Structural Integrity Program, Airplane Requirements,"

United States Air Force, Wright-Patterson Air Force Base, Dayton, Ohio, Sept. 1, 1972.

19. R. WECK, "A Rational Approach to Standard for Welded Construction," *British Welding Journal*, Nov. 1966.

20. J. W. FISHER and B. T. YEN, "Design, Structural Details, and Discontinuities in Steel, Safety and Reliability of Metal Structures," *ASCE*, Nov. 2, 1972.

21. E. V. BRAVENEC, "Analysis of Brittle Fractures During Fabrication and Testing," *ASCE*, Mar. 1972.

22. T. J. DOLAN, "Preclude Failure: A Philosophy for Materials Selection and Simulated Service Testing," Wm. M. Murray Lecture, SESA Fall Meeting, Houston, Oct. 1969, published *SESA Journal of Experimental Mechanics*, Jan. 1970.

23. E. R. PARKER, *Brittle Behavior of Engineering Structures*, John Wiley, New York, 1957.

24. Welding Research Council, *Control of Steel Construction to Avoid Brittle Failure*, edited by M. E. Shank, M.I.T. Press, Cambridge, Mass., 1957.

25. W. J. HALL, H. KIHARA, W. SOETE, and A. A. WELLS, *Brittle Fracture of Welded Plate*, Prentice-Hall, Englewood Cliffs, N.J., 1967.

26. The Royal Institution of Naval Architects, *Brittle Fracture in Steel Structures*, edited by G. M. Boyd, Butterworth's, London, 1970.

27. C. F. TIPPER, *The Brittle Fracture Story*, Cambridge University Press, New York, 1962.

28. H. LIBOWITZ, ed., *Fracture, An Advanced Treatise*, Vols. I–VII, Academic Press, New York, 1969.

29. The Japan Welding Society, "Cracking and Fracture in Welds," in *Proceedings of the First International Symposium on the Prevention of Cracking in Welded Structures, Tokyo*, Nov. 8–10, The Japan Welding Society, Tokyo, 1971.

30. W. S. PELLINI, "Principles of Fracture—Safe Design," *Welding Journal (Welding Research Supplement)*, pt. I, Mar. 1971, pp. 91-S–109-S, and pt. II, April 1971, pp. 147-S–162-S.

31. S. T. ROLFE, "Fracture Mechanics in Bridge Design," *Civil Engineering ASCE*, Aug. 1972.

32. AMERICAN SOCIETY FOR TESTING AND MATERIALS, "Fracture Toughness Testing and Its Applications," *ASTM STP 381*, Philadelphia, 1964.

33. AMERICAN SOCIETY FOR TESTING AND MATERIALS, "Plane Strain Crack Toughness Testing of High Strength Metallic Materials," edited by W. F. Brown and J. E. Srawley, *ASTM STP 410*, Philadelphia, 1966.

34. American Society of Civil Engineers, "Safety and Reliability of Metal Structures," ASCE Specialty Conference held in Pittsburgh, Nov. 2–3, published by ASCE, New York, 1972.

35. AMERICAN SOCIETY FOR TESTING AND MATERIALS, "Fracture Mechanics—Sixteenth Symposium," *ASTM STP 868*, Philadelphia, 1983.

36. AMERICAN SOCIETY FOR TESTING AND MATERIALS, "Elastic Plastic Fracture Test Methods—The User's Experience," *ASTM STP 856*, Philadelphia, 1985.

37. R. B. MADISON and G. R. IRWIN, "Dynamic $K_c$ Testing of Structural Steel," *Journal of the Structural Division, ASCE, 100, No. ST 7*, Proc. Paper 10653, July 1974, pp. 1331–49.

# 2 Stress Analysis for Members with Cracks

## 2.1. Introduction

Most structural members have discontinuities of some type, for example, holes, fillets, notches, and the like. If these discontinuities have well-defined geometries, it is usually possible to determine a stress-concentration factor, $K_T$, for these geometries and to account for the local elevation of stress using the well-known relation $\sigma_{max} = K_T \cdot \sigma_{nom}$ between the local maximum stress and the applied nominal stress. In fact, for most structures, the designer usually relies on the ductility of the material to redistribute the load around a mild stress concentration and, hence, to ignore any effects of the stress concentration. For mild stress concentrations (holes, smooth fillets) in ductile materials, this is a satisfactory procedure. However, if the stress concentration is severe, for example, approaching a sharp crack in which the radius of the crack tip approaches zero, this procedure is not satisfactory.

To illustrate this point, let us analyze the stress concentration at the edge of an ellipse as shown in Figure 2.1. This is given as

$$K_T = \frac{\sigma_{max}}{\sigma_{nom}} = 1 + \frac{2a}{b} \tag{2.1}$$

Thus for an ellipse,

$$\sigma_{max} = \sigma_{nom}\left(1 + \frac{2a}{b}\right) \tag{2.2}$$

For very sharp cracks,

$$a \gg b \quad \text{and} \quad \sigma_{max} \simeq \sigma_{nom}\left(\frac{2a}{b}\right) \tag{2.3}$$

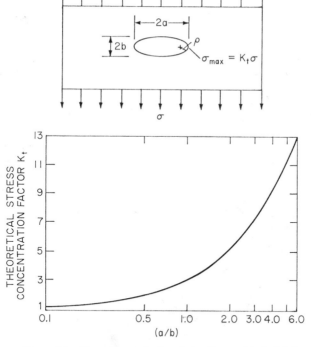

**Figure 2.1** Stress-concentration factor for an elliptical hole.

As shown in Figure 2.1, $K_T$ becomes very large as $a/b$ becomes large. The radius at the end of the major axis can be approximated by $\rho = b^2/a$. For sharp cracks,

$$a \gg \rho \quad \text{and} \quad \sigma_{max} \simeq \sigma_{nom} \cdot 2\sqrt{\frac{a}{\rho}} \tag{2.4}$$

and

$$K_T = \frac{\sigma_{max}}{\sigma_{nom}} = 2\sqrt{\frac{a}{\rho}} \tag{2.5}$$

Thus, for sharp cracks, $\rho \to 0$ and $K_T \to \infty$, and the use of the stress-concentration approach becomes meaningless. Consequently, an analytical method that is different from the stress-concentration approach is needed to analyze the behavior of structures or machine components that contain cracks or sharp imperfections. The first analysis of fracture behavior for components containing cracks was developed by Griffith. Presently, the fracture behavior for such components can be analyzed best by using fracture-mechanics technology. The Griffith[1-4] fracture criterion and some of the basic fracture-mechanics analyses, equations, and relationships are presented in this chapter.

## 2.2. The Griffith Theory

The first analysis of fracture behavior of components that contain sharp discontinuities was developed by Griffith.[1] The analysis was based on the assumption that incipient fracture in ideally brittle materials occurs when the magnitude of the elastic energy supplied at the crack tip during an incremental increase in crack length is equal to or greater than the magnitude of the elastic energy at the crack tip during an incremental increase in crack length. This energy approach can be presented best by considering the following example. Consider an infinite plate of unit thickness that contains a through-thickness crack of length $2a$ and that is subjected to uniform tensile stress, $\sigma$, applied at infinity. The total potential energy of the system, $U$, may be written as

$$U = U_0 - U_a + U_\gamma \tag{2.6}$$

where $U_0$ = elastic energy of the uncracked plate.

$\quad\quad U_a$ = decrease in the elastic energy caused by introducing the crack in the plate.

$\quad\quad U_\gamma$ = increase in the elastic-surface energy caused by the formation of the crack surfaces.

Griffith used a stress analysis that was developed by Inglis[2] to show that

$$U_a = \frac{\pi \sigma^2 a^2}{E} \tag{2.7}$$

Moreover, the elastic-surface energy, $U_\gamma$, is equal to the product of the elastic-surface energy of the material, $\gamma_e$, and the new surface area of the crack:

$$U_\gamma = 2(2a\gamma_e) \tag{2.8}$$

Consequently, the total elastic energy of the system, $U$, is

$$U = U_0 - \frac{\pi \sigma^2 a^2}{E} + 4a\gamma_e \tag{2.9}$$

The equilibrium condition for crack extension is obtained by setting the first derivative of $U$ with respect to crack length, $a$, equal to zero. The resulting equation can be written as

$$\sigma\sqrt{a} = \left(\frac{2\gamma_e E}{\pi}\right)^{1/2} \tag{2.10}$$

which indicates that crack extension in ideally brittle materials is governed by the product of the applied nominal stress and the square root of the crack length and by material properties. Because $E$ and $\gamma_e$ are material properties, the right-hand side of Equation (2.10) is equal to a constant value that is characteristic of a given ideally brittle material. Consequently, Equation (2.10) indicates that crack extension in such materials occurs when the product $\sigma\sqrt{a}$ attains a constant critical value. The value of this constant can be determined experimentally by measuring the fracture stress for a large plate that contains a through-thickness crack of known length and that is subjected to remotely applied uniform tensile stress. This value can also be measured by using other specimen geometries, which is what makes this approach to fracture analysis so powerful. Values of this constant for other geometries are discussed later in this chapter.

Equation (2.10) can be rearranged in the form

$$\frac{\pi\sigma^2 a}{E} = 2\gamma_e \qquad (2.11)$$

The left-hand side has been designated the energy-release rate, $G$, and represents the elastic energy per unit crack surface area that is available for infinitesimal crack extension. The right-hand side of Equation (2.11) represents the material's resistance to crack extension, $R$.

In 1948 Irwin[3] suggested that the Griffith fracture criterion for ideally brittle materials could be modified and applied to brittle materials and to metals that exhibit plastic deformation. A similar modification was proposed by Orowan[4] at about the same time. The modification recognized that a material's resistance to crack extension is equal to the sum of the elastic-surface energy and the plastic-strain work, $\gamma_p$, accompanying crack extension. Consequently, Equation (2.11) was modified to

$$\frac{\pi\sigma^2 a}{E} = 2(\gamma_e + \gamma_p) \qquad (2.12)$$

Because the left-hand side is the energy-release rate, $G$, and because $\sigma\sqrt{\pi a}$ represents the intensity, $K_I$, of the stress field at the tip of a through-thickness crack of length $2a$, the following relation exists between $G$ and $K_I$ for plane-strain conditions:

$$\frac{\pi\sigma^2 a}{E} = G = \frac{K_I^2}{E}(1 - \nu^2) \qquad (2.13)$$

The energy-balance approach to crack extension defines the conditions required for instability of an ideally sharp crack. This approach is not applicable to analysis of stable crack extension such as occurs under cyclic-load fluctuation or under stress-corrosion-cracking conditions. However, the stress-intensity parameter, $K$, *is* applicable to stable crack extension, and therefore development of linear-elastic fracture-mechanics theory has assisted greatly in improving our understanding of subcritical crack extension and crack instability.

## 2.3. Linear-Elastic Fracture Mechanics

Linear-elastic fracture-mechanics technology is based on an analytical procedure that relates the stress-field magnitude and distribution in the vicinity of a crack tip to the nominal stress applied to the structural member, to the size, shape, and orientation of the crack or crack-like discontinuity, and to material properties.

The fundamental principle of fracture mechanics is that the stress field ahead of a sharp crack in a structural member can be characterized in terms of a single parameter, $K$, the stress-intensity factor, that has units of ksi$\sqrt{\text{in.}}$ (MPa·m$^{1/2}$). This parameter, $K$, is related to both the nominal stress level ($\sigma$) in the member and the size of the crack present ($a$). Thus all structural members, or test specimens, that have flaws can be loaded to various levels of $K$, analogous to the situation where unflawed structural or mechanical members can be loaded to various stress levels, $\sigma$.

Because failure of most structural or mechanical components is caused by the propagation of cracks, an understanding of the magnitude and distribution of the stress field in the vicinity of the crack front is essential to determine the safety and reliability of structures. Because fracture mechanics is based on a stress analysis, a *quantitative* evaluation of the safety and reliability of structures is possible.

To establish methods of stress analysis for cracks in elastic solids, it is convenient to define three types of relative movements of two crack surfaces.[5] These displacement modes (Figure 2.2) represent the local deformation in an infinitesimal element containing a crack front. The opening mode, Mode I, is characterized by local displacements that are symmetric with respect to the x-y and x-z planes. The two fracture surfaces are displaced perpendicular to each other in opposite directions. Local displacements in the sliding or shear mode, Mode II, are symmetric with respect to the x-y plane and skew symmetric with respect to the x-z plane. The two fracture surfaces slide over each other in a direction perpendicular to the line of the crack tip. The tearing mode, Mode III, is associated with local displacements that are skew symmetric with respect to both x-y and x-z planes. The two fracture surfaces slide over each other in a direction that is parallel to the line of the crack front. Each of these modes of deformation corresponds to a basic type of stress field in the vicinity of crack tips. In any problem the deformations at the crack tip can be treated as one or a combination of these local displacement modes. Moreover, the stress field at the crack tip can be treated as one or a combination of the three basic types of stress fields. Most practical design situations and failures correspond to Mode I displacements, hence Mode I is emphasized in this textbook.

By using a method that was developed by Westergaard,[6] Irwin[7] found

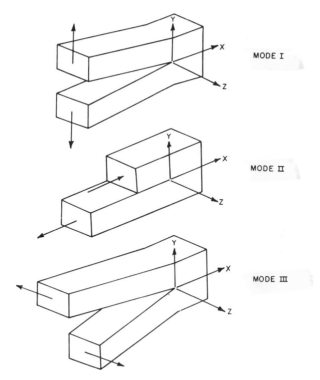

**Figure 2.2**   The three basic modes of crack surface displacements (Ref. 5).

that the stress and displacement fields in the vicinity of crack tips subjected to the three modes of deformation are given by

MODE I:

$$\sigma_x = \frac{K_I}{(2\pi r)^{1/2}} \cos \frac{\theta}{2} \left[ 1 - \sin \frac{\theta}{2} \sin \frac{3\theta}{2} \right]$$

$$\sigma_y = \frac{K_I}{(2\pi r)^{1/2}} \cos \frac{\theta}{2} \left[ 1 + \sin \frac{\theta}{2} \sin \frac{3\theta}{2} \right]$$

$$\tau_{xy} = \frac{K_I}{(2\pi r)^{1/2}} \sin \frac{\theta}{2} \cos \frac{\theta}{2} \cos \frac{3\theta}{2}$$

$$\sigma_z = \nu(\sigma_x + \sigma_y), \qquad \tau_{xz} = \tau_{yz} = 0$$

$$u = \frac{K_I}{G} \left[ \frac{r}{2\pi} \right]^{1/2} \cos \frac{\theta}{2} \left[ 1 - 2\nu + \sin^2 \frac{\theta}{2} \right] \qquad (2.14)$$

$$v = \frac{K_I}{G} \left[ \frac{r}{2\pi} \right]^{1/2} \sin \frac{\theta}{2} \left[ 2 - 2\nu - \cos^2 \frac{\theta}{2} \right]$$

$$w = 0$$

MODE II:

$$\sigma_x = -\frac{K_{II}}{(2\pi r)^{1/2}} \sin\frac{\theta}{2}\left[2 + \cos\frac{\theta}{2}\cos\frac{3\theta}{2}\right]$$

$$\sigma_y = \frac{K_{II}}{(2\pi r)^{1/2}} \sin\frac{\theta}{2}\cos\frac{\theta}{2}\cos\frac{3\theta}{2}$$

$$\tau_{xy} = \frac{K_{II}}{(2\pi r)^{1/2}} \cos\frac{\theta}{2}\left[1 - \sin\frac{\theta}{2}\sin\frac{3\theta}{2}\right]$$

$$\sigma_z = \nu(\sigma_x + \sigma_y), \qquad \tau_{xz} = \tau_{yz} = 0$$

$$u = \frac{K_{II}}{G}\left[\frac{r}{2\pi}\right]^{1/2}\sin\frac{\theta}{2}\left[2 - 2\nu + \cos^2\frac{\theta}{2}\right] \qquad (2.15)$$

$$v = \frac{K_{II}}{G}\left[\frac{r}{2\pi}\right]^{1/2}\cos\frac{\theta}{2}\left[-1 + 2\nu + \sin^2\frac{\theta}{2}\right]$$

$$w = 0$$

MODE III:

$$\tau_{xz} = -\frac{K_{III}}{(2\pi r)^{1/2}}\sin\frac{\theta}{2}$$

$$\tau_{yz} = \frac{K_{III}}{(2\pi r)^{1/2}}\cos\frac{\theta}{2}$$

$$\sigma_x = \sigma_y = \sigma_z = \tau_{xy} = 0 \qquad (2.16)$$

$$w = \frac{K_{III}}{G}\left[\frac{2r}{\pi}\right]^{1/2}\sin\frac{\theta}{2}$$

$$u = v = 0$$

where the stress components and the coordinates $r$ and $\theta$ are shown in Figure 2.3; $u$, $v$, and $w$ are the displacements in the $x$, $y$, and $z$ directions, respectively; $\nu$ is Poisson's ratio; and $G$ is the shear modulus of elasticity.

Equations (2.14) and (2.15) represent the case of plane strain ($w = 0$) and neglect higher-order terms in $r$. Because higher-order terms in $r$ are neglected, these equations are exact in the limit as $r$ approaches zero and are a good approximation in the region where $r$ is small compared with other $x$-$y$ planar dimensions. These field equations show that the distribution of the elastic-stress fields and of the deformation fields in the vicinity of the crack tip are invariant in all components subjected to a given mode of deformation and that the magnitude of the elastic-stress field can be described by single-term parameters, $K_I$, $K_{II}$, and $K_{III}$, that correspond to Modes I, II, and III, respectively. Consequently, the applied stress, the crack shape and size, and the structural configuration associated with structural components subjected to a given mode of deformation affect the value of the

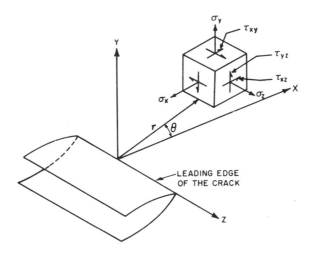

**Figure 2.3**   Coordinate system and stress components ahead of a crack tip.

stress-intensity factor but do not alter the stress-field distribution. Dimensional analysis of Equations (2.14), (2.15), and (2.16) indicates that the stress-intensity factor must be linearly related to stress and must be directly related to the square root of a characteristic length. Based on Griffith's original analysis of glass components with cracks and the subsequent extension of that work to more ductile materials, the characteristic length is the crack length in a structural member. Consequently, the magnitude of the stress-intensity factor must be directly related to the magnitude of the applied nominal stress, $\sigma$, and the square root of the crack length, $a$. In all cases, the general form of the stress-intensity factor is given by

$$K = \sigma \cdot \sqrt{a} \cdot f(g) \tag{2.17}$$

where $f(g)$ is a parameter that depends on the specimen and crack geometry and has been the subject of extensive investigations and research. Fortunately a number of relationships between the stress-intensity factor and various body configurations, crack sizes, orientations, and shapes, and loading conditions have been published,[5,8,9] and the more common ones are presented in the following section.

For unusual crack geometries, $K_I$ relations can be approximated by using known $K_I$ relations and "bounding" the real crack geometry. An example of this technique is given later in section 2.4.11.

One of the key aspects of the stress-intensity factor, $K_I$, is that it relates the *local* stress field ahead of a sharp crack in a structural member to the *global* (or nominal) stress applied to that structural member away from the crack. Specifically, Figure 2.3 shows the stresses ahead of a sharp crack. Most brittle fractures occur under conditions of Mode I loading

(Figure 2.2). Accordingly the stress of primary interest in Figure 2.3 and in most practical applications is $\sigma_y$. For $\sigma_y$ to be a maximum in Equation (2.14), let $\theta = 0$, and therefore

$$\sigma_y = \frac{K_I}{\sqrt{2\pi r}} \qquad (2.18)$$

Rearranging this expression shows that

$$K_I = \sigma_y\sqrt{2\pi r} \qquad (2.19)$$

At locations farther and farther from the crack tip (increasing r), the stress, $\sigma_y$, decreases. However, $K_I$ remains constant and describes the intensity of the stress *field* just ahead of a sharp crack. This same stress-intensity factor is also related to the global stress by

$$K_I = \sigma\sqrt{a} \cdot f(g) \qquad (2.20)$$

for various crack geometries as described through this chapter. Hence, $K_I$ describes the stress field intensity ahead of a sharp crack in any structural member (plates, beams, airplane wings, pressure vessels, etc.) as long as the correct geometrical parameter, $f(g)$, can be determined. Expressions for different crack geometries in various structural members are presented in the next section.

## 2.4. Stress-Intensity-Factor Equations

Stress-intensity-factor equations for the more common basic specimen geometries that are subjected to Mode I deformation are presented in this section. Extensive stress-intensity-factor equations for various geometries and loading conditions are available in the literature in tabular form.[5,8,9]

### 2.4.1. Through-Thickness Crack

The stress-intensity factor for an infinite plate subjected to uniform tensile stress, $\sigma$, and that contains a through-thickness crack of length $2a$ (Figure 2.4) is

$$K_I = \sigma\sqrt{\pi a} \qquad (2.21)$$

A tangent-correction factor having the form

$$\left(\frac{2b}{\pi a}\tan\frac{\pi a}{2b}\right)^{1/2} \qquad (2.22)$$

is used to approximate the $K_I$ values for plates of finite width, $2b$. Thus, the stress-intensity factor for a plate of finite width $2b$ that is subjected to uniform tensile stress, $\sigma$, and that contains a through-thickness crack of length $2a$ (Figure 2.4) is

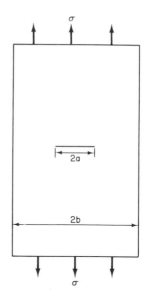

**Figure 2.4**  Finite-width plate containing a through-thickness crack.

$$K_{\mathrm{I}} = \sigma\sqrt{\pi a}\left(\frac{2b}{\pi a}\,\tan\frac{\pi a}{2b}\right)^{1/2} \tag{2.23}$$

The values for the tangent correction factor for various ratios of crack length to plate width are given in Table 2.1.  Equation 2.23 is accurate within 7 percent for $a/b \leqslant 0.5$.

TABLE 2.1.  Correction Factors
for a Finite-Width Plate Containing
a Through-Thickness Crack (Ref. 5)

| $a/b$ | $[2b/\pi a\cdot\tan\,\pi a/2b]^{1/2}$ |
|-------|----------------------------------------|
| 0.074 | 1.00 |
| 0.207 | 1.02 |
| 0.275 | 1.03 |
| 0.337 | 1.05 |
| 0.410 | 1.08 |
| 0.466 | 1.11 |
| 0.535 | 1.15 |
| 0.592 | 1.20 |

### 2.4.2.  Double-Edge Crack

The stress-intensity factor for double-edge-notched infinite plate that is subjected to uniform tensile stress (Figure 2.5) is given by

$$K_{\mathrm{I}} = 1.12\sigma\sqrt{\pi a} \tag{2.24}$$

where the constant 1.12 represents a free-surface correction factor for edge notches that are perpendicular to the applied tensile stress.

The tangent-correction factor, Equation (2.22), can be used to obtain good estimates of the stress-intensity-factor values for double-edge-notched specimens having a finite width, $2b$. However, more precise solutions are available in the literature[5,8,9] that incorporate the effects of variations in specimen length to width ratio and the decay of the free-surface-correction factor as the ratio of crack length to specimen width increases.

### 2.4.3. Single-Edge Notch

The stress-intensity-factor equation for a semi-infinite single-edge-notched specimen (Figure 2.6) is given by Equation (2.24). Moreover, the surface correction factors of 1.12 used to calculate the stress-intensity factor for a double-edge-notched specimen of finite width is used also to calculate the stress-intensity factor for a single-edge-notched specimen of finite width. However, an additional correction factor is necessary to account for bending stresses caused by lack of symmetry in the single-edge-notched specimen.

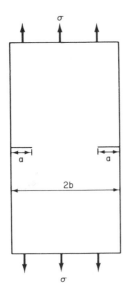
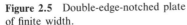
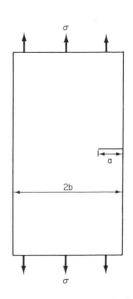

**Figure 2.5** Double-edge-notched plate of finite width.

**Figure 2.6** Single-edge-notched plate of finite width.

The stress-intensity-factor equation that incorporates the various correction factors for a single-edge-notched specimen is given by[5,10,11]

$$K_I = 1.12\, \sigma\sqrt{\pi a} \cdot k\!\left(\frac{a}{b}\right) \qquad (2.25)$$

The values of the function $k(a/b)$ are tabulated in Table 2.2 for various values of the ratio of crack length to specimen width. Note also that for $a/b = 1.0$, the notch depth is one-half the width of the plate, and the correction factor is becoming quite large.

**TABLE 2.2.    Correction Factors for a Single-Edge-Notched Plate (Ref. 5)**

| $a/b$ | $k(a/b)$ |
|-------|----------|
| 0.10 | 1.03 |
| 0.20 | 1.07 |
| 0.30 | 1.15 |
| 0.40 | 1.22 |
| 0.50 | 1.35 |
| 0.60 | 1.50 |
| 0.70 | 1.69 |
| 0.80 | 1.91 |
| 0.90 | 2.20 |
| 1.00 | 2.55 |

### 2.4.4.  Cracks Growing from Round Holes

The stress-intensity factor for cracks growing from a circular hole in infinite plates (Figure 2.7) is given by

$$K_I = \sigma\sqrt{\pi a} \cdot f\left(\frac{a}{r}\right)$$

(2.26)

where $r$ = radius of hole.

$a$ = crack length from one side of the hole

Note that as $a/r$ approaches zero, $f(a/r)$ for either one or two cracks approaches a value of about 3. This is the value of the stress-concentration factor, $K_T$, at a round hole. Thus, the $K_I$ factor can be thought of as being equal to $K_I \simeq K_T \sigma\sqrt{\pi a}$ for very short cracks near a stress concentration where the local stress is elevated by the stress concentration.

As the crack grows and $(a/r)$ becomes large, the effect of the stress concentration (in this case the hole) decreases, and the expression approaches that of a through-thickness crack in an infinite plate.

### 2.4.5.  Cracks Growing from Elliptical Notches

Figure 2.8 shows the relations between the theoretical stress-concentration factor, $K_T$, and the notch geometry for a long, narrow elliptical

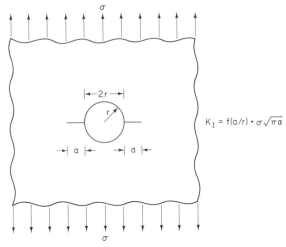

$$K_I = f(a/r) \cdot \sigma \sqrt{\pi a}$$

| a/r | f(a/r), ONE CRACK | f(a/r), TWO CRACKS |
|-----|------------------|--------------------|
| 0.00 | 3.39 | 3.39 |
| 0.10 | 2.73 | 2.73 |
| 0.20 | 2.30 | 2.41 |
| 0.30 | 2.04 | 2.15 |
| 0.40 | 1.86 | 1.96 |
| 0.50 | 1.73 | 1.83 |
| 0.60 | 1.64 | 1.71 |
| 0.80 | 1.47 | 1.58 |
| 1.0 | 1.37 | 1.45 |
| 1.5 | 1.18 | 1.29 |
| 2.0 | 1.06 | 1.21 |
| 3.0 | 0.94 | 1.14 |
| 5.0 | 0.81 | 1.07 |
| 10.0 | 0.75 | 1.03 |
| ∞ | 0.707 | 1.00 |

**Figure 2.7**   Cracks growing from a round hole (Ref. 5).

notch (approaching a crack) in an infinite plate; that is, $K_T = 2\sqrt{a_N/\rho}$. For short cracks growing from the tip of an elliptical notch, $K_I$ can be defined as

$$K_I = K_T \cdot \sigma \sqrt{\pi \Delta a_F} \tag{2.27}$$

where $K_T = 2\sqrt{a_N/\rho}$ for $\rho \ll a_N$ and $\Delta a_F$ is the length of the short crack that is growing from the tip of the ellipse (Figure 2.9).

The stress-concentration effect of an elliptical hole can be neglected when the parameter $\Delta a_F/\sqrt{a_N\rho}$ is equal to or greater than 0.25, as shown in Figure 2.9.[12,13] Thus, there is a definite limit to the design for which Equation (2.27) applies.

This expression generally is true regardless of the ratio of minor to major axis of the ellipse, that is, the $b/a_N$ ratio. However, the curve shown in Figure 2.9 is specifically for $b/a_N$ equal to 0.5. Curves for other ratios vary only slightly and are presented in Reference 12.

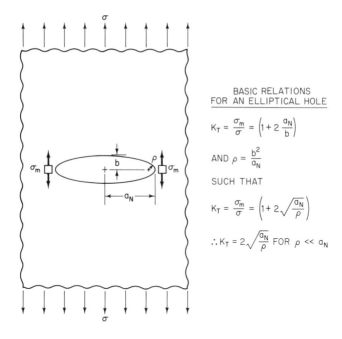

**Figure 2.8**  Basic relations between theoretical stress-concentration factor, $K_t$, and notch geometry for a long, narrow, elliptical notch (approaching a crack) contained in an infinite plate.

The $K_I$ value at the tip of a sharp crack growing from a notch in the center of an infinite plate is also defined as

$$K_I = \sigma\sqrt{\pi a_T} \cdot f_1(\lambda, \delta) \qquad (2.28)$$

where the terms are as shown in Figure 2.9. For $\Delta a_F/\sqrt{a_N\rho} \geq 0.25$, $K_I = \sigma\sqrt{\pi a_T}$ because $f_1(\lambda, \delta) \simeq 1.0$.

A similar relation also holds for cracks growing from a semielliptical notch at the edge of a plate, as shown in Figure 2.10. The curve shown in Figure 2.10 is for $b/a_N$ equal to 0.5. Curves for other ratios of $b/a_N$ differ slightly and are presented in Reference 12.

The $K_I$ value at the tip of a sharp crack growing from a notch at the edge of an infinite plate is

$$K_I = 1.12\sigma\sqrt{\pi a_T} \cdot k\left(\frac{a}{b}\right) \cdot f_2(\lambda, \delta) \qquad (2.29)$$

where the terms are defined as shown in Figure 2.10. For

$$\Delta a_F/\sqrt{a_N\rho} \geq 0.25, \ K_I = 1.12\sigma\sqrt{\pi a_T} \cdot k(a/b) \qquad (2.30)$$

because $f_2(\lambda, \delta)$ approaches unity.

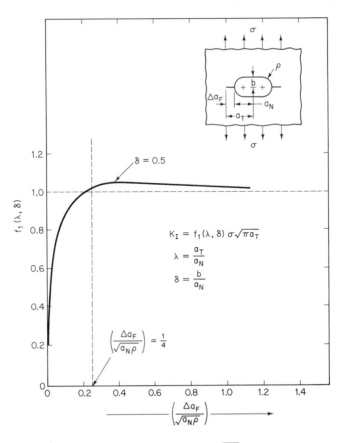

**Figure 2.9** Relation between $f_1(\lambda, \delta)$ and $\Delta a_F/\sqrt{a_N\rho}$ for cracks growing from an elliptical notch with $b/a_N = 0.5$ in an infinite plate.

### 2.4.6. Single Crack in Beam in Bending

The stress-intensity factor for a beam in bending that contains an edge crack (Figure 2.11) is given by

$$K_I = \frac{6M}{B(W-a)^{3/2}} \cdot g\left(\frac{a}{W}\right) \tag{2.31}$$

where $M$ is the moment per unit thickness and the values of $g(a/W)$ are presented in Table 2.3 for various values of the ratio of crack length, $a$, to beam depth, $W$.

**TABLE 2.3.   Stress-Intensity-Factor Coefficients for Notched Beams (Ref. 5)**

| $a/W$ | 0.05 | 0.1 | 0.2 | 0.3 | 0.4 | 0.5 | 0.6 (and larger) |
|-------|------|-----|-----|-----|-----|-----|------------------|
| $g(a/W)$ | 0.36 | 0.49 | 0.60 | 0.66 | 0.69 | 0.72 | 0.73 |

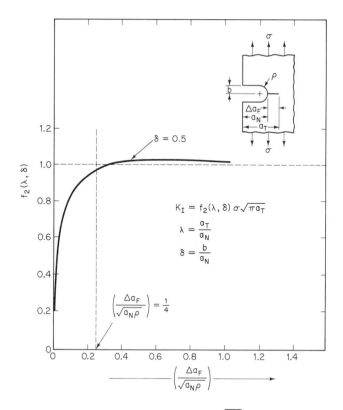

**Figure 2.10**  Relation between $f_2(\lambda, \delta)$ and $\Delta a_F/\sqrt{a_N \rho}$ for cracks growing from an elliptical notch with $b/a_N = 0.5$ in. at the edge of an infinite plate.

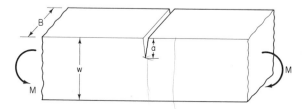

**Figure 2.11**  Edge-notched beam in bending.

### 2.4.7. Embedded Elliptical or Circular Crack in Infinite Plate

The stress-intensity factor at any point along the perimeters of elliptical or circular cracks embedded in an infinite body that is subjected to uniform tensile stress (Figure 2.12) is given by[14]

$$K_I = \frac{\sigma \sqrt{\pi a}}{\Phi_0} \left( \sin^2 \beta + \frac{a^2}{c^2} \cos^2 \beta \right)^{1/4} \qquad (2.32)$$

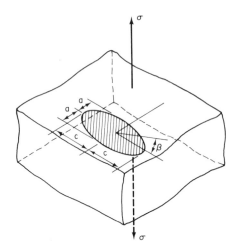

**Figure 2.12** Elliptical crack in an infinite body subjected to uniform tension (Ref. 5).

where $K_I$ corresponds to the value of the stress-intensity factor for a point on the perimeter of the crack whose location is defined by the angle $\beta$, and $\Phi_0$ is the elliptic integral

$$\Phi_0 = \int_0^{\pi/2} \left[ 1 - \left( \frac{c^2 - a^2}{c^2} \right) \sin^2 \theta \right]^{1/2} d\theta \tag{2.33}$$

The stress-intensity factor for an embedded elliptical crack reaches a maximum at $\beta = \pi/2$ and is given by the equation

$$K_I = \sigma \sqrt{\pi \frac{a}{Q}} \tag{2.34}$$

where $Q = \Phi_0^2$ and $\Phi_0$ is the elliptic integral given by Equation (2.33). $Q$ has been designated a shape factor because its value depends on the values of $a$ and $c$, as well as the ratio of $\sigma/\sigma_{ys}$. A graphic representation of this dependence is shown in Figure 2.13 for various crack shapes and values of the ratio of nominal applied stress to yield stress, $\sigma_{ys}$. The use of different curves for different values of $\sigma/\sigma_{ys}$ is intended to account for the effects of plastic deformation in the vicinity of the crack tip on the stress-intensity-factor value and is incorporated into the value of $Q$.

For circular cracks where $a = c$ and $Q \approx 2.4$, Equation (2.34) becomes

$$K_I = 0.65\sigma\sqrt{\pi a} = 1.15\sigma\sqrt{a} \tag{2.35}$$

The exact expression for an embedded circular crack is

$$K_I = \frac{2}{\sqrt{\pi}} \sigma\sqrt{a} = 1.13\sigma\sqrt{a} \tag{2.36}$$

and the two values are within 2 percent of each other.

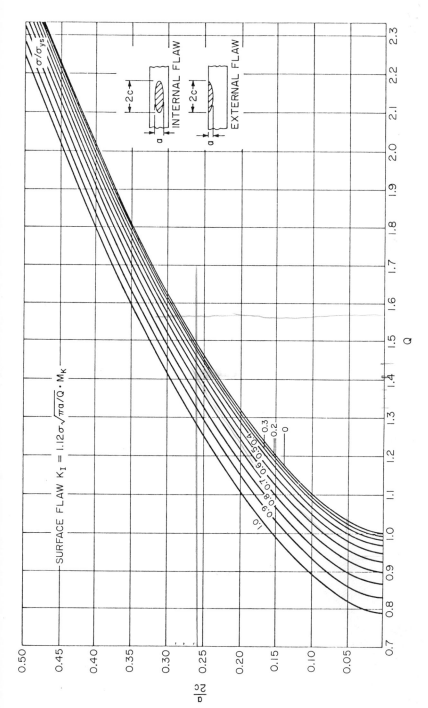

**Figure 2.13**  Effect of $a/2c$ ratio and $\sigma/\sigma_{ys}$ ratio on flaw-shape parameter $Q$.

### 2.4.8. Surface Crack

The stress-intensity factor for a part-through thumbnail crack in a plate subjected to uniform tensile stress (Figure 2.14) can be calculated by using Equations (2.32) and (2.33) and a free-surface-correction factor equal to 1.12. The stress-intensity factor for $\beta = \pi/2$ (which is the location of maximum stress intensity) is given by

$$K_I = 1.12\sigma\sqrt{\pi\frac{a}{Q}} \cdot M_K \tag{2.37}$$

where $Q = \Phi_0^2$ and is presented graphically in Figure 2.13. This equation is identical to Equation (2.34) except for the 1.12 multiplier that corresponds to the front free surface correction factor and $M_K$ which corresponds to the back free-surface correction factor. $M_K$ is approximately 1.0 as long as the crack depth, $a$, is less than one-half the wall thickness, $t$. As $a$ approaches $t$, $M_K$ approaches approximately 1.6 and a useful approximation is

$$M_K = 1.0 + 1.2\left(\frac{a}{t} - 0.5\right) \tag{2.38}$$

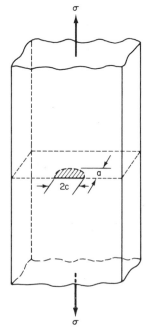

**Figure 2.14**  Plate containing a semi-elliptical surface crack.

### 2.4.9. Crack in WOL and CT Type Specimens

The stress-intensity-factor equation for wedge-opening-loading (WOL)–type specimens (Figure 2.15) can be presented in the form[15]

$$K_I = \frac{P}{B\sqrt{a}} \cdot f\left(\frac{a}{W}\right) \qquad (2.39)$$

where $P$ is the applied load, $B$ is the specimen thickness, and $f(a/W)$ is a measure of the compliance of the specimen. For a constant value of specimen height, $H$, to specimen width, $W$, the function $f(a/W)$ can be expressed in a polynomial form as a function of the dimensionless crack length $a/W$. Two WOL specimen geometries have been used extensively. These geometries are referred to as the T-type WOL specimen having $H/W = 0.972$ and the compact-tension (CT) specimen having $H/W = 1.2$. The dimensions for 1-in.-thick specimens for each of these two geometries are shown in Figures 2.16 and 2.17, respectively. Various sizes of a given WOL specimen geometry are proportional specimens.

The $f(a/W)$ for the T-type WOL specimen can be represented as[15,16]

$$f\left(\frac{a}{W}\right) = \left[ 30.96\left(\frac{a}{W}\right) - 195.8\left(\frac{a}{W}\right)^2 + 730.6\left(\frac{a}{W}\right)^3 \right.$$
$$\left. - 1186.3\left(\frac{a}{W}\right)^4 + 754.6\left(\frac{a}{W}\right)^5 \right] \qquad (2.40)$$

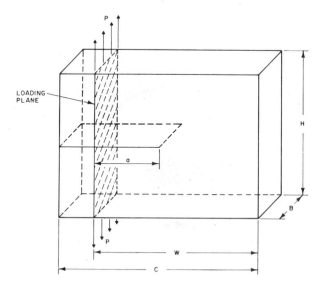

**Figure 2.15**  Idealized model of wedge-opening-loading specimen.

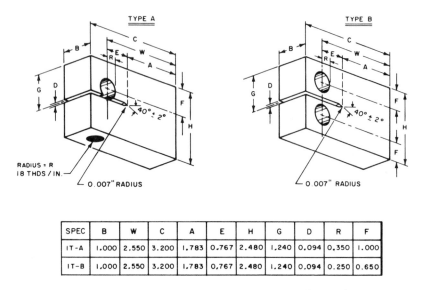

| SPEC | B | W | C | A | E | H | G | D | R | F |
|------|-----|-----|-----|-----|-----|-----|-----|-----|-----|-----|
| IT-A | 1.000 | 2.550 | 3.200 | 1.783 | 0.767 | 2.480 | 1.240 | 0.094 | 0.350 | 1.000 |
| IT-B | 1.000 | 2.550 | 3.200 | 1.783 | 0.767 | 2.480 | 1.240 | 0.094 | 0.250 | 0.650 |

**Figure 2.16**  Two geometries for 1-in.-thick T-type WOL specimens.

In the range $0.25 < a/W < 0.75$, the polynomial is accurate to within 0.5 percent of the experimental compliance.

The $f(a/W)$ for the compact tension specimen Figure 2.17 can be represented as[15,17]

$$f\left(\frac{a}{W}\right) = \left[0.2960\left(\frac{a}{W}\right) - 1.855\left(\frac{a}{W}\right)^2 + 6.557\left(\frac{a}{W}\right)^3 \right.$$
$$\left. - 10.17\left(\frac{a}{W}\right)^4 + 6.389\left(\frac{a}{W}\right)^5\right] \tag{2.41}$$

The polynomial is accurate within 0.25 percent for $a/W$ between 0.3 and 0.7.

### 2.4.10. Cracks with Wedge Forces and Internal Pressure

The stress-intensity factor for a crack that is subjected to eccentric line forces, $p$, per unit thickness, on its surfaces (Figure 2.18) is given by the equations

$$K_{I@1} = \frac{p}{\sqrt{\pi a}}\left(\frac{a + x}{a - x}\right)^{1/2} \tag{2.42}$$

$$K_{I@2} = \frac{p}{\sqrt{\pi a}}\left(\frac{a - x}{a + x}\right)^{1/2} \tag{2.43}$$

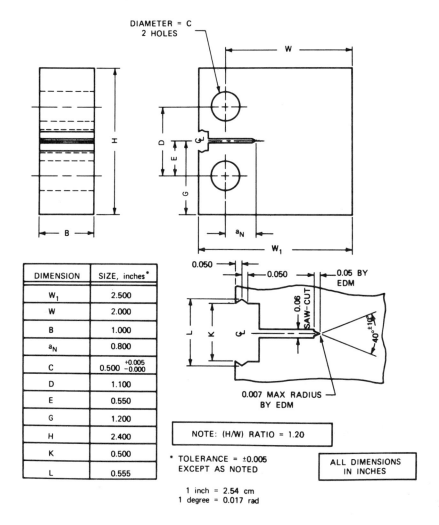

**Figure 2.17**  1-in.-thick compact tension (CT) specimen.

For centrally applied point forces, namely, $x = 0$, Equations (2.42) and (2.43) reduce to

$$K_{I@1} = K_{I@2} = \frac{p}{\sqrt{\pi a}} \qquad (2.44)$$

which indicates that as the crack grows, $K_I$ decreases. Thus, under the influence of a concentrated force on the crack surface, the rate of crack

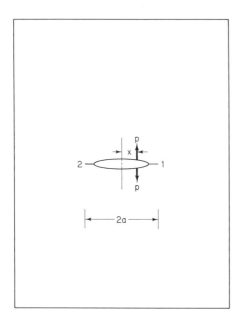

**Figure 2.18** Crack with eccentrical point forces.

propagation should decrease as the crack length increases and the crack may be arrested.

The stress-intensity factor for a crack that is subjected to an internal pressure $p$ (Figure 2.19) is given by the equation

$$K_I = p\sqrt{\pi a} \qquad (2.45)$$

### 2.4.11. Estimating Stress-Intensity Factors

In some circumstances, the particular crack of interest has an irregular shape. For most of these cases, the stress-intensity factor can be estimated from equations for simple and familiar shapes such as embedded tunnel, circular, or elliptical cracks.

Estimating stress-intensity factors for irregularly shaped cracks can be made easier if one remembers some basic relationships and trends. A tunnel crack is a special case of the equation for an embedded elliptical crack (Equation 2.34) with the crack-shape parameter, $Q$, equal to 1.0. Furthermore, the equation for an embedded circular crack is also a special case of Equation (2.34) for $a = c$ ($Q \approx 2.4$). Thus, as the crack shape changes from a circular crack front to a straight crack front, the value of $Q$ changes from 2.4 to 1.0 and the magnitude of the stress-intensity factor, $K_I$, increases by a factor of about 1.5. A corollary to these observations is that, for a tunnel crack that is curved, $K_I$ is higher for the convex front of the crack than for the concave front.

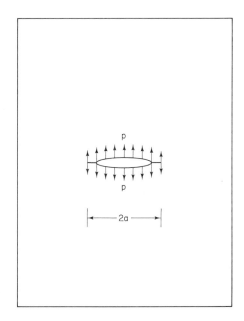

**Figure 2.19**   Crack subjected to internal pressure.

A study of the crack-shape parameter, $Q$ (Figure 2.13), shows that when $a/2c$ is less than about 0.15 (i.e., when the crack length on the surface, $2c$, is larger than about six times its depth, $a$), the crack-shape parameter approaches 1.0 and the crack behavior approaches that for a tunnel crack where $a/2c \to 0$. An approximate relationship between the change in crack geometry, $a/2c$, and the change in the crack-shape parameter, $Q$, for

$$2.4(\text{circular crack, } a/2c = 0.5) > Q > 1.1(\text{long crack, } a/2c \approx 0.15)$$

$$(2.46)$$

is a ratio of about 1.0 to 3.5. That is, for every 0.1 decrease in the magnitude of $a/2c$, the value of the crack-shape parameter decreases by about 0.35. This observation is derived from Figure 2.13, where the relationship between $a/2c$ and $Q$ is shown to be approximately linear in the range $2.4 > Q > 1.1$ with $\Delta Q/\Delta(a/2c) \approx 3.5$.

The application of the preceding information to estimate the stress-intensity factor along the perimeter of an irregularly shaped crack can be demonstrated by analyzing the embedded crack shown in Figure 2.20. The crack has an irregular shape and is subjected to a uniform tensile stress, $\sigma$, that is perpendicular to the crack plane.

The $K_I$ in region 1 (where the crack is denoted as $a_1$ in Figure 2.20), $K_{I1}$, for the crack can be represented by a circular crack having a radius $a_1$ because the crack is essentially circular in shape. The $K_{I1}$ value for this perimeter which opens into a tunnel crack should be slightly higher than

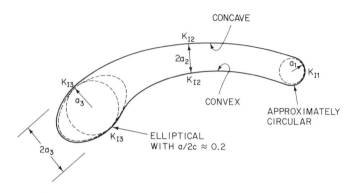

**Figure 2.20** Estimating stress-intensity factors for an embedded crack having an irregular shape.

the $K_I$ for a circular crack of radius $a_1$ and should be less than the $K_I$ value for a tunnel crack having a width equal to $2a_1$. Therefore,

$$\sigma\sqrt{\frac{\pi a_1}{Q}} < K_{I1} \ll \sigma\sqrt{\pi a_1} \qquad (2.47)$$

where $Q = 2.4$ or

$$0.65\,\sigma\sqrt{\pi a_1} < K_{I1} \ll \sigma\sqrt{\pi a_1} \qquad (2.48)$$

Thus,

$$K_{I1} \approx 0.75\sigma\sqrt{\pi a_1} \qquad (2.49)$$

should be a good estimate for that portion of the crack perimeter.

The crack front in region 2 corresponds closely to a tunnel crack and, therefore,

$$K_{I2} \approx \sigma\sqrt{\pi a_2} \qquad (2.50)$$

The elliptical crack front in region 3 can be represented by a circular crack of radius $a_3$ or a tunnel crack of width $2a_3$. Because of the general elliptical shape, the $K_I$ for this region is better described for a tunnel crack than for a circular crack. Therefore,

$$0.65\sigma\sqrt{\pi a_3} \ll K_{I3} < \sigma\sqrt{\pi a_3} \qquad (2.51)$$

A good estimate for $K_{I3}$ is

$$K_{I3} \approx 0.9\sigma\sqrt{\pi a_3} \qquad (2.52)$$

The crack front in region 3 can be approximated also by an elliptical crack with a minor axis equal to $a_3$ and for example a crack geometry, $a/2c$, of about 0.2. The decrease in the value of $a/2c$ from 0.5 for a circular crack to a value of 0.2 for the ellipse is equal to 0.3. Thus, the decrease in the value of $Q$ from 2.4 for a circular crack is 1.05 (i.e., 0.3 ×

$0.35/0.1 = 1.05$). Consequently, the value of $Q$ for the ellipse is 1.35 (i.e., $2.4 - 1.05 = 1.35$), and

$$K_{I3} \approx \sigma\sqrt{\frac{\pi a_3}{1.35}} = 0.86\sigma\sqrt{\pi a_3} \qquad (2.53)$$

This value is a good estimate of $K_{I3}$ especially for the lower portion of the crack front which fits the ellipse more closely than the upper portion. Because the upper portion of the perimeter has less curvature than the ellipse, the value for $K_{I3}$ in that region should be increased slightly. Consequently,

$$K_{I3} \approx 0.9\sigma\sqrt{\pi a_3} \qquad (2.54)$$

should be a very good estimate for this region.

## 2.5. Crack-Tip Deformation

The stress-field equations, Equations (2.14), (2.15), and (2.16), show that the elastic stress in the vicinity of a crack tip where $r \ll a$ can be very large. In reality, such high stress magnitudes do not occur because the material in this region undergoes plastic deformation, thus creating a plastic zone that surrounds the crack tip. Figure 2.21 is a schematic presentation

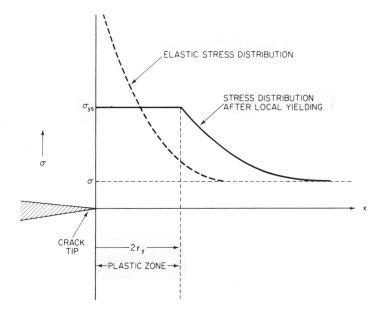

**Figure 2.21**  Distribution of the $\sigma_y$ stress component in the crack-tip region.

of the change in the distribution of the $y$ component of the stress caused by the localized plastic deformation in the vicinity of the crack tip.

The size of the plastic zone, $r_y$, can be estimated from the stress-field equations by treating the problem as one of plane stress and setting the $y$ component of stress, $\sigma_y$, equal to the yield strength, $\sigma_{ys}$, which results in[18]

$$r_y = \frac{1}{2\pi}\left(\frac{K}{\sigma_{ys}}\right)^2 \tag{2.55}$$

Irwin[19] suggested that the plastic-zone size under plane-strain conditions can be obtained by considering the increase in the tensile stress for plastic yielding caused by plane-strain elastic constraint. Under these conditions, the yield strength is estimated to increase by a factor of $\sqrt{3}$. Consequently the plane-strain plastic-zone size becomes

$$r_y = \frac{1}{6\pi}\left(\frac{K}{\sigma_{ys}}\right)^2 \tag{2.56}$$

The plastic zone along the crack front in a thick specimen is subjected to plane-strain conditions in the center portion of the crack front where $w = 0$ and to plane-stress conditions near the surface of the specimen where $\sigma_z = 0$. Consequently, Equations (2.55) and (2.56) indicate that the plastic zone in the center of a thick specimen is smaller than at the surface of the specimen. A schematic representation of the variation of the plastic-zone size along the front of a crack in a thick specimen is shown in Figure 2.22.

McClintock and Hult[20] investigated the effect of small plastic-zone size on the stress-field distribution in an elastic–perfectly plastic material that is subjected to Mode III deformation. Their results showed that the elastic-stress-field distribution in the vicinity of the plastic zone was identical to the elastic-stress-field distribution in the vicinity of a crack in a perfectly elastic material whose tip was placed at the center of the plastic zone. Irwin[18] suggested that the effect of small plastic zones corresponds to an apparent increase of the elastic crack length by an increment equal to $r_y$. This plastic-zone correction factor is valid for small plastic-zone sizes and should not be applied when large plastic deformations occur.

## 2.6. Superposition of Stress-Intensity Factors

Components that contain cracks may be subjected to one or more different types of Mode I loads such as uniform tensile loads, concentrated tensile loads, or bending loads. The stress-field distributions in the vicinity of the crack tip that is subjected to these loads are identical and are represented by Equation (2.1). Consequently, the total stress-intensity factor can be obtained by algebraically adding the stress-intensity factors that correspond

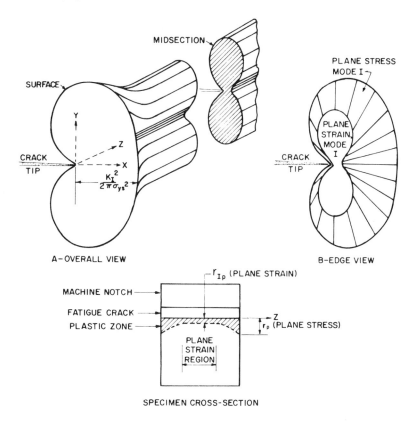

**Figure 2.22**  Schematic representation of plastic zone ahead of crack tip.

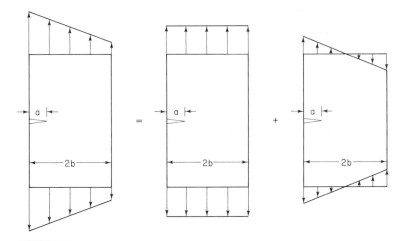

**Figure 2.23**  Superposition solution for tensile and bending stresses applied to a single-edge notched plate.

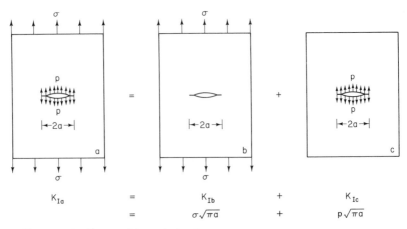

**Figure 2.24**  Superposition solution for a crack subjected to tensile stress and internal pressure.

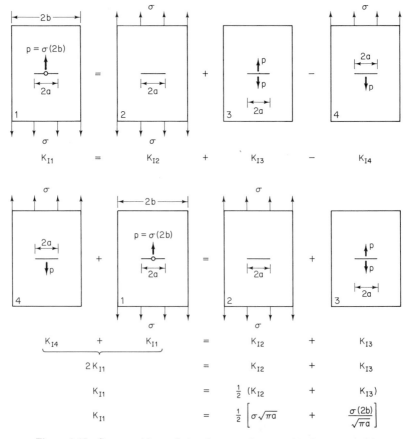

**Figure 2.25**  Superposition solution for a crack emanating from a pinhole.

to each load. Examples that illustrate the use of the superposition procedure to calculate the stress-intensity factor are presented in Figures 2.23 through 2.25.

Some components may be subjected to loads that correspond to various modes of deformation. Because the stress-field distributions in the vicinity of a crack, Equations (2.1), (2.2), and (2.3), are different for different modes of deformation, the stress-intensity factors for different modes of deformation cannot be added. Under these loading conditions, the total energy-release rate, $G$, rather than the stress-intensity factors, can be calculated by algebraically adding the energy-release rate for the various modes of deformation.

## 2.7. Crack-Tip Opening Displacement (CTOD) and the Dugdale Model

The tensile stresses applied to a body that contains a crack tend to open the crack and to displace its surfaces in a direction normal to its plane. For small crack-tip displacements and small plastic deformation at the crack tip, the stress and strain fields in the vicinity of the crack tip can be described by linear-elastic analyses. Under these loading conditions, the fracture instability can be predicted by using the plane-strain critical stress-intensity factor, $K_{Ic}$. As the size of the plastic zone and of the displacements at the crack tip increase, the stress and strain distributions in that neighborhood can be characterized better by using elastic-plastic analyses rather than linear-elastic analyses. Wells[21] argued that the opening displacement at the crack tip reflects the strain distribution in that region. He also proposed that fracture would initiate when the strains in the crack-tip region reach a critical value which can be characterized by a critical crack-tip opening displacement. This crack-tip opening displacement can be investigated by using an analytical model proposed by Dugdale.[22]

Dugdale proposed a simple analytical model that accounts, in part, for the plastic deformation at the tip of a stationary crack. This model, which sometimes is referred to as the strip-yield model, can be used to calculate the crack-tip opening displacement. In his model, Dugdale considered a through-thickness crack in an infinite plate that is subjected to a tensile stress normal to the plane of the crack, Figure 2.26. The crack is considered to have a length equal to $2a + 2r_y$. At each end of the physical crack, $2a$, there is a plastic zone of length $r_y$ that is subjected to yield-point stresses that tend to close the crack or, in reality, to prevent the crack from opening.

The plastic energy absorption rate and the crack-opening displacement for the Dugdale model were calculated by Goodier and Field.[23] For plane stress conditions, the crack-opening displacement, $\delta(a)$, is given by the equation

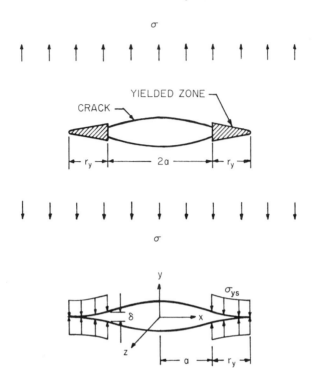

**Figure 2.26** The Dugdale crack.

$$\delta(a) = \frac{2(a + r_y)}{\pi E} \sigma_{ys} \left\{ \cos \phi \ln \left[ \frac{\sin (\beta - \phi)}{\sin^2 (\beta + \phi)} \right] \right.$$
$$\left. + \cos \beta \ln \left[ \frac{(\sin \beta + \sin \phi)^2}{(\sin \beta - \sin \phi)} \right] \right\} \qquad (2.57)$$

where  $\phi = \cos^{-1} [x/(a + r_y)]$.

$\beta = \frac{\pi}{2} \left( \frac{\sigma}{\sigma_{ys}} \right)$.

$\frac{r_y}{a} = \sec \beta - 1$.

$\sigma_{ys}$ = yield strength for the material.

$E$ = modulus of elasticity.

The crack-opening displacement at the tip of the physical crack is given by the equation

$$\delta = \frac{8\sigma_{ys} a}{\pi E} \ln \sec \beta \qquad (2.58)$$

or

$$\delta = \frac{8\sigma_{ys}a}{\pi E} \ln \sec \left( \frac{\pi}{2} \frac{\sigma}{\sigma_{ys}} \right) \qquad (2.59)$$

Using a series expansion for $\ln \sec [(\pi/2)(\sigma/\sigma_{ys})]$, this expression becomes

$$\delta = \frac{8\sigma_{ys}a}{\pi E} \left[ \frac{1}{2} \left( \frac{\pi}{2} \frac{\sigma}{\sigma_{ys}} \right)^2 + \frac{1}{12} \left( \frac{\pi}{2} \frac{\sigma}{\sigma_{ys}} \right)^4 + \frac{1}{45} \left( \frac{\pi}{2} \frac{\sigma}{\sigma_{ys}} \right)^6 + \cdots \right] \qquad (2.60)$$

For nominal stress values less than $\sigma_{ys}$, a reasonable approximation for $\delta$, using only the first term in the series, is

$$\delta = \frac{\pi \sigma^2 a}{E \sigma_{ys}} \qquad (2.61)$$

or

$$\delta = \frac{K_I^2}{E \sigma_{ys}} \qquad (2.62)$$

The strain-energy-release rate associated with crack extension is given by the equation[23]

$$G = 2G_e \left( \frac{\tan \beta}{\beta} - \frac{\ln \sec \beta}{\beta^2} \right) \qquad (2.63)$$

where $G_e = (\pi \sigma^2 a)/E$ is the strain-energy-release rate for a Griffith crack.

Combining Equations (2.58) and (2.63) results in the relation between the crack-tip opening displacement, $\delta$, and the energy release rate, $G$,

$$\delta = \frac{G}{\sigma_{ys}} \left( \frac{\beta \tan \beta}{\ln \sec \beta} - 1 \right)^{-1} \qquad (2.64)$$

For small values of $\beta$, that is, for low applied stress (in comparison with the yield strength of the material), Equation (2.64) demonstrates that $\delta$ can be approximated[19] by the relation

$$\delta = \frac{G}{\sigma_{ys}} \qquad (2.65)$$

Irwin[19] noted that this linear relationship between the crack-tip opening displacement, $\delta$, at the crack tip and the ratio of the strain-energy-release rate and the yield strength, $G/\sigma_{ys}$, is applicable even when the net section stresses approach the yield strength of the material. Combining equations (2.62) and (2.65) shows the relation between $G$ and $K_I$ is

$$G = \frac{K_I^2}{E} \qquad (2.66)$$

Additional discussion on the use of CTOD in design is presented in Chapter 17.

## 2.8. Relation Between Stress-Intensity Factor $(K_I)$ and Fracture Toughness $(K_{Ic})$

One of the underlying principles of fracture mechanics is that unstable fracture occurs when the stress-intensity factor at the crack tip, $K$, reaches a critical value. For Mode I deformation and for small crack-tip plastic deformation (plane-strain conditions), the critical stress-intensity factor for fracture instability is designated $K_{Ic}$. $K_{Ic}$ represents the inherent ability of a material to withstand a given stress-field intensity at the tip of a crack and to resist progressive tensile crack extension. Thus, $K_{Ic}$ represents the fracture toughness of a particular material, whereas $K_I$ represents the stress intensity ahead of a sharp crack in any material. Procedures used to measure $K_{Ic}$ are presented in Chapter 3. The fracture toughness behavior of structural metals as a function of temperature and rate of loading are presented in Chapter 4. The design relationship between $K_I$ and $K_{Ic}$ and a discussion of the use of fracture mechanics in design are presented in Chapter 6.

In general, $K_I$ should be kept below $K_{Ic}$ at all times to prevent fracture in the design of members with flaws in the same manner that $\sigma$ is kept below $\sigma_{ys}$ to prevent yielding in the design of members without flaws.

## References

1. A. A. GRIFFITH, "The Phenomena of Rupture and Flaw in Solids," *Transactions, Royal Society of London, A-221,* 1920.
2. C. E. INGLIS, "Stresses in a Plate due to the Presence of Cracks and Sharp Corners," *Proceedings, Institute of Naval Architects, 60,* 1913.
3. G. R. IRWIN, "Fracture Dynamics," in *Fracturing of Metals,* American Society of Metals, Cleveland, 1948.
4. E. OROWAN, "Fracture Strength of Solids," in *Report on Progress in Physics,* Vol. 12, Physical Society of London, 1949.
5. C. P. PARIS and G. C. SIH, "Stress Analysis of Cracks," in "Fracture Toughness Testing and Its Applications," *ASTM STP 381,* American Society for Testing and Materials, Philadelphia, 1965.
6. H. M. WESTERGAARD, "Bearing Pressures and Cracks," *Transactions, ASME, Journal of Applied Mechanics,* 1939.
7. G. R. IRWIN, "Analysis of Stresses and Strains Near the End of a Crack Traversing a Plate," *Transactions, ASME, Journal of Applied Mechanics, 24,* 1957.
8. H. TADA, P. C. PARIS, and G. R. IRWIN, ed., *Stress Analysis of Cracks Handbook,* Del Research Corporation, Hellertown, Pa., 1973.
9. G. C. SIH, *Handbook of Stress-Intensity Factors for Researchers and Engineers,* Institute of Fracture and Solid Mechanics, Lehigh University, Bethlehem, Pa., 1973.
10. O. L. BOWIE, "Rectangular Tensile Sheet with Symmetric Edge Cracks," *Transactions, ASME, Journal of Applied Mechanics, 31,* Series E. No. 2, June 1964.

11. B. Gross, J. E. Srawley, and W. F. Brown, Jr., "Stress Intensity Factors for a Single Edge Notch Tension Specimen by Boundary Collocation of a Stress Function," *NASA TN D-2395*, Aug. 1964.

12. S. R. Novak and J. M. Barsom, "Brittle Fracture ($K_{Ic}$) Behavior of Cracks Emanating from Notches," *Cracks and Fracture, ASTM STP No. 601*, American Society for Testing and Materials, Philadelphia, 1976, pp. 409–447.

13. J. C. Newman, Jr., "An Improved Method of Collocation for the Stress Analysis of Cracked Plates with Various Shaped Boundaries," *NASA Technical Note, NASA TN D-6376*, Aug. 1971.

14. G. R. Irwin, "The Crack Extension Force for a Part Through Crack in a Plate," *Transactions, ASME, Journal of Applied Mechanics, 29*, No. 4, 1962.

15. W. K. Wilson, "Stress Intensity Factors for Compact Specimens Used to Determine Fracture Mechanics Parameters," *Research Report 73-1E7-FMPWR-R1*, Westinghouse Research Laboratories, Pittsburgh, July 27, 1973.

16. W. K. Wilson, "Analytical Determination of Stress Intensity Factors for Manjoine Brittle Fracture Specimen," *Research Report AEC, WERL 0029-3*, Westinghouse Research Laboratories, Pittsburgh, Aug. 26, 1965.

17. W. K. Wilson, "Stress Intensity Factors for Compact Tension Specimen," *Research Memorandum 67-1D6-BTLFR-M1*, Westinghouse Research Laboratories, Pittsburgh, June 12, 1967.

18. G. R. Irwin, "Plastic Zone Near a Crack and Fracture Toughness," 1960 Sagamore Ordnance Materials Conference, Syracuse University, 1961.

19. G. R. Irwin, "Linear Fracture Mechanics, Fracture Transition, and Fracture Control," *Engineering Fracture Mechanics, 1*, No. 2, Aug. 1968.

20. F. A. McClintock and J. Hult, "Elastic-Plastic Stress and Strain Distribution Around Sharp Notches in Repeated Shear," Ninth International Congress of Applied Mechanics, New York, 1956.

21. A. A. Wells, "Unstable Crack Propagation in Metals—Cleavage and Fast Fracture," Cranfield Crack Propagation Symposium, 1, Sept. 1961, p. 210.

22. D. S. Dugdale, "Yielding of Steel Containing Slits," *Journal of Mechanics and Physics of Solids, 8*, 1960, p. 8.

23. J. H. Goodier and F. A. Field, "Plastic Energy Dissipation in Crack Propagation, *Fracture of Solids*, Wiley Interscience, New York, 1963, p. 103.

# Experimental Determination of $K_{Ic}$ and Other $K_c$ Values

## 3.1. General Overview

In Chapter 2 we described various analytical relationships for determining stress-intensity factors in elastic bodies with different shaped cracks. These stress-intensity factors ($K_I$, $K_{II}$, or $K_{III}$ for the opening, edge-sliding, or tearing modes of crack extension, as described in Chapter 2) are a function of load, crack size, and geometry. Ideally, a $K_I$ stress-intensity factor associated with a specific crack geometry (e.g., three of the most common geometries are the edge crack, surface crack, and through-thickness crack, as shown in Figure 3.1) can be used to model a particular crack geometry in an actual structure.

These particular stress-intensity values, $K_I$, are calculated for different load levels in the same general manner as particular stresses, $\sigma$, are calculated for different load levels in uncracked members subjected to tension. That is, for a member loaded in tension, the stress is calculated as $\sigma = P/A$ for various loads, $P$. In an analogous fashion, the stress-intensity is calculated as $K_I = C\sigma\sqrt{a}$ for various nominal stress levels, $\sigma$. Note that the calculations for stress-intensity factors, $K_I$, $K_{II}$, or $K_{III}$ (analogous to the calculation of stress, $\sigma$) are the same for any structural material as long as the general boundary conditions described in Chapter 2 are satisfied.

However, because actual structural materials have certain limiting characteristics (e.g., yielding in ductile materials or fracture in brittle materials), there are limiting values of both $\sigma$ and $K_I$, namely, $\sigma_{ys}$ (the yield strength) and $K_{Ic}$, $K_{Id}$, or $K_c$ (critical stress-intensity factors).

The critical stress-intensity factor, $K_c$, at which unstable crack growth occurs for conditions of *static* loading at a particular temperature actually depends on specimen thickness or constraint, as shown in Figure 3.2. The

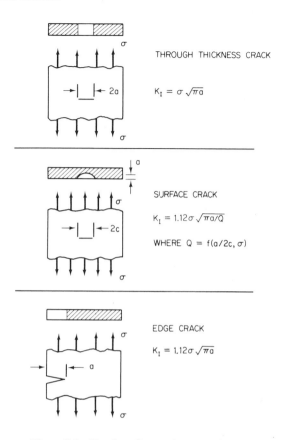

**Figure 3.1** $K_I$ values for crack geometries.

limiting value of $K_c$ for a particular test temperature and slow loading rate is $K_{Ic}$ for plane-strain (maximum constraint) conditions. Under conditions of *dynamic* (or impact) loading, the critical value is defined as $K_{Id}$.

After considerable study and experimental verification, a special ASTM committee on fracture testing, Committee E-24, prepared a "Standard Test Method for Plane Strain Fracture Toughness of Metallic Materials" (ASTM Designation: E-399—83),[1] which is used to measure $K_{Ic}$. This test method covers the determination of the plane-strain fracture toughness, $K_{Ic}$, of metallic materials by tests using a variety of fatigue-cracked specimens and is restricted to linear-elastic behavior.

In the following sections we shall describe the background of the $K_{Ic}$ test method and the details of the particular testing procedures. In Chapter 17, we shall describe the various elastic-plastic fracture test methods, that is, *J*-integral, *R*-curve,[2] and crack-tip opening displacement (CTOD).

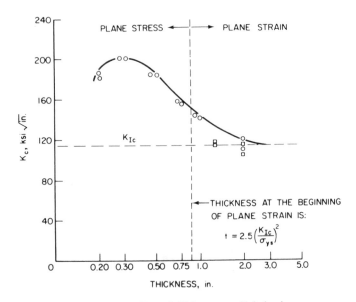

**Figure 3.2**   Effect of thickness on $K_c$ behavior.

## 3.2. Background of $K_{Ic}$ Test Method

A detailed review of the overall background and development of the E-399—83 test method for $K_{Ic}$ is presented in *ASTM STP 463*, "Review of Developments in Plane Strain Fracture Toughness Testing."[3] In one of the papers in *STP 463*, "Progress in Fracture Testing of Metallic Materials," J. G. Kaufman[3] stated that early in the development of the test methods, it was established that elastic fracture mechanics was the best analysis by which the resistance of materials to unstable crack growth could be described. Although it was realized that most real structural materials do not behave in a purely elastic manner on fracturing, it was hoped that it would be possible with appropriate modification to take into account the finite sizes of structural members and test specimens. If the small crack-tip plasticity could be accounted for, it was hoped that laboratory specimens could be used to recreate and describe the situation of unstable crack growth occurring in a large structure. Thus, the early work of the ASTM committee was directed toward work on the elastic-fracture problem, even though most structural materials fracture in an inelastic or elastic-plastic mode.

Early in this program it was realized that the critical stress-intensity factor for unstable crack growth, $K_c$, was a thickness-dependent property as shown by test data for 7075–T6 and –T651 sheet and plate materials (Figure 3.3). Over a certain range of thickness for this material (as well as other structural materials), the critical combination of load and the crack length at instability, that is, $K_c$, decreases with an increase in thickness,

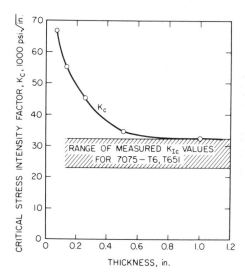

**Figure 3.3**  Test data for 7075–T6 and –T651 aluminums (Ref. 1).

reaching a rather constant minimum value, $K_{Ic}$, when plane-strain conditions are approached. In the fully plane-strain condition, the crack instability leading to complete fracture almost immediately follows crack initiation. It was also observed that the *initial* crack growth, whether stable or unstable, took place at about the same level of stress intensity for a given material regardless of specimen thickness. This initial crack growth was referred to as *pop-in* and led to the belief that a lower level of critical stress-intensity factor associated with plane-strain conditions, $K_{Ic}$, could be determined reproducibly. Thus, although much of the early work on fracture was for thin sheet specimens, it became clear that not only was thickness an important variable, but because fracture toughness decreased with increasing thickness to an apparently constant value ($K_{Ic}$ at a given temperature and loading rate), this value of $K_{Ic}$ might be considered to be a material property in the same manner that yield strength is considered to be a material property. If this were the case, then particular $K_{Ic}$ levels could be established for various structural materials in the same manner that yield strength values are established for various structural materials, as long as the thickness of plate was large enough to establish plane-strain conditions. The $K_{Ic}$ value would also be a minimum value for conditions of maximum structural constraint (for example, stiffeners, intersecting plates, etc.) that might lead to plane-strain conditions even though the individual structural members might be relatively thin.

Therefore, emphasis on the early ASTM E-24 fracture committee work shifted to the development of methods to determine $K_{Ic}$, because this might be a more fundamental property that would lead to a firmer basis for the development and handling of the overall fracture problem. Knowledge that $K_{Ic}$ provided a ''conservative'' approach to the fracture problem and that the material behavior in structural failures, even in relatively thin members, might well be controlled by plane-strain fracture conditions led to further emphasis on this particular approach. The committee felt that once the plane-strain problem was solved, attention could then be redirected to the thin-section problem ($K_c$ or $R$-curve analysis) and the inelastic-behavior problem ($J_{Ic}$ or CTOD). This was the case, and the latter techniques are described in Chapter 17.

Numerous specimen designs, test methods, and so on were considered, and there was considerable research during the 1960s and 1970s to study the various parameters that might affect plane-strain fracture behavior. The effects of notch acuity, stress level during fatigue precracking, plate thickness, fracture appearance, and others were all investigated and resulted in the development of a standardized, plane-strain $K_{Ic}$ test method, using either the notch-bend specimen or the compact tension (CT) specimen.

Using the recommended test specimens, which are shown in Figures 3.4 and 3.5, round robin test programs were conducted to establish that the recommended test procedure did indeed give reproducible results. The results of that program indicated that reproducible $K_{Ic}$ values could be obtained by different laboratories within about 15 percent.

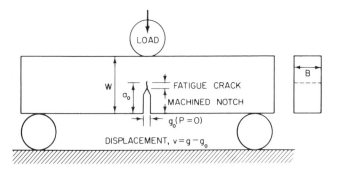

**Figure 3.4**   Essentials of bend specimen for $K_{Ic}$ test.

Thus, the test procedure was made a standard ASTM test method, E-399—83.[1] In subsequent sections in this chapter we shall describe the details of this test procedure as used to measure $K_{Ic}$ for particular structural materials at a given loading rate and temperature.

## 3.3. Specimen Size Requirements

The dominant advantage of measuring the notch toughness of structural materials in terms of $K_{Ic}$ is that this $K_{Ic}$ value can be compared directly to the various $K_I$ levels calculated for a particular structure of different design loadings. Prior to the development of fracture mechanics, the engineer had

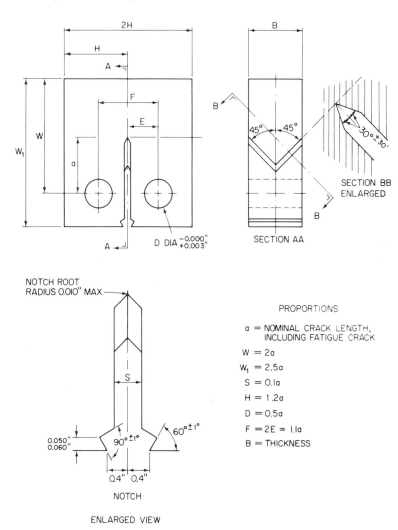

**Figure 3.5**   Compact-tension fracture-toughness specimen.

no such direct comparison of the fracture resistance of a structural material to the fracture behavior of that same material as measured in a laboratory test. The only comparison was that of empirical relations or service experience.

However, to model the analytical development of $K_I$ and the various structural cases properly, $K_{Ic}$ should be determined fairly accurately. Hence, the ASTM standard test method for $K_{Ic}$ is very restrictive with respect to specimen size requirements in order to obtain elastic plane-strain behavior. This restriction, while being necessary to use the linear-elastic methods of analysis described in Chapter 2, limits the applicability of the $K_{Ic}$ approach to relatively brittle structural materials, or to low testing temperatures that are below normal service temperatures, or to very high rates of loading.

In *ASTM STP 410*,[4] Brown and Srawley point out that the accuracy with which $K_{Ic}$ describes the fracture behavior of real materials depends on how well the stress-intensity factor represents the conditions of stress and strain inside the actual fracture process zone. The fracture process zone is the extremely small region just ahead of the tip of a crack where crack extension would originate (Figure 3.6). In this sense $K_I$ is exact only in the case of zero plastic strain, which occurs only in ideally brittle materials. For most structural materials, a sufficient degree of accuracy may be obtained if the plastic zone ahead of the crack tip is small in comparison with the region around the crack in which the stress-intensity factor yields a satisfactory approximation of the exact elastic-stress field. That is, as shown schematically in Figure 3.7, a small amount of yielding (plastic zone) ahead of the crack tip does not change the overall distribution of the stress field away from the crack front. Any loss in accuracy of the analysis associated with increasing the relative size of the plastic zone is gradual, and it is not possible to prescribe theoretical limits on the applicability of linear elastic fracture mechanics. Obviously, the decision of what is sufficient accuracy depends on the particular application. However, because a *standardized ASTM test method must be reproducible,* the specimen size requirements were chosen

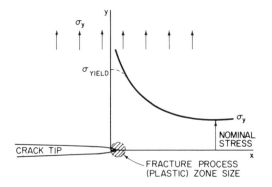

**Figure 3.6**   Fracture process zone size ahead of a crack.

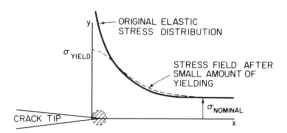

**Figure 3.7**   Schematic showing small effect of yielding on nominal stress field.

so that there should be essentially no question regarding this point. As discussed later, there are situations where satisfactory engineering decisions can be made even though the accuracy of the analysis is not as good as specified in the E-399—83 test method. These situations are discussed in Chapter 5 and 6, namely, correlations and design procedures.

Figure 3.8 shows the elastic-stress-field distribution ahead of a crack as developed in Chapter 2. The extent of the plastic zone ahead of the crack front can be estimated by using the following expression for stress in the $y$ direction, $\sigma_y$:

$$\sigma_y = \frac{K_I}{\sqrt{2\pi r}} \cos\frac{\theta}{2}\left(1 + \sin\frac{\theta}{2} \sin\frac{3\theta}{2}\right)$$

for $\theta = 0$ (along the $x$ axis)

$$\sigma_y = \frac{K_I}{\sqrt{2\pi r}}$$

Letting $\sigma_y = \sigma_{\text{yield stress}}$ ($\sigma_{ys}$), which is the 0.2 percent offset yield strength of the material at a particular temperature and loading rate, the extent of yielding ahead of the crack is

$$r_y = \frac{1}{2\pi}\left(\frac{K_I}{\sigma_{ys}}\right)^2$$

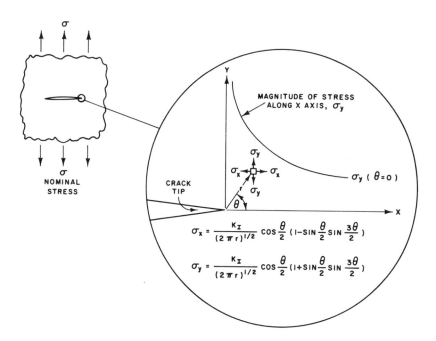

**Figure 3.8**   Elastic-stress-field distribution ahead of crack.

At instability, $K_I = K_c$, and the limiting value of $r_y$, or the plastic zone, is

$$r_y = \frac{1}{2\pi}\left(\frac{K_c}{\sigma_{ys}}\right)^2$$

This value of $r_y$ is estimated to be the plastic-zone radius at instability under plane-stress conditions.

As shown in Figure 3.9, this value is assumed to occur at the surface of a plate where the lateral constraint is zero and plane-stress conditions exist. Because of the increase in the tensile stress for plastic yielding under plane-strain conditions (Chapter 2), the plastic-zone radius at the center, where the constraint is greater and plane-strain conditions exist, is equal to one-third of this value, or

$$r_{y(\text{plane strain})} \simeq \frac{1}{6\pi}\left(\frac{K_{Ic}}{\sigma_{ys}}\right)^2$$

That is, the plain-strain yield strength is assumed to be equal to $\sqrt{3}$ times greater than the uniaxial (plane-stress) yield strength. Thus the relative

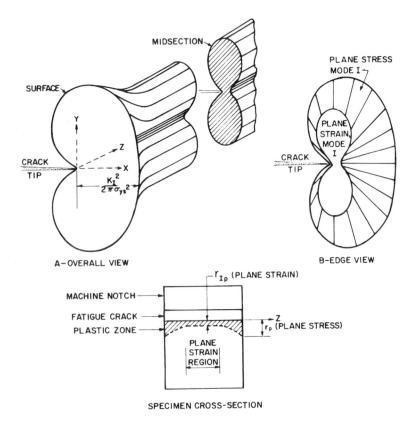

**Figure 3.9**   Schematic of plastic-zone size.

plastic-zone size ahead of a sharp crack is proportional to the $(K_{Ic}/\sigma_{ys})^2$ value of a particular structural material.

In establishing the specimen size requirements for $K_{Ic}$ tests, the specimen dimensions should be large enough compared with the plastic zone, $r_y$, so that any effects of the plastic zone on the $K_I$ analysis can be neglected. The pertinent dimensions of plate specimens for $K_{Ic}$ testing are crack length $(a)$, thickness (denoted as $B$ in the ASTM standard), and the remaining uncracked ligament length ($W - a$, where $W$ is the overall specimen depth). After considerable experimental work,[4] the following minimum specimen size requirements to ensure elastic plane-strain behavior were established:

$$a \geqslant 2.5\left(\frac{K_{Ic}}{\sigma_{ys}}\right)^2$$

$$B \geqslant 2.5\left(\frac{K_{Ic}}{\sigma_{ys}}\right)^2$$

$$W \geqslant 5.0\left(\frac{K_{Ic}}{\sigma_{ys}}\right)^2$$

The following calculation shows that for specimens meeting this requirement, the specimen thickness is approximately 50 times the radius of the plane-strain plastic-zone size:

$$\frac{\text{specimen thickness}}{\text{plastic-zone size}} = \frac{B}{r_y} \simeq \frac{2.5(K_{Ic}/\sigma_{ys})^2}{(1/6\pi)(K_{Ic}/\sigma_{ys})^2} \simeq 2.5(6\pi) \simeq 47$$

Thus, the restriction that the plastic zone be "contained" within an elastic-stress field certainly appears to be satisfied. In fact, during the development of the recommended test method, there was considerable debate about whether or not this requirement was too conservative. As described later, test results indicated that the requirement was very conservative. However, because it was felt that there should be no ambiguity regarding an ASTM standard, Brown and Srawley, in *ASTM STP—463*,[3] stated that

it has to be accepted that it will frequently not be possible to determine $K_{Ic}$ for a given material in any meaningful sense, usually because the material is not available in sufficient thickness, but sometimes for another reason. The concept of $K_{Ic}$ is fundamentally incompatible with a test method that could be applied to all available forms of materials, and neglect of this fact can only result in misrepresentation of the relative merits of different materials. The issue, therefore, is whether or not the ASTM Test Method could be made less restrictive with regard to specimen dimensions without significant risk that spurious results would be obtained in some cases. In our opinion there is no convincing evidence that the provisions of ASTM Method E 399-70 T are unduly restrictive; on the contrary, there is reason to suppose that they are not restrictive enough for some materials which are of technological importance. The test method should be applicable to any material that is available in sufficient bulk, and its provisions would be more than enough in other cases, but it is not feasible to discriminate between the more and the less favorable materials. It follows that the discovery of any number of favorable cases does not justify any relaxation of the provisions of the test method, though the discovery of a single unfavorable case may make it necessary to further restrict the test method. There is a further consideration which should not be neglected. Although the direct application of the ASTM Method E 399-70 T is restricted unavoidably to those materials that can be obtained in sufficient bulk, it does serve as a primary reference method for other fracture toughness tests that are based more empirically but less restricted in scope of application. This is an important reason why any margin of error in the provisions should be on the side of restrictiveness rather than uncertainty about the meaning of the results.

Accordingly, the preceding specimen size requirements were adopted into ASTM Test Method E-399—83, "Standard Method of Test for Plane Strain Fracture Toughness of Metallic Materials."

By adhering to the test specimen dimensions ($a$, $B$, $W$), the following two essential conditions are satisfied:

1. The test specimen is large enough so that linear-elastic behavior of the material being tested occurs over a large enough stress field so that any effect of the plastic zone ahead of the crack can be neglected.
2. There is a triaxial tensile stress field present such that the shear stress is very low compared to the maximum normal stress and a plane-strain opening mode behavior would be expected (Mode I, Chapter 2).

Note that before a $K_{Ic}$ test specimen can even be machined, the $K_{Ic}$ value *to be obtained* must already be known or at least estimated. Two general means of sizing test specimens before the $K_{Ic}$ value is even known are as follows:

1. Overestimate the $K_{Ic}$ value on the basis of experience with similar materials and judgment based on other types of notch-toughness tests. In Chapter 5 we shall describe various empirical correlations with other types of notch-toughness tests that can be used, such as the CVN impact test specimen.
2. Use specimens that have as large a thickness as possible, namely, a thickness equal to that of the plates to be used in service.

*Fortunately* for the structural designer (because such an individual would like his or her structure to be built from materials that do *not* exhibit elastic-plane-strain behavior) but *unfortunately* for the materials engineer responsible for determining $K_{Ic}$ values, many low- to medium-strength structural materials in the section sizes of interest for most large structures (such as ships, bridges, pressure vessels) are of insufficient thickness to maintain plane-strain conditions under slow loading and at normal service temperatures. In those cases, the linear-elastic analysis used to calculate $K_{Ic}$ values is invalidated by general yielding and the formation of large plastic zones. Under these conditions, which occur in the transition range approaching plane stress (mixed mode) and general yielding, alternative methods must be used for fracture analysis, as described in Chapters 5 and 15–17. Nonetheless, the basis for all subsequent fracture criteria and fracture control rests on a knowledge (or estimate) of how much a material exceeds $K_{Ic}$-type behavior at the service temperature and loading rate.

## 3.4. $K_{Ic}$ Test Procedure

The general test procedure to determine $K_{Ic}$ as well as a commentary on the significance of various aspects of that test method are presented in this section. The steps in conducting a $K_{Ic}$ test are as follows:

**1. Determine critical specimen size dimensions.**

$$a = \text{crack depth} \geqslant 2.5\left(\frac{K_{Ic}}{\sigma_{ys}}\right)^2$$

$$B = \text{specimen thickness} \geqslant 2.5\left(\frac{K_{Ic}}{\sigma_{ys}}\right)^2$$

$$W = \text{specimen depth} \geqslant 5.0\left(\frac{K_{Ic}}{\sigma_{ys}}\right)^2$$

Methods to estimate the probable $K_{Ic}$ value were discussed in Section 3.3.

**2. Select a test specimen.**    A variety of fatigue-cracked test specimens can be used to determine the plain-strain fracture toughness ($K_{Ic}$) of metallic materials as described in ASTM Specification E-399—83. These test specimens include

Three-point bend specimen (Figure 3.10)

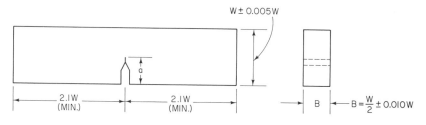

**Figure 3.10**  Three-point bend specimen.

Compact tension specimen (Figure 3.11)

Arc-shaped specimen (Figure 3.12)

Disk-shaped compact specimen (Figure 3.13)

The three-point bend specimen and the compact tension specimen have been widely used for years for testing plate materials from various applications. The arc-shaped specimen is generally used for material taken from a cylindrical geometry, for example, pressure vessels or piping. The disk-shaped compact specimen is relatively new and has been used for specimens machined from circular blanks, for example, from round bars or from cores drilled from actual structures in the field.

The initial machined crack length, $a$, should be $0.45W$ so that the crack can be extended by fatigue to approximately $0.5W$. As stated previously, the $K_{Ic}$ measurement capacity is limited primarily by the thickness of the material available since $a$ and $W$ can generally be made quite large, subject only to machining and testing capabilities. Thus selection of the specimen thickness, $B$, is usually made first.

**3. Fatigue-crack the test specimen.**     The purpose of notching the test specimen is to simulate an ideal plane crack with essentially zero root radius to agree with the assumption made in the $K_I$ analysis. Because a fatigue crack is considered to be the sharpest crack that can be reproduced in the laboratory, the machine notch is extended by fatigue. The fatigue

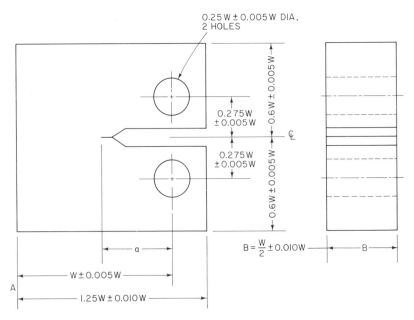

**Figure 3.11**  Compact tension specimen.

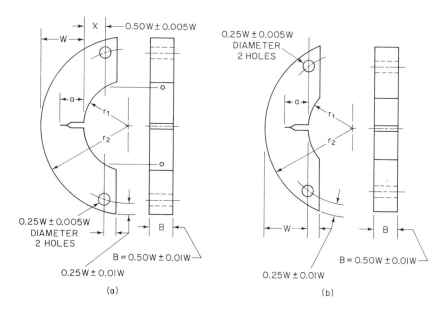

**Figure 3.12**   Arc-shaped specimen. (a) $X/W = 0.5$; (b) $X/W = 0$.

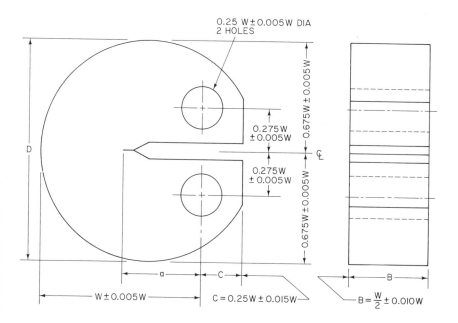

**Figure 3.13**   Disk-shaped compact specimen.

crack should extend at least $0.05W$ ahead of the machined notch to eliminate any effects of the geometry of the machined notch. Examples of notch geometry are presented in Figure 3.14. It has been found that a chevron notch has several advantages compared with a straight crack front. The chevron notch has been found to keep the crack in plane and to ensure that it extends well beyond the notch root $(0.05W)$. Thus the machining operation for the original chevron notch is not so critical as for a straight machined crack front. If a straight machined crack front is used, it is difficult to produce a fatigue crack that is uniform unless an extremely sharp uniform machined notch is used, such as might be obtained by electrodischarge machining (EDM).

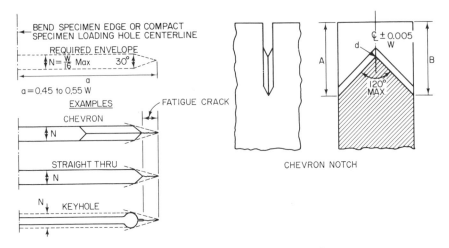

**Figure 3.14**   Notch geometry for $K_I$ test specimens.

To ensure that the plastic-zone size during the final fatigue cycle is less than the plastic-zone size during actual $K_{Ic}$ testing, the last 2.5 percent of the overall length of notch plus fatigue crack is loaded at a maximum stress-intensity level during fatigue, $K_{f(max)}$, such that $K_{f(max)}/E \leq 0.002^{1/2}$. $K_{f(max)}$ is the maximum $K_I$ level to which the specimen is subjected during cyclic loading and is calculated using the general expression for $K_Q$ given later. $K_Q$ is a conditional calculation of $K_{Ic}$ based on the actual fracture test results, but the expression for $K_Q$ is a general one for $K_I$. Figure 3.15 shows the general relation among $K_{f(max)}$, $\Delta K_f$, and $K_Q$ for corresponding values of load.

$K_{f(max)}$ must not exceed 60 percent of the $K_Q$ value determined from the actual fracture test results. Fatigue cracking should be considered as a special type of critical machining operation because cracks produced at high stress-intensity levels (high values of $K_{f(max)}$) can significantly affect the subsequent fracture test results. Control of the plastic-zone size during

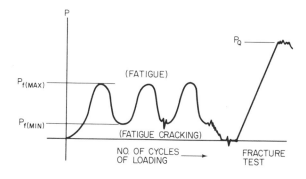

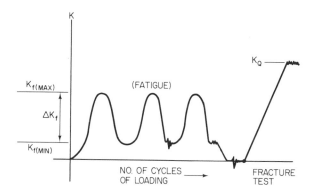

**Figure 3.15**  Relation among $P_{max}$, $K_{max}$, $\Delta K_{fatigue}$, and $K_Q$.

fatigue cracking is particularly important when the fatigue cracking is done at room temperature and the actual $K_{Ic}$ test is conducted at lower temperatures. In this case, $K_{f(max)}$ at room temperature must be kept to very low values so that the plastic-zone size corresponding to $K_Q$ at low temperatures is smaller than the plastic-zone size corresponding to $K_{f(max)}$ at room temperature.

Results of a limited study of the effects of stress level during fatigue cracking of air-melted (AM) 18Ni8Co3Mo maraging steel are summarized in Table 3.1 and Figure 3.16. The bend test results presented in Figure 3.16 showed that as long as the nominal stress for fatigue cracking was less than about 20 percent of the yield strength the $K_{Ic}$ values averaged 105 ksi$\sqrt{\text{in}}$. Stressing at about 30 percent of the yield strength increased the apparent $K_{Ic}$ value to 120 ksi$\sqrt{\text{in}}$.

In terms of $K_{f(max)}/E$, stressing at 25 percent of the yield strength corresponded to a $K_{f(max)}/E$ value of about 0.0012, which is less than the specification limit of 0.002. However, values of $K_{f(max)}$ and $K_{f(max)}/K_{Ic}$ presented in Table 3.1 show that even the restriction that $K_{f(max)} \leq 0.6K_Q$ may not be sufficient to ensure that fatigue cracking has no effect on the subsequent $K_{Ic}$ test results.

**TABLE 3.1  Analysis of $K_{Ic}$ Results for 18Ni8Co3Mo AM Steel (193-ksi yield strength)**

| Specimen Number | Nominal Stress for Fatigue Cracking (psi) | Number of Fatigue Cycles | $K_Q$ (ksi$\sqrt{in.}$) | Nominal Stress (ksi) | $K_{f(max)}$ | $\dfrac{K_{f(max)}}{K_{Ic}}$ |
|---|---|---|---|---|---|---|
| 18N-2 | 57,000 | 8,000 | 119 | 118 | 58 | 0.55 |
| 18NF-8a | 57,000 | 6,400 | 121 | 115 | 60 | 0.57 |
| 18NF-11 | 57,000 | 6,700 | 121 | 116 | 60 | 0.57 |
| 18NF-12 | 45,000 | 15,300 | 113 | 90 | 57 | 0.54 |
| 18NF-1 | 33,000 | 31,900 | 106 | 106 | 33 | 0.31 |
| 18NF-7 | 31,000 | 33,600 | 104 | 124 | 36 | 0.34 |
| 18NF-5 | 24,000 | 67,400 | 103 | 102 | 24 | 0.23 |
| 18NF-3 | 28,000 | 70,800 | 99 | 101 | 27 | 0.26 |
| 18NF-10 | 11,000 | 156,000 | 106 | 106 | 11 | 0.10 |
| 18N-3 | 18,000 | 123,000 | 107 | 108 | 18 | 0.17 |

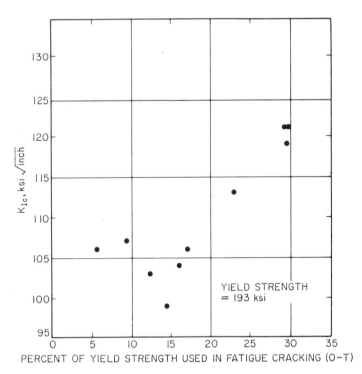

**Figure 3.16**  Effect of fatigue cracking on $K_{Ic}$.

This behavior is similar to that observed by Brown and Srawley[3] and confirms their statement that fatigue cracking of $K_{Ic}$ specimens should be conducted at the lowest practical stress level.

**4. Obtain test fixtures and displacement gauge.**  Recommended test fixtures for the various types of $K_{Ic}$ specimens are described in Reference 1. These fixtures were developed to minimize friction and have been used successfully by numerous laboratories. Other fixtures can be used as long as good alignment is maintained and frictional errors are minimized.

A key item in the $K_{Ic}$ test is the accurate measurement of some quantity that can be related to the beginning of crack extension from the fatigue crack, that is, unstable crack growth. The basic measurement selected is the relative displacement of two points located symmetrically on opposite sides of the crack plane, as shown in Figure 3.17. An extremely sensitive and highly linear displacement gauge that has no lost motion between the gauge and the locating portions on the specimen has been developed and is shown in Figure 3.18.

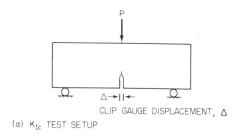

CLIP GAUGE DISPLACEMENT, $\Delta$

(a) $K_{Ic}$ TEST SETUP

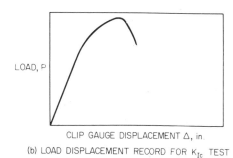

LOAD, P

**Figure 3.17**  Schematic showing displacement measurement for $K_{Ic}$ test. (a) $K_{Ic}$ test setup; (b) load displacement record.

CLIP GAUGE DISPLACEMENT $\Delta$, in.

(b) LOAD DISPLACEMENT RECORD FOR $K_{Ic}$ TEST

**5. Test procedure**

a. *Test setup*.  Center the specimen in the loading fixtures to ensure concentricity of loading. Seat the displacement gauge such that a continuous load-displacement record will be obtained throughout the

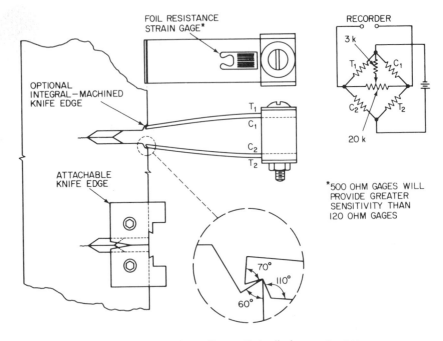

**Figure 3.18**   Double-cantilever clip-in displacement gauge.

test. Typical test setups for three-point bend and compact tension specimens are shown in Figures 3.19 and 3.20, respectively.

b. *Loading rate.* Load the specimen at a rate such that the rate of increase of stress intensity, $K_I$, is within the range from 30 to 150 ksi$\sqrt{\text{in.}}$/min (0.55 to 2.75 M Pa $\cdot$ m$^{1/2}$/sec). Detailed loading rates for each type of specimen are given in the E-399—83 Test Method.

c. *Test record.* A test record consisting of an autographic plot of the output of the load-sensing transducer versus the output of the displacement gauge should be obtained. The initial slope of the linear portion should be between 0.7 and 1.5. It is conventional to plot the load along the vertical axis, as in an ordinary tension test record. Select a combination of load-sensing transducer and autographic recorder so that the maximum load can be determined from the test record with an accuracy of $\pm 1$ percent. With any given equipment, the accuracy of readout will be greater the larger the scale of the test record. Continue the test until the specimen can sustain no further increase in load. A schematic test record is presented in Figure 3.21.

d. *Measurements.* Measurements of the specimen dimensions and fracture surfaces and features should be made as required to calculate $K_Q$ ($B, S, W, a$; Figures 3.10 through 3.13). The crack length, $a$, should

**Figure 3.19**   Three-point bend test setup.

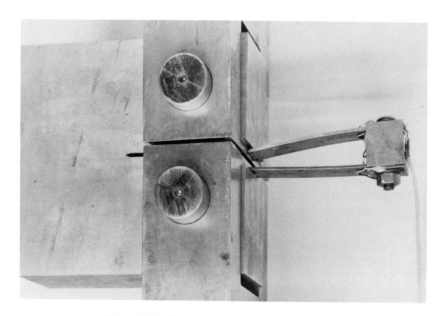

**Figure 3.20**   Compact-specimen $K_{Ic}$ test setup.

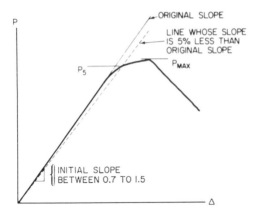

P

ORIGINAL SLOPE

LINE WHOSE SLOPE
IS 5% LESS THAN
ORIGINAL SLOPE

$P_5$

$P_{MAX}$

INITIAL SLOPE
BETWEEN 0.7 TO 1.5

Δ

**Figure 3.21**   Schematic $P$–Δ test record.

be measured to the nearest 0.5 percent at the center of the crack front and midway between the center and the end of the crack front on each side (Figure 3.22). Use the average of these three measurements as the crack length in subsequent calculations provided that the length of either surface trace is within 10 percent of the average crack length.

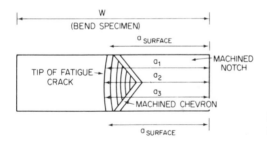

W
(BEND SPECIMEN)

$a$ SURFACE

TIP OF FATIGUE
CRACK

$a_1$
$a_2$
$a_3$

MACHINED
NOTCH

MACHINED CHEVRON

$a$ SURFACE

**Figure 3.22**   Various crack-length measurements to be made.

**6. Analysis of $P$–Δ records.**   If a material exhibited perfectly elastic behavior until fracture, the load-displacement curve would be merely a straight line until fracture. However, even very brittle structural materials exhibit some nonlinear behavior, and thus the idealized perfectly elastic load-displacement curve is rarely seen. The principal types of load-displacement curves observed are presented in Figure 3.23, which shows that considerable variation in behavior occurs for different structural materials. This is not unexpected, however, because of the wide variety of materials used in different structural applications.

To establish that a valid $K_{Ic}$ has been determined (that is, one that has satisfied the various restrictions to ensure that plane-strain behavior was obtained), it is necessary first to calculate a conditional result, $K_Q$, which involves a construction on the test record. Then it must be determined whether or not this $K_Q$ value is consistent with the size and yield strength

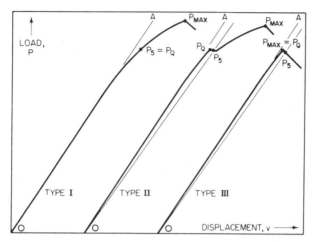

**Figure 3.23**   Types of load-displacement curves illustrating procedure for determination of $K_{Ic}$.

of the specimen according to the general requirements that

$$a \geq 2.5\left(\frac{K_Q}{\sigma_{ys}}\right)^2$$

$$B \geq 2.5\left(\frac{K_Q}{\sigma_{ys}}\right)^2$$

$$W \geq 5.0\left(\frac{K_Q}{\sigma_{ys}}\right)^2$$

If this $K_Q$ value meets the foregoing requirements, as well as other subsequent requirements, then $K_Q = K_{Ic}$. If not, the test is invalid, and whereas the results may be used to *estimate* the crack toughness of a material, they are not valid ASTM standard values.

As noted in Figure 3.23, the general types of $P$–$\Delta$ curves are not perfectly elastic but do exhibit different degrees of nonlinearity. Various criteria to establish the load corresponding to $K_{Ic}$ were considered, such as initial deviation from linearity, maximum load, specified offset, and a secant offset. After considerable experimentation, a 5 percent secant offset was chosen to define $K_{Ic}$ as the critical stress-intensity factor at which the crack reaches an effective length equal to 2 percent greater than that at the beginning of the test. Although somewhat arbitrary, this is analogous to defining the 0.2 percent offset yield strength for materials that do not have a well-defined yield point.

The procedure consists of drawing a secant line from the origin (slight nonlinearity at the very beginning of a record can be ignored) with a slope 5 percent less than that of the tangent *OA* to the initial part of the record.

The load, $P_5$, is the load at the intersection of the secant with the test record (Figure 3.24).

Define $P_Q$ according to the following procedure. If the load at every point on the $P-\Delta$ record which precedes $P_5$ is lower than $P_5$, then $P_Q$ is $P_5$ (Figure 3.23, Type I). If, however, there is a maximum load preceding $P_5$ that is larger than $P_5$, then this load is $P_Q$ (Figure 3.23, types II and III). Thus $P_Q = P_5$ in the example shown in Figure 3.24. If $P_{max}/P_Q$ is greater than 1.10, the test is not a valid test because it is possible that $K_Q$ is not representative of $K_{Ic}$.

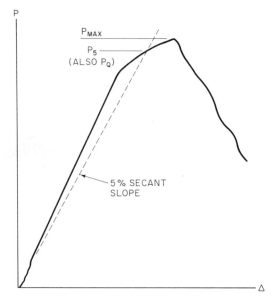

**Figure 3.24**   Schematic showing $P_Q = P_{5\%}$ and $P_{max}$.

For relatively tough structural materials, $P_{max}$ is usually $>1.10P_5$ as shown in typical $P-\Delta$ records for metals that are too tough to exhibit plane-strain behavior (Figure 3.25). For these tough materials, a "roundhouse" $P-\Delta$ curve is obtained, indicating general-yielding–type behavior rather than plane-strain behavior. Plane-strain behavior does not exist for these cases, and other test methods, such as those presented in Chapter 17, must be used.

**7. Calculation of conditional $K_{Ic}$ ($K_Q$).**   After determining $P_Q$ for either the bend specimen or the compact-tension specimen, calculate $K_Q$ using the following expression:

*Bend specimen (Figure 3.10).*

$$K_Q = \frac{P_Q S}{B \cdot W^{3/2}} \cdot f\left(\frac{a}{W}\right)$$

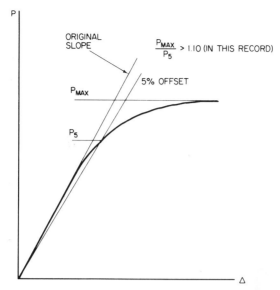

**Figure 3.25**  $P$–$\Delta$ test record for structural material with high level of notch toughness.

where $P_Q$ = load as determined, 1b.
   $B$ = thickness of specimen, in.
   $S$ = span length, in.
   $W$ = depth of specimen, in.
   $a$ = crack length as determined, in.

To facilitate calculation of $K_Q$, values of the power series $f(a/W)$ in the foregoing expression are tabulated in the following table for specific values of $a/W$ for bend specimens:

| $a/W$ | $f(a/W)$ | $a/W$ | $f(a/W)$ |
|-------|----------|-------|----------|
| 0.450 | 2.29 | 0.500 | 2.66 |
| 0.455 | 2.32 | 0.505 | 2.70 |
| 0.460 | 2.35 | 0.510 | 2.75 |
| 0.465 | 2.39 | 0.515 | 2.79 |
| 0.470 | 2.43 | 0.520 | 2.84 |
| 0.475 | 2.45 | 0.525 | 2.89 |
| 0.480 | 2.50 | 0.530 | 2.94 |
| 0.485 | 2.54 | 0.535 | 2.99 |
| 0.490 | 2.58 | 0.540 | 3.04 |
| 0.495 | 2.62 | 0.545 | 3.09 |
|       |      | 0.550 | 3.14 |

*Compact specimen (Figure 3.11).*   For the compact-tension specimen, calculate $K_Q$ in units of psi$\sqrt{\text{in.}}$ as follows:

$$K_Q = \frac{P_Q}{B \cdot W^{1/2}} \cdot f\left(\frac{a}{W}\right)$$

where $P_Q$ = load as determined, lb.
    $B$ = thickness of specimen, in.
    $W$ = width of specimen, in.
    $a$ = crack length as determined, in.

To facilitate calculation of $K_Q$, values of the power series $f(a/W)$ in the foregoing expression are tabulated here for specific values of $a/W$ for compact-tension specimens:

| $a/W$ | $f(a/W)$ | $a/W$ | $f(a/W)$ |
|-------|----------|-------|----------|
| 0.450 | 8.34 | 0.500 | 9.66 |
| 0.455 | 8.46 | 0.505 | 9.91 |
| 0.460 | 8.58 | 0.510 | 9.96 |
| 0.465 | 8.70 | 0.515 | 10.12 |
| 0.470 | 8.83 | 0.520 | 10.29 |
| 0.475 | 8.96 | 0.525 | 10.45 |
| 0.480 | 9.07 | 0.530 | 10.63 |
| 0.485 | 9.23 | 0.535 | 10.82 |
| 0.490 | 9.37 | 0.540 | 10.93 |
| 0.495 | 9.51 | 0.545 | 11.17 |
|       |      | 0.550 | 11.36 |

*Arc-Shaped specimen (Figure 3.12) and disk-shaped compact specimen (Figure 3.13).* Similar relations for the arc-shaped specimen and the disk-shaped specimen are given in ASTM Specification E-399—83.

**8. Final check for $K_{Ic}$.** Calculate $2.5(K_Q/\sigma_{ys})^2$, where $\sigma_{ys}$ = 0.2 percent offset yield strength in tension. If this quantity is less than both the thickness and the crack length of the specimen, then $K_Q$ is equal to $K_{Ic}$. Otherwise, it is necessary to use a larger specimen to determine $K_{Ic}$ in order to satisfy this requirement. The dimensions of the larger specimen can be estimated on the basis of $K_Q$. If $K_Q$ is invalid, the strength ratio, $R$, is a useful comparative measure of the toughness of materials when the test specimens tested are of the same size and that size is insufficient to produce a valid $K_{Ic}$.

For the bend specimen,

$$R = \frac{\text{bending moment at failure}}{\text{bending moment at yielding}}$$

$$R = \frac{(P/2)(S/2)}{\sigma_{ys}[(W - a)^2/6]B} = \frac{6PW}{\sigma_{ys}(W - a)^2 B}$$

where $S = 4W$. For the CTS specimen,

$$R = \frac{2P(2W + a)}{B(W - a)^2 \sigma_{ys}}$$

If $R < 1$, the specimen has failed before yielding, and the $K_Q$ value should be equal to $K_{Ic}$; that is, the behavior is linear elastic. For $R$ values slightly above 1, the same conclusion is probably true because the extent of yielding is small. For larger values of $R$, for example, 1.5 or above, the behavior is increasingly elastic plastic, indicating ductile-type behavior.

## 3.5. Typical $K_{Ic}$ Test Results

Examples of several types of $P$–$\Delta$ records for materials that exhibit behaviors ranging from almost perfectly elastic to general yielding are presented in Figures 3.26 through 3.29. Comparison of the various materials, specimen sizes, and test results are presented in Table 3.2. Comments on each of these records are as follows:

### 3.5.1. Very-High-Strength Aluminum (Figure 3.26)

The $P$–$\Delta$ record for an extremely high-strength aluminum alloy with a yield strength of 71 ksi, Table 3.2, is presented in Figure 3.26.

The test record is the general Type III (Figure 3.23) with essentially linear-elastic behavior to $P_{max}$, which occurs before $P_5$. $P_5$ and $P_{max}$ are essentially identical, and $K_{Ic}$ is 19.8 ksi$\sqrt{\text{in}}$. The required $B$ is only 0.2 in. $[B = 2.5(K_{Ic}/\sigma_{ys})^2]$ and thus the specimen thickness used (1.37 in.) is obviously much greater than necessary. Materials with this low a $(K_{Ic}/\sigma_{ys})^2$ ratio, 0.08, are not widely used in structural applications. Note that even for this very brittle material, a very slight nonlinearity in the $P$–$\Delta$ record occurs just prior to $P_{max}$, Figure 3.26.

### 3.5.2. 18Ni Air-Melt Maraging Steel (Figure 3.27)

As shown in Figure 3.27, the $P$–$\Delta$ record for this material is also representative of Type III types (Figure 3.23). The interpretation of the record is straightforward and $K_Q = K_{Ic} = 113$ ksi$\sqrt{\text{in}}$. The $(K_{Ic}/\sigma_{ys})^2$ ratio of $\left(\frac{113}{190}\right)^2 = 0.35$ is more representative of high-strength low-toughness structural materials than the value of 0.08 for the structural material presented in Figure 3.26.

### 3.5.3. 12Ni Vacuum-Melt Maraging Steel (Figure 3.28)

This test record is representative of Type I materials with relatively high levels of toughness. These types of materials exhibit various amounts

**TABLE 3.2  Typical Load-Displacement Results for Various Structural Materials**

| Material and Figure No. for $P$-$\Delta$ Record | $\sigma_{ys}$ (ksi) | $\sigma_T$ (ksi) | $K_{Ic}$ Test Specimen Dimensions (in.) $B$ | $a$ | $W$ | $P_5$ (lb) | $P_{max}$ (lb) | $P_Q$ (lb) | $K_Q$ (ksi$\sqrt{in.}$) | $K_{Ic}$ (ksi$\sqrt{in.}$) | Minimum Req'd $B = 2.5\left(\dfrac{K_{Ic}}{\sigma_{ys}}\right)^2$ (in.) |
|---|---|---|---|---|---|---|---|---|---|---|---|
| 7001-T75 very-high-strength aluminum (Figure 3.26) | 70.6 | 80.5 | 1.37 | 1.08 | 2.00 | 3,140 | 3,150 | 3,150 | 19.8 | 19.8 valid | 0.20 |
| 18Ni maraging steel (Figure 3.27) | 190.0 | 196.0 | 1.24 | 0.95 | 3.50 | 22,950 | 22,950 | 22,950 | 113.0 | 113.0 valid | 0.88 |
| 12Ni maraging steel (Figure 3.28) | 183.0 | 191.0 | 1.00 | 0.46 | 3.00 | 55,000 | 80,150 | 55,000 | 143.0 | Invalid | $B_{est} = 1.5(2.5)\left(\dfrac{K_Q}{\sigma_{ys}}\right)^2 = 2.3$  $B_{est} = 2.5\left(\dfrac{K_{max}}{\sigma_{ys}}\right)^2 = 3.2$ |
| A517 steel (Figure 3.29) | 110.0 | 121.0 | 2.00 | 2.60 | 6.00 | 47,800 | 66,000 | 47,800 | 150.0 | Invalid | $B_{est} = 1.5(2.5)\left(\dfrac{K_Q}{\sigma_{ys}}\right)^2 = 7.0$  $B_{est} = 2.5\left(\dfrac{K_{max}}{\sigma_{ys}}\right)^2 = 8.8$ |

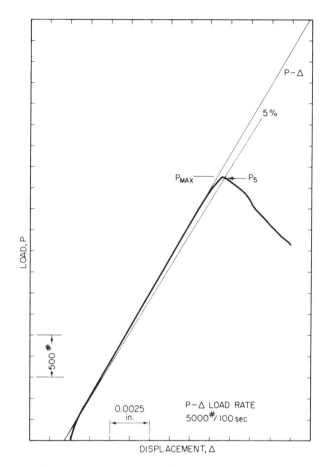

**Figure 3.26**   Actual $P$–$\Delta$ test record for a very-high-strength aluminum.

of nonlinear behavior between $P_5$ and $P_{max}$, and in this case $P_{max}/P_5$ is quite large, equal to 1.45, clearly violating the requirement that $P_{max}/P_5 \leqslant 1.10$.

It should be noted, however, that from a structural design viewpoint, such behavior is very desirable because considerable yielding occurs before fracture. The fact that the $K_{Ic}$ test procedure is limited to linear-elastic behavior has hampered the widespread application of fracture mechanics, particularly for specification purposes.

Using either a $K_{max}$ calculated from $P_{max}$ or a 1.5 amplification factor applied to $K_Q$, the probable minimum specimen thickness for plane-strain behavior in this material would be 2–3 in. (Table 3.2). If material of this thickness were available, additional $K_{Ic}$ tests could be conducted. If not, alternative methods of fracture analysis using either correlations (Chapter 5) or other elastic-plastic fracture mechanics approaches (Chapter 17) would need to be used to determine the fracture toughness of this material.

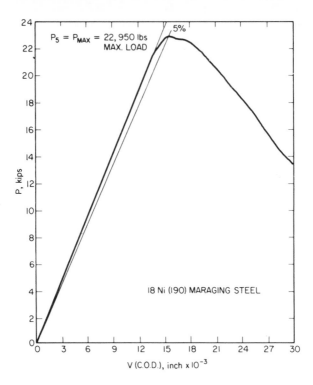

Figure 3.27 Actual $P$–$\Delta$ test record for an 18Ni air-melt maraging steel.

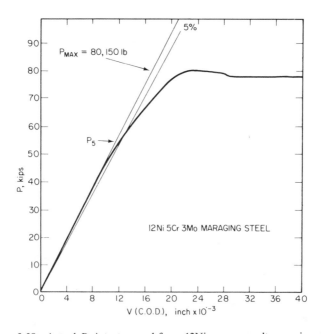

Figure 3.28 Actual $P$–$\Delta$ test record for a 12Ni vacuum-melt maraging steel.

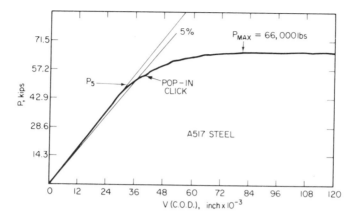

**Figure 3.29**   Actual $P-\Delta$ test record for an A517 structural steel.

### 3.5.4. A517 Structural Steel (Figure 3.29)

The test record for a 100-ksi yield-strength structural steel is shown in Figure 3.29 and indicates another example of inelastic behavior where $P_5$ is considerably less than $P_{max}(P_{max}/P_5 = 66/47.8 = 1.38$, which is well above the minimum allowable of 1.1). As sometimes happens while conducting $K_{Ic}$ tests, an apparent pop-in occurred (this behavior was verified on a duplicate specimen by interrupting the test and heat tinting to establish the crack growth by pop-in) at a load of about 56 kips, giving an "apparent" $K_{Ic}$ of 177 ksi$\sqrt{\text{in}}$. $K_Q$ was 150 ksi$\sqrt{\text{in}}$., and thus it could be estimated that a lower-bound value of $K_{Ic}$ was in the range 150–177 ksi$\sqrt{\text{in}}$. It would appear that a specimen thickness of 7–8 in. is necessary to provide sufficient constraint for plane-strain behavior in this steel plate.

## 3.6. Low-Temperature $K_{Ic}$ Testing

The fracture toughness of many structural materials is a function of temperature as well as loading rate. The effect of temperature on $K_{Ic}$ can be determined by conducting any of the $K_{Ic}$ test specimens described in the preceding sections at any test temperature following the ASTM E-399 standard method of testing. Obviously, it is more difficult to conduct the test because of the need for some cooling (or heating) medium compared with room temperature, but the test procedure and analysis is identical to that for room temperature. Numerous $K_{Ic}$ tests have been conducted at various testing temperatures, and actual test results on various structural materials are presented in Chapter 4.

## 3.7. Effect of Loading Rate on Fracture Toughness

$K_{Ic}$ tests are conducted at "slow" loading rates such that the time to maximum load is in the range of about 1–2 min. Specifically, the loading rate is specified to be within the range 30–150 ksi$\sqrt{\text{in.}}$/min. Because some structural materials are strain-rate sensitive, their fracture toughness at faster loading rates can be quite different from that measured in a "slow" $K_{Ic}$ test. Low-strength structural steels exhibit a large change in fracture toughness for different loading rates as shown in Figure 3.30. This figure shows $K_{Ic}$ test results conducted according to E-399 ($K_{Ic}$ tests), $K_{Ic}$ tests conducted at intermediate strain rates ($K_{Ic}$ ($t$)), and $K_{Ic}$ tests conducted at impact loading rates, referred to as $K_{Id}$ (dynamic). Note that the shift between the slow-bend test results and the dynamic-load results is over 150°F and that the difference in loading rates between slow and impact is about six orders of magnitude.

As a general rule, slow loading rates are conducted at strain rates of approximately $10^{-5}$ in./in./sec; that is, the maximum load is reached in about 1 min, as in a standard tension test. Intermediate-loading-rate tests, $K_{Ic}$ ($t$), are usually conducted at strain rates of about $10^{-3}$ in./in./sec or time to maximum load of about 1 sec. Dynamic tests usually are conducted at loading rates of 10 in./in./sec with time to maximum load of about 0.001 sec. Examples of test results are presented in Figure 3.30.

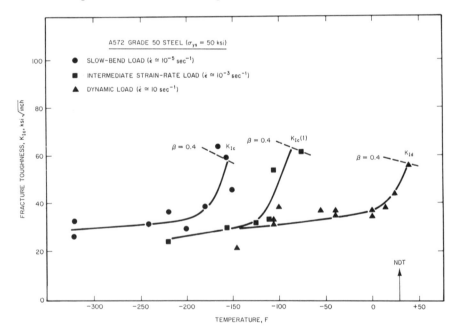

**Figure 3.30** Effect of temperature and strain rate on fracture toughness of A572 Grade 50 steel ($\sigma_{ys}$ = 50 ksi).

The more common reference to loading rate is

$$\dot{K} = \frac{K_{\text{critical}}}{t}$$

where $\dot{K}$ = stress-intensity factor rate, ksi$\sqrt{\text{in.}}$/sec.

$K_{\text{critical}}$ = critical stress-intensity factor ($K_{\text{Ic}}$, $K_{\text{Id}}$, $K_{\text{a}}$, etc., as defined in the next paragraph), ksi$\sqrt{\text{in.}}$

$t$ = time, in seconds, required to reach $K_{\text{critical}}$.

The different $K_{\text{Ic}}$ test results presented in Figure 3.30, as well as other rapid-loading-rate test results, are defined as follows:

$K_{\text{Ic}}$ = plane-strain fracture-toughness value from specimen tested at *slow* loading rates as described in ASTM E-399.

$K_{\text{Ic}}$ $(t)$ = plane-strain fracture-toughness value from specimen tested at *rapid-loading* rates as described by ASTM Committee E24.01.06 Report. The time to failure, $t$, in seconds, should be indicated in the parentheses ( ). This method does not include impact or quasi-impact testing (e.g., free-falling or swinging masses).

$K_{\text{Id}}$ = plane-strain fracture-toughness value from specimen tested at impact or dynamic loading rates (e.g., free-falling or swinging masses). As of 1986, there is no ASTM standard for this test, although there have been numerous dynamic tests conducted on structural steels, as described in Chapter 4.

$K_{\text{Ia}}$ = plane-strain crack-arrest fracture-toughness value for a *running* crack as measured in a crack-line-wedge–loaded compact-type specimen. This test is described in the ASTM document "Proposed Test Method for Crack Arrest Fracture Toughness of Metallic Materials."

$K_{\text{c}}$, $K_{\text{c}}$ $(t)$, $K_{\text{d}}$, $K_{\text{a}}$ = plane-stress fracture-toughness values from specimens tested as described ($K_{\text{Ic}}$, $K_{\text{Ic}}$ $(t)$, $K_{\text{Id}}$, $K_{\text{Ia}}$), but for specimens where conditions of plane *strain* (maximum constraint) are not met. The effect of constraint or thickness on fracture toughness is described in Chapter 4.

It should be reemphasized that the various critical $K$ values are the critical stress-intensity factors for particular structural materials obtained by testing these materials at different loading rates and also at various temperatures. Depending upon the chemical composition and the metallurgical processing of structural materials, these critical $K$ values will be different for different materials and even for materials of the same composition.

In contrast, $K_I$ or $K$ values as described in Chapter 2 are calculated values of the stress-intensity factor in various idealized structural members with cracks and do *not* depend on material properties. $K_I$ values have a similar relationship to $K_{Ic}$, $K_{Ic}$ ($t$), $K_{Id}$, $K_{Ia}$, and other values as stress ($\sigma$) values have to yield strength ($\sigma_{ys}$) values. $K_I$ and $\sigma$ are calculated values that are a function of geometry and loading only, whereas $K_{Ic}$, $K_{Id}$, $\sigma_{ys}$, and others are material properties that can be obtained only by testing specific materials at specific temperatures and loading rates.

For some structural materials, particularly the low-strength structural steels, there is a continual change in fracture toughness with increasing loading rate, as shown in Figure 3.31. The rate of change of $K_I$ with respect to time $\dot{K}$, is given in ksi$\sqrt{\text{in.}}$/sec. "Slow" loading rates, that is, those prescribed in the standard method of $K_{Ic}$ testing, are around 1 ksi$\sqrt{\text{in.}}$/sec, whereas those loading rates generally obtained in $K_{Id}$ testing are around $10^5$ ksi$\sqrt{\text{in.}}$/sec. Other structural materials, for example, aluminums, titaniums, and very-high-strength steels (yield strengths of 150 ksi and higher), generally do not exhibit loading-rate effects. Thus, for these materials, there generally would be no difference between $K_{Ic}$ and $K_{Id}$ values tested at the same temperature.

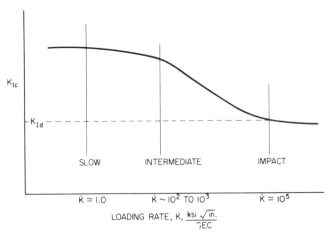

**Figure 3.31**   Effect of loading rate on $K_{Ic}$.

## 3.8.  Rapid-Load Plane-Strain Fracture-Toughness $K_{Ic}$ ($t$) Testing

The intermediate-loading-rate test results, $K_{Ic}$ ($t$), are referred to as rapid-load plane-strain fracture-toughness values. The time to failure, $t$, in seconds, is to be indicated in the parentheses ($t$). Although the tests must be conducted in testing machines capable of achieving the maximum load in the desired testing time, the specimens are prepared in the same manner as the "slow" $K_{Ic}$ test specimens are prepared. Load-deflection and load-time curves must be recorded for each test, and the instrumentation (load and displacement transducers) must have a frequency response sufficient to record the test signals. Because the time of a rapid-load test is much shorter than that of the conventional "slow" E-399 test, specialized testing machines and instrumentation generally are required.

The load-deflection curves obtained during testing are analyzed to eliminate data exhibiting excessive inertia or mechanical response effects peculiar to dynamic tests. Also, the time to critical load must be determined so that the test time ($t$) can be determined and a stress-intensity factor rate, $\dot{K} = K_c/t$, calculated. Restrictions on the test records to eliminate dynamic effects are necessary, and the yield strength used in the analysis must be for the temperature and loading rate used in the fracture test.

Other than these changes, all the criteria of ASTM E-399 for "slow" $K_{Ic}$ testing (except loading rate, obviously) apply to the rapid-load plane-strain fracture-toughness test. These and other special requirements for rapid-load plane-strain fracture-toughness $K_{Ic}$ ($t$) testing are described in an ASTM Report by Committee E24.01.06—"Task Group on Dynamic $K_{Ic}$ ( ) Testing."

Figure 3.30 shows typical rapid-load $K_{Ic}$ (1) test results compared to both slow $K_{Ic}$ results and impact $K_{Id}$ results. Note that because these are $K_{Ic}$ (1), the time to maximum load is about one second. Other examples of rapid-load $K_{Ic}$ ($t$) results for other structural steels are presented in Chapter 4. Note that, as would be expected, the test results fall between the $K_{Ic}$ and $K_{Id}$ results.

## 3.9. Crack-Arrest Fracture Toughness of Metallic Materials—$K_{Ia}$

This test method[5] measures the critical stress-intensity factor at which a fast-running crack will arrest, $K_{Ia}$. It is essentially a "dynamic" fracture-toughness value in that $K_{Ia}$ is measured for a propagating brittle fracture and thus would be for loading rates at the high end of Figure 3.31, for example, $\dot{K} \simeq 10^5$ ksi$\sqrt{in.}$/sec. Thus one might expect the results to be similar to those measured with the impact or dynamic-loading-rate $K_{Id}$ test method described in the following section, and on the basis of limited test results this is true. However, because the dynamic effects present in $K_{Id}$ testing are considered to be negligible in $K_{Ia}$ testing, an ASTM Standard for crack arrest $K_{Ia}$ testing is being prepared, whereas a $K_{Id}$ test method was not in preparation as of 1986. This $K_{Ia}$ document is the "Proposed Test Method for Crack Arrest Fracture Toughness of Metallic Materials."

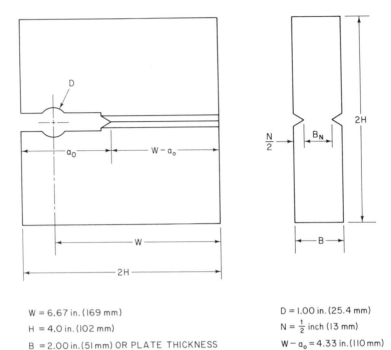

W = 6.67 in. (169 mm)

H = 4.0 in. (102 mm)

B = 2.00 in. (51 mm) OR PLATE THICKNESS

D = 1.00 in. (25.4 mm)

N = $\frac{1}{2}$ inch (13 mm)

W − $a_0$ = 4.33 in. (110 mm)

**Figure 3.32** Crack-arrest test specimen that is satisfactory for mild- and intermediate-strength steels.

The $K_{Ia}$ test is conducted using a compact-type specimen, Figure 3.32. Face notches are used to keep the crack in a straight plane, resulting in a net specimen thickness of $B_N$. During testing, a wedge is forced into a split pin, Figure 3.33, which develops an opening force across the crack face of the compact-type specimen. This force eventually causes a crack to initiate and propagate. Initially the dynamic effects are large because of the abrupt acceleration of the initial, stationary crack. After the crack initiates, propagates, and begins to arrest, these dynamic effects are assumed to be negligible and a static elastic-stress analysis is used to calculate $K_{Ia}$, on the basis of the load and crack length at arrest. For example, the $K$ value at arrest is

$$K_a = \frac{V \cdot E \cdot f\left(\dfrac{a}{W}\right)}{W^{1/2}} \left(\frac{B}{B_N}\right)^{1/2}$$

where     $V$ = measured displacement at arrest, Figure 3.34, which is correlated with load using a compliance calibration.

$E$ = modulus of elasticity

$f(a/W)$ = function of $a/W$ given in the Proposed Test Method.

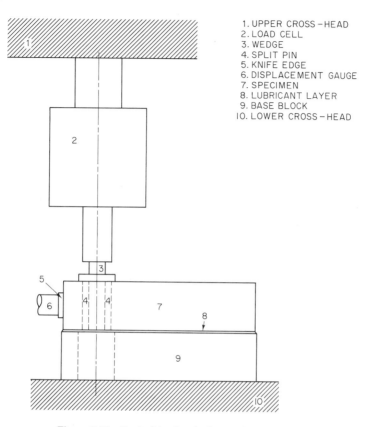

1. UPPER CROSS–HEAD
2. LOAD CELL
3. WEDGE
4. SPLIT PIN
5. KNIFE EDGE
6. DISPLACEMENT GAUGE
7. SPECIMEN
8. LUBRICANT LAYER
9. BASE BLOCK
10. LOWER CROSS–HEAD

**Figure 3.33**  Typical load train for crack-arrest tests.

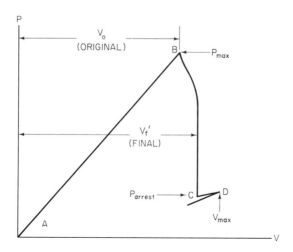

**Figure 3.34**  Typical load-displacement diagram for $K_{Ia}$ crack-arrest tests.

$a$ = crack length at arrest.
$W$ = specimen depth.
$B$ = specimen thickness.
$B_N$ = net specimen thickness at crack plane.
See Fig. 3.32 for specimen layout.

As with the E-399 $K_{Ic}$ test method, numerous criteria must be met before the calculated $K_a$ value can be designated to be a crack-arrest toughness, $K_{Ia}$. Thus, all preliminary calculations are referred to as $K_f$ or $K$ at the "final" conditions. These and other restrictions are described in the ASTM "Proposed Test Method for Crack Arrest Fracture Toughness of Metallic Materials." An example of $K_a$ values for A36 steel is presented in Figure 3.35. These values are referred to as plane-stress $K_a$ values rather than as plane-strain $K_{Ia}$ because the thickness requirement specified in the ASTM Test Method was not met.

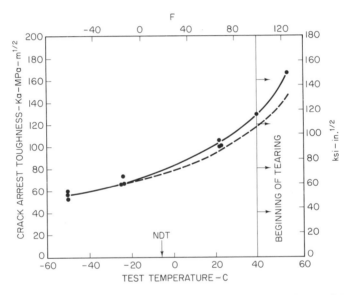

**Figure 3.35** Effect of test temperature on the crack-arrest toughness of 50.8-mm (2-in.)-thick A36 steel tested in the L-T direction.

## 3.10. Dynamic Fracture-Toughness Testing for $K_{Id}$

The test specimen most widely used for determining $K_{Id}$ values is a fatigue-cracked three-point bend specimen essentially identical to the $K_{Ic}$ slow-bend specimen (Figure 3.36). The specimen is loaded in three-point bending by a striking tup mounted on a free-falling weight. Either the specimen or the tup (or both) is instrumented to measure applied load as a function of time.

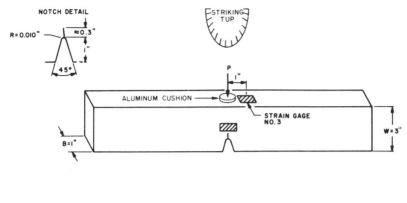

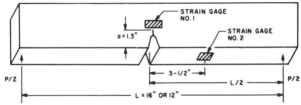

**Figure 3.36**   Fatigue-cracked $K_{Id}$ (dynamic $K_{Ic}$) bend test specimen.

The following dynamic test procedure was developed by Shoemaker and Rolfe[6] at the U.S. Steel Research Laboratory. Another similar procedure was developed by Madison and Irwin[7] at Lehigh. Both methods would be expected to give essentially the same results.

Strain gauges are applied to the bend specimen as shown in Figure 3.36. Gauge 1 is used as a crack detector to determine the time of crack initiation. The nominal strain at gauge 2 is recorded as the crack initiates and is used to calculate the corresponding nominal elastic stress at the point. Using the elementary strength of materials formula, $\sigma = My/I$, the moment and corresponding equivalent static load necessary to give this stress are calculated. This equivalent static load, which occurs at the time the crack initiates, is used in the following equation to determine the $K_{Id}$ value for the test conditions studied,

$$K_{Id} = f\left(\frac{a}{W}\right)\frac{6Ma^{1/2}}{BW^2}$$

where       $M$ = applied bending moment.
            $a$ = crack length.
            $B$ = specimen thickness.
            $W$ = specimen width.
            $S$ = span (see Figure 3.36),
$f(a/W) = A_0 + A_1(a/W) + A_2(a/W)^2 + A_3(a/W)^3 + A_4(a/W)^4.$

|            | $A_0$  | $A_1$   | $A_2$    | $A_3$    | $A_4$    |
|------------|--------|---------|----------|----------|----------|
| $S/W = 8$  | +1.96  | -2.75   | +13.66   | -23.98   | +25.22   |
| $S/W = 4$  | +1.93  | -3.07   | +14.53   | -25.11   | +25.80   |

To calculate dynamic $K_{Ic}$ values, $K_{Id}$, it is assumed that during the dynamic tests the strain distribution is the same as that obtained from an equivalent static load. This assumption implies that any inertial effects present do not significantly change the mode of deflection from that obtained during static loading. To minimize inertial effects, a low-impact velocity is attained by dropping a 1600-lb weight a distance of about 9 in. In addition, the tup of the drop weight is impacted against a soft aluminum or lead pad positioned on the test specimen. This dampening pad eliminates elastic "ringing" waves in the specimen and increases the loading time during the dynamic test.

In the initial stages of test development, strain gauges were used in positions 1 and 2 (Figure 3.36). A typical test record is shown in Figure 3.37. Gauge 1 was used to determine the time at which the crack extended. As the crack extended, it broke gauge 1, which gave an output discontinuity as shown by the upper strain-time trace. The peak strain from gauge 2, which occurred at approximately the same time (lower trace), was used to calculate an equivalent static load necessary to cause a peak strain of the same amount. The record shown in Figure 3.37(b) indicates a strain of 200 $\mu$in./in. at gauge 2 when the crack extended.

Subsequently, strain gauges were used in positions 2 and 3 (Figure 3.36). The ratio of the output of gauge 3 to that of gauge 2, which increased with increasing crack length, was determined from static tests. During a dynamic test, the time at which the crack extended was determined as the time when the ratio of the strains of the two gauges increased from its initial value.

During the development of the dynamic testing procedure, two different specimen span lengths (different stiffnesses and natural frequencies), two different sets of gauge positions, and different materials for dampening the impact blow were used to check for possible inertial effects. In no instance was it possible to find consistent significant deviations from the mean of the $K_{Id}$ data as any one of these testing variables was changed. The aluminum pads gave loading times shorter than those for the lead pads, but no difference in $K_{Id}$ was obtained.

As tougher material behavior and increased $K_{Id}$ values were obtained with increased temperature, the dynamic input energy became marginal. This led to strain-time records which had strain plateaus of 0.5-msec duration and greater. When the drop height was increased to increase the input energy, elastic "ringing" waves were encountered. Therefore, this test

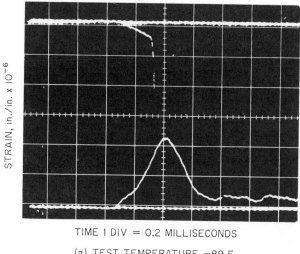

TIME 1 DIV = 0.2 MILLISECONDS

(a) TEST TEMPERATURE −89 F.

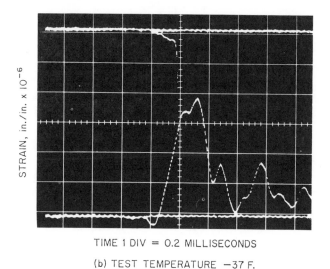

TIME 1 DIV = 0.2 MILLISECONDS

(b) TEST TEMPERATURE −37 F.

**Figure 3.37**   Typical strain-time records for dynamic tests. (a) Test temperature, −89°F; (b) test temperature, −37°F.

method appears to be restricted to a maximum $K_{Id}/\sigma_{yd}$ ratio of approximately 0.7.

Madison and Irwin[7] developed a similar test method at Lehigh that uses an instrumented tup, and a record of load versus loading time is obtained from signals from strain gauges mounted on the loading tup. Aluminum loading cushions are used to minimize any inertial effects.

Using a 3-in.-deep by 12-in.-long by $\frac{1}{2}$ to 2-in.-thick plate (span for three-point bending is 10 in.), they used the following relation to determine $K$:

$$K = \frac{1.5PS\sqrt{a}}{BW^2}\left[1.93 - 3.12\,\frac{a}{W} + 14.68\left(\frac{a}{W}\right)^2\right.$$
$$\left. - 25.30\left(\frac{a}{W}\right)^3 + 25.90\left(\frac{a}{W}\right)^4\right]$$

where $P$ = fracture load from strain gauge records on instrumented tup.

$\quad\quad\;\; a$ = effective crack length at fracture.

$\quad\quad\;\; S$ = support span = $3.33W$.

The effect of plasticity is estimated by increasing the crack length, $a$, by the plastic-zone radius, $r_y$, where

$$r_y \cong \frac{1}{2\pi}\left(\frac{K_{Id}}{\sigma_{yd}}\right)^2$$

The yield stress corresponding to the test temperature and loading rate is estimated by

$$\sigma_{yd_{\substack{(@\text{ test temp.} \\ \&\text{ loading rate})}}} \cong \left[\sigma_{ys_{\substack{(@\text{ room temp. \&} \\ \text{static loading rate})}}} + \frac{174,000}{\log(2 + 10^{10}t)(T + 459)} - 27.4\text{ ksi}\right]$$

where $\sigma_{ys_{\substack{(@\text{ room temp. \&} \\ \text{static loading rate})}}}$ = 0.2% offset yield strength, ksi.

$\quad\quad\;\; T$ = specimen temperature, in degrees Fahrenheit.

$\quad\quad\;\; t$ = load rise time for test—time from start of load to fracture, in seconds.

Static and some elevated-strain-rate yield-strength data have been obtained for seven steels from room temperature to $-320°F$.[8] In Figure 3.38, these values are plotted in terms of the rate-temperature parameter, $T \ln A/\dot{\varepsilon}$, suggested by Bennett and Sinclair,[9] where $T$ is the absolute temperature in degrees Rankine, $A$ is the frequency factor taken to be a constant of $10^8$/sec, $\dot{\varepsilon}$ is the strain rate, and ln is the natural logarithm to the base $e$. This rate-temperature parameter, based on the Arrhenius rate equation, was shown to correlate yield-strength behavior of body-centered cubic (bcc) materials for different temperatures and strain rates. This parameter appears to be useful to estimate yield behavior for low-strength alloy steels, for which significant changes in yield strength are observed for changes in temperature and strain rate. Because the effects of rate and temperature on the absolute yield strength decrease with steels of increasing yield strength, this parameter is used to estimate the yield behavior at the test conditions primarily for low-strength steels.

Madison and Irwin[7] compared their testing method with the ASTM Standard Method of Test for $K_{Ic}$. Their observations are summarized in Table 3.3.

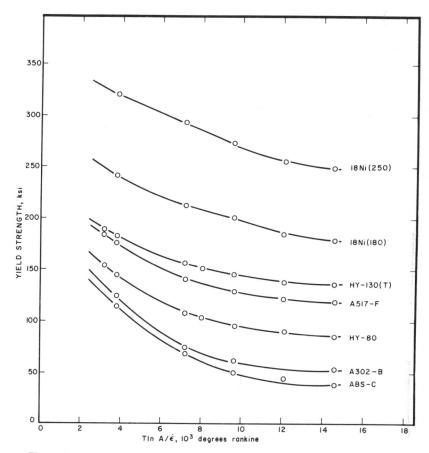

**Figure 3.38**  Yield strength for seven steels in terms of the rate-temperature parameter, $T \ln A/\dot{\epsilon}$.

**TABLE 3.3    Comparison of Madison and Irwin $K_{Id}$ Test Method
with Recommended ASTM $K_{Ic}$ Test Method**

| Testing Factors | ASTM $K_{Ic}$ Test Method | Madison and Irwin $K_{Id}$ Test Method |
| --- | --- | --- |
| Specimen support | Roller supports | Fixed simple supports |
| Fatigue precrack | $K_{max} < 60\% \; K_{Ic}$ | Crack growth rate at or below $1 \times 10^{-6}$ in./cycle |
| Support span | $S \geqslant 4W$ | $S = 3.33W$ |
| Plastic-zone size ($r_y$) adjustment | None | Plane stress $r_y$ used |
| Depth of initial crack | $0.5W$ | $0.33W$ |
| Measurement of slow stable growth | Not included | Not included |

Although the $K_{Id}$ test method is yet to be standardized, $K_{Id}$ values are being widely used in analysis, design, and specifications, as discussed throughout this text. Typical $K_{Id}$ results, compared with $K_{Ic}$ test results as a function of temperature, are presented in Chapter 4.

## 3.11. Other Standard Fracture-Mechanics Test Specimens

Several other standardized fracture-mechanics test methods are described briefly in the following sections. The $R$-curve, $J$-integral, and CTOD test methods are for elastic-plastic behavior and are described more completely in Chapter 17.

### 3.11.1. ASTM E-561—81: "Standard Practice for R-Curve Determination"

This practice covers the determination of resistance to fracturing of metallic materials by using the center-cracked tension panel (CCT), the compact tension (CT) specimen, or the crack-line wedge-loaded specimen (CLWL). The test result in an $R$-curve is a continuous record of toughness in terms of $K_R$ plotted against crack extension in the material as a crack is driven under a continuously increasing stress-intensity factor, $K$. See Chapter 17 for additional description.

### 3.11.2. ASTM E740–80: "Standard Practice for Fracture Testing with Surface-Crack Tension Specimens"

This practice covers the design, preparation, and testing of surface-cracked tension (SCT) specimens. It relates specifically to testing under continuously increasing load and excludes cyclic and sustained loadings. Results are obtained either in terms of the residual strength, $\sigma_r$, or the SCT specimen fracture toughness $K_{Ic} = \sigma_r\sqrt{\pi a} \cdot M/\theta$ as described in the standard practice. The $\sigma_r$ value determined is the residual strength of a specimen having a semielliptical or circular-segment fatigue crack in one surface. This value depends on the crack dimensions and the specimen thickness as well as on the characteristics of the material.

Metallic materials that can be tested are not limited by strength, thickness, or toughness. However, tests of thick specimens of tough materials may require a tension test machine of extremely high capacity.

### 3.11.3. ASTM E-813—81: "Standard Test for $J_{Ic}$, A Measure of Fracture Toughness"

This method covers the determination of $J_{Ic}$, which can be used as a toughness value at the initiation of crack growth for metallic materials.

The recommended specimens are generally bend type (three-point bend or compact-tension specimens) that contain deep initial cracks. The loading rate is slow, and environmentally assisted cracking is assumed to be negligible.

The recommended three-point bend specimen is a single edge-notched beam having an initial normalized crack size, $a_0/W$, of 0.5 or more. Overall span-to-width ratio, $S/W$, is set at 4.

The recommended compact specimen is a crack-line loaded type specimen that is pin loaded in tension. The specimen configuration has fixed planar dimensional proportionality with an initial normalized crack size, $a_0/A$, of 0.5 or more.

Specimen dimension requirements are based upon the ratio of $J$-integral to material effective yield strength. Therefore, in specimen design, it is helpful to have a general idea of the expected results in advance. See Chapter 17 for an additional description.

### 3.11.4. British Standard 5762:
### "Methods for Crack-Tip Opening-Displacement
### (CTOD) Testing"

This British standard specifies the method for carrying out crack-tip opening-displacement tests on metallic materials. It specifies the requirements for test pieces, test equipment, analysis of data, and recording results. The test method and specimen configuration are similar to those specified for $K_{Ic}$ testing. The test is carried out on material of the full thickness of interest. If this thickness is sufficient to give a valid $K_{Ic}$ and the other requirements of this standard are met, the result can be presented in those terms. If not, it can be presented in terms of CTOD in accordance with the methods specified in this standard. See Chapter 17 for further description.

## References

1. "Standard Method of Test for Plane-Strain Fracture Toughness of Metallic Materials," *ASTM Designation E-399—83*, Vol. 03.01, 1985, *ASTM Annual Standards,* American Society for Testing and Materials.
2. "Standard Practice for R-Curve Determination," *ASTM E-561—81*, Vol. 03.01, *1985 ASTM Annual Standards*, AMERICAN SOCIETY FOR TESTING AND MATERIALS, Philadelphia.
3. W. F. BROWN, JR., EDITOR, "Review of Developments in Plane Strain Fracture Toughness Testing," *ASTM STP 463*, 1970.
4. W. F. BROWN, JR., and J. E. SRAWLEY, "Plane Strain Crack Toughness Testing of High Strength Metallic Materials," *ASTM STP 410*, 1967.
5. P. B. CROSLEY and E. J. RIPLING, "Plane-Strain Crack Arrest Characterization of Steels," *ASME Paper No. 75-PVP-32*, presented at Second National Congress on Pressure Vessels and Piping, San Francisco, June 23–27, 1975.
6. A. K. SHOEMAKER and S. T. ROLFE, "Static and Dynamic Low-Temperature $K_{Ic}$ Behavior of Steels," *Transactions ASME*, Sept. 1969.

7. R. B. MADISON and G. R. IRWIN, "Dynamic $K_c$ Testing of Structural Steel," *Journal of the Structural Division, ASCE, 100*, No. ST 7, Proc. Paper 10653, July 1974, pp. 1331–1349.

8. D. P. CLAUSING, "Tensile Properties of Eight Constructional Steels Between 70 and $-320°F$," *Journal of Materials, ASTM, 4*, No. 2, June 1969.

9. P. E. BENNETT and G. M. SINCLAIR, "Parameter Representation of Low-Temperature Yield Behavior of Body-Centered-Cubic Transition Metals," *Journal of Basic Engineering, Transactions ASME*, Series D, *88*, No. 2, June 1966.

10. K. D. IVES and J. M. BARSOM, "Recent Developments in Dynamic Evaluation of $K_{Ic}$ and Analysis of Long Fractures," *Instrument Society of America, 12*, No. 4, 1973.

# 4

# Effect of Temperature, Loading Rate, and Plate Thickness on Fracture Toughness

## 4.1. Introduction

In Chapter 3, the ASTM Standard Test Method for determining the critical plane-strain stress-intensity factor, $K_{Ic}$, was described in detail. Also test methods for determining the fracture toughness under conditions of rapid load as well as dynamic loading were described. The fracture toughness under rapid-load conditions is referred to as $K_{Ic}$ $(t)$, where the time to maximum load is given in the parentheses. The dynamic crack toughness was called $K_{Id}$, the critical plane-strain stress-intensity factor under conditions of impact loading. Furthermore, it was stated that the $K_{Ic}$, $K_{Ic}$ $(t)$, and $K_{Id}$ tests are frequently conducted at various temperatures to determine the "static," "intermediate," and "dynamic" fracture toughness of various structural materials as a function of temperature.

The fact that the inherent fracture toughness of many structural materials increases with increasing test temperature is well known. This increase has been measured using various notch-toughness specimens such as the Charpy V-notch impact specimen, and it is certainly reasonable to expect a similar increase using fracture-mechanics–type test specimens. What is not so widely known is the fact that the same inherent fracture toughness can decrease significantly with increasing loading rate, that is, comparing $K_{Id}$ test results with $K_{Ic}$ test results. Also, testing plates thinner than those required for plane-strain values may result in plane-stress $K_c$ toughness values that are higher than the $K_{Ic}$ values. Thus, before the engineer can use fracture-toughness values in design (as described in Chapter 6), the critical fracture-toughness value for the particular service temperature, loading rate, and plate thickness must be known. In this chapter we shall describe the general effects of these three variables on the fracture toughness of various structural materials. Finally, test results will be presented for commonly used structural materials.

## 4.2. Plane-Strain Transition-Temperature Behavior

As discussed briefly in Chapter 1, there are numerous types of fracture-toughness specimens used to determine the notch toughness of structural materials. For most structural steels, these specimens—such as the Charpy V-notch (CVN) impact specimen, the dynamic tear (DT) test specimen, the crack-tip opening displacement (CTOD) specimen, the precracked impact (PCI) specimen, and so on—are tested at several temperatures to determine the notch-toughness behavior in the transition-temperature range. This temperature range is the region throughout which the inherent notch toughness of a material changes from brittle to ductile. It should be noted that certain structural materials, for example, aluminums, titaniums, and many very-high-strength steels ($\sigma_{ys} > 150$ ksi) do not exhibit a transition-temperature behavior.

The transition behavior is shown schematically in Figure 4.1 for a typical structural steel and indicates that temperature can have a significant effect on the notch-toughness behavior of certain structural materials, primarily structural steels. Examples of CVN, DT, and $K_{Ic}$ test data for an A517 steel are presented in Figure 4.2 and illustrate this marked change in notch toughness with increasing temperature.

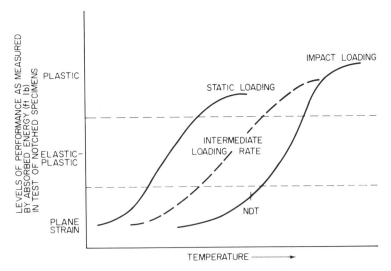

**Figure 4.1**  Schematic showing relation between notch-toughness test results and levels of structural performance for various loading rates.

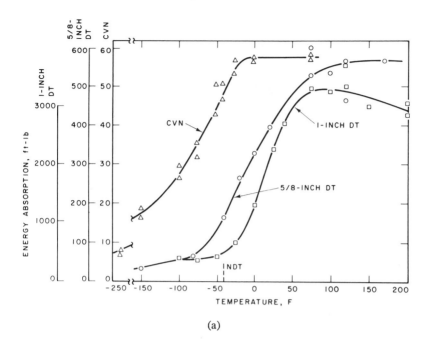

(a)

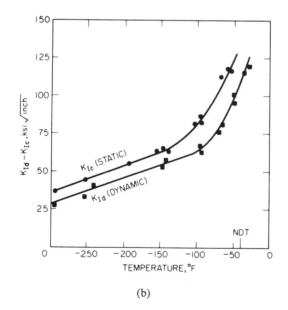

(b)

**Figure 4.2**   CVN, DT, $K_{Ic}$, and $K_{Id}$ test results for an A517 steel ($\sigma_{ys} = 100$ ksi).

This increase in notch toughness with increasing temperature is well known to most design engineers and has led to the *transition-temperature* design approach for structural steels. In this approach some level of notch toughness is selected as representing an "adequate" level of toughness for the particular structure. In Chapter 15 we shall describe various methods to establish an "adequate" level of toughness. One level that has been widely used is the 15-ft-lb CVN impact test value based on the analysis of the World War II ship fractures. The temperature at which the material exhibits this toughness level is required to be at or below the minimum service temperature. Figure 4.3 is a schematic showing two steels with different "15-ft-lb transition temperatures" and their relation to a particular service temperature. In this example, steel A would be considered to be satisfactory because its 15-ft-lb transition temperature would be below the service temperature, whereas steel B would not meet the criteria. This transition-temperature approach has been widely used by materials engineers to establish various toughness specifications.

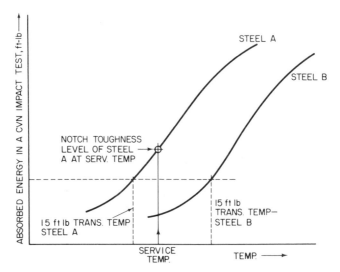

**Figure 4.3**  Schematic showing relation between 15-ft-lb transition temperatures and service temperature.

In a similar manner, the notch toughness of many structural materials exhibits an inherent change in the plane-strain fracture toughness with change in temperature, as shown in Figure 4.4.[1] The test results presented in Figure 4.4 satisfy the ASTM requirement for minimum specimen thickness in plane-strain fracture-toughness testing described in Chapter 3 as given by

$$a \text{ and } B \geqslant 2.5\left(\frac{K_{Ic}}{\sigma_{ys}}\right)^2$$

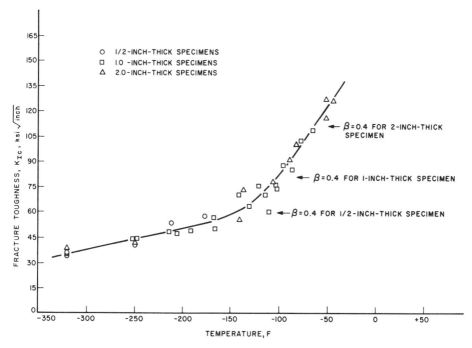

**Figure 4.4**  Plane-strain fracture-toughness transition behavior as a function of temperature for A517 Grade F steel.

where $B$ = specimen thickness.

$a$ = crack length.

$\sigma_{ys}$ = 0.2 percent yield strength at the test temperature.

This ASTM requirement can be expressed in terms of Irwin's plane-strain $\beta$ value[2] as follows:

$$\beta_{\text{Ic}} = \frac{1}{B}\left(\frac{K_{\text{Ic}}}{\sigma_{ys}}\right)^2 \leq 0.4$$

That is, if $\beta_{\text{Ic}}$ is 0.4, or less, the specimen size is sufficiently large to ensure plane-strain behavior as defined in Chapter 3.

All data points in Figure 4.4 define a single, common function, regardless of whether the points were obtained from tests that satisfied the criterion of $\beta$ = 0.4 or less. Because the maximum load at fracture for all data points shown in Figure 4.4 was within the ASTM 5 percent secant-intercept requirement for plane-strain behavior and because all data points define a single, common function, it would appear that all data points are valid $K_{\text{Ic}}$ values even though a few points did not satisfy the $\beta$ criterion. Actually, if only the points that are completely valid according to Test Method E-399 are considered, there is still a distinct change in the slope of the curve between $-100°F$ and $-150°F$.

Thus, the curve defined by these data points demonstrates that a $K_{\text{Ic}}$ temperature transition exists that is independent of specimen geometry; that is, the rate of increase in $K_{\text{Ic}}$ with temperature does not remain constant but increases markedly in a particular temperature range for a particular test specimen and testing conditions. In this case, the change in rate of increase of $K_{\text{Ic}}$ occurs in the temperature range $-150°$ to $-100°$F and is associated with the onset of change in the microscopic fracture mode.

Figure 4.5 shows the variation of the parameter $(K_{\text{Ic}}/\sigma_{ys})^2$, which is proportional to the plastic-zone size, as a function of temperature. Note that a distinct transition occurs in the same $-100°$F to $-150°$F temperature range as was shown in Figure 4.4. Above $-40°$F, the magnitude of the plastic zone appears to increase asymptotically. This behavior indicates that valid $K_{\text{Ic}}$ values probably cannot be obtained at temperatures greater than $-40°$F for this particular material regardless of specimen size.

It should be noted that early in the development of fracture mechanics, it was thought that there was no plane-strain transition. That is, it was believed that any marked increase in $K_{\text{Ic}}$ with temperature such as is shown

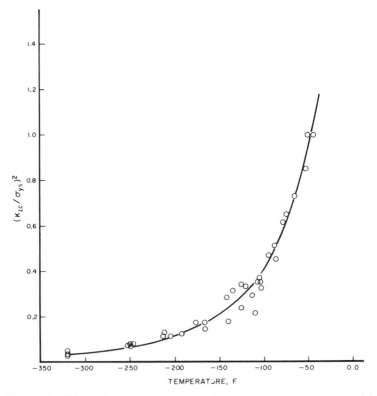

**Figure 4.5** Effect of temperature on the plastic-zone parameter for material presented in Figure 4.4.

in Figure 4.6 was in reality a loss in through-thickness constraint and actually a plane-strain to plane-stress, or thickness, transition. Furthermore, it was believed that as long as thicker and thicker specimens were used, the change in toughness at higher temperatures could be obtained by extrapolation from $K_{Ic}$ test results conducted at lower temperatures, as shown in Figure 4.6. These test results, as well as test results using extremely thick test specimens conducted by Greenburg et al.[3] (Figure 4.7), show that there is indeed a plane-strain transition with increasing temperature. In summary, $K_{Ic}$ can undergo a nonlinear change with temperature *independent* of any other factors such as loss of constraint which is described in Section 4.3.

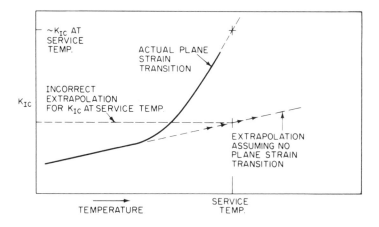

**Figure 4.6**    Effect of plane-strain transition on incorrect extrapolation for $K_{Ic}$.

The transition in plane-strain behavior of several structural steels was presented in Figures 4.4 and 4.7. Additional plane-strain transition behavior is presented in Figures 4.8 and 4.9 for A36 and A572 Grade 50 steels, respectively. Note that in all cases, the change in $K_{Ic}$ throughout the transition region is about a factor of 2. For example, the values in Figure 4.4 are about 60 ksi$\sqrt{\text{in}}$. at $-150°F$ and about 120–130 ksi$\sqrt{\text{in}}$. at $-50°F$. Similarly, the values shown in Figure 4.9 change from about 30 ksi$\sqrt{\text{in}}$. at the transition to about 60 ksi$\sqrt{\text{in}}$. at the point $K_{Ic}$ values can no longer be measured because of loss of constraint.

Over this transition in plane-strain behavior, there is little or no change in the *macro*scopic fracture appearance or specimen geometry. That is, the fracture surface always appears to be cleavage, and there are no shear lips. On a *micro*scopic level, the fracture surface is beginning to develop isolated areas of ductility, referred to as microvoid coalescence. It is this start of the change in *micro*scopic fracture behavior that results in the plane-strain transition.

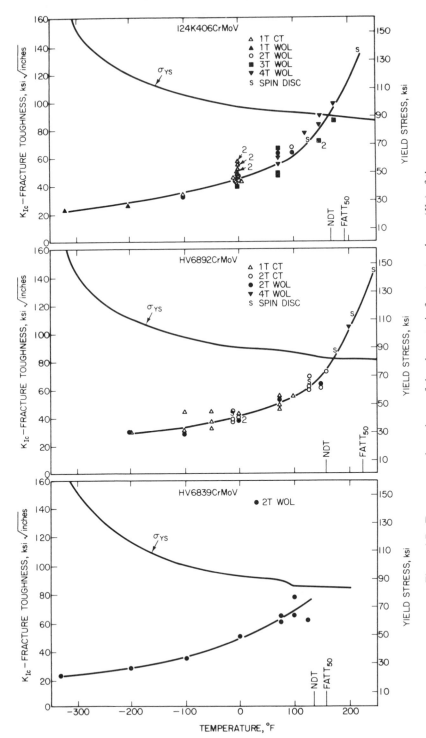

**Figure 4.7** Temperature dependence of the plane-strain fracture toughness ($K_{Ic}$) of three CrMoV alloy forgings.

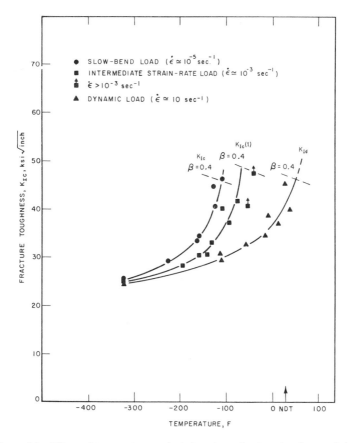

**Figure 4.8**   Effect of temperature and strain rate on fracture toughness of A36 steel.

Beyond the plane-strain transition (which is only about a factor of 2) is the plane-strain to plane-stress transition. As will be shown later, in Chapter 17, this transition results in a *ten*fold change in toughness and is dependent on two factors. These two factors are

1. A significant change in the microstructural behavior as the fracture mode becomes almost all microvoid coalescence or shear.
2. A loss in mechanical constraint resulting in lateral contraction.

These changes are synergistic and result in a very large change in toughness as is shown in Figure 4.10. This figure is taken from Chapter 17 and shows the plane-strain transition behavior of the A572 steel presented in Figure 4.9 along with the plane-stress transition results that will be presented in Chapter 17. Note that the change in plane stress values is

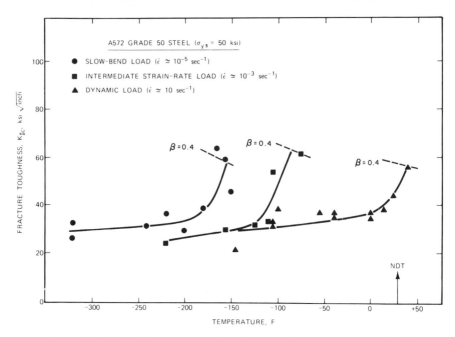

**Figure 4.9** Effect of temperature and strain rate on fracture toughness of A572 Grade 50 steel ($\sigma_{ys}$ = 50 ksi).

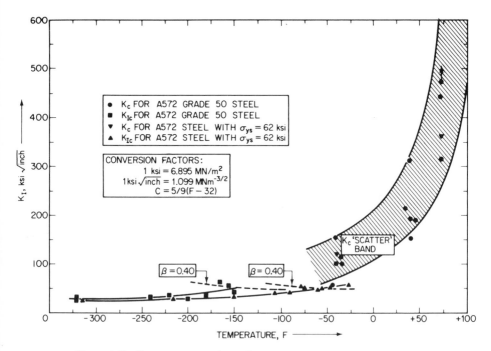

**Figure 4.10** Summary comparison of $K_c$ and $K_{Ic}$ behavior obtained from 1.5-in-thick plates of A572 steel.

from about 50 ksi$\sqrt{\text{in.}}$ to 500 ksi$\sqrt{\text{in.}}$, that is, an order of magnitude change in toughness.

Thus, a plane-strain transition does exist and is the beginning of metallurgical changes that increase the $K_{\text{Ic}}$ behavior. However, it is the plane-stress transition that is a synergistic effect of metallurgical changes and mechanical (loss of constraint) changes that results in very large changes in fracture toughness.

## 4.3. Effect of Thickness (Constraint) and Notch Acuity on Fracture Toughness

Ahead of a sharp crack, the lateral constraint (which increases with increasing plate thickness) is such that through-thickness stresses are present.

Because these through-thickness stresses must be zero at each surface of a plate, they are less for thin plates compared with thick plates. For very thick plates, the through-thickness stresses are large, and a triaxial tensile state of stress occurs ahead of the crack. This triaxial state of stress reduces the apparent ductility of the steel by decreasing the shear stresses. Because yielding is restricted, the constraint ahead of the notch is increased and thus the notch toughness is reduced. This decrease in notch toughness is controlled by the thickness of the plate, even though the inherent metallurgical properties of the material may be unchanged. Thus the notch toughness decreases for thick plates compared with thinner plates of the same material. This behavior is shown schematically in Figure 4.11, which indicates that the minimum toughness of a particular material, $K_{\text{Ic}}$, is reached when the thickness of the specimen is large enough so that the state of stress is plane strain. In Figure 4.12, actual test results are presented for a high-strength maraging steel that illustrate this behavior.

For thicknesses greater than some value related to the toughness and yield strength of individual materials, maximum constraint occurs and plane-strain, $K_{\text{Ic}}$, behavior results. In Chapter 3 it was shown that this limiting thickness has been defined to be $B \geq 2.5(K_{\text{Ic}}/\sigma_{ys})^2$, as given in the ASTM Standard Method of Test for $K_{\text{Ic}}$. Conversely, as the thickness of the plate is decreased, *even though the inherent metallurgical characteristics of the steel are not changed,* the notch toughness increases, and plane-stress, $K_c$, behavior exists.

Figure 4.13 shows the shear lips at the surface of fracture test specimens machined from a single plate with different thicknesses. The percentage of shear lips as compared with the total fracture surface is a qualitative indication of notch toughness. A small percentage of shear-lip area indicates a relative brittle behavior. A comparison of the fracture surfaces in Figure 4.13 shows that thinner plates are more resistant to brittle fracture than thick plates in that the percentage of shear lips is larger for the thinner specimens compared with the thicker ones.

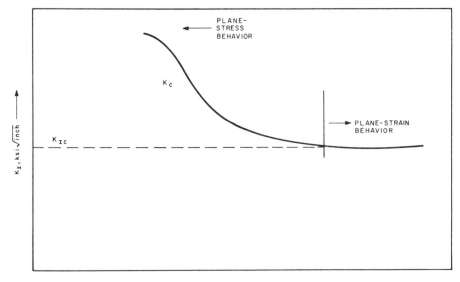

**Figure 4.11**  Effect of thickness on $K_c$.

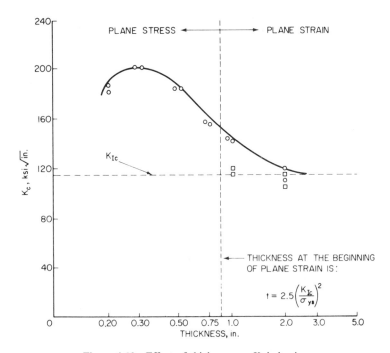

**Figure 4.12**  Effect of thickness on $K_c$ behavior.

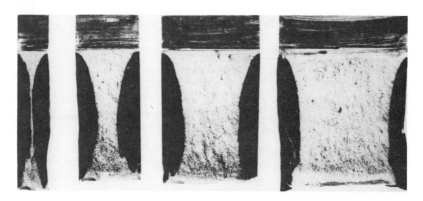

**Figure 4.13**   Effect of specimen thickness ($\frac{1}{2}$, 1, $1\frac{1}{2}$, and 2 in.) on toughness as determined by size of shear lips.

Pellini[4] has described the physical significance of constraint and plate thickness on fracture toughness in terms of plastic flow, as shown in Figure 4.14. This figure shows that the introduction of a circular notch in a bar loaded in tension causes an elevation of the stress-strain, or flow, curve. He describes the plastic flow of the smooth tensile bar as "free" flow, or that normally observed in conventional stress-strain curves, where lateral contraction is not constrained during the initial loading.

In the notched bar, however, the reduced section deforms inelastically while the ends of the specimen are still loaded elastically. Since the amount of elastic contraction (Poisson's ratio) is small compared to the inelastic contraction of the reduced section, a *restriction* to plastic flow is developed.

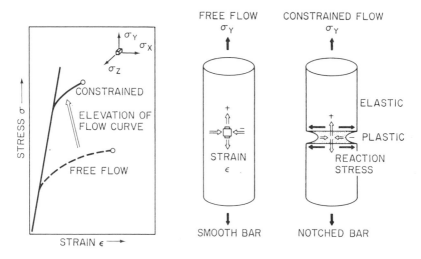

**Figure 4.14**   Origins of constraint effects.

This restriction is in the nature of a reaction-stress system such that the $\sigma_x$ and $\sigma_z$ stresses restrict or constrain the flow in the $\sigma_y$ (load) direction. Thus the uniaxial stress state of the smooth bar is changed to a triaxial tensile stress system in the notched bar.

For a triaxial state of stress, where the three principal stresses, $\sigma_x$, $\sigma_y$, and $\sigma_z$, are equal, there are no shearing stresses. This results in almost complete constraint against plastic flow, thus increasing the elastic stresses at the tip of a crack to extremely high values, compared with the lower "free"-flow stresses in an unrestrained tension specimen. In the case of most notched specimens, $\sigma_y > \sigma_x$ or $\sigma_z$, but the stresses are not equal. Thus some shearing stresses do occur, and there is some nonlinear behavior.

Figure 4.15 is a schematic description of the state of stress at the tip of a through-thickness crack in a sharply notched specimen loaded in tension. To satisfy compatibility conditions, the plastic "cylinder" (plastic-zone region defined in Chapter 2) that is developed at the crack tip must increase in diameter with an increase in stress in the $y$ direction due to load. However, this can happen only if through-thickness lateral contraction in the $z$ direction occurs. This lateral contraction is constrained by the elastically stressed material surrounding the "cylinder" and leads to the triaxial state of stress which raises the flow stress. Furthermore, the material behind the notch is unstressed because of the free surface of the notch and adds to the lateral constraint ahead of the notch. Therefore, *as the plate thickness is increased, the constraint increases, and the flow stress curve* is raised, as shown schematically in Figure 4.15.

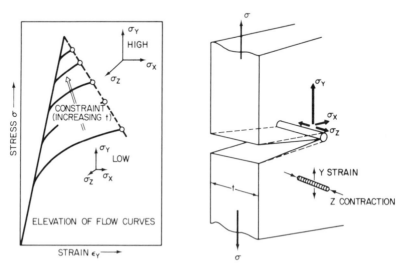

**Figure 4.15** Constraint conditions for through-thickness cracks. Increasing section size increases constraint because through-thickness flow must evolve with increased opposition of the larger volume of surrounding elastically loaded metal.

In summary, the constraint ahead of a sharp crack is increased by increasing the plate thickness. Thus the critical stress intensity, $K_c$, for a particular structural material tested at a particular temperature and loading rate decreases with increasing specimen thickness (Figures 4.11 and 4.12). Beyond some limiting thickness, maximum constraint is obtained, and the critical stress-intensity factor reaches the minimum plane-strain value, $K_{Ic}$. This maximum constraint occurs when the plate thickness is sufficiently large in a notched specimen of the particular material being tested at a particular test temperature and loading rate. As described previously, the limiting thickness for plane-strain behavior has been established by the ASTM standard test method as

$$B \geq 2.5\left(\frac{K_{Ic}}{\sigma_{ys}}\right)^2$$

For dynamic loading, the limiting thickness would be

$$B \geq 2.5\left(\frac{K_{Id}}{\sigma_{yd}}\right)^2$$

where $B$ = thickness of test specimen.

$K_{Ic}$ = critical plane-strain stress-intensity factor under conditions of static loading described in ASTM Method E-399—"Standard Method of Test for Plane Strain Fracture Toughness of Metallic Materials"—as described in Chapter 3.

$\sigma_{ys}$ = static tensile yield strength obtained in "slow" tension test as described in ASTM Test Method E-8—"Standard Methods of Tension Testing of Metallic Materials."

$K_{Id}$ = critical plane-strain stress-intensity factor as measured by "dynamic" or "impact" test; the test specimen is similar to a $K_{Ic}$ test specimen but is loaded rapidly as described in Chapter 3; there was no standardized test procedure as of 1986.

$\sigma_{yd}$ = dynamic tensile yield strength obtained in "rapid" tension test at loading rates comparable to those obtained in $K_{Id}$ tests; although extremely difficult to measure, a good engineering approximation based on experimental results of structural steels is

$$\sigma_{yd} = \sigma_{ys} + (20\text{--}30 \text{ ksi})$$

This limiting constraint condition for $K_{Ic}$ or $K_{Id}$ is established for a crack tip of "infinite" sharpness, namely, $\rho = 0$. This "infinite" sharpness is obtained by fatigue cracking the test specimens at low-stress levels, as described in Chapter 3. As a test specimen is loaded, some local plastic flow will occur at the crack tip, and the crack tip will be blunted slightly. For a brittle material, that is, any structural material tested at a temperature and loading rate where it has very low crack toughness, the degree of crack blunting is very small. Consequently unstable crack *extension* will occur under conditions of continued crack sharpness. In essence, the material fractures under elastic loading and exhibits plane-strain behavior under conditions of maximum constraint.

However, if the inherent toughness of the structural material is such that it is *not* brittle at the particular testing temperature and loading rate (e.g., the start of elastic-plastic behavior), then an increase in plastic deformation at the crack tip occurs and the crack tip is "blunted." As a result, the limit of plane-strain constraint is exceeded. At temperatures above this test temperature, the inherent toughness begins to increase rapidly with increasing test temperature because the effects of the crack blunting and relaxation of plane-strain constraint are synergistic. That is, the crack blunting leads to a relaxation in constraint which causes increased plastic flow, which leads to additional crack blunting. Thus elastic-plastic behavior begins to occur rapidly at increasing test temperatures once this plane-strain constraint (thickness plus notch acuity) is exceeded. Figures 4.5 through 4.9 show examples of the very rapid change in $K_{Ic}$ (after the plane-strain transition) for structural steels once the limit of plane-strain behavior ($\beta_{Ic} = 0.4$) has been reached.

Pellini[4] has described this behavior in terms of a constraint relaxation which changes the flow curve as shown schematically in Figure 4.16(a). The degree of crack-tip blunting establishes the particular flow-stress curve, leading to plane-strain, elastic-plastic, or plastic behavior. For example, the dashed curves *A* and *B* represent flow curves of unconstrained material (as in standard tension tests) leading to plastic or elastic-plastic behavior, depending on the inherent ductility of the structural material. Curve *C* represents the flow curve of a notched fully constrained material leading to failure under conditions of plane strain. If the material is tested at a temperature above the limit of plane strain such that partial crack-tip blunting occurs, a partial relaxation of constraint occurs, leading to elastic-plastic behavior. If the material is tested at still higher temperatures, considerable crack-tip blunting occurs, and considerable relaxation of constraint occurs, leading to plastic behavior.

Figure 4.16(b) is a schematic of the metal-grain structure ahead of the crack tip and indicates the microscopic behavior of the particular structural material. The dark line tracings within grains indicate slip on crystal planes, which is necessary to produce deformation. This deformation of individual grains is necessary to provide for growth of the plastic zone $r_y$ (approximated by the dashed circle) ahead of the crack tip. Depending on the inherent metallurgical structure of the metal, continued loading either increases slip and deformation (elastic-plastic or plastic behavior) or leads to the development of cracks and/or voids (plane-strain behavior).

A material that is brittle at a particular temperature will develop microcracks or voids before the plastic-zone size is very large, resulting in rapid

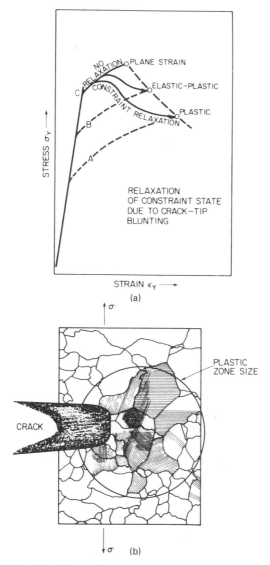

**Figure 4.16**   Relaxation of constraint state due to crack-tip blunting. (a) Relaxation of plane-strain constraint due to metal-grain flow, which causes crack-tip blunting; (b) effect of crack-tip blunting ahead of crack.

or unstable crack growth, that is, brittle fracture. As the test temperature is increased for successive test specimens, the metal becomes more ductile and slip occurs before microcracks occur, resulting in larger plastic-zone sizes. This is the beginning of the plane-strain transition where the individual grains begin to undergo large amounts of plasticity (*microscopic plasticity*), but the overall specimen is still elastic (*macroscopic plane strain*).

This transition in behavior from elastic (plane-strain) to plastic behavior occurs over the region known as the transition-temperature region. This region can be measured with various types of test specimens, but the $K_{Ic}$, CVN, $K_{Ic}$ $(t)$, $K_{Id}$, and DT test specimens are the most common types used. Note that $K_{Ic}$ or $K_{Id}$ specimens cannot be used to measure the entire range of behavior since they require essentially elastic plane-strain behavior to satisfy the restrictions placed on the analysis described in Chapter 2. Fracture-mechanics–type specimens ($J_{Ic}$, CTOD, and $R$-curve) that *can* be used to obtain quantitative estimates of the toughness in the elastic-plastic region are described in Chapter 17.

Because the $J_{Ic}$ and $R$-curve elastic-plastic tests are complex (as well as expensive) to conduct and interpret, they are not yet widely used in specifications or fracture-control plans. However, the CTOD test method has been used in various applications such as the Alaskan pipeline and North Sea offshore drilling rigs. In other cases, for example, nuclear pressure vessels and steel bridges, the engineer has used the results of auxiliary test specimens, such as the CVN or DT specimens, coupled with empirical correlations as discussed in Chapter 5, to describe the various levels of performance quantitatively. Examples of criteria and fracture-control plans are presented in Chapters 15 and 16.


## 4.4. Effect of Temperature and Loading Rate on $K_{Ic}$ and $K_{Id}$

In general, the crack toughness of structural materials, particularly steels, increases with increasing temperature and decreasing loading rate. These two general types of behavior are shown schematically in Figures 4.17 and 4.18. Figure 4.17 shows that both $K_{Ic}$ and $K_{Id}$ increase with increasing test temperature but that, for any given temperature, the crack toughness measured in an impact test, $K_{Id}$, generally is lower than the crack toughness measured in a static test, $K_{Ic}$. Figure 4.18 shows that, at a constant temperature, crack-toughness tests conducted at higher loading rates generally result in lower toughness values.

This effect of temperature and loading rate is similar to that obtained with Charpy V-notch impact test results, as shown schematically in Figure 4.19. Figure 4.8 showed the general effect of loading rate and temperature on the behavior of $K_{Ic}$ and $K_{Id}$, for an A36 structural steel ($\sigma_{ys}$ = 36 ksi). Figure 4.20 shows the same general effect for impact CVN tests and slow-bend CVN tests of the same A36 structural steel. Thus the transition from brittle to ductile behavior begins at lower temperatures for specimens tested at slow loading rates compared with specimens tested at impact loading rates.

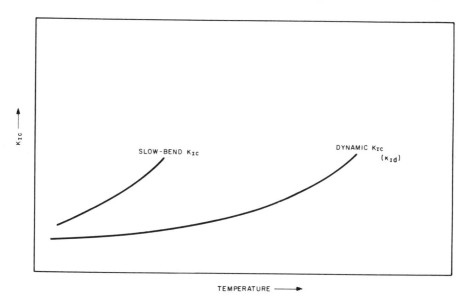

**Figure 4.17**   Schematic showing effect of temperature and loading rate on $K_{Ic}$.

Data obtained for various structural steels demonstrate that a true $K_{Ic}$ temperature transition exists that is independent of geometry. This is the plane-strain transition as was described in Section 4.2. This change in $K_{Ic}$ is not related to a loss in constraint as the test temperature is increased but is an inherent characteristic of many structural materials.

The rate of increase in $K_{Ic}$ with temperature does not remain constant but increases markedly above a given test temperature. Fractographic anal-

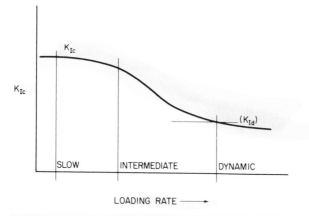

**Figure 4.18**   Schematic showing effect of loading rate on $K_{Ic}$.

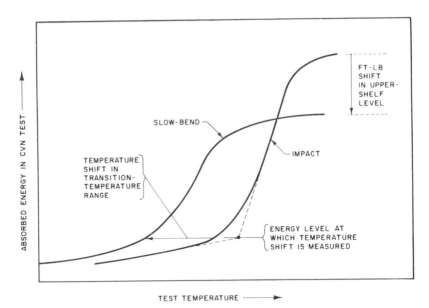

**Figure 4.19** Schematic representation of shift in CVN transition temperature and upper-shelf level due to strain rate.

yses show that the fracture-toughness transition temperature is associated with the onset of change in the microscopic-fracture mode at the crack tip. At the low end of the transition-temperature range, the mode of fracture initiation is cleavage, and at the upper end, the fracture initiation mode is ductile tear. In the transition-temperature region, a continuous change in fracture mode occurs.

Thus, there are *two* transitions in fracture behavior with temperature, namely, the $K_{Ic}$ temperature transition just described, which is referred to as the plane-strain transition with temperature, and a transition from plane-strain to plane-stress (elastic-plastic) behavior as the constraint at the crack tip decreases. Plane strain generally refers to the *macroscopic* state of stress. In the plane-strain transition region, the *microscopic* fracture behavior can change with increasing temperature from 100 percent cleavage or quasi-cleavage to a mixture of quasi-cleavage, tear dimples, and large flat tear areas. Thus, the *microscopic* fracture behavior can change while the *macroscopic* state of stress is still one of plane strain. This change in the *microscopic* fracture behavior leads to the rapid increase in the plane-strain fracture toughness. Figure 4.21 illustrates the *macroscopic* plane-strain transition for 100-ksi yield-strength structural steel. Figure 4.22 shows the fracture surfaces of selected specimens in this transition-temperature region and illustrates the *macroscopic* fracture behavior under *macroscopic* plane-strain conditions. Figure 4.23 is a series of fractographs showing the changes in *microscopic* behavior for these same test specimens. At $-250°F$, the

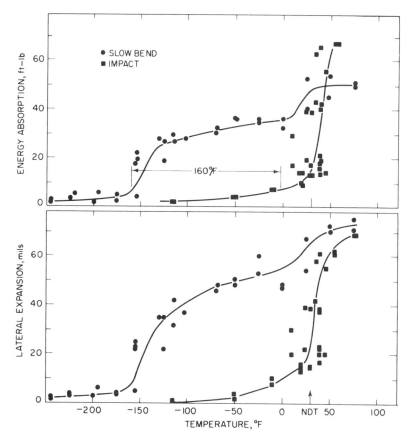

**Figure 4.20**   Charpy V-notch energy absorption and lateral expansion for impact and slow-bend tests of standard CVN specimens.

fracture surfaces are 100 percent quasi-cleavage and, at +75°F, are 100 percent tear dimples. At −140°F, the fractures were approximately 90 percent quasi-cleavage, and at −50°F (approximately the limit of the plane-strain transition at which point the plane-strain *constraint* at the crack tip decreased significantly), the fractures were approximately 90 percent tear dimples.

Similar behavior exists for CVN specimens,[1] leading to the conclusion that the transition-temperature behavior in $K_{Ic}$ (plane-strain transition) and Charpy tests reflects predominantly a transitional change in the microscopic mode of fracture from quasi-cleavage at very low temperatures to tear dimples at the upper-shelf region of the CVN test results.

Under the combined effects of both transitions, that is, the plane-strain $K_{Ic}$ transition and the transition from plane strain to plane stress, the rate of change of $K_c$ as a function of temperature should be *greater* than

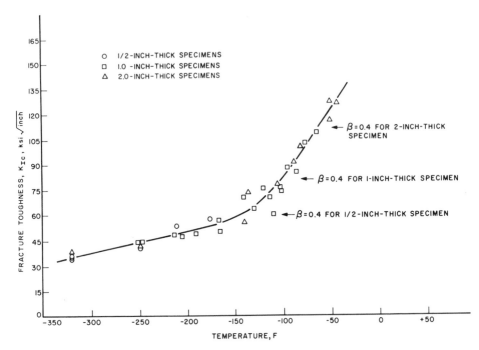

**Figure 4.21**   Plane-strain fracture-toughness transition behavior as a function of temperature.

the rate shown in Figure 4.21. Data obtained from $\frac{1}{2}$-in.-thick specimens of the same material described in Figure 4.21 substantiate this assumption. The test results from these $\frac{1}{2}$-in.-thick specimens are presented in Figure 4.24 and compared with the test results for the 1- and 2-in.-thick valid $K_{Ic}$ test results of Figure 4.21. Approximate values of $K_{Ic}$ were calculated from

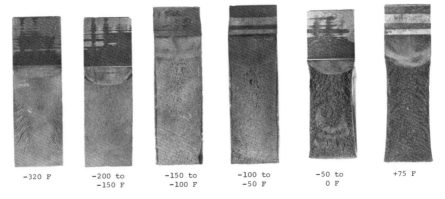

**Figure 4.22**   Fracture surfaces of $K_{Ic}$ specimens at various temperatures.

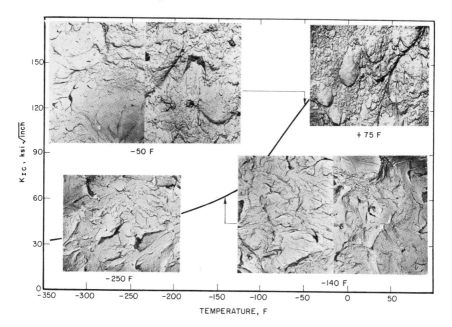

**Figure 4.23**    Fractographs of $K_{Ic}$ specimens at various temperatures.

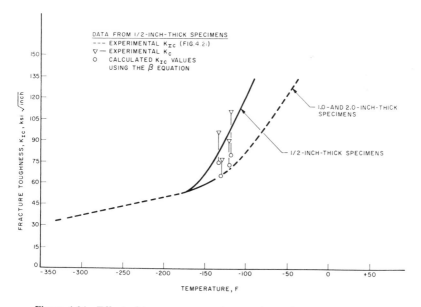

**Figure 4.24**    Effect of temperature on plane-strain to plane-stress transition.

the plane-stress results obtained on the $\frac{1}{2}$-in.-thick specimens by using Irwin's $\beta$ relation between $K_{Ic}$ and $K_c$ that is applicable only when the differences between $K_c$ and $K_{Ic}$ are small.[2] This relation is as follows:

$$K_c^2 = K_{Ic}^2[1 + 1.4\beta_{Ic}^2]$$

where

$$\beta_{Ic} = \frac{1}{B}\left(\frac{K_{Ic}}{\sigma_{ys}}\right)^2$$

The calculated $K_{Ic}$ values agree quite well with those obtained from 1- and 2-in.-thick specimens (Figure 4.24), demonstrating the fact that the $K_c$ plane-stress transition occurs before the plane-strain transition and that Irwin's $\beta$ concept does relate $K_c$ and $K_{Ic}$ values.

Above some temperature that depends on the particular structural material (for this A517 steel it was approximately $-40°F$), valid $K_{Ic}$ results according to the standard ASTM test method described in Chapter 3 cannot be obtained. For these cases, auxiliary test methods and correlations (such as described in Chapter 5) must be used to estimate $K_{Ic}$ or $K_{Id}$. For high levels of elastic-plastic behavior, methods of fracture analyses as described in Chapter 17 can also be used to describe the fracture behavior of structural materials.

## 4.5. Significance of Initiation and Propagation (Arrest) Fracture Toughness

The general difference in initiation and propagation behavior of low- to medium-strength structural steels as related to fracture-toughness test results is shown schematically in Figure 4.25. The curve labeled "static" refers to the fracture toughness obtained in a standard ASTM E-399 $K_{Ic}$ test under conditions of slow loading. [The curve for rapid or intermediate loading-rate tests, $K_{Ic}(t)$, would be shifted slightly to the right of the static curve.] The impact curve is from a $K_{Id}$ or other dynamic test under conditions of impact loading. The difference between these two is the temperature shift, which is a function of yield strength for structural steels as will be discussed in Section 4.7.

In region $I_s$ for the static curve (Figure 4.25), the crack initiates in a cleavage mode from the tip of the fatigue crack. In region $II_s$ the fracture toughness that will result in initiation of unstable crack propagation increases with increasing temperature. This increase in the crack-initiation toughness corresponds to an increase in the size of the plastic zone and in the zone of ductile tear (shear) at the tip of the crack prior to unstable crack extension. In region $III_s$ the static fracture toughness is quite large and somewhat difficult to define (that is, elastic-plastic fracture-mechanics tests are required— Chapter 17), but the fracture initiates by ductile tear (shear).

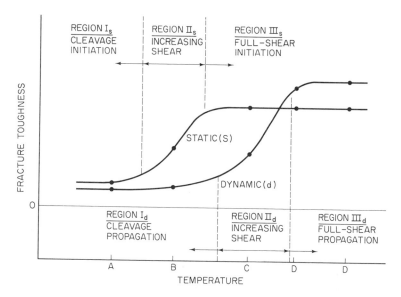

**Figure 4.25**  Schematic showing relation between static and dynamic fracture toughness.

The ductile tear zone at the tip of a statically loaded crack is confined to the zone of plastic deformation along the crack front. The maximum size for this plastic zone is restricted by the ASTM E-399 test requirement to ensure valid plane-strain test results. Deviations from this requirement for elastic plane-strain conditions toward elastic-plastic conditions usually result in high initiation fracture-toughness values. The ductile tear zone at the tip of the crack under elastic-plastic conditions is usually very small and is difficult to delineate by visual examination.

In an actual steel structure loaded at temperature A, initiation may be static and propagation dynamic. However, there is no apparent difference between the two because both initiation and propagation are by cleavage. If a similar structure is loaded slowly to failure at temperature B, there will be some localized shear and a reasonable level of static fracture-toughness at the initiation of failure. However, for rate-sensitive structural materials, such as structural steels used in bridges, offshore rigs, or ships, once the crack has initiated, the notch toughness is characterized by the dynamic toughness level on the impact curve. Thus the fracture appearance for the majority of the fracture service is cleavage. If the structure is loaded slowly to fracture initiation at temperature C, the initiation characteristics will be full-shear initiation with a high level of plane-stress crack toughness, $K_c$. However, the fracture surface of the running crack may still be predominately cleavage but with some amount of shear as shown in the lower impact curve at temperature C in Figure 4.25. Figure 4.26 shows the fracture

**Figure 4.26**   Fracture surfaces of full-thickness ($B = 1.5$ in.) 4-T compact-tension specimens of A572 Grade 50 steel tested under load-control conditions using a total-unload/reload loading sequence.

surfaces and representative toughness levels (CVN and $K_c$) for an A572 Grade 50 steel. The specimen tested at $-42°F$ exhibits some small amount of shear initiation, that is, at a temperature slightly below B, Figure 4.25. The specimen tested at $+38°F$ exhibits increasing shear initiation (between B and C). The specimen tested at $+72°F$ exhibits full shear initiation (temperature C) but still exhibits a large region of cleavage propagation, Figure 4.26. Thus ductile crack propagation (dynamic) would only occur at temperature D, which is essentially dynamic upper-shelf CVN impact behavior (i.e., 80 percent or above shear fracture appearance).

Schematic $P$–$\Delta$ records as would be obtained from an E-399 slow-bend test are presented in Figure 4.27 for each of the three fracture surfaces shown in Figure 4.26. Note that although all of them are shown to exhibit some nonlinear or elastic-plastic behavior, the characteristics of these records depend on the size and shape of the crack and of the specimen.

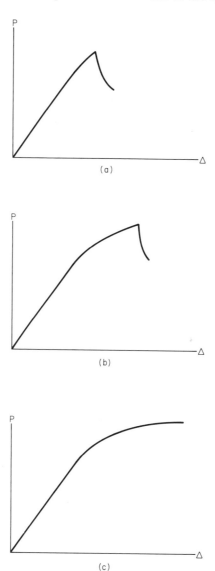

**Figure 4.27**    Schematic $P$–$\Delta$ curves corresponding to fracture surfaces (A), (B), and (C), Figure 4-26.

## 4.6. Representative Fracture-Toughness Results for Structural Steels

Low-strength structural steels (for example, steels with yield strengths less than 140 ksi) generally are temperature- and loading-rate sensitive. That is, as described previously, these steels exhibit a considerable increase in fracture toughness with increasing test temperature or decreasing loading rate. Examples of $K_{Ic}$, rapid-load $K_{Ic}$ (0.03 to 3 sec), and impact $K_{Id}$ fracture-toughness results for various structural steels are presented in Figures 4.28 through 4.37. The general characteristics of these steels are presented in Table 4.1. These steels have yield strengths ranging from 40 to 250 ksi and, although believed to be representative, are presented for general information only and should not be used in any design without independent investigation of actual materials. Table 4.1 presents a general description of the steels whose fracture-toughness results are presented in Figures 4.28

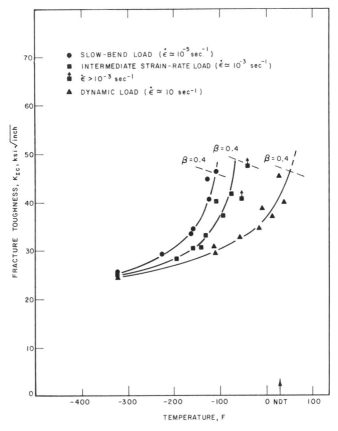

**Figure 4.28**   Effect of temperature and loading rate on fracture toughness of an A36 steel.

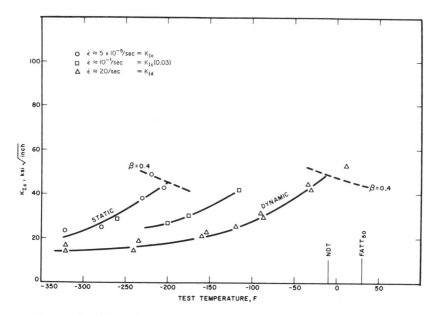

**Figure 4.29**   Effect of temperature and loading rate on fracture toughness of an ABS-C steel.

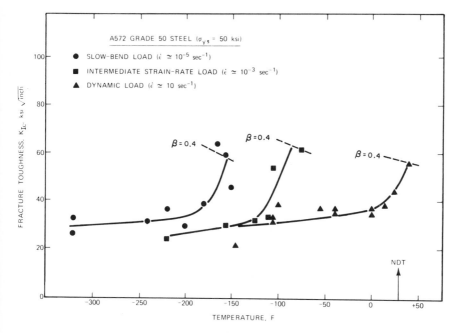

**Figure 4.30**   Effect of temperature and loading rate on fracture toughness of an A572 steel.

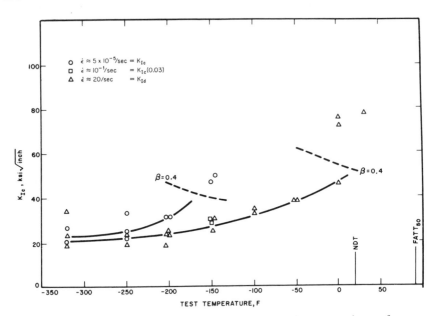

**Figure 4.31**  Effect of temperature and loading rate on fracture toughness of an A302 Grade B steel.

through 4.37. Also shown on these figures are the NDT temperatures determined from drop weight tests.

$K_{Ic}$, rapid-load $K_{Ic}$ $(t)$, and $K_{Id}$ values were determined as described in Chapter 3. The validity of these values was determined according to ASTM Test Method E-399. The $K_{Ic}/\sigma_{ys}$ ratio was evaluated at the testing

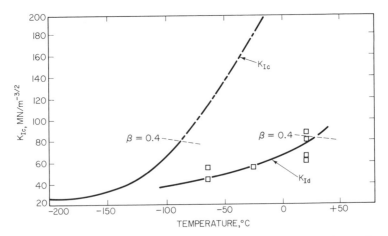

**Figure 4.32**  Effect of temperature and loading rate on fracture toughness of a C-Mn steel. (From *Dynamic Fracture Toughness Measurements*, by A. P. Glover, F. A. Johnson, J. C. Radon, and C. E. Turner, The Welding Institute, 1976.)

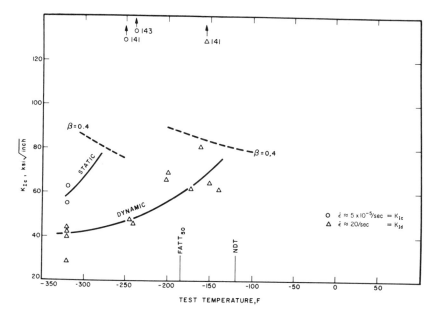

**Figure 4.33**   Effect of temperature and loading rate on fracture toughness of an HY-80 steel.

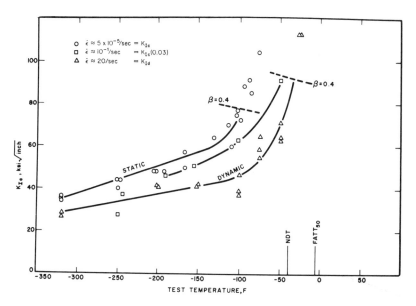

**Figure 4.34**   Effect of temperature and loading rate on fracture toughness of an A517 Grade F steel.

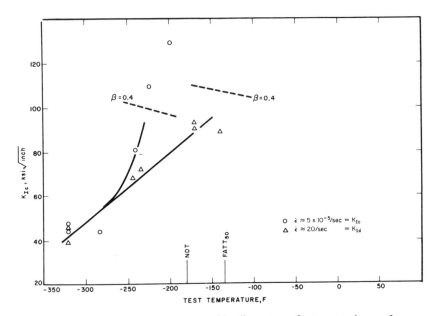

**Figure 4.35** Effect of temperature and loading rate on fracture toughness of an HY-130 steel.

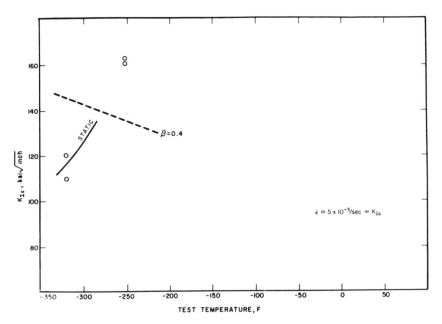

**Figure 4.36** Effect of temperature and loading rate on fracture toughness of an 18Ni (180) maraging steel.

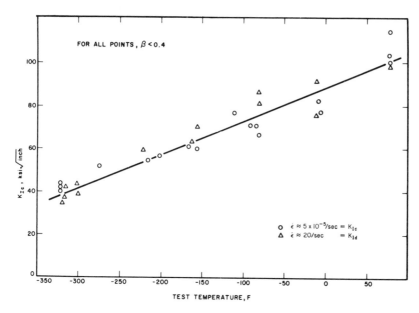

FOR ALL POINTS, $\beta < 0.4$

$\circ$  $\dot{\epsilon} \approx 5 \times 10^{-5}/\text{sec} = K_{\text{Ic}}$
$\triangle$  $\dot{\epsilon} \approx 20/\text{sec}$   $= K_{\text{Id}}$

$K_{\text{Ic}}$, ksi $\sqrt{\text{inch}}$

TEST TEMPERATURE, F

**Figure 4.37**  Effect of temperature and loading rate on fracture toughness of an 18Ni (250) maraging steel.

temperature and loading rate. For the rapid-load $K_{\text{Ic}}$ and $K_{\text{Id}}$ tests, strain rates were calculated for a point on the elastic-plastic boundary as determined by Irwin and using the following equation:

$$\dot{\varepsilon} = \frac{2\sigma_{ys}}{tE} \qquad (4.1)$$

where $\sigma_{ys}$ = yield strength for the test temperature and loading rate.

$\quad\quad t$ = loading time for the test, that is, the time stated in the brackets for the rapid-load $K_{\text{Ic}}$ $(t)$ test results.

$\quad\quad E$ = elastic modulus of the material tested.

The rate of application of $K$ is

$$\dot{K} = \frac{K_{\text{critical}}}{t}$$

where $K_{\text{critical}}$ = the $K_{\text{Ic}}$, $K_{\text{Ic}}(t)$, or $K_{\text{Id}}$ value.

$\quad\quad t$ = loading time for the test.

The dynamic $K_{\text{Id}}$ tests (falling weight) were conducted at crack-tip strain rates varying between 10 and 40/sec, and the rapid-load $K_{\text{Ic}}$ (0.03) tests were conducted at strain rates of approximately $10^{-1}/\text{sec}$. Two steels, A36 and A572 Grade 50 (Figures 4.28 and 4.30), were tested at strain rates of approximately $10^{-3}/\text{sec}$. These would be called rapid-load $K_{\text{Ic}}$ (3) test

TABLE 4.1   General Characteristics of Steels Described in Figures 4.28 to 4.37

| Figure No. | Steel Grade* | Classification | Steel Quality | Steel Condition | Minimum Yield Strength (ksi)** |
|---|---|---|---|---|---|
| 4.28 | ASTM A36 | Carbon | Structural | As rolled† | 36 |
| 4.29 | ABS Grade C | Carbon | Structural | As rolled† | ‡ |
| 4.30 | ASTM A572 Grade 50 | HSLA§ | Structural | As rolled† | 50 |
| 4.31 | ASTM A302 Grade B | Alloy | Pressure vessel | As rolled† | 50 |
| 4.32 | British steel | C-Mn | Structural | Normalized and tempered | 75 |
| 4.33 | HY-80 | Armor | Armor | Quenched and tempered | 80 |
| 4.34 | ASTM A517 Grade F | Alloy | Pressure vessel | Quenched and tempered | 100 |
| 4.35 | HY-130 | Armor | Armor | Quenched and tempered | 130 |
| 4.36 | 18Ni (180) | High strength | High strength | Maraging | 180 |
| 4.37 | 18Ni (250) | High strength | High strength | Maraging | 250 |

* ASTM = American Society for Testing and Materials, ABS = American Bureau of Shipping. The Grade C steel was deleted from ABS Specifications in 1974.

† These steels are usually supplied in the as-rolled condition. However, they are sometimes supplied in the normalized condition.

‡ No specified yield strength. Tensile-strength range: 58 to 71 ksi.

§ HSLA = High-strength low-alloy steel.

** Conversion factor: 1 ksi = 6.895 MPa

results. The static $K_{Ic}$ tests were conducted at strain rates of approximately $10^{-5}$/sec.

The thickness requirement for plane-strain behavior is determined by the ratio of plastic-zone size and specimen thickness. This ratio is defined as

$$\beta = \frac{(K_{Ic}/\sigma_{ys})^2}{B} \qquad (4.2)$$

where $K_{Ic}$ = measured critical stress-intensity factor.

$\sigma_{ys}$ = yield strength.

$B$ = specimen thickness.

At present, the ASTM Standard Method of Testing requires that $\beta$ be equal to or less than 0.4 to ensure that crack extension occurs under plane-strain conditions. In Figures 4.28 through 4.37, the dashed lines represent constant values of 0.4 for either static or dynamic loading of the 1-in.-thick plates. Data points falling below these dashed lines are valid $K_{Ic}$ values. The "$K_{Ic}$" values measured above these lines might be affected by crack-tip conditions in which the stress state is changing from plane strain to plane stress. Hence, these values cannot be considered valid by current ASTM standards, although they appear to be extensions of the valid results and are useful approximations for engineering purposes.

For the strain rates used in this investigation, the $K_{Ic}$ values decreased with increasing loading rate for a constant testing temperature. Furthermore, the $K_{Ic}$ values obtained at the intermediate loading rate were always between those obtained by static and by dynamic loading. The most significant effect of increased loading rate on these five steels was the increase in the threshold temperature below which plane-strain behavior occurred. This increased range of plane-strain behavior was due to both an increase in yield strength and a reduction in $K_{Ic}$ caused by the increase in strain rate.

Only a few static fracture-toughness ($K_{Ic}$) values were obtained for the 18Ni (180) maraging steel (Figure 4.36). Meaningful dynamic test records could not be obtained because elastic "ringing" was encountered at the high loads necessary to obtain the high $K_{Id}$ values. No significant difference between static and dynamic $K_{Ic}$ values was observed for the 18Ni (250) maraging steel (Figure 4.37).

In general, these results indicate that because of the increase in temperature range over which valid $K_{Ic}$ behavior occurs, loading rate is an especially significant variable in $K_{Ic}$ testing, particularly for those steels having yield strengths less than about 140 ksi.

The best method of comparing the resistance to fracture of steels having different yield strengths is to evaluate the crack-toughness performance in terms of $K_{Ic}/\sigma_{ys}$. Because this ratio is both a measure of the critical crack-tip plastic-zone size and a critical flaw-size parameter, the larger this ratio, the better the resistance to fracture. Thus steels having different yield strengths but the same $K_{Ic}/\sigma_{ys}$ ratio should have the same resistance to fracture in terms of critical flaw size.

The crack-toughness performance, $K_{Ic}/\sigma_{ys}$, of ABS-C steel is shown in Figure 4.38. In this steel, and others, the $K_{Ic}/\sigma_{ys}$ values increased with increasing temperature, and the dynamic $K_{Id}/\sigma_{yd}$ values were less than the static $K_{Ic}/\sigma_{ys}$ values at a given temperature. Similar results for A517 steel are presented in Figure 4.39 and show a similar type of behavior.

All the results show that these structural materials undergo an increase in toughness with increasing temperature in the plane-strain transition region. Then, as the inherent toughness increases, and as the constraint at the crack tip decreases, these materials undergo a very rapid increase in toughness as they begin to exhibit elastic-plastic behavior. At this point, the $K_I$ analysis is no longer valid because of the excessive yielding at the crack tip, and $K_{Ic}$, $K_{Ic}$ ($t$), or $K_{Id}$ values cannot be determined. Although this type of behavior is very desirable in structural materials, it makes the application of fracture mechanics in materials testing very difficult.

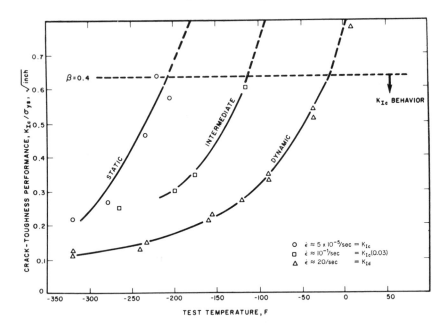

**Figure 4.38**   Crack-toughness performance for ABS-C steel (see Figure 4.29).

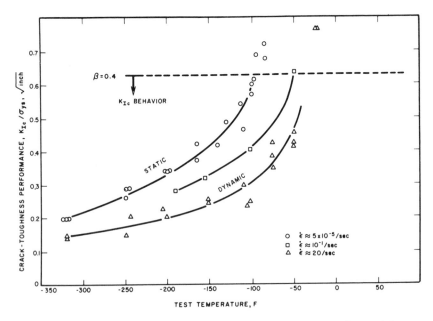

**Figure 4.39**   Crack-toughness performance for A517 steel (see Figure 4.34).

## 4.7. Loading-Rate Shift for Structural Steels

### 4.7.1. CVN Temperature Shift

The $K_{Ic}$, $K_{Ic}$ $(t)$, and $K_{Id}$ test results presented in Section 4.6 demonstrated the general effect of loading rate on $K_{Ic}$. A similar effect of loading rate exists for CVN specimens tested in three-point slow-bend and standard-impact loading. The general effect of a slow loading rate (compared with standard-impact loading rates for CVN specimens) is to shift the CVN curve to the left and to lower the upper-shelf values. This behavior was shown schematically in Figure 4.19.

Slow-bend and impact CVN test results for nine steels having yield strengths in the range 40–250 ksi are presented in Figures 4.40 through 4.48. The shifts in the transition temperature are related to yield strength, as shown in Figure 4.49. These results are similar to those observed by Roberts et al.[5]

For low-strength steels, the rate of change of absorbed energy as a function of temperature is greater in the impact test than in the slow-bend tests. Thus the magnitude of the temperature shift caused by high-strain-rate testing should be measured, at the same energy level, from the onset of the dynamic temperature transition to the onset of the transition on the static curve as shown in Figure 4.19. This onset of the dynamic temperature transition is defined arbitrarily by the intersection of tangent lines drawn

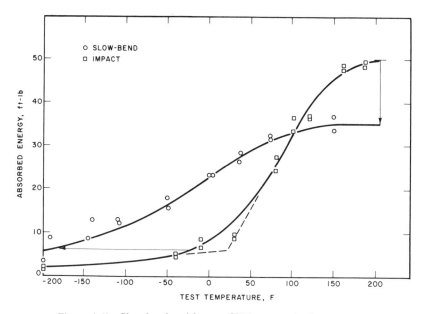

**Figure 4.40**   Slow-bend and impact CVN test results for A36 steel.

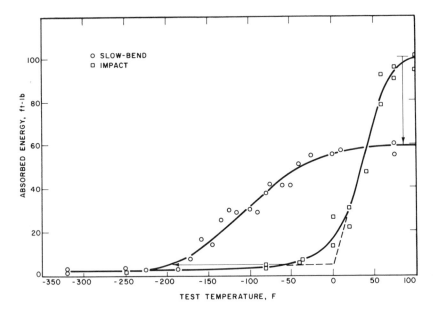

**Figure 4.41**   Slow-bend and impact CVN test results for ABS-C steel.

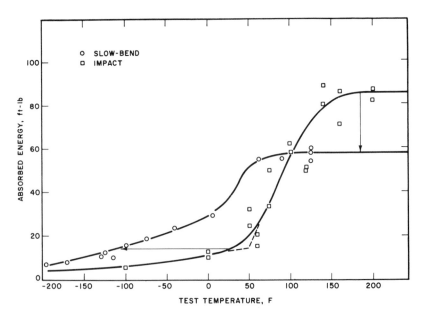

**Figure 4.42**   Slow-bend and impact CVN test results for A302 Grade B steel.

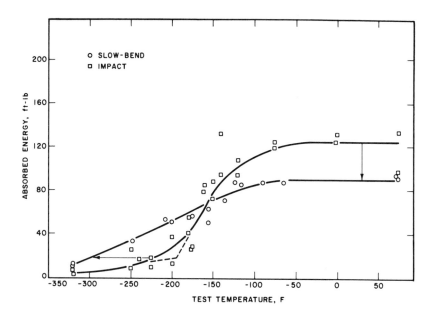

**Figure 4.43**  Slow-bend and impact CVN test results for HY-80 steel.

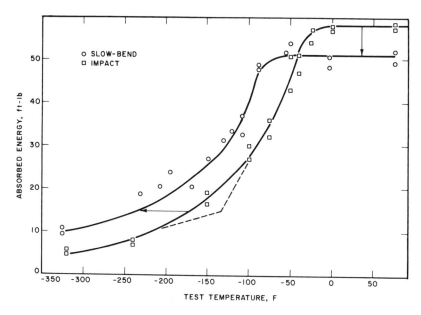

**Figure 4.44**  Slow-bend and impact CVN test results for A517-F steel.

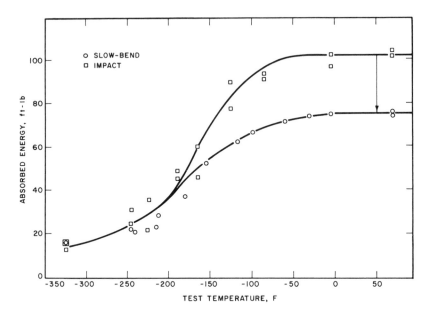

**Figure 4.45**  Slow-bend and impact CVN test results for HY-130 steel.

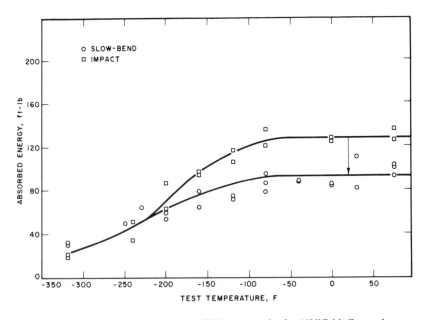

**Figure 4.46**  Slow-bend and impact CVN test results for 10NiCrMoCo steel.

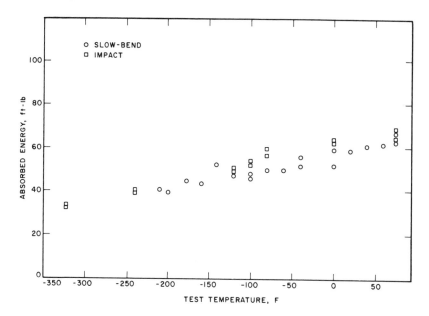

**Figure 4.47**   Slow-bend and impact CVN test results for 18Ni (180) steel.

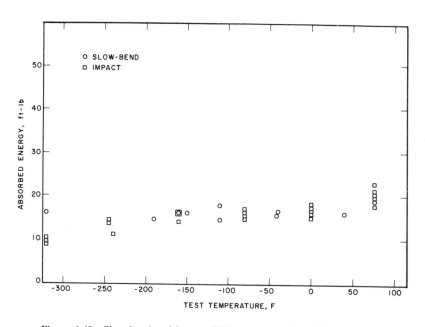

**Figure 4.48**   Slow-bend and impact CVN test results for 18Ni (250) steel.

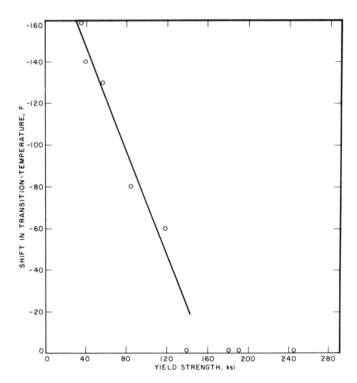

**Figure 4.49** Effect of yield strength on shift in transition temperature between impact and slow-bend CVN tests.

from the lower shelf level and the transition region (Figure 4.19). The loading-rate shift has been verified by Roberts et al.[5] using the 15-ft-lb level so the presence of the shift does not depend on the particular means used to measure it. However, the onset of dynamic transition seems to be the best reference point from which to measure strain-rate effects because this point is located in the energy-absorption region where a change in the microscopic mode of fracture starts to occur at the initial crack front for both static and dynamic testing.[1] Also, because the onset of the static temperature transition occurs at a lower temperature than that marking the onset of the dynamic temperature transition and because the static upper energy-absorption shelf is usually of lower magnitude than that measured in the dynamic test, the static and dynamic energy-absorption curves usually intersect. Thus, measurements of the temperature shift at temperatures above that defined by the onset of dynamic temperature transition may underestimate the magnitude of the shift. Below this reference temperature, the slopes with respect to the temperature axis of both the static and the dynamic CVN energy become very small; consequently, it is difficult to measure the magnitude of the shift between the two curves in the lower-shelf region.

Extreme care should be exercised when using the test results from slow-bend CVN specimens because, in many cases, the transition from brittle to ductile crack initiation occurs between very low values of absorbed energy (e.g. 5 and 12 foot pound) above which the specimen is subjected to general yielding conditions. Under these conditions, the brittle to ductile crack initiation transition may be missed and the only obvious transition would be that corresponding to an increased resistance to crack propagation which is closely related and may coincide with the toughness transition observed under impact loading.

### 4.7.2.  $K_{Ic}$–$K_{Id}$ Impact-Loading-Rate Shift

The maximum difference in $K_{Ic}$ and $K_{Id}$ fracture behavior for a given steel occurs between static loading and full-impact loading that correspond to strain rates on the order $10^{-5}$ sec$^{-1}$ and 10 sec$^{-1}$, respectively, as was shown in Figures 4.28 through 4.37. Moreover, various investigations[8-10] showed that the temperature shift between static loading ($\dot{\varepsilon} \approx 10^{-5}$ sec$^{-1}$) and impact loading ($\dot{\varepsilon} \approx 10$ sec$^{-1}$) decreases as the room temperature yield strength of the steel increases in the same manner as the shift in CVN results shown in Figure 4.49. The magnitude of the temperature shift between slow loading and impact loading (in both CVN and $K_{Ic}$–$K_{Id}$ tests) in steels of various yield strengths can be approximated by

$$T_{shift} = 215 - 1.5\sigma_{ys} \tag{4.3}$$

for 36 ksi $< \sigma_{ys} <$ 140 ksi (250 MPa $< \sigma_{ys} <$ 965 MPa) and

$$T_{shift} = 0$$

for $\sigma_{ys} >$ 140 ksi

where $T_{shift}$ = absolute magnitude of the shift in the transition temperature between slow loading and impact loading, °F.

$\sigma_{ys}$ = room temperature yield strength, ksi.

### 4.7.3.  $K_{Ic}$–$K_{Ic}$ (t) Intermediate-Loading-Rate Shift

Sufficient data are available to show that Equation (4.3) can be used to predict the shift from a static ($\dot{\varepsilon} \approx 10^{-5}$ sec$^{-1}$) $K_{Ic}$ transition curve to one caused by impact loading ($\dot{\varepsilon} \approx 10$ sec$^{-1}$). $K_{Ic}$ transition curves obtained at intermediate loading rates ($10^{-3}$ sec$^{-1} < \dot{\varepsilon} <$ 10 sec$^{-1}$) are always between those obtained under static and under impact loading. Unfortunately, a systematic investigation of the effect of loading rate on $K_{Ic}$ transition curves is yet to be conducted. Some data have been obtained by Shoemaker et al., and Barsom,[6,7] at an intermediate loading rate that corresponded to a strain rate of about $10^{-3}$ sec$^{-1}$ at the elastic-plastic boundary in the vicinity of the crack tip. Figures 4.28, 4.29, and 4.30 show these data for ASTM A36, ABS-B, and A572 Grade 50 steels, respectively. The data suggested that the shift between a $K_{Ic}$ curve obtained under static load ($\dot{\varepsilon} \approx 10^{-5}$ sec$^{-1}$) and under an intermediate loading rate ($\dot{\varepsilon} \approx 10^{-3}$ sec$^{-1}$) was approximately equal to 25 percent of the total shift, Equation (4.2), between the curves obtained under static loading and under impact loading ($\dot{\varepsilon} \approx 10$ sec$^{-1}$).

### 4.7.4. Predictive Relationship for Temperature Shift

Based on the preceding observations for $K_{Ic}$ transition behavior under static, intermediate, and impact rates of loading, a generalized characterization for the dependence of temperature shift, $T_{shift}$, strain rate, $\dot{\varepsilon}$, and yield strength, $\sigma_{ys}$, for a material can be formulated, Figure 4.50. For intermediate strain rates in the range $10^{-3}$ sec$^{-1} \leq \dot{\varepsilon} \leq 10$ sec$^{-1}$ and for steels having yield strengths of less than about 140 ksi (965 MPa), the relationship among $T_{shift}$, $\dot{\varepsilon}$, and $\sigma_{ys}$, Figure 4.51, can be approximated by

$$T_{shift} = (150 - \sigma_{ys})(\dot{\varepsilon})^{0.17} \qquad (4.4)$$

where $T_{shift}$ is in °F, $\sigma_{ys}$ is in ksi, and $\dot{\varepsilon}$ is in sec$^{-1}$.

The plane-strain fracture-toughness value for a given steel tested at a given temperature and for strain rate equal to or less than $10^{-4}$ sec$^{-1}$ is constant and is equal to the static $K_{Ic}$ value. Although the plane-strain

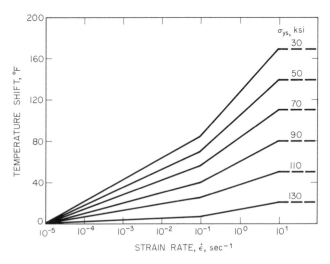

**Figure 4.50**  Shift in transition temperature for various steels and for strain rates greater than $10^{-5}$ sec$^{-1}$.

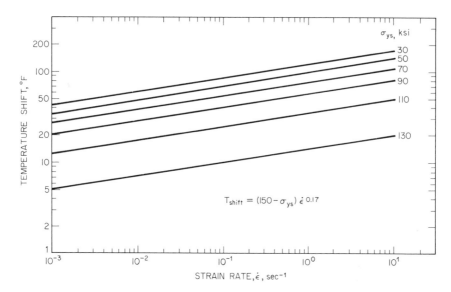

**Figure 4.51**    Shift in transition temperature for various steels and for strain rates in the range $10^{-3} \text{ sec}^{-1} \leq \dot{\varepsilon} \leq 10 \text{ sec}^{-1}$.

fracture-toughness behavior for a given steel tested at a given temperature and at a strain rate greater than 10 sec$^{-1}$ is yet to be established, it is generally assumed that the value is almost constant and is independent of the rate of loading.[8-13]

### 4.7.5. Significance of Temperature Shift

Because of this shift, increasing the loading rate can decrease the fracture-toughness value at a particular temperature for steels having yield strengths of less than 140 ksi. The change in fracture-toughness values for loading rates varying from slow to impact rates is particularly important to those applications where the actual loading rates are slow or intermediate. Many types of structures fall into this classification.

For example, for a design stress of about 20 ksi (138 MN/m$^2$), a $K_{\text{Ic}}$ of 40 ksi$\sqrt{\text{in.}}$ (44 MN/m$^{3/2}$) for a given steel would correspond to the tolerance of a through-thickness flaw of approximately 3 in. (76 mm). If a structure were loaded statically, this size flaw could be tolerated at extremely low temperatures. If the structure were loaded dynamically, however, the temperature at which this size flaw could cause failure would be much higher. That is, using the test results for an ABS-C ship steel presented in Figure 4.29 as an example, the static $K_{\text{Ic}}$ of 40 ksi$\sqrt{\text{in.}}$ occurs at $-225°$F, whereas the dynamic $K_{\text{Id}}$ of 40 ksi$\sqrt{\text{in.}}$ occurs at $-50°$F. Similar observations could be made for other steels as shown in Figures 4.30 to 4.35. Note that this difference between the static and dynamic results decreases with increasing yield strength, as indicated in the equation for $T_{shift}$.

If the loading rates of structures are closer to those of slow loading than to impact loading, a considerable difference in the behavior of these structures would be expected. Thus, not only should the effects of temperature and constraint (plate thickness) on structural steels be established, but more important, *the maximum loading rates that will occur in the actual structure being analyzed under operating conditions should be established.* As discussed in Chapter 16, the loading-rate shift has been used to establish the AASHTO material-toughness requirements for bridge steels.[7]

The results presented in Figure 4.49 show that the greatest shifts in transition temperature occurred for the low-strength steels and decreased with increasing yield strength up to strength levels of about 140 ksi. Above about 140 ksi, no shift in transition temperature was observed.

A similar shift in $K_{Ic}$ behavior as a function of loading rate was described in Figures 4.28 through 4.37.[6] To determine whether the shift in CVN test values is the same as the shift in $K_{Ic}$ test results, the dynamic $K_{Ic}$ test results were shifted by an amount equal to the CVN transition-temperature shifts and compared with the actual slow-bend $K_{Ic}$ test results (Figures 4.52 through 4.57). In general, the measured values agreed quite well with the predicted values. It should be emphasized that a prediction can be made either from slow-bend $K_{Ic}$ values to dynamic values ($K_{Id}$), or vice versa.

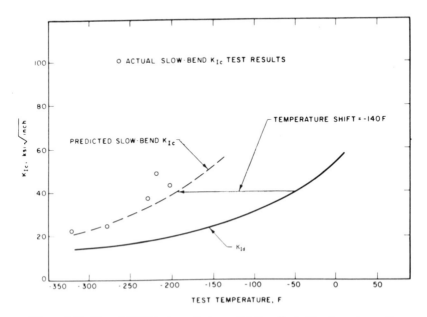

**Figure 4.52**  Use of CVN test results to predict the effect of loading rate on $K_{Ic}$ for ABS-C steel.

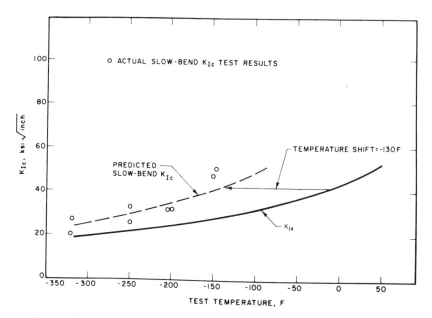

**Figure 4.53**   Use of CVN test results to predict the effect of loading rate on $K_{Ic}$ for A302 Grade B steel.

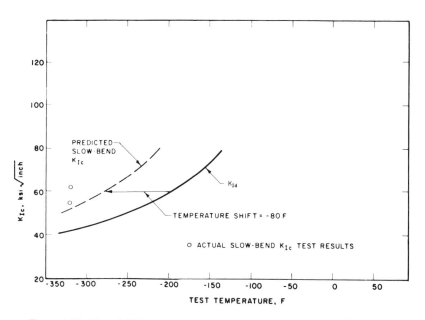

**Figure 4.54**   Use of CVN test results to predict the effect of loading rate on $K_{Ic}$ for HY-80 steel.

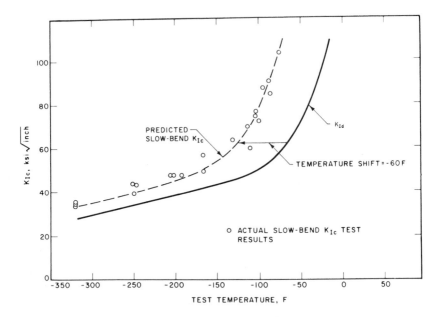

**Figure 4.55** Use of CVN test results to predict the effect of loading rate on $K_{Ic}$ for A517 Grade F steel.

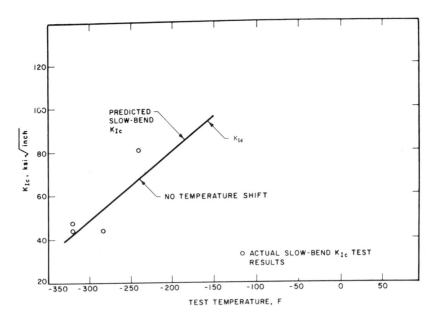

**Figure 4.56** Use of CVN test results to predict the effect of loading rate on $K_{Ic}$ for HY-130 steel.

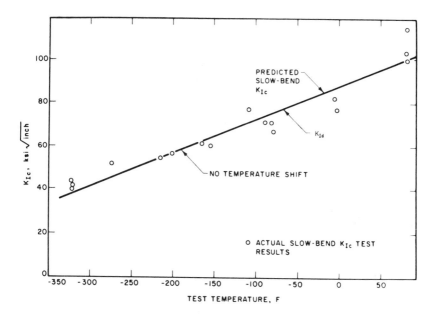

**Figure 4.57**    Use of CVN test results to predict the effect of loading rate on $K_{Ic}$ for 18Ni (250) steel.

Because the dynamic $K_{Ic}$ curves ($K_{Id}$) were better defined, the results shown in Figures 4.52 through 4.57 were from dynamic $K_{Id}$ values to slow-bend $K_{Ic}$ values. Because dynamic $K_{Id}$ tests are extremely difficult to conduct and analyze, the presently developed prediction procedure should be quite useful in obtaining a first-order approximation of the effects of loading rate on the $K_{Ic}$ behavior of steels by adjusting experimentally obtained slow-bend $K_{Ic}$ values. Further application of this observation is described in Chapter 5 on Correlations Between Various $K_{Id}$ Values and Fracture-Toughness Test Results.

## References

1. J. M. Barsom and S. T. Rolfe, "$K_{Ic}$ Transition-Temperature Behavior of A517-F Steel," *Journal of Engineering Fracture Mechanics, 2,* 1971, p. 341.
2. G. R. Irwin, "Crack-Toughness Testing of Strain-Rate Sensitive Materials," *Journal of Engineering for Power, Transactions of the ASME, Series A, 86,* 1964, p. 444.
3. H. D. Greenberg, E. T. Wessel, and W. H. Pryle, "Fracture Toughness of Turbine-Generator Rotor Forgings," *Engineering Fracture Mechanics, 1,* 1970, pp. 653–674.
4. W. S. Pellini, "Design Options for Selection of Fracture Control Procedures in the Modernization of Codes, Rules, and Standards," in *Proceedings: Joint*

*United States–Japan Symposium on Application of Pressure Component Codes, Tokyo, Japan*, Mar. 13–15, 1973.

5. R. ROBERTS, G. R. IRWIN, G. V. KRISHNA, and B. T. YEN, "Fracture Toughness of Bridge Steels—Phase II Final Report," *Fritz Engineering Laboratory Report No. 379.2*, Lehigh University, Bethlehem, Pa., June 1974.

6. A. K. SHOEMAKER and S. T. ROLFE, "The Static and Dynamic Low-Temperature Crack-Toughness Performance of Seven Structural Steels," *Engineering Fracture Mechanics 2*, 1971, pp. 319–339.

7. J. M. BARSOM, "Development of the AASHTO Fracture-Toughness Requirements for Bridge Steels," *Journal of Engineering Fracture Mechanics, 7*, No. 3, Sept. 1975, p. 605.

8. J. M. BARSOM, "Relationship Between Plane-Strain Ductility and $K_{Ic}$," *Transactions of the ASME, Journal of Engineering for Industry, 93*, Series B, No. 4, Nov. 1971, p. 1209.

9. J. M. BARSOM, J. F. SOVAK, and S. R. NOVAK, "AISI Project 168—Toughness Criteria for Structural Steels: Fracture Toughness of A36 Steel," *Research Laboratory Report 97.021-001(1)*, May 1, 1972 (available from the American Iron and Steel Institute).

10. J. M. BARSOM, J. F. SOVAK, and S. R. NOVAK, "AISI Project 168—Toughness Criteria for Structural Steels: Fracture Toughness of A572 Steels," *Research Laboratory Report 97.021-001(2)*, Dec. 29, 1972 (available from the American Iron and Steel Institute).

11. J. M. BARSOM, and S. T. ROLFE, "The Correlations Between $K_{Ic}$ and Charpy V-Notch Test Results in the Transition-Temperature Range," *Impact Testing of Metals, ASTM STP 466*, 1970, p. 281.

12. J. EFTIS and J. M. KRAFFT, "A Comparison of the Initiation with the Rapid Propagation of a Crack in a Mild Steel Plate," *Transactions of the ASME, Journal of Basic Engineering, 87*, Series D, Mar. 1965, p. 257.

13. G. T. HAHN, R. G. HOAGLAND, A. R. ROSENFIELD, and R. SEJNOHA, "Rapid Crack Propagation in a High Strength Steel," *Metallurgical Transactions, 5*, No. 2, Feb. 1974, p. 321.

# Correlations Between Fracture Mechanics and Other Fracture-Toughness Test Results

## 5.1. General Overview

In previous chapters we have introduced the concepts and analyses of fracture mechanics; described the test methods used to measure the critical material parameters, $K_{Ic}$, $K_{Ic}$ $(t)$, $K_{Ia}$, and $K_{Id}$; and demonstrated the general effects of temperature and loading rate on these parameters. These topics are the basis of fracture mechanics and are well founded on theory.

*However, the inherent fracture toughness of most structural materials is such that, using fracture mechanics test methods, $K_{Ic}$, $K_{Ic}$ $(t)$, and $K_{Id}$ values often cannot be measured at the service temperatures and loading rates for most large complex structures.* In these situations, it may be possible to conduct elastic-plastic fracture-mechanics tests as described in Chapter 17, and this is being done to a greater extent. However, for preliminary design analyses, and for quality control and failure analyses, a need exists for correlations between small-scale inexpensive fracture-toughness tests (for example, the Charpy V-notch impact specimen) and the larger, more expensive fracture-mechanics–type specimens ($K_{Ic}$, $J_{Ic}$, and so on). With such correlations, the small-scale inexpensive fracture test results can be used to estimate the various critical fracture mechanics material values (for example, $K_{Ic}$, $K_{Id}$). These values, in turn, can be used in design to predict failure stresses or critical flaw sizes as will be described in Chapter 6. Accordingly, although correlations are generally empirical in nature, they are extremely useful to the engineer.

To use fracture mechanics effectively throughout all phases of design, fabrication, quality control, and inspection, the engineer often must rely on auxiliary test methods and empirical correlations to obtain estimates of $K_{Ic}$ or $K_{Id}$ for most structural materials. In fact, many recently developed structural codes and specifications that have material-toughness requirements are based on fracture-mechanics principles, but the *specific* toughness tests specified for material purchase or quality control are in terms of *auxiliary* test specimens such as the CVN impact test specimen.

For example, the American Association of State Highway and Transportation Officials (AASHTO) material requirements for bridge steels[1] are based on concepts of fracture mechanics but are specified in terms of Charpy V-notch impact test results. Toughness requirements for thick-walled nuclear pressure-vessel steels are based on minimum dynamic toughness values, $K_{Id}$ (actually $K_{IR}$ for critical reference values[2]). However, the actual material-toughness requirements for steels used in these pressure vessels are specified using NDT (nil-ductility transition) values and CVN impact values using lateral expansion measurements. Proposed toughness requirements for welded ship hulls were developed using $K_{Id}/\sigma_{yd}$ values, but the proposed material procurement values are in terms of NDT and DT (dynamic tear) values. Thus empirical correlations as well as engineering judgment and experience are used to *translate* fracture-mechanics guidelines or controls into actual material-toughness specifications.

The cost of machining, fatigue precracking, and testing a $K_{Ic}$ or $K_{Id}$ specimen and the size requirements necessary to ensure valid $K_{Ic}$ or $K_{Id}$ test results render the $K_{Ic}$ or $K_{Id}$ test impractical as a quality-control test. Consequently the need exists to correlate $K_{Ic}$ or $K_{Id}$ data with notch-toughness test results obtained with smaller and less costly specimens.

For years, many specimens which are smaller and less costly than $K_{Ic}$ specimens have been used to measure the notch toughness of steels. Each of these specimens has limitations and shortcomings. However, correlations between $K_{Ic}$ and $K_{Id}$ data and various notch-toughness test results can be developed if the limitations of the specimen have been thoroughly investigated and understood.

In this chapter, we shall review the various theoretical relationships between the various fracture mechanics parameters and describe briefly those fracture-toughness test methods for which various correlations with either $K_{Ic}$ or $K_{Id}$ have been developed. We shall then describe some of the most widely used correlations between these various auxiliary test methods and $K_{Ic}$ or $K_{Id}$ as well as other fracture mechanics tests, that is, $J_{Ic}$ and CTOD.

## 5.2. Basic Relations
## Among Fracture-Mechanics Parameters

Several theoretical relationships between the various fracture-mechanics parameters exist. In the linear-elastic regime, the critical-strain energy-release rate, $G_{Ic}$ (Chapter 2), is equal to $J_{Ic}$, the critical toughness value at the initiation of crack growth for metallic materials, which is described in Chapter 17. Furthermore, because

$$K_{Ic} = \sqrt{\frac{EG_{Ic}}{1 - v^2}} \qquad (5.1)$$

for plane strain, it follows that

$$K_{Ic} = \sqrt{\frac{EJ_{Ic}}{1 - v^2}} \qquad (5.2)$$

for plane strain. Also

$$K_c \cong \sqrt{EJ_c}$$

for plane stress, except that $J_c$ is not defined by an ASTM standard at the present time. Also, in Chapter 2 it was shown that

$$G = \frac{\pi \sigma^2 \cdot a}{E} \qquad (5.3)$$

for a wide plate with a center crack of length $2a$.

In Chapter 17, it will be shown that the crack-tip opening displacement, $\delta$, is equal to

$$\delta = \frac{\pi \sigma^2 a}{E\sigma_{ys}} \qquad (5.4)$$

for a wide plate with center crack of length $2a$. Thus

$$G = \delta \cdot \sigma_{ys} \qquad (5.5)$$

Because

$$K_I = \sigma \sqrt{\pi a}$$

for this case,

$$\delta = \frac{K_I^2}{E\sigma_{ys}} \qquad (5.6)$$

Also, because $E \cong \sigma_{ys}/\varepsilon_{ys}$,

$$\frac{\delta}{\varepsilon_{ys}} = \left(\frac{K_I}{\sigma_{ys}}\right)^2 \qquad (5.7)$$

At the onset of crack instability, where $K_I$ equals $K_{Ic}$ and $\delta$ is equal to the critical value, $\delta_c$,

$$\frac{\delta_c}{\varepsilon_{ys}} = \left(\frac{K_{Ic}}{\sigma_{ys}}\right)^2 \qquad (5.8)$$

Equation (5.6) can be rewritten as follows:

$$K_{Ic} = \sqrt{E\sigma_{ys} \cdot \delta_c} \qquad (5.9)$$

Because actual specimens are neither completely plane strain nor plane stress, a constraint factor, $m$, is introduced into Equation (5.9) to account for the actual state of stress in the specimen:

$$K_{Ic} = \sqrt{mE\sigma_{ys} \cdot \delta_c} \qquad (5.10)$$

Finite-element analyses by Wellman[3] have shown that the constraint factor $m$ varies between 1.2 for plane stress and 1.6 for plane strain. Because most test results fall between these two extremes, a value of $m = 1.4$ is recommended unless the state of stress is clearly either plane stress or plane strain. Furthermore, Wellman[3] has shown that the use of the flow stress

$$\sigma_{flow} = \frac{\sigma_{ys} + \sigma_{ult}}{2} \qquad (5.11)$$

gives a better correlation between $K_c$ and CTOD values than the use of $\sigma_{ys}$. Hence the suggested relation between CTOD and $K_c$ is

$$K_c = \sqrt{1.4E\sigma_{flow}\,\delta_c} \qquad (5.12)$$

Because $K_c = \sqrt{EJ_c}$, it follows that

$$J_c = 1.4\sigma_{flow}\,\delta_c \qquad (5.13)$$

## 5.3. Other Fracture-Toughness Test Specimens

### 5.3.1. General Background

Throughout the years, there have been many fracture-toughness test specimens developed throughout the world. However, only those that are both widely used and that have been correlated with either $K_{Ic}$ or $K_{Id}$ values are discussed in this chapter. For a more complete discussion of the various fracture test specimens that have been developed, the reader is referred to technical papers[4-10] and to Chapter 17 for a discussion of CTOD, $R$-curve, and $J$-integral.

### 5.3.2. CVN Impact or Slow-Bend Test Specimens

Prior to the development of fracture mechanics, the Charpy V-notch impact fracture-toughness test specimen was the one most widely used to determine the toughness behavior of structural materials. In fact, even today it is widely used throughout the world, not only as a general reference specimen but in many actual toughness specifications. The specimen is a standard ASTM specimen (E-23—"Standard Methods for Notched Bar Impact Testing of Metallic Materials").

Figure 5.1(a) shows the dimensions of a CVN specimen, and Figure 5.1(b) is a photograph of a specimen before and after testing. Figure 5.1(c)

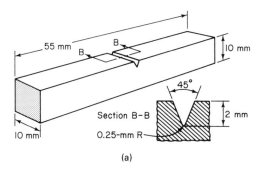

(a)

(b)

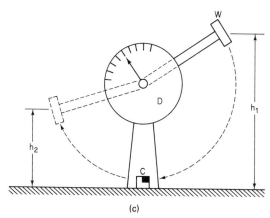

(c)

**Figure 5.1**    Charpy V-notch impact test specimen and testing procedure.

is a schematic showing the impact hammer $W$ dropping from height $h_1$, impacting the specimen at $C$, and rising to the maximum final height $h_2$. The energy absorbed by the specimen, related to the height differential $h_1-h_2$, is recorded on dial $D$.

Using the CVN impact specimen, notch toughness is measured in terms of

1. Energy absorbed, ft-lb, or
2. Lateral expansion (equivalent to notch contraction), mils, or
3. Fracture appearance, percent shear.

Examples of these three measurements for a low-strength structural steel are presented in Figure 5.2. Note that the absorbed energy and the lateral expansion measurements undergo a change over the same general temperature range. The percent fibrous fracture does not always conform to this behavior, especially in the slow-bend tests.

Examples of CVN impact curves of absorbed energy versus temperature for various structural materials are presented in Figure 5.3. It should be

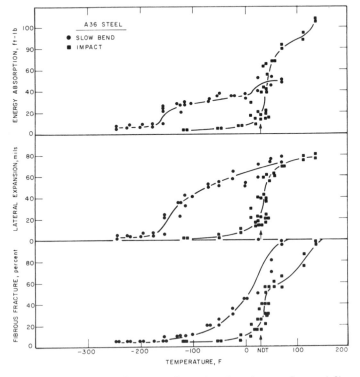

**Figure 5.2** Charpy V-notch energy absorption, lateral expansion, and fibrous fracture for impact and slow-bend test of standard CVN specimens.

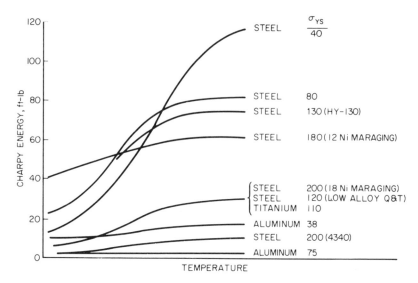

**Figure 5.3**  Charpy V-notch energy versus temperature behavior of various materials.

reemphasized that, like fracture-mechanics fracture-toughness test results, there is no single unique CVN curve for any single composition but that these curves depend also on the thermomechanical processing history. Approximately 15–20 specimens of a single material are tested at various temperatures, and a smooth curve is fitted to the results. Because of the relative small changes in behavior with temperature of the very-high-strength steels, the titaniums, and the aluminums, the CVN impact specimen is rarely used to evaluate the toughness behavior of these materials.

Although there are some valid criticisms of the CVN impact test when compared with $K_{Ic}$ or $K_{Id}$ tests, for example, it has a blunt notch (root radius of 0.01 in.), small size, and does not differentiate between initiation and propagation energies, it is still widely used because it is very fast to conduct, inexpensive, and simple to use. Furthermore, it has many years of correlation with service performance, and, as noted later, there are several empirical correlations between CVN impact test results and $K_{Ic}$ or $K_{Id}$.

The CVN slow-bend specimen is identical to the CVN impact test specimen shown in Figure 5.1. The quantities measured are also identical, and the only difference is the loading rate. The CVN slow-bend specimen is loaded in three-point bending at a loading rate comparable to that used in conducting standard tension tests. The area under the load-displacement curve is measured to determine the total energy to fracture. Typical results are compared with CVN impact results in Figure 5.4 for a low-strength

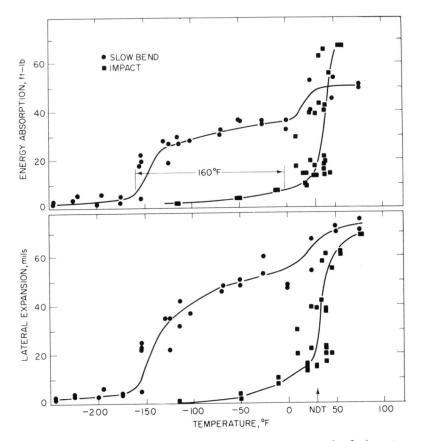

**Figure 5.4**  Charpy V-notch energy absorption and lateral expansion for impact and slow-bend tests of standard CVN specimens.

structural steel. The effect of the slower loading rate on a strain-rate–sensitive material is to cause the start of the transition in behavior from brittle to ductile behavior to occur at a lower temperature, compared with impact results.

Extreme care should be exercised when using the test results from slow-bend CVN specimens because, in many cases, the transition from brittle to ductile crack initiation occurs between very low values of absorbed energy (e.g., 5 and 12 foot pound) above which the specimen is subjected to general yielding conditions. Under these conditions, the brittle to ductile crack initiation transition may be missed and the only obvious transition would be that corresponding to an increased resistance to crack propagation which is closely related and may coincide with the toughness transition observed under impact loading.

### 5.3.3. Precracked CVN Impact or Slow-Bend Test Specimens

The *precracked* Charpy V-notch specimen[11] was introduced to eliminate the machined notch in the standard Charpy V-notch specimen. The notch in a regular CVN specimen is precracked under cyclic loading so that the actual test is conducted on a CVN specimen that has a sharp crack, rather than the 0.01-in. notch of the standard CVN specimen.

The specimens can be tested either slowly or under impact conditions, in the same manner as the regular CVN test specimen is tested.

By instrumenting a standard CVN impact testing machine with strain gauges whose output is recorded with an oscilloscope, measurements of fracture load can be obtained and values of $K_{Id}$ are estimated using expressions presented in Chapter 2. It should be noted, however, that the specimen size used (0.4- × 0.4-in. cross section) rarely meets the size requirements for $K_{Id}$ values presented in Chapter 3.

The precracked slow-bend CVN specimen exhibits problems similar to the slow-bend V-notch specimen. Consequently, extreme care should be exercised when using the data, especially for establishing the plane-strain brittle-to-ductile crack initiation fracture toughness transition.

### 5.3.4. Nil-Ductility Transition (NDT) Temperature Test Specimen

The NDT test specimen is a standardized ASTM test specimen and is described in ASTM E-208—"Standard Method for Conducting Drop Weight Test to Determine Nil-Ductility Transition Temperature of Ferritic Steels." Standard dimensions of the specimen and details of the crack starter are presented in Figure 5.5. The crack starter is a brittle weld in which a saw cut is made to localize the fracture in the center of the weld bead. The specimen is loaded in three-point bending by a falling weight at various temperatures. As the weight first hits the specimen, the brittle weld bead cracks, creating a small very sharp semicircular surface crack. As the weight continues to fall, the material being tested is loaded dynamically in the presence of the small weld crack. A mechanical stop is placed under the center of the specimen to limit the amount of deformation to which the specimen is subjected. The NDT temperature is defined as the highest temperature at which a standard specimen breaks in a brittle manner; that is, the material is said to have "nil ductility." Above the NDT temperature, the material has sufficient ductility to deflect to the mechanical stop before fracturing, that is, deflect inelastically. Thus, a series of about six to eight specimens are tested at various temperatures to establish the NDT temperature of a materal, usually within ±10°F.

The NDT test is used primarily for structural steels having yield strengths less than 140 ksi that undergo a brittle to ductile transition.

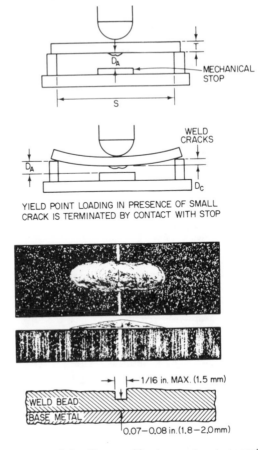

YIELD POINT LOADING IN PRESENCE OF SMALL
CRACK IS TERMINATED BY CONTACT WITH STOP

Figure 5.5    Nil-ductility transition-temperature test specimen.

### 5.3.5. Dynamic Tear (DT) Test Specimen

The DT test specimen is a sharply notched specimen impact loaded along one edge in three-point bending (Figure 5.6) and is standardized as described in ASTM E-604—"Standard Test Method for Dynamic Tear Energy of Metallic Materials."[12] The specimen has a very sharp pressed notch, is standardized, and the results have been related to structural behavior of dynamically loaded structures by Pellini.[13]

A complete transition curve is usually obtained, similar to that obtained for CVN impact specimens, as shown in Figure 5.7. Because the DT specimen is slightly thicker than the CVN specimen and has a sharper notch (additional constraint), because the material that has been subjected to plastic deformation caused by the pressed knife loses some of its original ductility, and because of residual tensile stresses in this plastically deformed

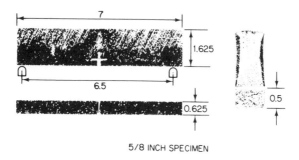

5/8 INCH SPECIMEN

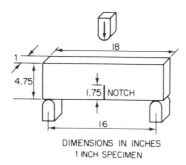

DIMENSIONS IN INCHES
1 INCH SPECIMEN

**Figure 5.6**  Dynamic tear test specimen.

region, the transition behavior usually occurs at higher temperatures compared with impact CVN or $K_{Id}$ test results.

## 5.4. $K_{Ic}$–CVN Upper-Shelf Correlation

One of the most widely used fracture-toughness tests in material development, specifications, and quality control is the Charpy V-notch impact test. Accordingly, Barsom and Rolfe[14] suggested relationships between $K_{Ic}$ and upper-shelf CVN test results on the basis of the results of various investigations by Clausing,[15] Holloman,[16] and Gross[17]. Clausing[15] showed that the state of stress at fracture initiation in the CVN impact specimen is plane strain, which is the state of stress in a thick $K_{Ic}$ specimen. Holloman[16] has shown that for the dimensions used in the CVN specimen, the maximum possible lateral stress is obtained, indicating a condition approaching maximum constraint. Tests by Gross[17] on CVN specimens of various thicknesses showed that the transition temperature for a standard CVN specimen is identical to the transition temperature for a CVN specimen of twice the standard width, substantiating the observation that the standard CVN test specimen has considerable constraint at the notch root. These observations, combined with the conclusion of Barsom and Rolfe[14] that the effect of temperature

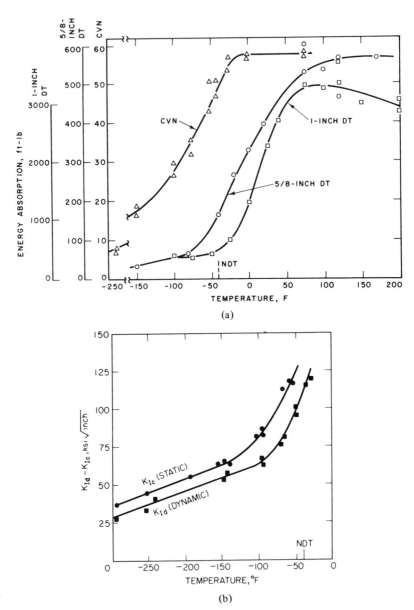

**Figure 5.7** Relation among NDT, CVN, DT, $K_{Ic}$, and $K_{Id}$ for an A517 steel.

and rate of loading on CVN and $K_{Ic}$ values is the same, suggested that it should be possible to establish empirical correlations between $K_{Ic}$ and CVN test results. The development of the $J$-integral concept,[18,19] as discussed in Chapter 17, suggests a possible theoretical basis for such empirical correlations.

The upper-shelf $K_{Ic}$–CVN correlation shown in Figure 5.8 was developed empirically[14,20] from results obtained on 11 steels having yield strengths in the range 110–246 ksi (Table 5.1). The $K_{Ic}$ values for these steels ranged from 87 to 246 ksi$\sqrt{in.}$, and the CVN impact values ranged from 16 to 89 ft-lb. Since the development of this correlation, additional test results have verified the correctness of this relation for upper-shelf values.

At the upper shelf, the effects of loading rate and notch acuity are not so critical as in the transition-temperature region. Thus, the differences in the $K_{Ic}$ and CVN test specimens (namely, loading rate and notch acuity) are not that significant, and a reasonable correlation would be expected.

Furthermore, in a discussion of a paper on the *J*-integral as a fracture criterion,[21] P. C. Paris states that "This paper (*J*-integral) finally explains the reasonableness of the Rolfe-Novak-Barsom correlation[14,20] of upper shelf Charpy values, CVN, with $K_{Ic}$ numbers; that is,

$$\left(\frac{K_{Ic}}{\sigma_{ys}}\right)^2 = 5\left(\frac{CVN}{\sigma_{ys}} - 0.05\right) \tag{5.14}$$

where $K_{Ic}$ = critical plane-strain stress-intensity factor at slow loading rates, ksi$\sqrt{in.}$

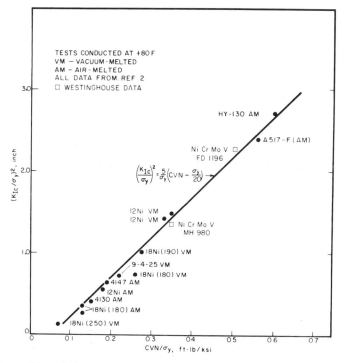

**Figure 5.8**  Relation between $K_{Ic}$ and CVN values in the upper-shelf region.

**TABLE 5.1   Longitudinal Mechanical Properties of Steels Investigated for Room Temperature $K_{Ic}$–CVN Correlation**

| Steel and Melting Practice* | Yield Strength, 0.2% Offset (ksi) | Tensile Strength (ksi) | Elongation in 1 in. (%) | Reduction of Area (%) | Charpy V-Notch Energy Absorption at +80°F (ft-lb) | $K_{Ic}$ (ksi$\sqrt{\text{in.}}$) |
|---|---|---|---|---|---|---|
| A517-F, AM | 110 | 121 | 20.0 | 66.0 | 62 | 170 |
| 4147, AM | 137 | 154 | 15.0 | 49.0 | 26 | 109 |
| HY-130, AM | 149 | 159 | 20.0 | 68.4 | 89 | 246 |
| 4130, AM | 158 | 167 | 14.0 | 49.2 | 23 | 100 |
| 12Ni-5Cr-3Mo, AM | 175 | 181 | 14.0 | 62.2 | 32 | 130 |
| 12Ni-5Cr-3Mo, VIM | 183 | 191 | 15.0 | 61.2 | 60 | 220 |
| 12Ni-5Cr-3Mo, VIM | 186 | 192 | 17.0 | 67.1 | 65 | 226 |
| 18Ni-8Co-3Mo (200 Grade), AM | 193 | 200 | 12.5 | 48.4 | 25 | 105 |
| 18Ni-8Co-3Mo (200 Grade), AM | 190 | 196 | 12.0 | 53.7 | 25 | 112 |
| 18Ni-8Co-3Mo (190 Grade), VIM | 187 | 195 | 15.0 | 65.7 | 49 | 160 |
| 18Ni-8Co-3Mo (250 Grade), VIM | 246 | 257 | 11.5 | 53.9 | 16 | 87 |

*AM signifies electric-furnace air-melted; VIM signifies vacuum-induction-melted.

$\sigma_{ys}$ = 0.2% offset yield strength at the upper shelf temperature, ksi.
CVN = standard Charpy V-notch impact test value at upper shelf, ft-lb.

This equation relating CVN, a limit-load-relating energy parameter, to $K_{Ic}$ is not only now acceptable but is, for us, in agreement with the $J$ failure criteria.'' Thus, in addition to empirically relating $K_{Ic}$ to CVN test results over a wide range of strength and toughness levels, the $K_{Ic}$–CVN relation shown in Figure 5.8 appears to have some theoretical basis as expressed in the development of the $J$-integral[21] and has been substantiated by additional tests by other investigators.[22]

Because $K_{Ic}$ is a static test, and the CVN impact test is an impact test, the relationship presented in Equation (5.1) is limited to steels having yield strengths greater than 100 ksi. However, because this correlation is an upper-shelf correlation, where the slow-bend and impact CVN values are constant for steels of various yield strengths, Equation (5.1) may be applicable to steels having yield strengths < 100 ksi. Because the upper-shelf CVN impact results are higher than the upper-shelf CVN slow-bend results, substituting the dynamic yield strength, $\sigma_{yd}$, into Equation (5.14) may give a better correlation for steels having a yield strength < 100 ksi.

# 5.5. $K_{Ic}$–CVN Correlation in the Transition-Temperature Region

### 5.5.1. Correlations That Account for Loading Rate

To establish correlations between $K_{Ic}$ and CVN test results in the transition-temperature region, the effects of both notch acuity and loading rate should be considered. $K_{Ic}$ values and CVN values in the transition-temperature region can be correlated (1) when the test results for slow-bend $K_{Ic}$ specimens are related to the test results for slow-bend fatigue-cracked CVN specimens and (2) when the test results for dynamic $K_{Ic}$ specimens are related to the test results for dynamic fatigue-cracked CVN impact specimens (Figure 5.9). The correspondence between $K_{Ic}$ and CVN energy-absorption values obtained at a particular test temperature and at the same strain rate for both $K_{Ic}$ and CVN can be approximated by[14,23,24] either

$$\frac{K_{Ic}^2}{E} = A \text{ (CVN slow-bend)} \tag{5.15}$$

or

$$\frac{K_{Id}^2}{E} = A \text{ (CVN impact)} \tag{5.16}$$

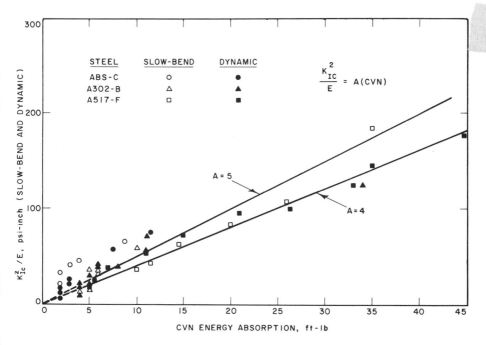

**Figure 5.9**  Correlation between $K_{Ic}$ and CVN test results for slow-bend and dynamic loading.

The constant of proportionality, $A$, for *precracked* CVN specimens is approximately 4. Thus the following relations exist for precracked CVN impact (PCI) or precracked CVN slow-bend (PSB) or specimens.

$$\frac{K_{Ic}^2}{E} = 4 \text{ (PSB)} \tag{5.17}$$

$$\frac{K_{Id}^2}{E} = 4 \text{ (PCI)} \tag{5.18}$$

where the PSB and PCI specimens are as described in Section 5.3.3 and where $E$ is Young's modulus, in pounds per square inch.

The constant of proportionality, $A$, incorporates the effects of specimen size as well as notch acuity. Thus by changing the value of $A$ in Equations (5.15) and (5.16), it is possible to correlate $K_{Ic}$ data and CVN energy-absorption values obtained by testing V-notched specimens. The constant of proportionality, $A$, for $K_{Ic}$ and slow-bend V-notched Charpy specimens, and $K_{Id}$ and impact V-notched Charpy specimens (as well as at intermediate loading rates) for low-strength structural steels was determined to be equal to about 5.

Thus the recommended expressions for $K$–CVN correlations are

$$\frac{K_{Ic}^2}{E} = 5 \text{ (CVN)}_{\text{slow bend (psi-in., ft-lb)}} \tag{5.19}$$

and

$$\frac{K_{Id}^2}{E} = 5 \text{ (CVN)}_{\text{impact (psi-in., ft-lb)}} \tag{5.20}$$

where the CVN specimens are the standard *notched* specimens shown in Figure 5.1(a).

### 5.5.2. Two-Stage CVN–$K_{Id}$–$K_{Ic}$ Correlations

Engineering estimates of $K_{Ic}$ at any strain rate can be predicted by using impact CVN data in conjunction with Equation (5.20) and then shifting the curve to lower temperatures. This approach has been used in investigating the effects of irradiation for steels used in nuclear reactors as well as in other studies.[25]

The magnitude of the temperature shift between dynamic ($\dot{\varepsilon} \approx 10 \text{ sec}^{-1}$) and slow-bend ($\dot{\varepsilon} \approx 10^{-5} \text{ sec}^{-1}$) curves was given in Chapter 4 by

$$\begin{aligned} T_{\text{shift}} &= 215 - 1.5\sigma_{ys} \quad &\text{for } 36 \text{ ksi} < \sigma_{ys} < 140 \text{ ksi} \\ T_{\text{shift}} &= 0 \quad &\text{for } \sigma_{ys} > 140 \text{ ksi} \end{aligned} \tag{5.21}$$

The magnitude of the temperature shift between dynamic ($\dot{\varepsilon} \approx 10 \text{ sec}^{-1}$) and intermediate strain rate of $10^{-3} \text{ sec}^{-1}$ was found to be equal to about 75 percent of the shift between dynamic and slow-bend curves.[23,24]

The general use of this procedure to estimate $K_{Ic}$ values in the transition-temperature region from CVN impact results is summarized as follows:

1. Obtain standard nonprecracked CVN impact test results in the transition-temperature region.

2. Calculate $K_{Id}$ values at each test temperature using

$$\frac{K_{Id}^2}{E} = 5\,(\text{CVN}) \tag{5.20}$$

   where the calculated $K_{Id}$ values are at the same temperature as each CVN value.

3. Use the following equation for temperature shift to determine the temperature shift between $K_{Id}$ and $K_{Ic}$ values:

$$T_{shift} = 215 - 1.5\sigma_{ys} \qquad \text{for } 36\,\text{ksi} < \sigma_{ys} < 140\,\text{ksi}$$

4. Shift the $K_{Id}$ values at each temperature by the temperature shift calculated in step 3 to obtain $K_{Ic}$ (static) values as a function of temperature.

This procedure is limited to the lower end of the transition curve where the CVN impact value in ft-lb is less than or equal to about one-half the yield strength in ksi. As the upper-shelf region is approached, where loading rate and notch acuity do not have so great an influence on the fracture-toughness behavior, the upper-shelf correlation may be used, even though it was developed primarily for steels with yield strengths greater than 80 ksi.

For intermediate strain rates in the range $10^{-3}\,\text{sec}^{-1} \leq \dot{\varepsilon} \leq 10\,\text{sec}^{-1}$ and for steels with yield strengths less than about 140 ksi, the relationship among $T_{shift}$, $\dot{\varepsilon}$, and $\sigma_{ys}$ can be approximated by

$$T_{shift} = (150 - \sigma_{ys})(\dot{\varepsilon})^{0.17} \tag{5.22}$$

where $T_{shift}$ is in °F.

   $\sigma_{ys}$ is the standard yield strength in ksi.

   $\dot{\varepsilon}$ is the strain rate in $\text{sec}^{-1}$.

The plane-strain fracture-toughness value for a given steel tested at a given temperature and for a strain rate equal to or less than $10^{-4}\,\text{sec}^{-1}$ is constant and equal to the static $K_{Ic}$ value.

### 5.5.3. Correlations That Do Not Account for Loading Rate

Several correlations between *slow-bend* $K_{Ic}$ values and *impact* CVN values have been developed. In these correlations, the effect of strain rate is not considered.

One of the first correlations by the authors was

$$\frac{K_{Ic}^2}{E} = 2\,(\text{CVN})^{3/2} \ (\text{psi-in., ft-lb}) \tag{5.23}$$

Corten and Sailors[22] developed a similar correlation as follows:

$$K_{Ic} = 15.5\,(CVN)^{1/2}\,(ksi\sqrt{in.}, ft\text{-}lb) \qquad (5.24)$$

Roberts and Newton[26] analyzed these and other similar correlations and arrived at a recommended lower-bound relation as follows:

$$K_{Ic} = 9.35\,(CVN)^{0.63}\,(ksi\sqrt{in.}, ft\text{-}lb) \qquad (5.25)$$

Because $K_{Ic}$–CVN impact relations do not account for strain-rate effects, these correlations would be expected to be quite conservative and not applicable to low strength steels.

Of the various correlations in the transition-temperature region, the recommended procedure for determining $K_{Ic}$ values from CVN impact values is the two-stage method involving the temperature shift, as described in Section 5.5.2.

## 5.6. $K_{Id}$ Value at NDT Temperature

In the drop weight NDT test, a specimen is subjected to a crack initiated from a brittle weld bead under impact-loading conditions (Figure 5.10). After examining the crack shape obtained in this type of test, Irwin[27] et al. proposed an analysis for determining a dynamic crack-toughness value, $K_{Id}$, from the drop weight test results. Assuming that at the NDT temperature the plate surface reached the dynamic yield stress, $\sigma_{yd}$, corresponding to the testing temperature and that the pop-in crack geometry was of an $a/2c$ ratio of 1 to 4, where $a$ is crack length and $c$ is half the crack width, Irwin[27] et al. arrived at the following relationship for a part-through-thickness crack:

$$K_{Id} = 0.78(\sqrt{in.})\sigma_{yd} \qquad (5.26)$$

Shoemaker and Rolfe[28] observed that the NDT temperature is close to the temperature at which a 1-in.-thick $K_{Ic}$ specimen tested under impact loading ceases to satisfy the ASTM requirements for valid $K_{Ic}$ tests. The thickness requirement for valid $K_{Ic}$ tests is given by

$$B \geqslant 2.5\left(\frac{K_{Ic}}{\sigma_{ys}}\right)^2 \qquad (5.27)$$

where   $B$ = specimen thickness.

$\quad\quad\sigma_{ys}$ = yield strength.

This relationship can be used to represent this observation by Shoemaker and Rolfe[28] concerning the dynamic $K_{Ic}$ value at NDT temperature. The resulting equation is

$$K_{Id} = 0.64(\sqrt{in.})\sigma_{yd} \qquad (5.28)$$

where $K_{Id}$ = critical plane-strain stress-intensity factor at NDT temperature and under dynamic loading ($\dot{\varepsilon} \approx 10\ sec^{-1}$).

$\quad\quad\sigma_{yd}$ = dynamic yield strength at NDT temperature.

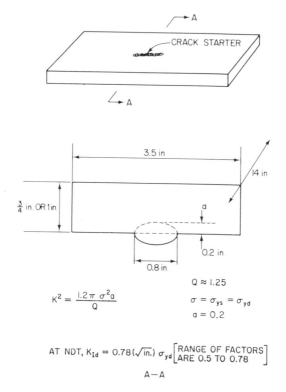

$$K^2 = \frac{1.2\pi\,\sigma^2 a}{Q}$$

$Q \approx 1.25$

$\sigma = \sigma_{ys} = \sigma_{yd}$

$a = 0.2$

AT NDT, $K_{Id} = 0.78(\sqrt{in.})\,\sigma_{yd}\left[\begin{array}{c}\text{RANGE OF FACTORS}\\ \text{ARE 0.5 TO 0.78}\end{array}\right]$

A—A

**Figure 5.10**  Cross section of NDT test specimen showing initial crack and fracture-mechanics analysis.

Pellini has estimated that the factor relating $K_{Id}$ and $\sigma_{yd}$ should be 0.5. However, the differences in the various factors are slight, and the suggested relationship between $K_{Id}$ and $\sigma_{yd}$ at the NDT temperature is

$$K_{Id} = 0.6(\sqrt{in.})\sigma_{yd} \tag{5.29}$$

where $K_{Id}$ = dynamic critical plane-strain stress-intensity factor at the NDT temperature, ksi$\sqrt{in}$.

$\sigma_{yd}$ = dynamic yield strength at the NDT temperature, ksi.

Using Equation (5.29), the calculated $K_{Id}$ values at NDT for an A36 steel and an A572 Grade 50 steel would be

$$\text{A36:}\quad K_{Id} = 0.6\sigma_{yd} \simeq 0.6(40\,+\,25) \simeq 39 \text{ ksi}\sqrt{in.}$$

$$\text{A572 Grade 50:}\quad K_{Id} = 0.6\sigma_{yd} \simeq 0.6(55\,+\,25) \simeq 48 \text{ ksi}\sqrt{in.}$$

The values of the dynamic yield strength, $\sigma_{yd}$, are approximately equal to the static yield strength plus 25 ksi, that is, $\sigma_{yd} \simeq \sigma_{ys}\,+\,25$ ksi.

The $K_{Id}$ test results presented in Figures 4.28 and 4.30 indicate measured values of about 40 and 50 ksi$\sqrt{in}$. compared with the calculated values of

39 and 48 ksi$\sqrt{\text{in.}}$, respectively. Thus Equation (5.29) appears to give a realistic approximation to $K_{Id}$ for low-strength structural steels at their NDT temperature.

## 5.7. Comparison of CVN Correlations and *J*-Integral, or CTOD, Relations to Predict $K_c$ Results

Throughout this chapter, we have presented various relations and correlations between fracture mechanics and small-scale test results. Several of these correlations are used to estimate plane-stress $K_c$ results for structural steels having yield strengths ranging from about 40 to 100 ksi. The relationships investigated include the following:

### Two-stage CVN–$K_{Id}$–$K_{Ic}$

$$K_{Id}^2 = 5\,(\text{CVN})E \tag{5.20}$$

$$T_{\text{shift}} = 215 - 1.5\sigma_{ys} \tag{5.21}$$

### Roberts-Newton lower-bound CVN–$K_{Ic}$

$$K_c = 9.35\,(\text{CVN})^{0.63} \tag{5.25}$$

### CTOD–$K_c$

$$K_c = \sqrt{1.4 E \sigma_{\text{flow}} \delta_c} \tag{5.12}$$

### *J*-integral

$$K_{Ic} = \sqrt{\frac{E J_{Ic}}{1 - v^2}} \tag{5.2}$$

where *J* values were determined as described in Chapter 17.

Wellman[3] studied the relationships between the foregoing correlations using the following steels:

| STEEL | $\sigma_{ys}$, ksi |
|-------|--------------------|
| ABS-B | 38 |
| A516  | 39 |
| A533  | 62 |
| A508  | 71 |
| A517  | 108 |

The results of that study are presented in Figures 5.11 through 5.15 and show the benefits of using correlations to estimate $K_c$ values.

In all cases, the Roberts-Newton lower-bound curve is indeed a lower

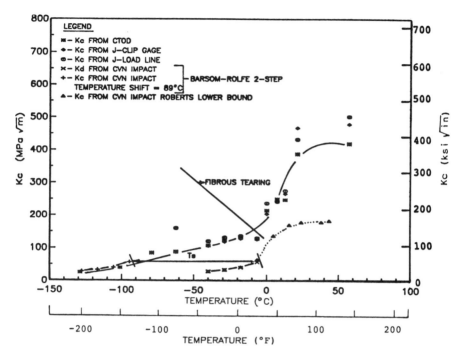

**Figure 5.11**   $K_c$–CVN–CTOD–$J$ correlations for an ABS-B steel [specimen size (1 × 2 × 8) lower-bound values].

bound. The two-stage CVN–$K_{Id}$–$K_{Ic}$ correlation appears to match the CTOD–$K_c$ relation quite well, but its usefulness is limited to the lower region of the transition-temperature curve.

The CTOD–$K_c$ and $K_c$–$J$ relations yield results that increase with temperature as might be expected. However, because it was not possible to conduct $K_c$ tests directly because of specimen-size limitations, the correctness of the $K_c$ values cannot be verified.

The high-strength steel (A517) behaved in a manner somewhat different from the lower-strength steels. Because of the higher yield strength, there is a much smaller loading-rate shift. Also because of the higher yield strength, the determination of stable cracking is much more difficult for the A517 steel than for the lower-yield-strength steels. This is analogous to the difficulty encountered in fracture appearance rating Charpy V-notch specimens of high-strength steel. However, the general appearance of the temperature-transition curve, with a lower shelf, a lower- and upper-transition region, and an upper shelf, is maintained but at lower toughness levels.

Figure 5.16 is a schematic showing the general regions of transition-temperature curves. Generally, they can be divided into four regions, which

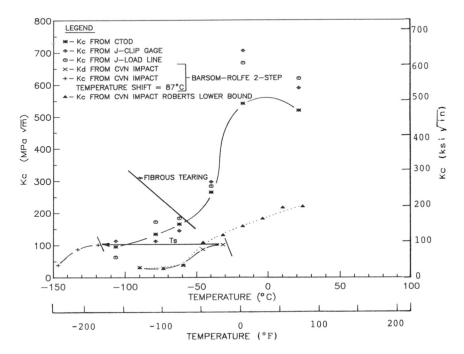

**Figure 5.12** $K_c$–CVN–CTOD–$J$ correlations for an A516 steel [specimen size (1.6 × 3.2 × 12) lower-bound values].

are (1) lower shelf, (2) lower transition, (3) upper transition, and (4) upper shelf.

The lower-shelf region is characterized by little or no change in toughness with changes in temperature. All fracture surfaces show brittle behavior both for initiation and propagation. Behavior is nearly linear elastic, and $K_{Ic}$ can be used to describe the behavior throughout most of this region. As shown in Figure 5.16, the lower-shelf region lies approximately 150°C below the NDT for the steels studied.

The lower-transition region is characterized by a brittle fracture with no visible evidence of prior stable cracking. The toughness increases steadily with increasing temperature from near the linear elastic $K_{Ic}$ values to the toughness at the initiation of a fibrous thumbnail visible to the naked eye. This toughness increase takes place over a temperature range of about 200°F. Finite element analysis shows that within this range (which varies with yield strength and specimen size), a plastic hinge develops in the three-point bend test specimen. The development of a plastic hinge demonstrates that the lower transition region is definitely a region of elastic-plastic fracture behavior. Because $K_{Ic}$ measures linear-elastic plane-strain behavior, and the $J_{Ic}$ measures the initiation of stable cracking, CTOD is the only fracture

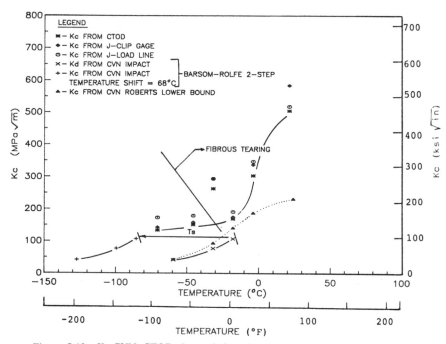

**Figure 5.13** $K_c$–CVN–CTOD–$J$ correlations for an A533 steel [specimen size $(1 \times 2 \times 8)$ lower-bound values].

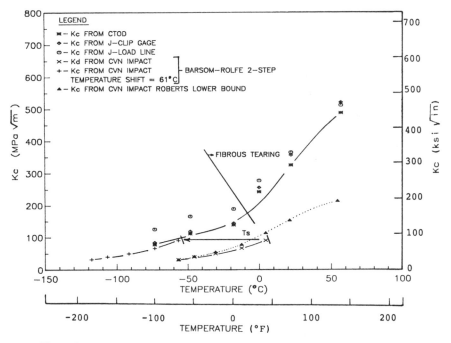

**Figure 5.14** $K_c$–CVN–CTOD–$J$ correlations for an A508 steel [specimen size $(1 \times 2 \times 8)$ lower-bound values].

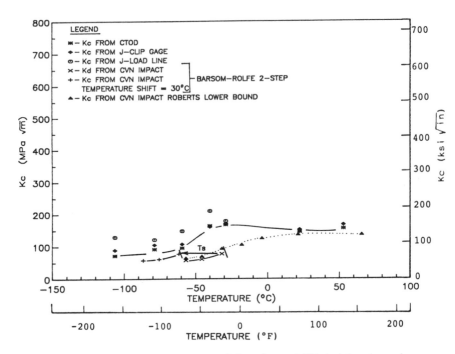

**Figure 5.15** $K_c$–CVN–CTOD–$J$ correlations for an A517 steel [specimen size $(1 \times 2 \times 8)$ lower-bound values].

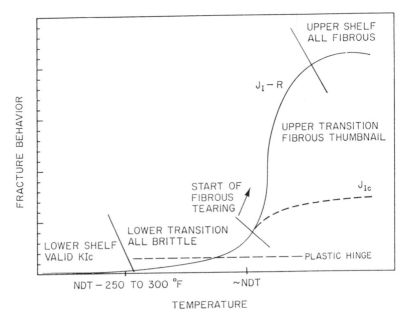

**Figure 5.16** Schematic CTOD—temperature-transition curve showing four regions.

parameter currently applicable in this lower-transition region. This observation has significant implications for the usefulness of the CTOD parameter because the service temperatures of most structures are in the lower-transition region.

In the upper-transition region, the failure initiates by stable ductile tearing recognizable by a coarse fibrous "thumbnail" detectable by unaided visual observation of the fracture surface. This ductile "thumbnail" initiation is followed by a fast brittle propagation. The significant increase in the CTOD values in the upper-transition and upper-shelf regions appears to represent increased resistance to ductile crack *propagation*. If the *initiation* of ductile crack propagation is plotted, there would be no significant increase in CTOD values with temperatures in the upper-transition and upper-shelf regions. This observation suggests that the difference between the behavior of the A517 steel tested and the lower-strength steels was related to a reduced resistance of this A517 steel plate to ductile crack propagation.

The upper-shelf region is characterized by fibrous ductile tearing over the entire surface. The exact location of the start of the upper shelf is somewhat ambiguous due to the data scatter in the upper-transition region. This ambiguity is avoided if the upper shelf is arbitrarily defined to start at the temperature at which all specimens show 100 percent fibrous tearing failures.

## 5.8.  $K_{Id}$ from Precracked CVN Impact Test Results

Although impact tests on precracked CVN test specimens using uninstrumented impact machines have been conducted for many years,[29] the *instrumented* precracked CVN test is relatively new.[11] Its growth in popularity results primarily from the fact that it has been used to estimate $K_{Id}$ values for nuclear pressure-vessel steels. The use of a lower-bound $K_{Id}$ curve for the design of thick-section nuclear pressure vessels (see Chapter 16), the fact that conducting full-thickness $K_{Id}$ tests (in some cases using 12-in.-thick test specimens) is extremely expensive, and the fact that space limitations restrict the engineer to small-size specimens in surveillance programs that monitor the effects of irradiation on these steels have led to the use of the instrumented precracked CVN impact test for fracture control.

The primary advantage of the instrumented precracked Charpy test is that it has many of the features of the standard CVN test (small specimen, simple test, high strain rate, large sampling capability). It also has some of the advantages of the dynamic tear test (sharp energy transition and low initiation energy). As yet, it does not have the established correlation with service performance that the standard CVN impact test specimen has, particularly for low-strength structural steels. By recording the loads during the precracked Charpy test, the total energy can be separated into initiation and propagation energies, and values of $K_{Id}$ can be measured under appropriate conditions. Disadvantages of the instrumented precracked Charpy test are that it cannot measure full-thickness behavior and that control of the pre-cracking procedure is difficult. Moreover, it does not appear to correlate with $K_{Id}$ as well as the CVN impact test does.[26]

As part of the old Atomic Energy Commission (AEC) Heavy Section Steel Technology (HSST) program on nuclear pressure vessel steels, the static[29] and dynamic fracture toughness of A533B steel ($\sigma_{ys} = 70$ ksi) has been measured as a function of temperature (using test specimens up to 12 in. thick, Figure 5.17).[30-33] As shown in Figure 5.18, $K_{Id}$ is less than $K_{Ic}$ at any given temperature. At the NDT temperature of this A533B steel ($+10°F$), 6-in.-thick specimens were required for valid $K_{Ic}$ tests, and 2-in.-thick specimens were required for valid $K_{Id}$ tests ($\dot{K} \simeq 10^4$ ksi$\sqrt{in.}$/sec). Valid $K_{Ic}$ values could not be obtained above 50°F even when using 12-in.-thick specimens, and valid $K_{Id}$ values could not be obtained above 125°F for 8-in.-thick specimens.

There are several points to be made concerning the results presented in Figure 5.18:

**Figure 5.17** Various sizes of compact-tension test specimens.

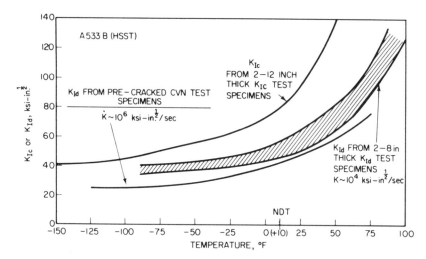

**Figure 5.18**  Comparison of static, dynamic, and instrumented precracked CVN impact fracture toughness as a function of temperature for A533 Grade B steel.

1. $K_{Ic}$ and $K_{Id}$ show a similar transition-temperature behavior as the CVN impact, dynamic tear, and precracked CVN curves for this steel (Figure 5.19).

2. A change in the microfracture mechanism from cleavage to fibrous tearing produces a large increase in toughness, and increasing the specimen thickness (constraint) has very little effect once this metallurgical transition occurs (this transition in the inherent toughness behavior of a material was discussed in Chapter 4).

3. Dynamic testing allows valid fracture-toughness values to be obtained at higher temperatures for a given thickness or allows use of a thinner specimen at a given temperature.

4. The $K_{Ic}$ and $K_{Id}$ test results presented in Figure 5.18 were very costly to obtain (in excess of $1 million).

The remaining curve in Figure 5.18 represents the dynamic fracture toughness of A533B steel determined from instrumented precracked Charpy impact tests.[33] The impact velocity of the Charpy hammer was approximately 200 in./sec, and this corresponds to $\dot{K} \simeq 10^6$ ksi$\sqrt{\text{in.}}$/sec. The higher strain rate of the Charpy impact test causes a greater increase in the yield strength of A533B steel and thus a slightly lower $K_{Id}$.

As discussed in Chapter 16, the Pressure Vessel Research Committee (PVRC) has proposed that a lower-bound fracture-toughness curve be used in the design of nuclear pressure vessels. Its recommended design curve essentially parallels the lower part of the scatter band for $K_{Id}$ at $\dot{K} \simeq 10^4$ ksi$\sqrt{\text{in.}}$/sec. It is apparent from Figure 5.18 that a conservative

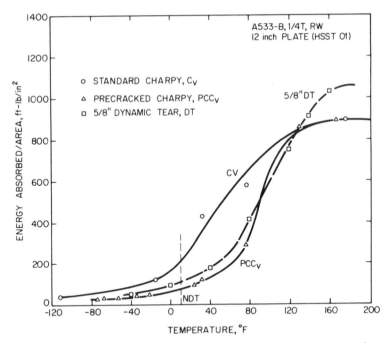

**Figure 5.19** Comparison of V-notch and precracked Charpy data with ⅝-in. dynamic tear data for an A533 Grade B steel.

lower-bound $K_{Id}$ design curve was obtained from instrumented precracked Charpy tests for temperatures up to 70°F. Most important of all, the $K_{Id}$ curve from the instrumented precracked Charpy test can be determined very inexpensively compared to the presently accepted techniques of testing full-thickness specimens capable of satisfying the size requirements presented in Chapter 3.

## References

1. *American Association of State Highway and Transportation Officials (AASHTO) Material Toughness Requirements*, Association General Offices, Washington, D.C., 1978.
2. "PVRC Recommendations on Toughness Requirements for Ferritic Materials," *WRC Bulletin 175*, New York, Aug. 1972.
3. G. W. Wellman and S. T. Rolfe, "Engineering Aspects of CTOD Fracture Toughness Testing," *WRC Bulletin 299*, Nov. 1984.
4. E. R. Parker, *Brittle Behavior of Engineering Structures*, John Wiley, New York, 1957.
5. Welding Research Council, *Control of Steel Construction to Avoid Brittle Failure*, edited by M. E. Shank, M.I.T. Press, Cambridge, Mass., 1957.

6. W. J. Hall, H. Kihara, W. Soete, and A. A. Wells, *Brittle Fracture of Welded Plate*, Prentice-Hall, Englewood Cliffs, N.J., 1967.

7. The Royal Institution of Naval Architects, *Brittle Fracture in Steel Structures*, edited by G. M. Boyd, Butterworth's, London, 1970.

8. H. Libowitz, ed., *Fracture, An Advanced Treatise*, Vols. I–VII, Academic Press, New York, 1967.

9. C. F. Tipper, *The Brittle Fracture Story*, Cambridge University Press, New York, 1962.

10. W. S. Pellini, "Principles of Fracture—Safe Design" (Parts I and II), *Welding Journal (Welding Research Supplement)*, Mar. 1971, pp. 91-S–109-S, and Apr. 1971, pp. 147-S–162-S.

11. R. A. Wullaert, D. R. Ireland, and A. S. Tetelman, "The Use of Precracked Charpy Specimen in Fracture Toughness Testing," presented at the Fracture Prevention and Control Symposium, WESTEC 72, Mar. 13–17, 1972, Los Angeles, Calif.

12. "Standard Test Method for Dynamic Tear Testing of Metallic Materials," *ASTM E-604—83*, Vol. 03.01, 1985.

13. W. S. Pellini, "Design Options for Selection of Fracture Control Procedures in the Modernization of Codes, Rules and Standards," *Proceedings of the Joint U.S.–Japan Symposium on Application of Pressure Component Codes*, Tokyo, Mar. 13–15, 1973.

14. J. M. Barsom and S. T. Rolfe, "Correlations Between $K_{Ic}$ and Charpy V-Notch Test Results in the Transition-Temperature Range," *Impact Testing of Metals, ASTM STP 466*, American Society for Testing and Materials, Philadelphia, 1970, pp. 281–302.

15. D. P. Clausing, "Effect of Plane-Strain Sensitivity on the Charpy Toughness of Structural Steels," *International Journal of Fracture Mechanics, 6*, No. 1, March 1970.

16. J. H. Holloman, "The Notched-Bar Impact Test," *Transactions, American Institute of Mining, Metallurgical, and Petroleum Engineers, 158*, 1944, pp. 310–322.

17. J. H. Gross, "Effect of Strength and Thickness on Notch Ductility," *Impact Testing of Metals, ASTM STP 466*, American Society for Testing and Materials, Philadelphia, 1970, p. 21.

18. J. A. Begley and J. D. Landes, "The *J* Integral as a Fracture Criterion," *Stress Analysis and Growth of Cracks, ASTM STP 514*, American Society for Testing and Materials, Philadelphia, 1972.

19. J. D. Landes and J. A. Begley, "The Effect of Specimen Geometry on $J_{Ic}$," *Stress Analysis and Growth of Cracks, ASTM STP 514*, American Society for Testing and Materials, Philadelphia, 1972.

20. S. T. Rolfe and S. R. Novak, "Slow-Bend $K_{Ic}$ Testing of Medium-Strength High-Toughness Steels," *Review of Developments in Plane Strain Fracture-Toughness Testing, ASTM STP 463*, American Society for Testing and Materials, Philadelphia, 1970, pp. 124–159.

21. P. C. Paris, discussion of paper by Begley and Landes: "The *J* Integral as a Fracture Criterion," *ASTM STP 514*, American Society for Testing and Materials, Philadelphia, 1972 (Ref. 18).

22. J. T. CORTEN and R. H. SAILORS, "Relationship Between Material Fracture Toughness Using Fracture Mechanics and Transition Temperature Tests," *T. & A. M. Report No. 346*, University of Illinois, Urbana, Aug. 1971. See also Paper No. 3, HSST Program Information Meeting, ORNL, March 25, 1971.

23. J. M. BARSOM, "The Development of AASHTO Fracture-Toughness Requirements for Bridge Steels," American Iron and Steel Institute paper, Washington, D.C., February 1975.

24. J. M. BARSOM, "Development of the AASHTO Fracture-Toughness Requirements for Bridge Steels," *Engineering Fracture Mechanics, 7*, No. 3, Sept. 1975.

25. J. R. HAWTHORNE and T. R. MAGER, "Relationship Between Charpy V and Fracture Mechanics $K_{Ic}$ Assessments of A533-B Class 2 Pressure Vessel Steel," *ASTM STP 514*, American Society for Testing and Materials, Philadelphia, 1972.

26. R. ROBERTS and C. NEWTON, "Report on Small-Scale Test Correlations with $K_{Ic}$ Data," *WRC Bulletin 299*, Nov. 1984.

27. G. R. IRWIN, J. M. KRAFFT, P. C. PARIS, and A. A. WELLS, "Basic Aspects of Crack Growth and Fracture," *NRL Report 6598*, Washington, D.C., Nov. 21, 1967.

28. A. K. SHOEMAKER and S. T. ROLFE, "The Static and Dynamic Low-Temperature Crack-Toughness Performance of Seven Structural Steels," *Engineering Fracture Mechanics, 2*, No. 4, June 1971.

29. C. E. HARTBOWER, "Crack Initiation and Propagation in the V-Notch Charpy Impact Specimen," *Welding Journal, 35*, No. 11, Nov. 1957, p. 494-s.

30. W. O. SHABBITS, W. H. PRYLE, and E. T. WESSEL, "Heavy Section Fracture Toughness Properties of A533 Grade B Class 1 Steel Plate and Submerged Arc Weldment," *WCAP-7414*, Westinghouse Electric Corporation, Pittsburgh, Dec. 1969.

31. P. B. CROSLEY and E. J. RIPLING, "Crack Arrest Fracture Toughness of A-533 Class 1 Pressure Vessel Steel," *HSSTP-TR-8*, Materials Research Laboratory, Glenwood, Ill., March 1970.

32. W. O. SHABBITS, "Dynamic Fracture Toughness Properties of Heavy Section A533 Grade B Class 1 Steel Plate," *WCAP-6723*, Westinghouse Electric Corporation, Pittsburgh, Dec. 1970.

33. W. L. SERVER and A. S. TETELMAN, "The Use of Precracked Charpy Specimens to Determine Dynamic Fracture Toughness," *Engineering Fracture Mechanics, 4*, No. 2, June 1972.

# Fracture-Mechanics Design

## 6.1. Introduction

"Design" is a term used in different ways by engineers. From a structural viewpoint, the term usually refers to the synthesis of various disciplines (statics, strength of materials, structural analysis, matrix algebra, and so on) to create a structure that is proportioned and then detailed into its final shape using a set of scale drawings. When the word "design" is used in this sense, designing to prevent brittle fracture usually refers to using an appropriate stress as well as to the elimination (as much as possible) of those structural details that act as stress raisers and that can be potential fracture-initiation sites, for example, weld discontinuities, mismatch, and intersecting plates. Unfortunately, large complex structures (either welded or bolted) rarely are designed (or fabricated) without these discontinuities, although good design and fabrication practices can minimize the size and number of these discontinuities.

From a materials viewpoint, the word "design" usually refers to the selection of a material and of the appropriate design stress level at the particular service temperature and loading rate that the structure will see. Thus from this viewpoint, designing to prevent brittle fracture refers to appropriate material selection as well as to selection of the appropriate allowable stress.

Both these approaches to design are valid, but they should not be confused. The first (and traditional) definition assumes that the designer starts with a given material and design stress level (often specified by codes) and thus the design involves the process of detailing and proportioning members to carry the given loads without exceeding the allowable stress. The allowable stress is usually a certain percentage of the yield strength for tension members and a certain percentage of the buckling stress for compression members. These "allowable" design stress levels assume "perfect" fabrication in that the structures are usually assumed to have no crack-like discontinuities, defects, or cracks. It is realized that stress concentrations or mild discontinuities will be present, but the designer assumes that his structural materials will yield locally and redistribute the load in the vicinity of these stress concentrations or mild discontinuities.

The second use of the term "design" refers to selection of materials and allowable stress levels based on the more appropriate realization of the fact that discontinuities in large complex structures may be present or may initiate under cyclic loads or stress-corrosion cracking and that some level of notch toughness is desirable. This aspect of design has recently been made more quantitative by the development of fracture mechanics as an engineering science.

*To provide a safe, fracture-resistant structure, both of these design approaches should be followed.* The designer must properly proportion the structure to prevent failure by either tensile overload or compressive instability *and* by unstable crack growth by brittle fracture. Historically, most design criteria have been established to prevent yielding (either in tension or compression) and to prevent buckling. The yielding and buckling modes of failure have received considerable attention because analytical design procedures were available and could be used in various theories of failure. The Euler buckling analysis and the maximum shearing stress theory of failure are among the most widely used analytical design principles to prevent failure by general yielding or buckling. The recent development of fracture mechanics as an analytical design tool finally "fills the gap" in the designers' various techniques for safe design of structures.

Numerous textbooks describe the various design techniques to prevent either general yielding or buckling by proper proportioning of the various structural members, and the reader is referred to the large number of design textbooks in the particular field of structures (bridges, ships, pressure vessels, aircraft, etc.) for these well-established design procedures.

This textbook deals primarily with the selection of materials and appropriate design stress levels for fracture-resistant structures. This design approach recognizes the fact that discontinuities may be present in large complex structures. It is true that the incidence of brittle fractures (as well as fatigue cracks) can be *reduced* by good detailing practices and minimizing discontinuities. Examples of design details to optimize fatigue strength in structures are presented in Reference 1.

The remainder of this chapter deals with a description of how fracture mechanics can aid the engineer in the initial selection of

1. Materials,
2. Design stress levels, and
3. Tolerable crack sizes for quality control or inspection

for the fracture-resistant design of *any* large complex structure, for example, bridges, ships, pressure vessels, aircraft, and earthmoving equipment. Thus, this textbook should be used in conjunction with textbooks that describe the more traditional methods of designing to prevent either the buckling or yielding modes of failure to ensure that all possible failure modes are considered.

Regardless of the type of structure to be considered, fracture mechanics design assumes that the engineer has established the following general information:

1. Type and overall dimensions of the structure (bridge, pressure vessel, etc.).
2. General size of the tension members (length, diameter, etc.).
3. Additional performance criteria (e.g., minimum weight, least cost, maximum resistance to fracture, specified design life, loading rate, operating temperature).
4. Stress and cyclic stress range, where crack growth can occur (as discussed in Chapters 7–14).

With this basic information, the designer can incorporate fracture-mechanics values at the service temperature and loading rate in the "design" of a fracture-resistant structure. The fatigue life can be estimated as well.

If the designer knows or can measure the critical value of notch toughness at the service conditions (that is, $K_{Ic}$, $K_c$, $K_{Id}$, and so on, as described in Chapter 3), or can estimate these values using the correlations described in Chapter 5, the philosophy of design using fracture mechanics is fairly straightforward. Basically, the designer should make sure that the applied stress intensity factor, $K_I$, is always less than the critical stress intensity factor that is appropriate for the particular structure at all times in the same manner that $\sigma$ is kept below $\sigma_{ys}$ to prevent yielding.

Fracture-mechanics "design" generally follows either a fail-safe or a safe-life principle, both of which have been used extensively in design. Fail-safe design assumes that if an individual member fails, the overall structure is still safe from total fracture. Conversely, safe-life principles assume that for the particular service loading the structure will last the entire design life of the structure and failure will not occur while the structure is in service. Structures with multiple-load paths are essentially fail-safe structures because of the structural redundancy. Safe-life structures may or may not have multiple-load paths, that is, redundancy. In either case, the objective is to keep the applied $K_I$ below the critical $K$.

Current design thinking recognizes the fact that discontinuities may be present in large complex structures or that they may initiate and grow during the service life of the structure. The structural designer, therefore, should not limit his or her analysis to the traditional approaches of design whereby factors of safety or indices of reliability for unflawed structures are used but rather should incorporate the fact that discontinuities can be present in his or her structure.

It should be emphasized that brittle fractures in structures are rare. Hence, fracture mechanics is not widely used, and this is actually a desirable situation. However,

1. When designs become more complex,
2. When the use of high-strength thick welded structural materials becomes more common compared with the use of lower-strength thin riveted plates,
3. When the fabrication and construction become more complex,
4. When the magnitude of loading increases, and
5. When actual factors of safety decrease because of the use of computer design,

the probability of brittle fracture in large complex structures increases, and the designer should be aware of what can be done to prevent brittle fractures.

In this chapter, design procedures using fracture-mechanics concepts will be limited to design considerations related to the *terminal* failure after possible crack extension had occurred by fatigue (Chapters 7, 8, 9, and 10), stress corrosion (Chapter 11), corrosion fatigue (Chapters 12 and 13), or weldment fatigue (Chapter 14). This particular chapter forms the basis for subsequent development of fracture criteria (Chapter 15) or fracture-control plans (Chapter 16) where these additional considerations are important. Also, the principles of design discussed in this chapter are applicable to nonplane-strain or elastic-plastic levels of toughness, although the design applications may be more complex. Chapter 17 discusses elastic-plastic behavior.

## 6.2. General Fracture-Mechanics Design Procedure for Terminal Failure

The critical stress-intensity factor for a particular material at a given temperature and loading rate is related to the nominal stress and flaw size as follows:

$$K_{Ic}, K_c, K_{Ic}(t), K_{Id}, \text{etc.} = C\sigma\sqrt{a} \qquad (6.1)$$

where $K_{Ic}$ or $K_{Id}$ = crack toughness of a material, $ksi\sqrt{in.}$, tested at a particular temperature and loading rate.

$C$ = constant, function of crack and specimen geometries, Chapter 2.

$\sigma$ = nominal applied stress, ksi.

$a$ = flaw size as a critical dimension for a particular crack geometry, in.

Thus, the maximum flaw size a structural member can tolerate at a particular stress level is

$$a = \left[ \frac{K_{Ic}, K_c, K_{Ic}\,(t),\ \text{or}\ K_{Id}}{C \cdot \sigma} \right]^2 \qquad (6.2)$$

Accordingly, the engineer can analyze the safety of a structure against failure by brittle fracture in the following manner:

1. Determine the values of the appropriate critical $K$ [$K_{Ic}$, $K_{Ic}\,(t)$, $K_c$, $K_{Id}$, etc.] and the corresponding yield strength at the service temperature and loading rate for the materials being considered for use in the structure. Note that for a complete analysis of welded structures, values for the weldment should be used also.
2. Select the most probable type of flaw that will exist in the member being analyzed and the corresponding $K_I$ equation. Figure 6.1 shows

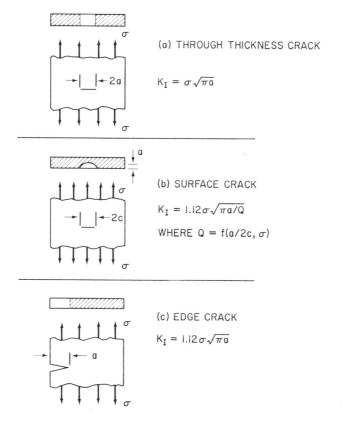

(a) THROUGH THICKNESS CRACK

$K_I = \sigma\sqrt{\pi a}$

(b) SURFACE CRACK

$K_I = 1.12\sigma\sqrt{\pi a/Q}$

WHERE Q = f(a/2c, $\sigma$)

(c) EDGE CRACK

$K_I = 1.12\sigma\sqrt{\pi a}$

**Figure 6.1**  $K_I$ values for various crack geometries. (a) Through-thickness crack, (b) surface crack, and (c) edge crack.

the fracture-mechanics models that describe three of the more common types of flaws occurring in structural members. Complex-shaped flaws can often be approximated by one of these models. Additional equations to analyze other crack geometries were described in Chapter 2.

3. Determine the stress–flaw-size relation at various possible design stress levels using the appropriate $K_I$ expression and the appropriate critical $K$ value.

As an example, the relation among stress, flaw size, and stress intensity factor, $K_I$, for the through-thickness crack geometry shown in Figure 6.1 is

$$K_I = \sqrt{\pi}\sigma\sqrt{a} \tag{6.3}$$

Assume that the material being analyzed has a $K_{Ic}$ value at the service temperature of 50 ksi$\sqrt{\text{in}}$. and a yield strength of 100 ksi. Substituting $K_I = K_{Ic}$, the possible combinations of stress and critical flaw size at failure are described by

$$K_{Ic} = 50 \text{ ksi}\sqrt{\text{in}}. = \sqrt{\pi}\sigma\sqrt{a} \tag{6.4}$$

Using this equation, values of the critical crack size ($2a$) for various stress levels are calculated as follows:

| $\sigma$ (ksi) | $2a$, in. |
|:---:|:---:|
| 10 | 16 |
| 20 | 4 |
| 30 | 1.8 |
| 40 | 1.0 |
| 50 | 0.64 |
| 60 | 0.44 |
| 70 | 0.32 |
| 80 | 0.24 |
| 90 | 0.20 |
| 100 | 0.16 |

These results are plotted in Figure 6.2. The curve labeled $K_{Ic}$ is the locus of points at which unstable crack growth (fracture) will occur. Thus, if the nominal design stress level is 30 ksi, then the maximum tolerable flaw size, or *critical crack size*, $2a$, is 1.8 in. Conversely, if the nominal design stress level is 60 ksi, then the crack size is 0.44 in. It can be seen that there is no single critical crack size for a particular structural material (at a given temperature and loading rate) but rather a "semi-infinite" number of critical crack sizes, depending on the nominal design stress level. Figure 6.3 shows the locus of values of stress and flaw size for a "design $K_I$" of $K_{Ic}/2$, that is, a factor of safety of 2 against fracture that is based on the critical stress-intensity factor using the traditional definition of factor of

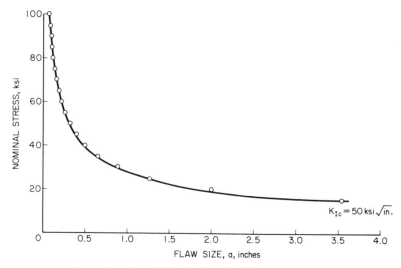

**Figure 6.2**  Stress–flaw-size relation for through-thickness crack in material having $K_{Ic}$ = 50 ksi$\sqrt{\text{in.}}$

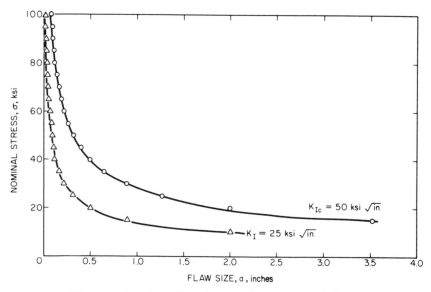

**Figure 6.3**  Stress–flaw-size relation for through-thickness crack showing $K_{I(design)}$ = 25 ksi$\sqrt{\text{in.}}$

safety.  In this case a structure having a flaw size of 0.22 in. loaded to a design stress of 30 ksi has a factor of safety of 2 against fracture.  Similarly the $K_I$ = 25 ksi$\sqrt{\text{in.}}$ curve is the locus of all stress levels and corresponding flaw sizes with a factor of safety of 2 against fracture.  The factor of safety also may be based on stress alone or on crack size alone rather than on the critical stress-intensity factor.

Obviously, for the high-toughness materials (compared with materials with lower notch toughness), the number of possible combinations of design stress and allowable flaw sizes that will not lead to failure is much larger. In Figure 6.4 the locus of failure points for a tougher material with a $K_{Ic}$ of 100 ksi$\sqrt{\text{in.}}$ is shown along with the locus of failure points of the material having a $K_{Ic}$ of 50 ksi$\sqrt{\text{in.}}$ For the material with a $K_{Ic}$ of 100 ksi$\sqrt{\text{in.}}$, the tolerable flaw sizes at all stress levels are considerably larger than are those for the material with the lower toughness, and the possibility of fracture is reduced considerably.

Thus it can be seen that to minimize the possibility of brittle fracture in a given structure, the designer has three *primary* factors that can be controlled:

1. Material toughness at the particular service temperature and loading rate (critical $K$), ksi$\sqrt{\text{in.}}$
2. Nominal stress level ($\sigma$), ksi.
3. Flaw size present in the structure ($a$), in.

All three of these factors affect the possibility of a brittle fracture occurring in structures. Other factors such as temperature, loading rate, residual stresses, and so on merely affect the three primary factors. It should be noted that flaws need not be present for brittle fractures to occur if the other factors are sufficiently severe and if the general level of constraint is high—for example, at the toe of welds that join intersecting thick plates.

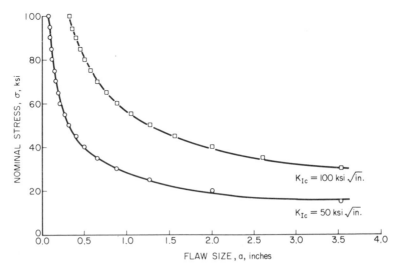

**Figure 6.4** Stress–flaw-size relation for through-thickness crack showing effect of higher $K_{Ic}$ (100 ksi$\sqrt{\text{in.}}$).

Under these conditions, several factors other than the fracture toughness would be the controlling factors and the application of fracture mechanics technology is not appropriate for analyzing the fracture initiating event.

Design engineers have known these facts for many years and have reduced the susceptibility of their structures to brittle fractures by applying these concepts to their structures qualitatively. The traditional use of "good design" practice, that is, by using appropriate stress levels and minimizing discontinuities, has led to the reduction of brittle fractures in many structures. In addition, the use of good fabrication practices, that is, eliminating or decreasing the flaw size by proper welding control and inspection, as well as the use of materials with good notch-toughness levels (for example, as measured with a Charpy V-notch impact test), has minimized the probability of brittle fractures in structures. This has been the traditional design approach to reducing brittle fractures in structures and has been used reasonably successfully. However, fracture mechanics offers the engineer the possibility of a *quantitative* approach to designing to prevent brittle fractures in structures containing crack-like imperfections.

In summary, the general relationship among material toughness ($K_{Ic}$ or $K_{Id}$), nominal stress ($\sigma$), and flaw size ($a$) is shown schematically in Figure 6.5. If the particular combinations of stress and flaw size in a structure ($K_I$) reach the $K_{Ic}$ level, fracture can occur. Thus, there are many combinations of stress and flaw size, $\sigma_f$ and $a_f$, that may cause fracture in a structure that is fabricated from a structural material having the particular value of $K_{Ic}$ (or $K_{Id}$) at a particular service temperature and loading rate. Conversely, there are many combinations of stress and flaw size, for example, $\sigma_0$ and $a_0$, that will not cause failure of a particular material.

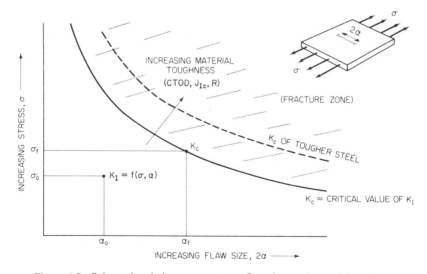

**Figure 6.5**  Schematic relation among stress, flaw size, and material toughness.

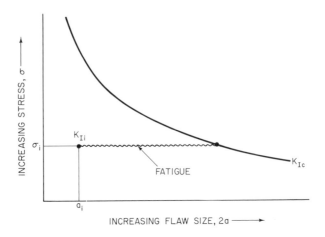

**Figure 6.6**   Schematic showing how $K_{Ii(initial)}$ can increase by fatigue to $K_{Ic}$.

As will be discussed later, $K_I$ can increase throughout the life of a structure because of crack growth by fatigue. This behavior is shown schematically in Figure 6.6 for $a$ increasing from $a_i$ to $a_f$ by fatigue.

A useful analogy for the designer is the relation among applied load ($P$), nominal stress ($\sigma$), and yield stress ($\sigma_{ys}$) in an unflawed structural member and among applied load ($P$), stress intensity factor ($K_I$), and critical stress intensity factor for fracture ($K_c$, $K_{Ic}$, or $K_{Id}$) in a structural member with a flaw. In an unflawed structural member, as the load is increased, the nominal stress increases until an instability (yielding at $\sigma_{ys}$) occurs. As the load is increased in a structural member with a flaw (or as the size of the flaw grows by fatigue or stress corrosion), the stress intensity factor, $K_I$, is increased until an instability (fracture at $K_c$, $K_{Ic}$, $K_{Id}$) occurs. Thus the $K_I$ level in a structure should always be kept below the appropriate critical $K$ value in the same manner that the nominal design stress ($\sigma$) is kept below the yield strength ($\sigma_{ys}$).

Another analogy that may be useful in understanding the fundamental aspects of fracture mechanics is the comparison with the Euler column instability (Figure 6.7). The stress level required to cause instability in a column (buckling) decreases as the $L/r$ ratio increases. Similarly, the stress level required to cause instability (fracture) in a flawed tension member decreases as the flaw size ($a$) increases. As the stress level in either case approaches the yield strength, both the Euler analysis and the $K_{Ic}$ analysis are invalidated because of yielding. To prevent buckling, the actual stress and $L/r$ values must be below the Euler curve. To prevent fracture, the actual stress and flaw size, $a$, must be below the $K_{Ic}$ line shown in Figure 6.7. Obviously, using a material with a high level of notch toughness (for example, a $K_{Ic}$ level of 100 ksi$\sqrt{\text{in.}}$ compared with 50 ksi$\sqrt{\text{in.}}$ as shown in

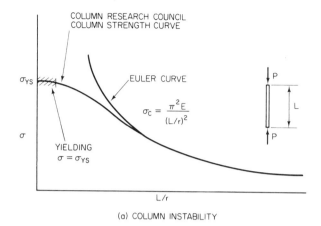

(a) COLUMN INSTABILITY

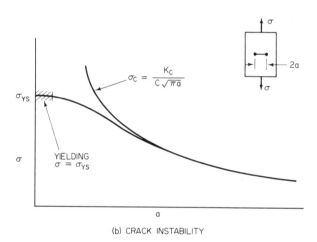

(b) CRACK INSTABILITY

**Figure 6.7**   (a) Column instability and (b) crack instability.

Figure 6.4) will increase the possible combinations of design stress and flaw size that a structure can tolerate without fracturing.

At this point, it should be reemphasized that (fortunately) the critical stress-intensity-factor value of many structural materials used in various types of structures cannot be measured directly at the service temperature and loading rates by using the existing test methods described in Chapter 3. Thus although concepts of fracture mechanics can be used to develop fracture-control guidelines and stress–flaw-size trade-offs for design use, the state of the art is such that in many cases actual fracture-toughness values cannot be measured for various structural applications at the service temperatures and loading rates. In these cases various empirical correlations with other notch-toughness tests can be used to approximate $K_{Ic}$, $K_{Ic}(t)$, or $K_{Id}$, as was described in Chapter 5. These approximations of $K_{Ic}$, $K_{Ic}(t)$, or $K_{Id}$ can then be used as described, but with less accuracy. Also elastic-plastic analyses can be used as described in Chapter 17.

## 6.3. Design Selection of Materials

Current methods of design and fabrication of large complex structures are such that engineers expect these structures to be able to tolerate yield stress loading in tension without failing, at least in local regions around points of stress concentration. Because yield stress loading may occur in the vicinity of structural discontinuities, the critical crack size, $a$, often is proportional to $(K_{Ic}/\sigma_{ys})^2$ or $(K_{Id}/\sigma_{yd})^2$.

Thus either the $K_{Ic}/\sigma_{ys}$ or $K_{Id}/\sigma_{yd}$ ratio becomes a good index for measuring the relative toughness of structural materials. Because for most structural applications it is desirable that the structure tolerate large flaws without fracture, the use of materials with as high a $K_{Ic}/\sigma_{ys}$, $K_c/\sigma_{ys}$, or $K_{Id}/\sigma_{yd}$ as is possible *consistent with economic considerations* is a desirable condition.

The question becomes, *How high must these ratios be to ensure satisfactory performance in large complex structures, where complete inspection for initial cracks and continuous monitoring of crack growth throughout the life of a structure are not always possible, practical, or economical.*

No simple answer exists because the answer is obviously dependent on the type of structure, frequency of inspection, access for inspection, quality of fabrication, design life of the structure, consequences of failure for a structural member, redundancy of load path, probability of overload, fabrication and material costs, and others. However, fracture mechanics does provide an engineering approach to evaluate this question rationally. Basic assumptions are that flaws do exist in structures, yield stress loading is possible in parts of the structure, and plane-strain conditions can exist. Under these very conservative conditions, the $K_c/\sigma_{ys}$ ratios for materials used in a particular structure are one of the primary controlling design parameters that can be used to define the relative safety of a structure against brittle fracture.

As an example of the use of the $K_{Ic}/\sigma_{ys}$ ratio as a material selection parameter, the behavior of a wide plate with a through-thickness center crack of length $2a$, Figure 6.1(a), is analyzed for materials having assumed levels of strength and notch toughness. Table 6.1 presents assumed values of $K_{Ic}$ for various steels having yield strengths that range from 40 to 260 ksi. For each of these steels, the critical crack size, $2a$, is calculated for four design stress levels, that is, 100 percent $\sigma_{ys}$, 75 percent $\sigma_{ys}$, 50 percent $\sigma_{ys}$, and 25 percent $\sigma_{ys}$. It should be emphasized that there is no single

TABLE 6.1    **Values of Critical Crack Size as a Function of Yield Strength and Notch Toughness**

| $\sigma_{ys}$ (ksi) | Assumed $K_{Ic}$ Values (ksi$\sqrt{\text{in.}}$) | Critical Flaw Size, $2a$ (in.) (Actual Design Stress Level, ksi, Is Shown in Parentheses) | | | |
|---|---|---|---|---|---|
| | | $\sigma = 100\%\sigma_{ys}$ | $\sigma = 75\%\sigma_{ys}$ | $\sigma = 50\%\sigma_{ys}$ | $\sigma = 25\%\sigma_{ys}$ |
| 260 | 80 | 0.06(260) | 0.11(195) | 0.24(130) | 0.96(65) |
| 220 | 110 | 0.16(220) | 0.28(165) | 0.64(110) | 2.55(55) |
| 180 | 140 | 0.39(180) | 0.68(135) | 1.54(90) | 6.16(45) |
| 180 | 220 | 0.95(180) | 1.69(135) | 3.80(90) | 15.22(45) |
| 140 | 260 | 2.20(140) | 3.90(105) | 8.78(70) | 35.13(35) |
| 110 | 170 | 1.52(110) | 2.70(82.5) | 6.08(55) | 24.33(27.5) |
| 80 | 200 | 3.98(80) | 7.07(60) | 15.92(40) | 63.66(20) |
| 40 | 100 | 3.98(40) | 7.07(30) | 15.92(20) | 63.66(10) |

Through-Thickness Crack in a Wide Plate, Figure 6.1(a)

$$K_{Ic} = \sqrt{\pi}\sigma_{design}\sqrt{a}$$

$$\therefore \quad a = \frac{1}{\pi}\left(\frac{K_{Ic}}{\sigma_{design}}\right)^2$$

$$2a = \frac{2}{\pi}\left(\frac{K_{Ic}}{\sigma_{design}}\right)^2$$

unique critical stress-intensity factor for any one steel at a given test temperature and rate of loading because the $K_{Ic}$, $K_c$, $K_{Ic}(t)$, or $K_{Id}$ for a given steel depends on the thermomechanical history, that is, heat treatment, rolling, and so on.

The results presented in Table 6.1 demonstrate the influence of strength level, notch toughness, and design stress level on the critical crack size in a wide plate. For example, the critical flaw size for the 260-ksi yield strength steel loaded to 50 percent of the yield strength (design stress of 130 ksi) is 0.24 in. If a design stress of 130 ksi were required for a particular structure, it would be preferable to use a lower-strength, higher-toughness material (e.g., the 180-ksi yield strength material) at a design stress of 75 percent $\sigma_{ys}$ (135 ksi). For either of the two 180-ksi yield-strength steels analyzed in Table 6.1, the critical crack sizes are much larger than 0.24 in., for example, 0.68 in. for the material with a $K_{Ic}$ of 140 ksi$\sqrt{\text{in.}}$ or 1.69 in. for the material with a $K_{Ic}$ of 220 ksi$\sqrt{\text{in.}}$

Obviously the tougher of these two materials would be a more fracture-resistant structural material, but it probably would be a more expensive material also. This point illustrates one of the basic aspects of fracture-resistant design that is not as obvious in other more traditional modes of design, namely, economics. That is, structural materials that are extremely notch tough at service temperatures and loading rates are available. However, because the cost and availability of these materials generally increase with their ability to perform satisfactorily under more severe operating conditions, the engineer usually does not want to specify more notch toughness than is required for the particular application. Thus the problem of fracture-resistant design is really one of optimizing structural performance consistent with safety and economic considerations. However, this is no different from the general definition of good engineering design, that is, an optimization of performance, safety, and cost.

Further analysis of Table 6.1 indicates that the traditional method of selecting a design stress as some percentage of the yield strength does not always give the same degree of safety and reliability against fracture as it is presumed to give for yielding. For example, assume that the design stress for the two steels having yield strengths of 220 ksi and 110 ksi is 50 percent $\sigma_{ys}$, or 110 ksi and 55 ksi, respectively. For the 220-ksi steel, the critical crack size is 0.64 in., whereas for the 110-ksi yield-strength steel, the critical crack size is 6.08 in. If the design stress for the lower-yield-strength steel were increased to 100 percent $\sigma_{ys}$ (110 ksi), the critical crack size would be 1.52 in. This is still significantly greater than the 0.64-in. size for the 220-ksi yield-strength steel with a design stress of 50 percent $\sigma_{ys}$.

For the lower-strength steels shown in Table 6.1 that have higher values of $K_{Ic}/\sigma_{ys}$, the critical crack sizes become extremely large, indicating that fracture-mechanics theory no longer applies in these cases. This is analogous to saying that for very low $L/r$ ratios, the calculated Euler buckling stress is well above the yield stress and the Euler buckling analysis no longer applies.

In summary, there may be situations where the designer should specify a *lower*-yield-strength material at a *higher* design stress level (as a percentage of the yield strength) actually to *improve* the overall safety and reliability of a structure from a fracture-resistant design viewpoint. Good design dictates that the engineer design to prevent failure of his or her structure against *all* possible modes of failure, including fracture.


## 6.4. Design Analysis of Failure of a 260-In.-Diameter Motor Case

An excellent example of the fact that specifying a percentage of yield strength is not always the best method of establishing the design stress level is the failure of a 260-in.-diameter motor case during hydrotest.[2,3] The motor case failed during hydrotest at a pressure of 542 psi, which was about 56 percent of proof pressure. The motor case was constructed of 250 Grade maraging steel plate joined primarily by submerged arc automatic welding. An investigation of the failure was conducted by a committee composed of members from industry and government.

Gerberich[3] summarized the failure as follows:

Although this motor case was designed to withstand proof pressures of 960 psi, it failed during hydrotest at a pressure of 542 psi. In this 240 ksi yield strength material, the failure occurred at a very low membrane stress of 100 ksi. The fracture was both premature and brittle with crack velocities approaching 5000 feet/second. The result of this 65-ft-high chamber literally flying apart is shown in Figure 6.8. Postfailure examination revealed that the fracture had originated in an area of two defects that had probably been produced by manual gas-tungsten arc weld repairs. After the crack initiated it branched into multiple cracks as shown in Figure 6.9, leading to complete catastrophic failure of the motor case.

The real lesson in this failure is the lack of design knowledge that went into the material selection. First, the chamber was 0.73 in. thick, which put it into the plane-strain regime for the high-strength material being considered. Grade 250 maraging steel which had a yield strength of about 240 ksi was chosen. For plane-strain conditions this was not a particularly good choice since the base metal had a plane-strain fracture toughness, $K_{Ic}$, of only 79.6 ksi$\sqrt{in}$. At the design stress of 160 ksi, a critical defect only 0.08 in. in depth could have caused catastrophic failure. Still, the chamber manufacturer thought that defects of this size could be detected. In retrospect, this was not a very judicious decision. Furthermore, the error was compounded by the fact that welding this material produced an even lower $K_{Ic}$ value ranging from 39.4 to 78.0 ksi$\sqrt{in}$, the toughness level depending on the location of the flaw in the weld.

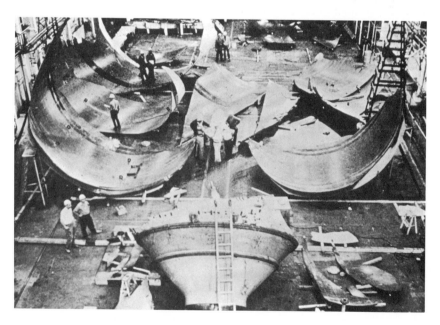

**Figure 6.8**  Failed motor case with pieces laid out in approximately the proper relation to each other. (Courtesy of J. E. Srawley, NASA Lewis Research Center.)

260-inch DIAMETER SOLID ROCKET MOTOR CASE

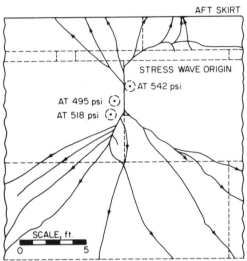

**Figure 6.9**   Map of fracture path about failure origin; dashed lines indicate welds (Ref. 2).

A postfailure analysis was run on several types of flaw configurations and weld positions which indicated the $K_{Ic}$ value to range from 38.8 to 83.1 ksi$\sqrt{\text{in}}$. with the average being 55.0 ksi$\sqrt{\text{in}}$. As the exact value of the fracture toughness at the failure origin in the chamber is not known, this average value is used as an estimate of $K_{Ic}$. Postfailure examination of the fracture origin indicated the responsible defect had an irregular banana shape that could best be approximated by an internal ellipse that was 0.22 in. in depth and 1.4 in. long. The critical dimension was the in-depth value of 0.22 in. since the crack would first propagate through the thickness from this dimension. To calculate the critical defect size from fracture-mechanics principles, a solution for an internal ellipse gives

$$K_{I} = \sigma(a)^{1/2}f\left(\frac{a}{c}\right) \qquad (6.5)$$

where

$$f\left(\frac{a}{c}\right) = \pi^{1/2}\left(\frac{1}{Q}\right)^{1/2}$$

and $a$ is the half-crack depth, $c$ is the half-crack length, and $f(a/c)$ is related to the complete elliptical integral of the second kind (Chapter 2). However, as $a/c \to 0$, $1/Q \to 1.0$ and $f(a/c) \to (\pi)^{1/2}$, and this equation approaches

$$K_{I} = \sigma(\pi a)^{1/2} \qquad (6.6)$$

For the shape of flaw under consideration, the difference between these two equations (6.5) and (6.6) is only 4 percent, and so for the sake of simplicity, the second equation is utilized. Based on the fracture toughness of 55.0

ksi$\sqrt{\text{in}}$., the second equation was utilized to make a plot of membrane stress versus defect size for crack instability. This is shown in Figure 6.10 as the curve for Grade 250 maraging steel. An intercept of the 100-ksi failure stress for the chamber predicts a critical defect size, $2a$, of 0.2 in., which is very close to the observed value also indicated in Figure 6.10. Besides the Grade 250 data, there is also shown here a curve for Grade 200 maraging steel. Although this material has a yield strength that is about 10 percent lower than Grade 250, its toughness is about triple the average $K_{Ic}$ value for the Grade 250, with $\underline{K_{Ic}}$ being about 150 ksi$\sqrt{\text{in}}$. for a member 0.7 in. thick. Using the 150-ksi$\sqrt{\text{in}}$. value for $K_{Ic}$ and the above equation (6.6), a curve for Grade 200 maraging steel was constructed for Figure 6.10. Significantly, the defect that failed the Grade 250 chamber would not have failed a Grade 200 chamber, and, in fact, yield stresses could have been reached without failure. At the failure stress of 100 ksi, it would have taken a flaw 1.42 in. in depth to burst a Grade 200 chamber. This would not have been possible since the thickness was only 0.73 in. Even at the design stress of 160 ksi, it would take a flaw 0.56 in. in depth to cause plane-strain fracture, and this is a very large flaw. In all probability, a flaw this large in this type of material would be arrested under plane-stress conditions anyway as soon as it grew through the thickness.

Thus, the Grade 200 maraging steel would have been a more reliable material for this chamber than the Grade 250 maraging steel. The proof of this is that there was a competition to make the 260-in.-diameter chamber for NASA, and the competitor used Grade 200 maraging steel successfully. Not only

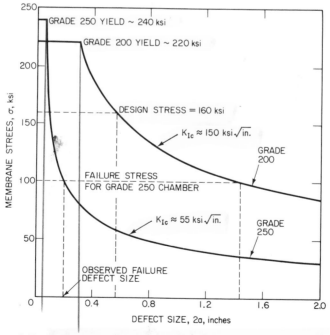

**Figure 6.10**    Design curves for critical defect in 260-in. motor case (Ref. 3).

were there two successful proof tests of Grade 200 chambers, but there were also two firings which developed thrusts of more than 6 million lb.

In summary, using a lower-strength level steel with higher toughness (a larger $K_{Ic}/\sigma_{ys}$ ratio), even at a design stress which was a higher percentage of the yield strength, would have been better.

A failure in an F-111 Wing Pivot Fitting occurred in a D6-AC steel heat treated to obtain a high-strength level so that the design stress level of $\frac{2}{3}\,\sigma_{ult}$ would lead to a weight reduction. However, this also led to a lower fracture-toughness level so that the *actual* degree of safety was lowered. In this case, it also seems preferable to have used a heat treatment that gave a higher toughness (but a lower $\sigma_{ult}$) and to have used a design stress somewhat higher than $\frac{2}{3}\,\sigma_{ult}$ so that the *actual* design stresses were the same but the critical crack size greater. This example plus the example of the 260-in. missile motor case illustrate that *both* the yielding and fracture modes of failure should be considered in structural design.

## 6.5. Design Example—Selection of a High-Strength Steel for a Pressure Vessel

As a simplified example of the desirability of using fracture-mechanics principles to *select* materials during the preliminary design stages, assume that a high-strength steel pressure vessel must be built to withstand 5000 psi of internal pressure, $p$, that the diameter, $d$, of the vessel is nominally 30 in., and that the wall thickness, $t$, must be equal to or greater than 0.5 in. The designer can use any yield-strength steel available, but in addition to satisfactory performance, cost and weight of the vessel are important factors that must be considered. The steels available for use in the vessel are shown in Table 6.2 along with their yield strengths and assumed $K_{Ic}$ values at the service temperature and loading rate.

TABLE 6.2   Yield-Strength and Crack-Toughness Values of Steels Used in Example Problem as Well as Assumed Costs*

| Steel | Yield Strength $\sigma_{ys}$ (ksi) | Assumed $K_{Ic}$ Values (ksi$\sqrt{\text{in.}}$) | Cost ($/lb) |
|-------|------|------|------|
| A | 260 | 80 | 1.40 |
| B | 220 | 110 | 1.40 |
| C | 180 | 140 | 1.00 |
| D | 180 | 220 | 1.20 |
| E | 140 | 260 | 0.50 |
| F | 110 | 170 | 0.15 |

*Assumed values for example only.

**Case I—*Traditional design approach.*** The traditional design approach is to assume "perfect" fabrication, and therefore there is no flaw. The factor of safety would be based on yielding only, e.g., $\sigma_{\text{design}} = \sigma_{ys}/2$ for a factor of safety of 2.

This design procedure is direct since

$$\sigma_{\text{design}} = \frac{\sigma_{ys}}{2} \quad \text{and} \quad t = \frac{pd}{2\sigma_{\text{design}}}$$

Knowing $t$ for each of the six steels studied in this example, the estimated weight per foot and cost per foot are

weight/ft = volume/ft × density

volume/ft = $\pi dt$ × 12

Therefore weight/ft = $\pi dt$ × 12 × density

and cost/ft = weight/ft × cost/lb

A typical calculation for steel $D$ with $\sigma_{ys}$ = 180 ksi is as follows:

$$\sigma_{\text{design}} = \frac{\sigma_{ys}}{2} = \frac{180}{2} = 90 \text{ ksi}$$

$$t = \frac{pd}{2\sigma_{\text{design}}} = \frac{(5000)(30)}{2(90,000)} = 0.83 \text{ in.}$$

volume/ft = (3.14)(30)(0.83) × 12 = 78.5(12) = 942 in.$^3$

weight/ft = volume/ft × density = 942 × 0.283 = 267 lb/ft

cost/ft = weight/ft × cost/ft = 267 × ($1.20) = $320/ft

The values for the other steels are calculated in a similar manner and are presented in Table 6.3. These results show the direct effect of reducing weight by using a higher-strength steel. The cost per foot is not necessarily directly related to yield strength because of the numerous factors involved in pricing of structural materials. However, weight is decreased directly by increasing the yield strength, but if fracture is possible, the overall safety and reliability of the vessel may be decreased, as will be illustrated in the

**TABLE 6.3   Comparison of Results for $\sigma_{\text{design}} = \sigma_{ys}/2$, Example Problem**

| Steel | Yield Strength $\sigma_{ys}$ (ksi) | $\sigma_{\text{design}}$ (ksi) | $t$ (in.) | Weight (lb/ft) | Cost ($/ft) |
|-------|------|------|------|------|------|
| A | 260 | 130 | 0.58 | 185 | 259 |
| B | 220 | 110 | 0.68 | 218 | 306 |
| C | 180 | 90  | 0.83 | 267 | 267 |
| D | 180 | 90  | 0.83 | 267 | 320 |
| E | 140 | 70  | 1.07 | 343 | 172 |
| F | 110 | 55  | 1.36 | 437 | 66  |

following examples. No $K_I$ calculations are made in case I because it is assumed that there are no cracks.

**Case II—*Fracture-mechanics design.*** In this case, it is assumed that a flaw is present. As a first step, the designer should estimate the maximum possible flaw size that can exist in the vessel wall, based on fabrication and inspection considerations. In this particular design example, assume that we want to prevent failures caused by a surface flaw of depth 0.5 in. and an $a/2c$ ratio of 0.25 is possible, as shown in Figure 6.11. This size flaw may be due to improper fabrication or crack growth by fatigue or stress corrosion. Obviously, the decision regarding the maximum possible flaw size will depend on many factors related to the design, fabrication, and inspection for the particular structure, but the designer must use the best information available and make the decision. Too often, this step in the design process has been overlooked or, at best, casually handled by assuming that the fabrication will be inspected and "all injurious flaws removed," to quote one specification.

The general relation among $K_{Ic}$, $\sigma$, and $a$ for a surface flaw as developed in Chapter 2 is

$$K_{Ic} = 1.12\sqrt{\pi}M_k\sigma\sqrt{\frac{a}{Q}}$$

where $M_k$ = magnification factor for deep flaws (similar to the tangent correction factor in Chapter 2), assumed in this example to vary linearly between 1.0 and 1.6 as $a/t$ (crack

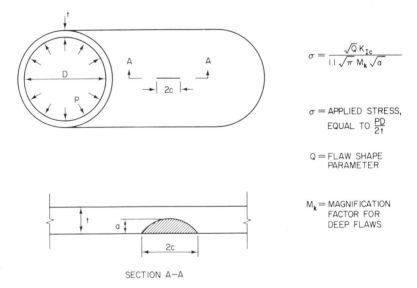

$$\sigma = \frac{\sqrt{Q}\,K_{Ic}}{1.1\sqrt{\pi}\,M_k\sqrt{a}}$$

$\sigma$ = APPLIED STRESS, EQUAL TO $\frac{PD}{2t}$

$Q$ = FLAW SHAPE PARAMETER

$M_k$ = MAGNIFICATION FACTOR FOR DEEP FLAWS

SECTION A–A

**Figure 6.11**   Section of pressure vessel with surface flaw for example problem.

depth/vessel thickness) varies from 0.5 to 1.0 (Figure 6.12); for $a/t$ values less than 0.5, $M_k \simeq 1.0$.

$Q$ = flaw shape parameter as shown in Figure 6.13.

$\sigma$ = applied hoop stress, ksi, equal to $pd/2t$ for the vessel section shown in Figure 6.11.

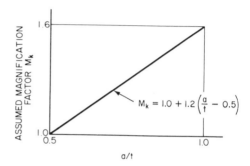

**Figure 6.12**  Assumed magnification factor, $M_k$, for example problem.

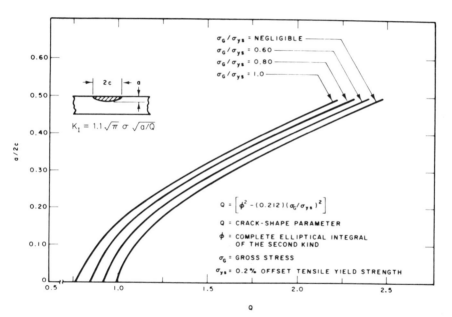

**Figure 6.13**  Crack-shape parameter, $Q$, for surface flaw.

Rearranging the preceding equation yields

$$\sigma = \frac{\sqrt{Q}K_{Ic}}{1.12\sqrt{\pi}M_k\sqrt{a}}$$

The design stress, $\sigma_{\text{design}}$ for a factor of safety of 2.0 against fracture is calculated by replacing $K_{Ic}$ with $K_{I\text{design}} = K_{Ic}/2$ as follows:

$$\sigma_{\text{design}} = \frac{\sqrt{Q}\left(\frac{K_{\text{Ic}}}{2}\right)}{1.12\sqrt{\pi}M_k\sqrt{a}}$$

and

$$t = \frac{pd}{2\sigma_{\text{design}}}$$

For $a = 0.5$, and the different values of $K_{\text{Ic}}$ for the steels shown in Table 6.2, the calculated *allowable* stress values, $\sigma$, for each of the steels being studied are found by (1) calculating the design stress based on the fracture resistance needed for an 0.5-in.-deep crack and then (2) calculating the vessel thickness required at that stress level. Because $M_k$ and $Q$ are functions of the design stress, an iterative procedure must be used.

A step-by-step calculation of the value for steel D (Table 6.2) ($\sigma_{ys} = 180$ ksi) is as follows. Given

$$\sigma = \frac{\sqrt{Q}\left(\frac{K_{\text{Ic}}}{2}\right)}{1.12\sqrt{\pi}M_k\sqrt{a}}$$

for steel D

1. $K_{\text{Ic}} = 220$ ksi$\sqrt{\text{in.}}$
2. $a = 0.5$ in.
3. Assume that $\sigma/\sigma_{ys} = 0.55$ and thus $Q = 1.4$ (Figure 6.13).
4. Assume that $M_k = 1.0$ (Figure 6.12) for the first trial.

Thus

$$\sigma = \frac{\sqrt{1.4}\left(\frac{220}{2}\right)}{1.12(1.77)(1.0)\sqrt{0.5}}$$

$$= 95 \text{ ksi}$$

Using $\sigma = 95$ ksi, solve for the wall thickness, $t$, required to contain the design pressure of 5000 psi (5 ksi):

$$t = \frac{pd}{2\sigma}$$

$$= \frac{(5)(30)}{2(95)} = 0.8 \text{ in.}$$

For

$$t = 0.8 \text{ in.}$$

$$\frac{a}{t} = \frac{0.5}{0.8} = 0.63$$

and

$$M_k = 1.15$$

also

$$\frac{\sigma}{\sigma_{ys}} = \frac{95}{180} = 0.53$$

and

$$Q = 1.4$$

Using $M_k = 1.15$ and $Q = 1.4$, a second iteration results in

$$\sigma = \frac{\sqrt{1.4}(110)}{1.1(1.77)(1.15)\sqrt{0.5}}$$

$$\sigma = 82.2 \text{ ksi}$$

$$t = \frac{pd}{2\sigma} = \frac{(5)(30)}{2(82.2)}$$

$$t = 0.91 \text{ in.}$$

For

$$t = 0.91 \text{ in.}$$

$$\frac{a}{t} = \frac{0.5}{0.91} = 0.55$$

and

$$M_k = 1.06$$

$$\frac{\sigma}{\sigma_{ys}} = \frac{82.2}{180} = 0.46 \quad \text{and} \quad Q = 1.42$$

Using $M_k = 1.06$ and $Q = 1.42$,

$$\sigma = \frac{\sqrt{1.42}(110)}{(1.12)(1.77)(1.06)\sqrt{0.5}}$$

$$\sigma = 90 \text{ ksi}$$

$$t = \frac{pd}{2\sigma} = \frac{5(30)}{2(90)}$$

$$t = 0.83 \text{ in.}$$

Assuming that $t$ will be $\simeq 0.85$ in.,

$$\frac{a}{t} = \frac{0.5}{0.85} = 0.59$$

and

$$M_k = 1.11$$

also

$$\frac{\sigma}{\sigma_{ys}} = \frac{85}{180} = 0.47$$

and

$$Q \simeq 1.42$$

therefore

$$\sigma = \frac{\sqrt{1.42}(110)}{1.12(1.77)(1.11)\sqrt{0.5}}$$

$$\sigma = 85.8 \text{ ksi}$$

$$t = \frac{(5)(30)}{2(85.8)}$$

$$t = 0.87 \text{ in.}$$

Because the calculated thickness is essentially equal to the assumed thickness, a final iteration should show that $t_{assumed} = t_{calculated}$.

As the fourth iteration, for $\sigma = 87$ ksi and $t = 0.86$ in., assume that

1. $\sigma/\sigma_{ys} = 87/180 = 0.48$, and therefore $Q = 1.42$.
2. $a/t = 0.5/.86$ and therefore $M_k = 1.10$.

Thus,

$$\sigma = \frac{\sqrt{1.42}(110)}{(1.12)(1.77)(1.10)\sqrt{0.5}}$$

$$\sigma = 87 \text{ ksi}$$

For a design stress of 87 ksi, the required wall thickness, $t$, is

$$t = \frac{pd}{2\sigma} = \frac{(5)(30)}{2(87)}$$

$$t = 0.86 \text{ in.}$$

This value essentially agrees with the initial value of thickness of 0.86 in. for the assumed value, and thus further trials are not required. Note that the convergence is fairly rapid and that the procedure can easily be programmed for a computer.

Wall thicknesses for the remaining steels in this example are calculated in a similar manner and are presented in Table 6.4. These results show that to withstand an internal pressure of 5000 psi in a 30-in.-diameter vessel having an 0.5-in.-deep surface flaw, the design stresses and wall thicknesses that should be used to give the same resistance to fracture vary considerably for the steels investigated. For example, the allowable design stress level

**TABLE 6.4    Comparison of Results for $K_1 = K_{Ic}/2$, Example Problem**

| Steel | Yield Strength $\sigma_{ys}$ (ksi) | $\sigma_{design}$ (ksi) | $K_{Ic}$ (ksi$\sqrt{in.}$) | $K_1$ (ksi$\sqrt{in.}$) | $t$ (in.) | Weight (lb/ft) | Cost ($/ft) |
|---|---|---|---|---|---|---|---|
| A | 260 | 35 | 80 | 40 | 2.14 | 685 | 959 |
| B | 220 | 48 | 110 | 55 | 1.56 | 499 | 699 |
| C | 180 | 61 | 140 | 70 | 1.23 | 394 | 394 |
| D | 180 | 87 | 220 | 110 | 0.86 | 275 | 330 |
| E | 140 | 95 | 260 | 130 | 0.79 | 253 | 127 |
| F | 110 | 72 | 170 | 85 | 1.04 | 333 | 50 |

for the 260-ksi yield-strength steel is only 35 ksi, and the required wall thickness is 2.14 in., whereas for a lower-strength, tougher steel having a 180-ksi yield strength (and a $K_{Ic}$ of 140 ksi$\sqrt{in.}$), the design stress is 61 ksi and the required wall thickness is 1.23 in. If an even tougher, 180-ksi yield-strength steel is selected, that is, steel D with a $K_{Ic}$ value of 220 ksi$\sqrt{in.}$, the design stress can be increased to 86 ksi and the wall thickness decreased to 0.86 in.

Because each of the vessels in this example is designed on the basis of *equivalent resistance to fracture in the presence of an 0.5-in.-deep surface flaw,* there would be an obvious savings in weight of the vessel by using a lower-strength, tougher steel compared with the 260-ksi yield-strength steel. That is, the weight is proportional to the wall thickness, and the required wall thicknesses for the lower-strength, tougher steels are *less* than for the higher-strength less tough steels.

To show that the highest-strength steel may *not* yield the least weight or most economical vessel, an approximate estimate of the weight per foot and assumed cost per foot of the vessel (neglecting the costs of forming, fabrication, etc.) are presented in Table 6.4. The weights per foot of vessel were calculated by estimating the volume of material as the cross-sectional area ($\pi dt$) times a 12-in. length and then multiplying by the density of steel, 0.283 lb/in$^3$.

These results show the significant effect of toughness on the allowable design stress and illustrate dramatically that high-strength materials with low values of notch toughness do not necessarily yield the least weight vessel when fracture is a possible mode of failure. When cost is the primary criterion, the advantage of the lower-strength, tougher materials based on the assumed prices is obvious.

**General analysis of cases I and II.**    The calculated thicknesses tabulated in Tables 6.3 and 6.4 are plotted as a function of yield strength in Figure 6.14. Comparison of the required thicknesses shows that for the two lowest-strength steels, yielding is the most likely mode of failure and the factor

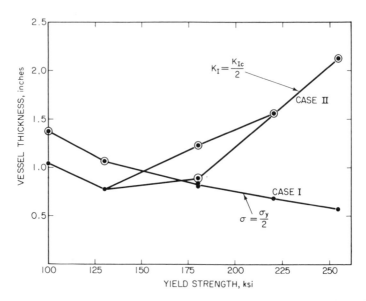

**Figure 6.14** Comparison of vessel thicknesses required for factor of safety of 2.0 against either yielding or fracture in example problem.

of safety against fracture will be greater than 2.0. Specifically for steel F, the required vessel thickness is 1.36 in., and the design stress ($\sigma_{ys}/2$) is 55 ksi; thus, the factor of safety against yielding is 2.0. The corresponding $K_I$ value is 63 ksi$\sqrt{\text{in.}}$, which gives a factor of safety against fracture of $\frac{170}{63}$ = 2.7. However, to have a factor of safety of at least 2 against *both* modes of failure, a wall thickness of 1.36 in. is required.

Conversely, the required thickness for the highest-strength steel is 2.14 in. based on a factor of safety against fracture of 2.0. However, the corresponding design stress value is 35 ksi, which gives a factor of safety against yielding of $\frac{260}{35}$ = 7.43.

The factors of safety against yielding or fracture are tabulated in Table 6.5 and illustrate the necessity of considering all possible modes of failure prior to selecting a final geometry (thickness) as well as selecting a particular material.

As a last step in this example, the required thicknesses for a factor of safety of at least 2.0 against *both* yielding and fracture are listed in Table 6.6. In addition, the corresponding weight per foot and cost per foot are tabulated for each material. Analysis of the results shows that steel D would be the optimum selection on the basis of minimum weight and steel F would be the least expensive steel. Note that for both cases, the factors of safety against yielding and fracture are 2.0 or greater. Thus, the vessels are compared on an equivalent performance basis and show the advantage of using fracture mechanics during the material selection process.

TABLE 6.5    Comparison of Factors of Safety Against Yielding and Fracture
for Example Problem

| Steel | Yield Strength, $\sigma_{ys}$ (ksi) | $t$ (in.) | Factor of Safety Against Yielding | Factor of Safety Against Fracture | Required Thickness to Satisfy Both Criteria (in.) |
|---|---|---|---|---|---|
| A | 260 | 0.58 | 2.0 | 0.35 | 2.14 |
|   |     | 2.14 | 7.43 | 2.0 |    |
| B | 220 | 0.68 | 2.0 | 0.64 | 1.56 |
|   |     | 1.56 | 4.58 | 2.0 |    |
| C | 180 | 0.83 | 2.0 | 1.12 | 1.23 |
|   |     | 1.23 | 2.95 | 2.0 |    |
| D | 180 | 0.83 | 2.0 | 1.77 | 0.86 |
|   |     | 0.86 | 2.07 | 2.0 |    |
| E | 140 | 1.07 | 2.0 | 3.01 | 1.07 |
|   |     | 0.79 | 1.47 | 2.0 |    |
| F | 110 | 1.36 | 2.0 | 2.51 | 1.36 |
|   |     | 1.04 | 1.53 | 2.0 |    |

TABLE 6.6    Weight and Cost of Steel for Factor of Safety of 2.0 or Greater
Against Both Yielding and Fracture, Example Problem

| Steel | Yield Strength $\sigma_{ys}$ (ksi) | $t$ (in.) | Weight (lb/ft) | Cost ($/ft) |
|---|---|---|---|---|
| A | 260 | 2.14 | 685 | 959 |
| B | 220 | 1.56 | 499 | 699 |
| C | 180 | 1.23 | 394 | 394 |
| D | 180 | 0.86 | 275 | 330 |
| E | 140 | 1.07 | 343 | 172 |
| F | 110 | 1.36 | 437 | 66 |

## 6.6. Significance of Loading Rate

For materials that exhibit loading-rate or strain-rate effects, such as structural steels having yield strengths less than about 140 ksi, the loading rate at a given temperature can affect the notch toughness significantly. Ideally, the loading rate in the particular notch-toughness test used to evaluate the behavior of a structural material should be essentially equal to the loading rate in the structure being analyzed. Toughness criteria can then be established by using either static, intermediate, or dynamic toughness test results for a particular structure.

By knowing a specific $K_{Ic}$ value for a particular structural material the critical crack size can be calculated for a given design stress level. If the structure is loaded statically, this size crack can be tolerated at all temperatures above an extremely low one, as shown schematically in Figure 6.15. If the structure is loaded dynamically, however, the minimum temperature at which this size crack can be tolerated would be much higher, possibly above the minimum service temperature.

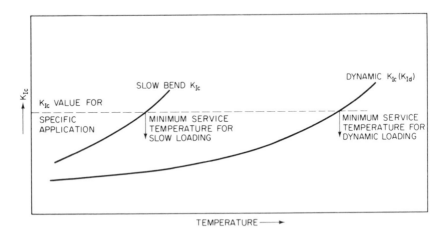

**Figure 6.15** Schematic showing effect of temperature and loading rate on $K_{Ic}$.

As a specific example for a structural steel commonly used in shipbuilding, test results for an ABS, Class C steel are presented in Figure 6.16. Assume that a fracture criterion stating that $K_{Ic}/\sigma_{ys} \geq 0.6$ is required for a particular application.

The test results presented in Figure 6.16 show that if the actual structure is loaded *statically* ($\dot{\varepsilon} \approx 5 \times 10^{-5}$/sec), this level of performance will be obtained at all service temperatures above $-220°F$. If the structure is loaded *dynamically* ($\dot{\varepsilon} \approx 20$/sec), however, this level of performance ($K_{Id}/\sigma_{yd} \geq 0.6$) will be obtained at all service temperatures above $-20°F$. (For an intermediate loading rate ($\dot{\varepsilon} \approx 10^{-1}$/sec), this level of performance will be obtained at all temperatures above $-120°F$.) Thus, the *actual service loading rate* has a very large influence on the structural performance at a given temperature.

Assume that this ABS-C steel is to be used in a low-temperature application at $0°F$ and that the possibility exists that a through-thickness crack is present in the vicinity of a weld, where the local stress is of yield-point magnitude. This situation is shown schematically in Figure 6.17, where the critical crack size, $2a$, is

$$K_{Ic} = \sigma\sqrt{\pi a}$$

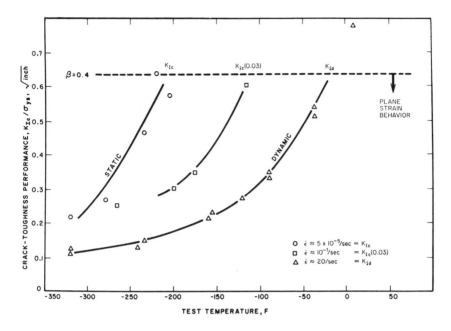

**Figure 6.16**  Crack-toughness performance for an ABS-C steel showing effect of loading rate.

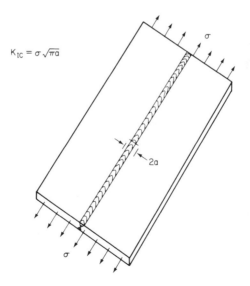

**Figure 6.17**  Schematic showing flaw in welded plate.

For
$$\frac{K_{\mathrm{Ic}}}{\sigma_{ys}} \sim 0.6 = \sqrt{\pi a}$$

$$a \simeq \frac{(0.6)^2}{\pi} \simeq 0.11 \text{ in.}$$

Therefore $2a = 0.22$ in.

If the structure is loaded slowly, this size flaw could be tolerated at extremely low temperatures, that is, $-220°$F. However, if the structure is loaded dynamically, this same size flaw can be tolerated only at temperatures above $-20°$F, and the possibility of a brittle fracture at a service temperature of only $0°$F becomes much greater.

Note that above a $K_{\mathrm{Id}}/\sigma_{yd}$ value of about 0.6, which occurs slightly above the NDT temperature of a structural steel, the relative toughness levels increase quite rapidly with temperature. This transition from plane-strain behavior to elastic-plastic and plastic behavior is the basis for many fracture criteria developed using the "transition-temperature approach." At a $K_{\mathrm{Ic}}/\sigma_{ys}$ or $K_{\mathrm{Id}}/\sigma_{yd}$ ratio of about 0.9, the minimum service temperatures for the ABS-C steel described in Figure 6.16 are approximately $-180°$F, $-80°$F, and $+20°$F for static, intermediate, and dynamic loading rates, respectively. For the same through-thickness crack situation (Figure 6.17), the critical crack size is

$$\frac{K_{\mathrm{Ic}}}{\sigma_{ys}} \sim 0.9 = \sqrt{\pi a}$$

$$a \simeq \frac{(0.9)^2}{\pi} \simeq 0.26$$

and $2a \sim 0.52$ in.

Thus, for only a $40°$F change in minimum service temperature above a $K_{\mathrm{Ic}}/\sigma_{ys}$ or $K_{\mathrm{Id}}/\sigma_{yd}$ ratio of 0.6, the critical crack size has more than doubled because of the rapid increase in toughness in this region. For dynamic loading, this would be about $40°$F above NDT. Accordingly, the temperature range through which either the $K_{\mathrm{Ic}}/\sigma_{ys}$ or $K_{\mathrm{Id}}/\sigma_{yd}$ ratio increases rapidly can be and has been used as a limiting service temperature, that is, a transition temperature where the transition is from plane-strain to elastic-plastic behavior.

The importance of using a test specimen that "models" the loading rate in the actual structure should be emphasized. That is, if the designer can be certain that his or her structure will always be loaded statically or that the structural material is *not* strain-rate sensitive, then static tests can be used to predict the service behavior. Conversely, if impact loading is the proper loading rate, then an impact test should be used to predict the service behavior.

Because of the large difference in behavior of structural materials that are strain-rate sensitive, there are two widely differing schools of thought regarding the possibility of dynamic loading of structures.

The first school of thought assumes that there can be highly localized regions where the notch toughness is low in all large structures. Examples of these regions might be microcracks in weld metal, a grain-coarsened region in the heat-affected zone of a weldment, arc strikes on the base plate, nonmetallic inclusions, and so on. These highly localized regions of low notch toughness are assumed to initiate microcracks under statically applied loads so that the surrounding material is suddenly presented with a moving dynamic (but small) crack. Thus, even though the applied structural loading rate may be slow, this school of thought believes that the crucial loading rate is that of the crack-tip pop-in. This sudden separation of a few metal grains ($<{\sim}0.01$ in. in size) is assumed to create dynamic loading rates. This implies that the dynamic toughness ($K_{Id}/\sigma_{yd}$) always controls fracture extension, irrespective of initiation conditions. Accordingly, this leads to the conclusion that the $K_{Id}/\sigma_{yd}$ parameter *always* should be used to establish the behavior for all structures regardless of the measured structural loading rates. This is often referred to as designing for crack-arrest behavior.

The second school of thought assumes that for those structures where the measured loading rates are slow, the static $K_{Ic}/\sigma_{ys}$ ratio is the controlling notch-toughness parameter. For those structures loaded at some intermediate loading rate (which applies to many structures such as bridges, ships, and offshore rigs), an "intermediate" ($K_{Ic}/\sigma_{ys}$) ratio is believed to be the controlling notch-toughness parameter. The latter viewpoint helps to explain the fact that many structures whose dynamic toughness ($K_{Id}/\sigma_{yd}$) is relatively low have been performing quite satisfactorily because the actual structural loading rate is slow or, at most, intermediate.

A realistic appraisal of these two schools of thought leads to the following observation. While it is true that there may exist highly localized regions of low toughness in complex structures, it seems hard to visualize that a dynamic stress field can be created by the localized extension of a small microcrack. A more realistic mechanism of creating a localized dynamic stress field under nominally static loading would appear to result from the sudden separation of secondary members such as a stiffener or a gusset plate. This sudden separation might change the local stress distribution over a reasonably large area rather suddenly and could create a large dynamic stress field surrounding a small moving crack that would be necessary for the dynamic properties ($K_{Id}/\sigma_{yd}$) to control.

Accordingly, whereas it would be conservative always to design on the basis of possible dynamic loading (use of $K_{Id}/\sigma_{yd}$ toughness values at the service temperature), it does not seem to be necessary unless extremely conservative toughness values are desired for a single load-path member. That is, if the failure of a single member can lead to collapse of the entire structure, then a conservative approach would be to design for the possibility of a dynamic crack, that is, use $K_{Id}$ values at the minimum service temperature rather than $K_{Ic}$ values. Another approach to ensure the safety and reliability of such structures can be achieved by decreasing the maximum stress and stress fluctuation levels. This approach would result in a significant increase in the fatigue life of the structural component and in tolerating larger cracks prior to failure. Thus, factors *other than* just material toughness are important in fracture control, as discussed in Chapter 16. For multiple-load-path structures, such as most bridges, the use of static $K_{Ic}/\sigma_{ys}$ or intermediate $K_{Ic}(t)/\sigma_{ys}(t)$ as the controlling toughness parameter appears quite realistic.

For example, the AASHTO material-toughness requirements for non-fracture critical tension members of A36 bridge steels having service temperatures down to 0°F require that the steel plates exhibit 15-ft-lb CVN impact energy at +70°F. Because bridges have been shown to be loaded at an intermediate rate of loading, this requirement is designed to ensure nonplane-strain behavior beginning at about −50°F, well *below* the minimum service temperature of 0°F (the development of this criterion is presented in Chapter 16). Thus (1) because of the intermediate loading rate to which bridges are subjected, (2) because the loadings are reasonably well known, (3) because bridges are usually highly redundant structures (multiple-load paths) such that failure of a single member rarely leads to failure of the entire structure (except for the Point Pleasant Bridge, which was *not* a multiple-load-path structure), and (4) because of the conservative AASHTO fatigue requirements for bridge details, these toughness requirements appear to be quite satisfactory. Based on satisfactory service experience and the fact that most bridges *are* multiple-load-path structures so that failure of a single member probably will not lead to failure of the entire structure, the consequences of a failure in a single member are probably minor—generally only the cost of repair or replacement of a portion of the structure. Tests conducted on nonredundant, welded bridge details indicated that the AASHTO toughness requirements were adequate even when the details were subjected to the total design fatigue life, the maximum design stress, an operating temperature of −30°F, and the maximum expected loading rate (as discussed in Chapter 16).

## 6.7. Fitness for Purpose

The general design philosophy using fracture mechanics is to keep $K_I < K_c$ at all times throughout the life of a structure. This behavior is shown schematically in Figure 6.5, which illustrates the general relation among stress, flaw size, and material toughness for a structural member with a center crack length of $2a$. If the member is subjected to fatigue, such that the crack can grow, thereby increasing $K_I$ (Figure 6.6), then $K_I$ at the end of the fatigue life of the structure should still be less than $K_c$ for a safe design.

This approach assumes that a crack is present in the structural member and is a conservative design approach. In Chapters 7 and 8, it is shown that a significant portion of the fatigue life can occur *prior to* crack growth, and certainly this initiation fatigue life should be considered in design.

Making $K_I$ much less than $K_c$ certainly appears to be a desirable situation. However, because design is an optimization of performance, safety, and cost, making a structure safer than it needs to be, at the expense of performance and cost, is not good engineering design.

The development of this concept has also been referred to as *fitness for purpose*. Wells[4] defines fitness for purpose as that design

> which is consciously chosen to be the right level of material and fabrication quality for each application, having regard to the risks and consequences of failure; it may be contrasted with the best quality that can be achieved within a given set of circumstances, which may be inadequate for some exacting requirements, and needlessly uneconomic for others which are less demanding. A characteristic of the fitness-for-purpose approach is that it requires to be defined beforehand according to known facts, and by agreement with purchasers which will subsequently seek to be national and eventually international.

The fitness-for-purpose approach certainly uses concepts of fracture mechanics to develop a set of principles for design and even to develop specific recommendations. However, the present state of the art is such that fitness for purpose is really an attempt to quantify what has often been referred to as sound engineering judgment or even common sense. Engineers have never applied the same set of rigid criteria to materials for critical structural applications (e.g., steels for nuclear pressure vessels) as they have to noncritical applications such as steels for truck bodies.

Fitness for purpose has been described briefly in the fracture mechanics design because of the importance of fracture mechanics to the fitness-for-purpose concept. However, it should be noted that other disciplines are also incorporated into the fitness-for-purpose approach, for example, risk analysis and reliability, nondestructive examination, and quality assurance as well as recent metallurgical advances such as electron microscopes and spectrographic analyses.

Reference 4, *Fitness for Purpose Validation of Welded Construction*, sponsored by the Welding Institute, the American Welding Society, and the Welding Research Council presents an extensive set of papers describing the fitness-for-purpose approach.

# References

1. J. W. Fisher, *Guide to 1977 AASHTO Fatigue Specifications, American Iron and Steel Construction*, New York, 1974.
2. J. E. Srawley and J. B. Esgar, "Investigation of Hydrotest Failure of Thiokol Chemical Corporation 260-Inch-Diameter SL-1 Motor Case," *NASA TMX-1194*, Cleveland, Jan. 1966.
3. W. W. Gerberich, "Fracture Mechanics Approach to Design-Application," presented in a *Short Course on Offshore Structures*, University of California, Berkeley, Calif., 1967.
4. A. A. Wells, "The Meaning of Fitness for Purpose and the Concept of Defect Tolerance," *Fitness for Purpose Validation of Welded Constructions*, The Welding Institute, London, Nov. 17–19, 1981, paper No. 33.

# 7
# Introduction
# to Fatigue

## 7.1. Introduction

The discussion in the preceding chapters described the fracture behavior of components subjected to a monotonically increasing load. However, most equipment and structural components are subjected to repeated fluctuating loads whose magnitude is well below the fracture load under monotonic loading. Examples of equipment and structures that are subjected to fatigue loading include pumps, vehicles, earthmoving equipment, drilling rigs, aircraft, bridges, ships, and offshore structures.

Fatigue is the process of cumulative damage in a benign environment that is caused by repeated fluctuating loads and, in the presence of an aggressive environment, is known as corrosion fatigue. Fatigue damage occurs only in regions that deform plastically under the applied fluctuating load. Thus, fatigue damage for components that are subjected to normally elastic stress fluctuations occurs at regions of stress (strain) raisers where the localized stress exceeds the yield stress of the material. After a certain number of load fluctuations, the accumulated damage causes the initiation and subsequent propagation of a crack or cracks in the plastically damaged regions. This process can and in many cases does cause the fracture of components. The more severe the stress concentration, the shorter the time to initiate a fatigue crack.

The number of cycles required to initiate a fatigue crack is the fatigue-crack-initiation life, $N_i$. The number of cycles required to propagate a fatigue crack to a critical size is called the fatigue-crack-propagation life, $N_p$. The total fatigue life, $N_T$, is the sum of the initiation and propagation lives,

$$N_T = N_i + N_p \qquad (7.1)$$

There is no simple or clear delineation of the boundary between fatigue-crack initiation and propagation. Furthermore, a preexisting crack in a structural component can reduce or eliminate the fatigue-crack-initiation life and, thus, decrease the total fatigue life of the component.

Concern with fatigue damage was recognized in Europe early in the nineteenth century.[1,2] In 1852, Wohler conducted comprehensive experiments on axles[2] subjected to tensile, bending, and torsional repeated-load fluctuations. This work is important because it formed the basis for the Goodman diagram,[2] which is the first developed methodology to predict the fatigue performance of components. It is a graphic representation of the fatigue limit, at any stress ratio, given as a function of the ultimate tensile strength. Fatigue was incorporated into design criteria near the end of the nineteenth century and has been studied since. However, the most significant developments have occurred since the 1950s. At present, fatigue is part of design specification for many engineering structures.

## 7.2. Factors Affecting Fatigue Performance

Many parameters affect the fatigue performance of structural components. They include parameters related to stress (load), geometry and properties of the component, and external environment. The stress parameters include state of stress, stress range, stress ratio, constant or variable loading, frequency, and maximum stress. The geometry and properties of the component include stress (strain) raisers, size, stress gradient, and metallurgical and mechanical properties of the base metal and weldments. The external environment parameters include temperature and aggressiveness of the environment. The effects of most of these parameters are discussed in the following chapters.

The primary factor that affects the fatigue behavior of structural components is the fluctuation in the localized stress or strain. Consequently, the most effective methods for increasing the fatigue life significantly are usually accomplished by decreasing the severity of the stress concentration and the magnitude of the applied nominal stress. In many cases, decrease in the severity of the stress concentration can be easily accomplished by using transition radii in fillet regions as for keyways and geometrical changes and by minimizing the size of weld imperfections.

Corrosion-fatigue behavior is affected by the same parameters that affect the fatigue behavior as well as factors that do not affect fatigue, for example, frequency and waveform and effects of the environment. Unfortunately, at present, the only means of determining the corrosion-fatigue behavior of materials is by conducting the tests on the actual material-environment system of interest. Corrosion fatigue is discussed in Chapters 12 and 13.

## 7.3. Fatigue Loading

Structural components are subjected to a variety of load (stress) histories. The simplest of these histories is the constant-amplitude cyclic-stress fluctuation shown in Figure 7.1. This type of loading usually occurs in machinery parts such as shafts and rods during periods of steady-state rotation. The most complex fluctuating-load history is a variable-amplitude random sequence as shown in Figure 7.2. This type of loading is experienced by many structures, including offshore drilling rigs, ships, aircraft, bridges, and earth-moving equipment.

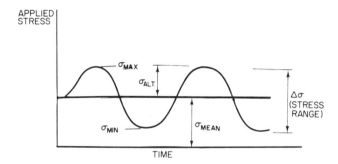

**Figure 7.1**   Terminology used in constant-amplitude fatigue. (From N. Willems, J. T. Easley, and S. T. Rolfe, *Strength of Materials*, McGraw-Hill, New York, 1981, Figure 14.5.)

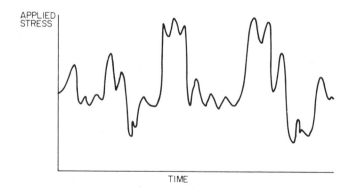

**Figure 7.2**   Random-stress loading.

### 7.3.1. Constant-Amplitude Loading

Constant-amplitude load histories can be represented by a constant load (stress) range, $\Delta P(\Delta\sigma)$; a mean stress, $\sigma_{mean}$; an alternating stress or stress amplitude, $\sigma_{amp}$; and a stress ratio, $R$, Figure 7.1. The stress range is the algebraic difference between the maximum stress, $\sigma_{max}$, and the minimum stress, $\sigma_{\min}$, in the cycle

$$\Delta\sigma = \sigma_{\max} - \sigma_{\min} \qquad (7.2)$$

The mean stress is the algebraic mean of $\sigma_{\max}$ and $\sigma_{\min}$ in the cycle

$$\sigma_{\text{mean}} = \frac{\sigma_{\max} + \sigma_{\min}}{2} \qquad (7.3)$$

The alternating stress or stress amplitude is half the stress range in a cycle

$$\sigma_{\text{amp}} = \frac{\Delta\sigma}{2} = \frac{\sigma_{\max} - \sigma_{\min}}{2} \qquad (7.4)$$

The stress ratio represents the relative magnitude of the minimum and maximum stresses in each cycle

$$R = \frac{\sigma_{\min}}{\sigma_{\max}} \qquad (7.5)$$

Thus, a complete reversal of load from a minimum compressive stress to an equal maximum tensile stress corresponds to an $R = -1$, as shown in Figure 7.3(a); a stress fluctuation from zero stress to any maximum tensile stress corresponds to an $R = 0$, Figure 7.3(b); and a stress fluctuation from a given minimum tensile load to a maximum tensile load would be characterized by a positive value for $R$ that is less than or equal to 1.0, $0 < R \leqslant 1.0$, as shown in Figure 7.3(c).

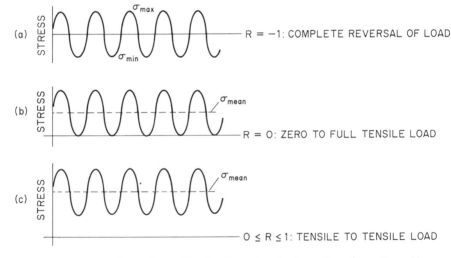

**Figure 7.3**  Comparison of $R$-ratios for various loadings: $R\ \sigma_{\min}/\sigma_{\max}$. (From N. Willems, J. T. Easley, and S. T. Rolfe, *Strength of Materials*, McGraw-Hill, New York, 1981, Figure 14.6.)

### 7.3.2. Variable-Amplitude Loading

Variable-amplitude random-sequence load histories are very complex functions in which the probability of the same sequence and magnitude of stress ranges recurring during a particular time interval is very small. Such histories lack a describable pattern and cannot be represented by an analytical function. Examples of this type of fatigue-load histories include wind loading on aircraft, wave loading on ships and offshore platforms, and truck loading on bridges.

Between the extremes of constant-amplitude cyclic-stress histories and variable-amplitude random-sequence stress histories, there is a multitude of stress patterns of varying degrees of complexity. Many of these histories can be described by analytic functions and represented by various parameters. Some simple variable-amplitude stress histories are those corresponding to a single cycle or multiple high-tensile-load cycles superimposed upon constant-amplitude cyclic-load fluctuations, Figure 7.4. Further discussion of variable-amplitude load histories is presented in Chapter 10.

## 7.4. Fatigue Testing

Fatigue tests are conducted on small laboratory specimens, on specimens that simulate actual structural components, or on actual components. The objective of such tests is to develop information on the fatigue behavior of a particular material, weldment, or geometry. This information can then be used to select the proper material, or design, or both, for a particular application. Ideally, the material condition, stress history, and environment for the test closely simulate the actual service conditions for the structure under consideration.

### 7.4.1. Small Laboratory Tests

Small laboratory test specimens usually have simple geometries and are tested to obtain basic material properties for base metal and weldments. They can be used to study fatigue-crack-initiation or -propagation life, but, in some cases, no distinction is made between the two life regions. Obtaining only the total life on small specimens complicates the use of the results to predict the behavior of actual structural components that have different size, shape, and surface and material conditions from those for the tested specimens. It is preferable to determine the initiation and propagation behaviors separately and then to combine them, as appropriate, to predict the behavior of an actual structural component.

Although there are many small specimens that have been used in laboratory tests, the following specimens and the corresponding test procedures will be discussed briefly because of their importance in the de-

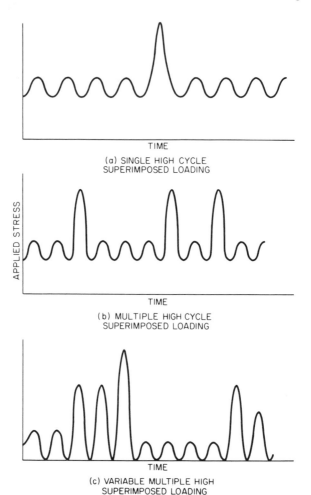

**Figure 7.4** Single or multiple high-cycle superimposed loading. (a) Single high-cycle superimposed loading, (b) multiple high-cycle superimposed loading, and (c) variable multiple high superimposed loading.

velopment of the present knowledge on fatigue behavior of materials and structural components.

### a.   Fatigue-crack-initiation tests

*Stress-life tests.*    Extensive laboratory fatigue investigations have been conducted to evaluate the fatigue behavior of metals by testing a series of specimens in the rotating beam test, the flexure test, or the axial load test. For the rotating beam test, a polished, round specimen with a reduced cross section is supported as a beam and is subjected to a bending moment while being rotated so that fibers of the specimen are subjected alternately to compression and tension stresses of equal magnitude.  For the flexure test, a specimen is bent back and forth as a beam instead of being rotated and, for the axial load test, the specimen is subjected to an alternating axial stress.  In the flexure and the axial load tests, the specimen may be polished or may be a rectangular specimen cut from a plate so that the mill surface is left intact.  In the rotating beam test, the stress ratio, $R$, which is the algebraic ratio of minimum to maximum stress, is $-1.0$, while in the other two tests the effect of various stress ratios may be investigated.  In addition, there is a stress gradient over the cross section of the flexure or rotating beam specimen, whereas the stress is uniformly distributed over the cross section of the axially stressed specimen.

In each of the tests, the specimen is subjected to alternating stresses that vary between fixed limits of maximum and minimum stress until failure occurs.  This procedure is repeated for other specimens at the same stress ratio but at different maximum stress values.  The results of the tests are plotted to form an $S–N$ diagram, where $S$ represents the maximum stress, $\sigma_{max}$, in the cycle and $N$ represents the number of cycles required to cause failure.  An $S–N$ diagram for polished specimens of A514 steel obtained from a series of rotating beam tests is shown in Figure 7.5.[3]  Because the number of cycles to failure span several orders of magnitude, and because most tests have been conducted under rotating bending, $R = -1.0$, where $\Delta\sigma$ is equal to $2\sigma_{max}$, the data are usually presented as semilog plot of $\sigma_{max}$ and the number of cycles to cause failure as shown in Figure 7.5.  At any

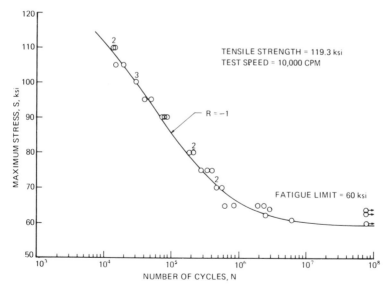

**Figure 7.5**   $S–N$ diagram for polished specimens of an A517 steel obtained from rotating-beam fatigue tests.

point on the curve, the stress value is the "fatigue strength"—the value of maximum stress that will cause failure at a given number of stress cycles and at a given stress ratio—and the number of cycles is the "fatigue life"— the number of stress cycles that will cause failure at a given maximum stress and stress ratio.

As shown in Figure 7.5, the fatigue strength of a structural steel decreases as the number of cycles increases until a "fatigue limit" is reached. If the maximum stress does not exceed the fatigue limit, an unlimited number of stress cycles can be applied at that stress ratio without causing failure. Tests on a large number of steels having tensile strengths up to 200,000 psi indicate that the fatigue limit of polished rotating beam specimens is about one-half the tensile strength. The fatigue limit of polished rotating beam specimens is about the same as that for axially loaded polished specimens, although the fatigue strength at a lower number of cycles is different.

Some materials, such as aluminum, do not exhibit a well-defined fatigue limit. Rather, the test results continue to decrease even beyond $10^7$ cycles of loading, as shown in Figure 7.6.[4]

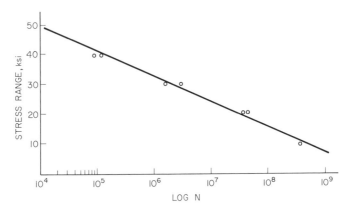

**Figure 7.6**  Typical $S$–$N$ fatigue test results for material with no fatigue limit. (From N. Willems, J. T. Easley, and S. T. Rolfe, *Strength of Materials*, McGraw-Hill, New York, 1981, Figure 14.14.)

The influence of the stress ratio on fatigue strength is illustrated by the $S$–$N$ curves for axially loaded polished specimens of A514 steel shown in Figure 7.7.[3] Stress ratios of 0, $-\frac{1}{2}$, and $-1$ correspond to the following respective loading conditions: zero to tension, half compression to full tension, and complete reversal of stress between equal compression and tension stresses. The curves show that the fatigue strength decreases significantly as the stress ratio decreases. However, the apparent effect of stress ratio on the fatigue behavior is an artifact of the method adopted for presenting the $S$–$N$ data in terms of $\sigma_{max}$ rather than in terms of the stress range, $\Delta\sigma$, which is the primary stress parameter affecting fatigue performance.

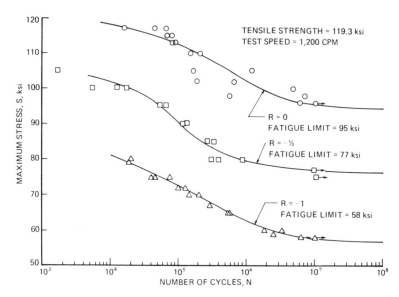

**Figure 7.7**  *S–N* diagrams for polished specimens of an A517 steel obtained from axial-load fatigue tests.

Representation of the data in Figure 7.7 in terms of $\Delta\sigma$ (rather than $\sigma_{max}$) versus $N$ shows that stress ratio has less effect on the fatigue behavior of the tested specimens than would be predicted from Figure 7.7.

These *S–N* curves can be used to construct a fatigue chart shown in Figure 7.8[3] for the polished A514 steel specimens.  Each curved line in the

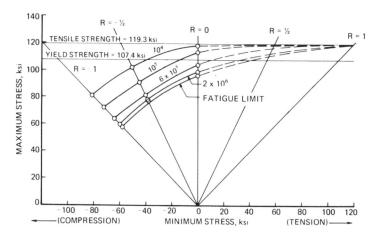

**Figure 7.8**  Fatigue chart for axially loaded polished specimens of an A517 steel.

chart represents the locus of all combinations of maximum and minimum stress at which failure will occur in the indicated number of cycles. Rays can be drawn from the origin in an obvious manner to represent various stress ratios. Such charts are convenient for determining fatigue strength at a stress ratio different from those at which the tests were conducted.

*Strain-life tests.* Because the nominal stresses in most structures are elastic, the zone of plastically deformed metal in the vicinity of stress concentrations is surrounded by an elastic stress field. The true strains, $\varepsilon$, and true stresses, $\sigma$, of the plastic zone are limited by the elastic displacements of the surrounding elastic stress field. Therefore, even when the structure is stress controlled, the localized plastic zones are approximately strain controlled. Consequently, to predict the effects of stress concentration on the fatigue-crack-initiation behavior of structures, the fatigue behavior of the localized plastic zones has been simulated by testing smooth specimens under strain-controlled conditions, Figure 7.9, such that the minimum cross section for the specimen is some fraction of the plastic-zone size. However, suitable correction factors must be used to account for differences in stress state, size, and strain gradient between the smooth specimen and the plastic zone for the structural detail of interest.[5] Assuming that such correction factors are available, fatigue data obtained by testing smooth specimens

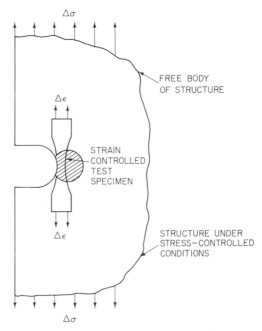

**Figure 7.9** Strain-controlled test specimen simulation for stress concentrations in structures.

under strain-controlled conditions can be used to predict the fatigue-crack-initiation behavior in structural components.

*Stress-strain hysteresis loops.*    The strain-life approach is based on the cyclic properties of the material under constant-strain fluctuations.  These properties are established by subjecting a smooth specimen to fully reversed cyclic plastic-strain amplitudes of constant magnitude.  Such a fully reversed cycle results in a stress-strain hysteresis loop[6] such as that shown in Figure 7.10.  The American Society for Testing and Materials (ASTM) has developed a tentative recommended practice for "Constant-Amplitude Low-Cycle Fatigue Testing," which is designated ASTM E-606—80.[7]

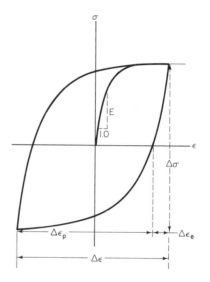

**Figure 7.10**   Schematic of a stress-strain hysteresis loop.

The total strain range, $\Delta\varepsilon$, for a hysteresis loop is equal to twice the strain amplitude, $\varepsilon_a$ (i.e., $\Delta\varepsilon = 2\varepsilon_a$), and the total stress range, $\Delta\sigma$, is equal to twice the stress amplitude $\sigma_a$ (i.e., $\Delta\sigma = 2\sigma_a$).  Moreover, the total strain amplitude can be represented as the sum of its elastic and plastic components, Figure 7.10, such that

$$\varepsilon_a = \frac{\Delta\varepsilon}{2} = \frac{\Delta\varepsilon_e}{2} + \frac{\Delta\varepsilon_p}{2} = \frac{\Delta\sigma}{2E} + \frac{\Delta\varepsilon_p}{2} \qquad (7.6)$$

because $\Delta\varepsilon_e = \dfrac{\Delta\sigma}{E}$, where $E$ is Young's modulus.

*Transient cyclic stress-strain behavior.*    The stress-strain behavior for metals depends on the initial condition (heat treated, cold worked, etc.) of the metal and on the test conditions used.  When exposed to fully reversed cyclic strain amplitudes of constant magnitude, the metal may exhibit

1. Cyclically stable behavior.

2. Cyclically hardening behavior.
3. Cyclically softening behavior.
4. Complex cyclic behavior.

These behaviors can be illustrated by observing the variation in stress as a smooth specimen is subjected to completely reversed, constant-amplitude strain fluctuations with a zero mean value, as shown in Figure 7.11.[6] If the stress magnitude required to apply the constant strain cycles remains constant, Figure 7.11(a), the metal is cyclically stable, which implies that

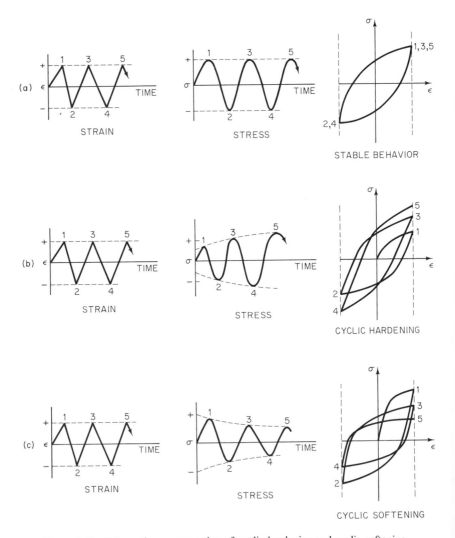

**Figure 7.11**   Schematic representation of cyclic hardening and cyclic softening.

the metal's cyclic stress-strain properties are identical to its monotonic stress-strain properties (engineering stress and strain properties obtained from a tension test). If the stress increases, Figure 7.11(b), the metal cyclically hardens; if it decreases, Figure 7.11(c), the metal cyclically softens. A metal exhibits complex cyclic behavior when cyclic softening, hardening, or stable behavior occur for different strain ranges.

*Cyclically stable stress-strain behavior.*    A metal subjected to constant cyclic-strain amplitude usually exhibits an initial transient behavior, but reaches an essentially cyclically stable stress-strain behavior that corresponds to a constant hysteresis loop. This stable behavior is usually reached in less than 50 percent of the total fatigue life of the specimen. The stabilized stress values are plotted at the corresponding strain values to construct a cyclic stress-strain curve similar to a monotonic stress-strain curve. This procedure is shown schematically in Figure 7.12(a).[6] Superposition of the cyclic and monotonic stress-strain curves, Figure 7.12(b),[6] shows how different steels may exhibit significantly different stress-strain behavior when cyclically loaded than when tested monotonically.

Another means of producing the cyclic stress-strain curve consists of subjecting a specimen to blocks of gradually decreasing and then increasing strain amplitudes as shown in Figure 7.13.[8] A maximum strain amplitude of $\pm 1.5$ to 2.0 percent is usually sufficient to stabilize cyclically the metal quickly without the danger of causing the specimen to neck, fail, or buckle before a stable state is achieved. The cyclic stress-strain curve is then determined by the locus of superimposed hysteresis loop tips.

*Fatigue-life behavior.*[9]    A log-log plot of the stable plastic-strain amplitude, $\Delta\varepsilon_p/2$, versus the number of reversals to failure, $2N_f$, generally results in a straight-line relationship, Figure 7.14, given by the equation

$$\frac{\Delta\varepsilon_p}{2} = \varepsilon_f'(2N_f)^c \tag{7.7}$$

where $\varepsilon_f'$ = fatigue-ductility coefficient.

$c$ = fatigue-ductility exponent.

$N_f$ = number of cycles; therefore, $2N_f$ is equal to the number of reversals.

Similarly, a log-log plot of the stable stress amplitude, $\Delta\sigma/2$, versus the number of reversals to failure, $2N_f$, results in a straight-line relationship, Figure 7.15, given by the equation

$$\frac{\Delta\sigma}{2} = \sigma_a = \sigma_f'(2N_f)^b \tag{7.8}$$

where $\sigma_a$ = true-stress amplitude.

$\sigma_f'$ = fatigue-strength coefficient.

$b$ = fatigue-strength exponent.

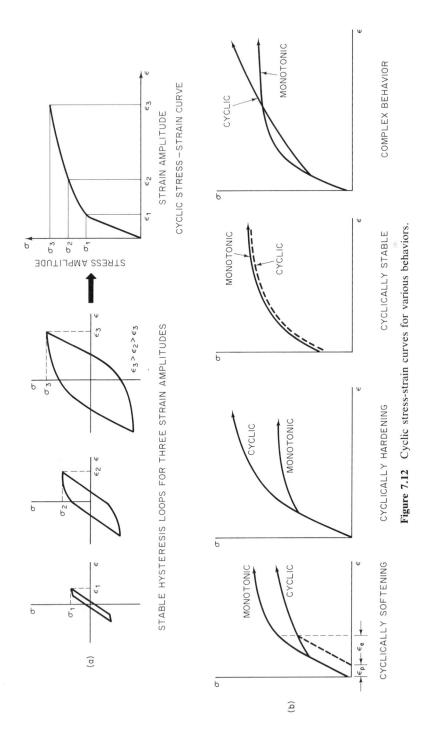

**Figure 7.12** Cyclic stress-strain curves for various behaviors.

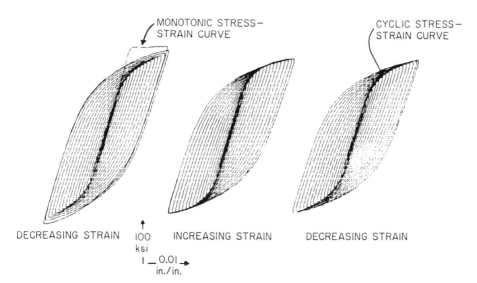

**Figure 7.13**    Stress-strain record of incremental step test on quenched and tempered SAE 4142, 380 BHN.

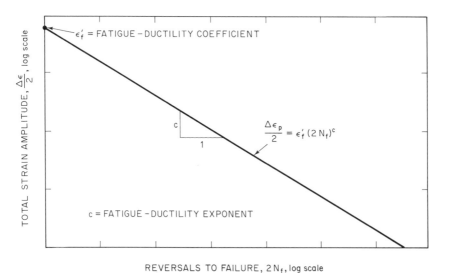

**Figure 7.14**    Fatigue ductility—life plot.

Dividing Equation (7.8) by Young's modulus, $E$, gives the elastic-strain amplitude in terms of the fatigue-strength coefficient and exponent and the fatigue life:

$$\frac{\Delta\varepsilon_e}{2} = \frac{\sigma_a}{E} = \frac{\sigma_f'}{E}(2N_f)^b \tag{7.9}$$

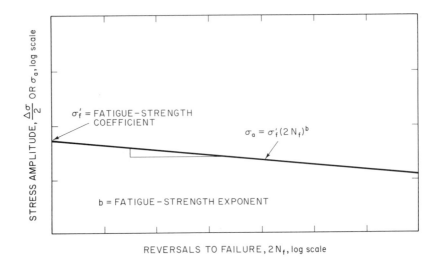

**Figure 7.15**  Fatigue strength—life plot.

Combining Equations (7.6), (7.7), and (7.9) results in the strain-life relationship

$$\frac{\Delta\varepsilon}{2} = \frac{\sigma_f'}{E}(2N_f)^b + \varepsilon_f'(2N_f)^c \tag{7.10}$$

which is represented schematically in Figure 7.16.

The transition-fatigue life, $2N_t$, obtained when the elastic and plastic

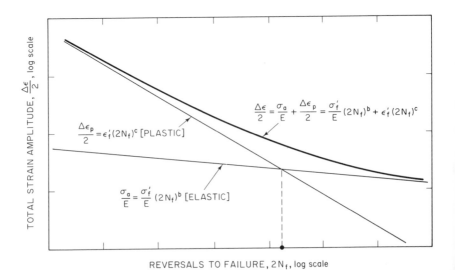

**Figure 7.16**  Total strain—life plot.

components of the total strain are equal (i.e., $\Delta\varepsilon_p/\Delta\varepsilon_e = 1$) is given by the relation

$$2N_t = \left(\frac{E\varepsilon_f'}{\sigma_f'}\right)^{1/(b-c)} \tag{7.11}$$

Equations (7.10) and (7.11) and Figure 7.16 show that the total fatigue lives less than $2N_t$ are governed primarily by the plastic-strain amplitude, whereas the total fatigue lives greater than $2N_t$ are governed primarily by the elastic-strain amplitude. Moreover, it is claimed that the fatigue-ductility coefficient, $\varepsilon_f'$, and the fatigue-strength coefficient, $\sigma_f'$, determined from regression analysis of fatigue data, can be approximated by the true-fracture ductility, $\varepsilon_f$, and the true-fracture strength, $\sigma_f$, respectively, obtained from a monotonic tension test.[9]* However, only general trends have been observed. These trends indicate that low-cycle fatigue life ($2N < 2N_t$) is governed primarily by the ductility of the material, whereas the high-cycle fatigue life ($2N_f > 2N_t$) is governed primarily by the static strength of the material.

True stresses ($\sigma$) and true strains ($\varepsilon$) are used in the foregoing relationships; however, engineering stresses ($S$) and engineering strains ($e$) are obtained from the hysteresis loops and may be converted as follows,

$$\sigma = S(1 + e) \tag{7.12}$$

$$\varepsilon = \ln(1 + e) \tag{7.13}$$

provided that uniform elongation is occurring. Since the engineering strains of the hysteresis loops are small ($e \leq 10^{-2}$), the error in assuming that $\sigma = S$ and $\varepsilon = e$ is small and is therefore usually neglected.

**b.    Fatigue-crack-propagation tests.**    Most fatigue-crack-propagation tests are conducted by subjecting a fatigue-cracked specimen to cyclic-load fluctuations. Figure 7.17 shows a specimen that has been used extensively to obtain fatigue-crack-propagation data.

Incremental increase of crack length is measured visually at low magnification, or ultrasonically, or by using electrical potential methods, and the corresponding number of elapsed load cycles is recorded. The data are presented as a linear plot of crack length, $a$, and the corresponding total number of load cycles, $N$, as shown in Figure 7.18. An increase in the magnitude of cyclic-load fluctuation results in a decrease of fatigue life of specimens having identical geometry (Figure 7.19). Furthermore, the fatigue life of specimens subjected to a fixed constant-amplitude cyclic-load fluctuation decreases as the length of the initial crack is increased (Figure 7.20). Consequently, under a given constant-amplitude stress fluctuation,

---

* There is no standard method of determining the true-fracture properties from monotonic tension tests of specimens with rectangular cross sections; various investigators use different methods.

**Figure 7.17** Compact-tension specimen used to determine propagation fatigue behavior (specimen length is about 5 in.). (From N. Willems, J. T. Easley, and S. T. Rolfe, *Strength of Materials*, McGraw-Hill, New York, 1981, Figure 14.10.)

most of the useful cyclic life is expended when the crack length is very small. Various $a$ versus $N$ curves can be generated by varying the magnitude of the cyclic-load fluctuation and/or the size of the initial crack. These curves reduce to a single curve when the data are represented in terms of crack-growth rate per cycle of loading, $da/dN$, and the fluctuation of the stress-intensity factor, $\Delta K_I$, because $\Delta K_I$ is a single-term parameter that incorporates the effect of changing crack length and cyclic-load magnitude. The parameter $\Delta K_I$ is representative of the mechanical driving force and is independent of geometry. The most commonly used presentation of fatigue-crack-growth data is a log-log plot of the rate of fatigue-crack growth per cycle of load fluctuation, $da/dN$, and the fluctuation of the stress-intensity factor, $\Delta K_I$.

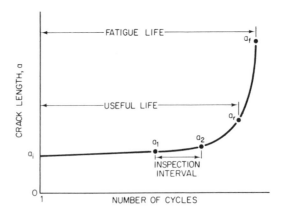

**Figure 7.18** Schematic representation of fatigue-crack-growth curve under constant-amplitude loading.

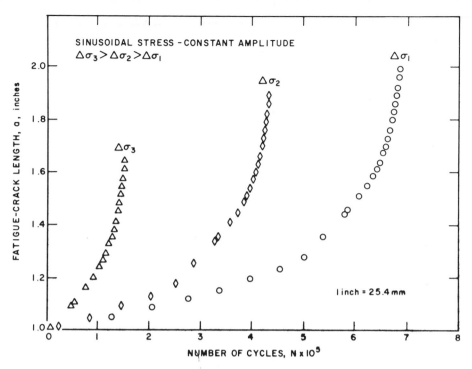

**Figure 7.19**   Effect of cyclic-stress range on crack growth.

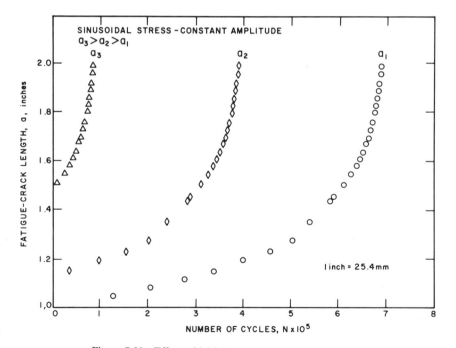

**Figure 7.20**   Effect of initial crack length on crack growth.

### 7.4.2. Tests of Actual or Simulated Structural Components

Ideally, fatigue tests are conducted on actual structural components under conditions that closely simulate the loading and environment for the actual structure. Figure 7.21 is a photograph of a full-scale fatigue test of a 25-ft-long welded beam subjected to four-point loading. Data obtained from such tests can be used directly in design. However, these tests are difficult to conduct, time consuming, and usually very costly. Consequently, simple specimens that simulate structural components such as shown in Figure 7.22 are more commonly tested. Generally, these specimens are tested to failure to obtain information on the total fatigue life of the components. Test data from simulated and actual welded structural components are presented in Chapter 14.

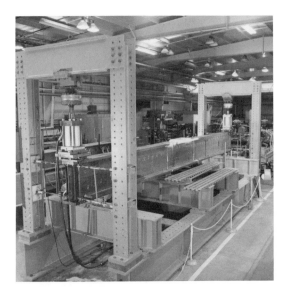

**Figure 7.21**  Full-scale fatigue test of welded beam used as structural component. Beam test setup (specimen length is about 25 ft.). (From N. Willems, J. T. Easley, and S. T. Rolfe, *Strength of Materials*, McGraw-Hill, New York, 1981, Figure 14.12.)

## 7.5. Some Characteristics of Fatigue Cracks

Initiation and propagation of fatigue cracks are caused by localized cyclic-plastic deformation. A fatigue crack initiates more readily and propagates more rapidly as the magnitude of the local cyclic-plastic deformation increases. Thus, a smooth component that is free of imperfections and geometrical discontinuities does not fail by fatigue when subjected to elastic-stress (strain) fluctuations. However, when the same component is subjected to

**Figure 7.22** Fatigue specimen used to determine fatigue behavior of welded attachment (specimen length is about 2 ft). (From N. Willems, J. T. Easley, and S. T. Rolfe, *Strength of Materials*, McGraw-Hill, New York, 1981, Figure 14.11.)

stress ranges approximately equal to or larger than the yield strength of the material, the plastic deformation causes the component to deform along slip planes that coincide with maximum shear stress—about 45° from the direction of load in the case of axial loading—which results in slip steps on the surface. These steps correspond to stress raisers that become the nucleation sites for fatigue cracks which initiate along the maximum shear planes and propagate normal to the maximum tensile-stress component. Because the initial surface was free of stress raisers, the plastically induced stress raisers can occur at various locations on the surface, resulting in possible initiation of multiple fatigue cracks as shown in Figure 7.23. The stochastic character of this process results in the possible initiation of multiple cracks and their propagation at different rates until one reaches a critical size causing failure of the component under the applied loads, Figure 7.23.

The probability for fatigue cracks to initiate at identical stress raisers subjected to the same stress range is equal. Consequently, multiple fatigue cracks may initiate in a uniformly loaded region of a component that contains stress raisers of equal severity.

Load distribution (e.g., three-point loading), imperfections (e.g., gouges), geometrical changes (e.g., change in cross section), as well as other factors can localize the initiation and propagation of fatigue cracks. Furthermore, the geometry of the component and the type of loading can significantly affect the location of fatigue-crack initiation, the rate of crack propagation, and the shape of the propagating crack. Figure 7.24[10] is a schematic representation of these observations for smooth and notched components with round, square, and rectangular cross sections under various loading conditions.

In most structural components, fatigue cracks initiate and propagate

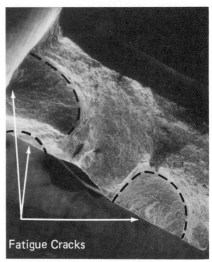

Figure 7.23   Multiple fatigue-crack initiation in a smooth specimen.

from stress raisers.  The stress raisers in unwelded components can be either surface imperfections or geometrical changes.  In welded components, the stress raisers can be embedded imperfections, such as gas pockets and entrapped slag, or weld terminations and weld toes, or geometrical changes.  The effects of these stress raisers on the fatigue behavior of weldments are discussed in Chapter 14.

A fatigue crack propagates normal to the primary tensile-stress component trying to assume a constant stress-intensity factor, $K$ (or strain-energy-release rate, $G$), along the entire crack front.  Thus, a fatigue crack that initiates from a point source in a uniform uniaxial tensile field propagates as a penny-shaped crack if the source is embedded and as a semicircular part-through crack if the source is on the surface, Figure 7.25.  A crack that initiates from an irregularly shaped imperfection in a uniaxial tensile field propagates at different rates to achieve the circular crack-front shape, Figure 7.26.

A constant-$K$ crack front in a pure bending stress field is a straight line as in a single-edge-notch specimen.  Thus, a fatigue crack that initiates from a point source on the tension surface of a specimen in bending initially assumes a semicircular shape and then propagates more rapidly along the

Figure 7.24   Schematic representation of marks on surfaces of fatigue fractures produced in smooth and notched components with round, square and rectangular cross sections, and in thick plates, under various loading conditions at high and low nominal stress. (From *Metals Handbook*, Vol. 10, 8th edition, American Society for Metals, Metals Park, OH 44073, 1975, p. 102. With permission.).

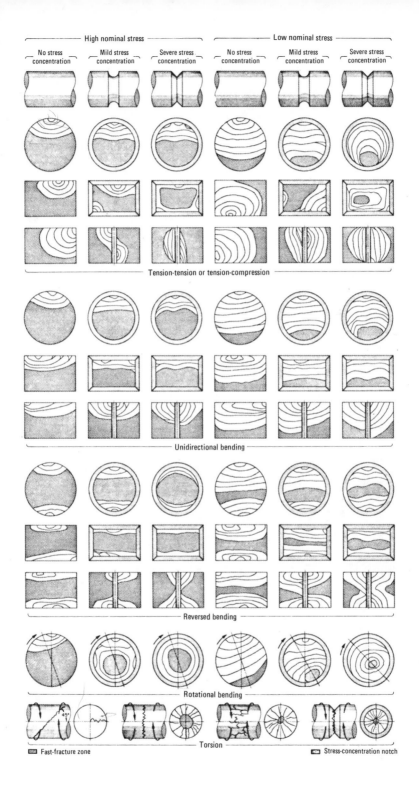

High nominal stress | Low nominal stress

No stress concentration | Mild stress concentration | Severe stress concentration | No stress concentration | Mild stress concentration | Severe stress concentration

Tension-tension or tension-compression

Unidirectional bending

Reversed bending

Rotational bending

Torsion

Fast-fracture zone | Stress-concentration notch

Figure 7.25   Semicircular surface crack propagating from a single source.

tension surface than in the depth direction until it reaches constant-$K$ straight-line crack front. The relative rate for crack propagation along the surface and in the depth direction depend on the stress gradient along the depth such that the higher the stress gradient, the slower the relative propagation in the depth direction. Fatigue cracks in complex stress fields like those possible when nominal, residual, and thermal stresses are superimposed can exhibit complex fronts with widely varying propagation rates at different points and at the same relative point for different crack sizes.

Figure 7.26   Growth of crack from ir-regularly shaped porosity maintaining a penny-shaped crack front.

Visually, a fatigue crack on a fracture surface appears smoother than the surrounding fracture regions, Figure 7.23. Microscopically, the fatigue region exhibits striations, Figure 7.27, which correspond to crack extension with each cycle of load. Usually, these striations are more distinct for aluminum than for steel.

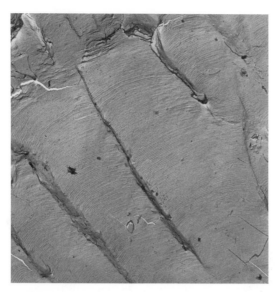

**Figure 7.27**    Fatigue striations on a fracture surface of an aluminum alloy.

Figure 7.28 is a scanning electron micrograph of a fatigue surface in A36 steel. Although the surface indicates the presence of striations, some of the striation-like features are caused by the composite properties of ferrite-pearlite steels.[11] Ferrite-pearlite steels behave as particulate composites such that the path of least resistance to fatigue-crack growth is through the

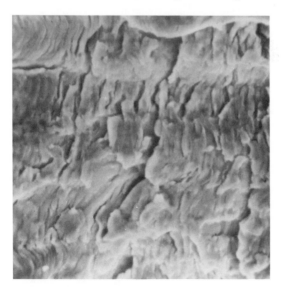

**Figure 7.28**    Scanning electron micrograph of the fracture surface in A36 steel, X2500.

ferrite matrix and the pearlite colonies tend to retard crack growth. Figure 7.29 is a light micrograph of a portion of the fatigue surface shown in Figure 7.28. The micrograph shows that plastic deformation under cyclic loading is more extensive in the ferrite matrix than in the pearlite colonies. It also shows that secondary fatigue cracks preferentially seek to propagate around a pearlite colony rather than through it. Such secondary cracks make it very difficult to delineate the true striations on the fatigue surface.

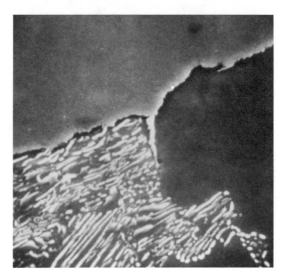

**Figure 7.29** Scanning electron micrograph of the fracture profile in A36 steel, X6000.

Finally, an aggressive environment may eliminate the striation from the surface of a corrosion fatigue crack, thus making it very difficult to distinguish between corrosion-fatigue crack extension under cyclic loading and stress-corrosion crack extension under static loading.

## References

1. R. CAZAUD, *Fatigue of Metals*, Chapman & Hall, London, 1953.
2. *Structural Steel Design*, edited by L. Tall, L. S. Beedle, and T. V. Galambos, Ronald Press, New York, 1964.
3. R. L. BROCKENBROUGH and B. G. JOHNSTON, *Steel Design Manual*, United States Steel Corporation, ADUSS 27-3400-04, Jan. 1981.
4. N. WILLEMS, J. T. EASLEY, and S. T. ROLFE, *Strength of Materials*, McGraw-Hill Book Company, New York, 1981.
5. L. F. COFFIN, JR., "Fatigue at High Temperature," *Fatigue at Elevated Temperatures, ASTM STP 520*, American Society for Testing and Materials, Philadelphia, 1973, pp. 5–34.

6. D. F. SOCIE and J. D. MORROW, "Review of Contemporary Approaches to Fatigue Damage Analysis," *Fracture Control Program Report No. 24*, College of Engineering, Urbana, Ill., Dec. 1976.

7. ASTM Standards, Section 3, Vol. 03.01, *Metals—Mechanical Testing, Elevated and Low-Temperature Tests*, 1983.

8. D. T. RASKE and J. D. MORROW, "Mechanics of Materials in Low Cycle Fatigue Testing," *Manual on Low Cycle Fatigue Testing, ASTM STP 465*, 1969, pp. 1–25.

9. *Fatigue Design Handbook, 4*, edited by J. A. Graham, Society of Automotive Engineers, Warrendale, Pa., 1968.

10. *Metals Handbook*, Vol. 10, 8th ed., American Society for Metals, Metals Park, Ohio, 1975.

11. J. M. BARSOM, "Fatigue-Crack Propagation in Steels of Various Yield Strengths," *Transactions of the ASME, Journal of Engineering for Industry*, Series B, 93, No. 4, Nov. 1971.

# Fatigue-Crack Initiation

## 8.1. General Background

The fatigue life of structural components is determined by the sum of the elapsed cycles required to initiate a fatigue crack and to propagate the crack from subcritical dimensions to the critical size. Consequently, the fatigue life of structural components may be considered to be composed of three continuous stages: (1) fatigue-crack initiation, (2) fatigue-crack propagation, and (3) fracture. The fracture stage represents the terminal conditions (i.e., the particular combination of $\sigma$, $a$, and $K_{Ic}$) in the life of a structural component, as described in Chapter 6. The useful life of cyclically loaded structural components can be determined only when the three stages in the life of the component are evaluated individually and the cyclic behavior in each stage is thoroughly understood.

Conventional procedures used to design structural components subjected to fluctuating loads provide a design fatigue curve which characterizes the basic unnotched fatigue properties of the material and a fatigue-strength-reduction factor. The fatigue-strength-reduction factor incorporates the effects of all the different parameters characteristic of the specific structural component that make it more susceptible to fatigue failure than the unnotched specimen, such as surface finish, geometry, defects, and so on. The design fatigue curves are based on the prediction of cyclic life from data on nominal stress (or strain) versus elapsed cycles to failure ($S$–$N$ curves), as determined from laboratory specimens. Such data are usually obtained by testing unnotched specimens and represent the number of cycles required to initiate a crack in the specimen plus the number of cycles required to propagate the crack from a subcritical size to a critical dimension. The dimensions of the critical crack required to cause failure depend on the magnitude of the applied stress and on the specimen size, as well as on the particular testing conditions used.

Figure 8.1 is a schematic $S–N$ curve divided into an initiation component and a propagation component. The number of cycles corresponding to the endurance limit represents initiation life primarily, whereas the number of cycles expended in crack initiation at a high value of applied alternating stress is negligible. As the magnitude of the applied alternating stress increases, the total fatigue life decreases and the percentage of the fatigue life to crack initiation decreases. Consequently, $S–N$–type data do not provide complete information regarding safe-life predictions in structural components, particularly in components having surface irregularities different from those of the test specimens and in components containing crack-like imperfections, because the existence of surface irregularities and crack-like imperfections reduces and may eliminate the crack-initiation portion of the fatigue life of structural components.

Many attempts have been made to characterize the fatigue behavior of metals.[1-3] The results of some of these attempts have proved invaluable in the evaluation and prediction of the fatigue strength of structural components. However, these fatigue-strength-evaluation procedures are subject to limitations, caused primarily by the failure to distinguish adequately between fatigue-crack initiation and fatigue-crack propagation.

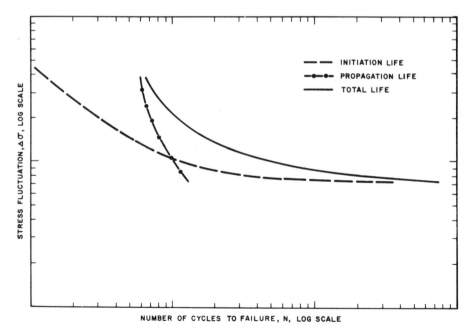

**Figure 8.1**   Schematic $S–N$ curve divided into initiation and propagation components.

## 8.2. Fracture-Mechanics Methodology

The development of fracture-mechanics methodology offers considerable promise in improving our understanding of fatigue-crack initiation, fatigue-crack propagation, and unstable crack propagation and in solving the problem of designing to prevent failures caused by fatigue. As described in Chapter 2, linear-elastic fracture-mechanics technology is based on an analytical procedure that relates the stress-field magnitude and distribution in the vicinity of a crack tip to the nominal stress applied to the structure, to the size, shape, and orientation of the crack or crack-like imperfection, and to the material properties. Figure 8.2 presents the $\sigma_x$ and $\sigma_y$ components of the elastic-stress field in the vicinity of a crack tip in a body subjected to tensile stresses normal to the plane of the crack (Mode I deformation).[4] These stress-field equations show that the distribution of the elastic-stress field in the vicinity of the crack tip is invariant in all structural components subjected to Mode I deformation and that the magnitude of the elastic-stress field can be described by a single parameter, $K_I$, designated the stress-intensity factor. Consequently, the applied stress, the crack shape, size, and orientation, and the structural configuration associated with structural components subjected to Mode I deformation affect the value of the stress-intensity factor but do not alter the stress-field distribution. Relationships

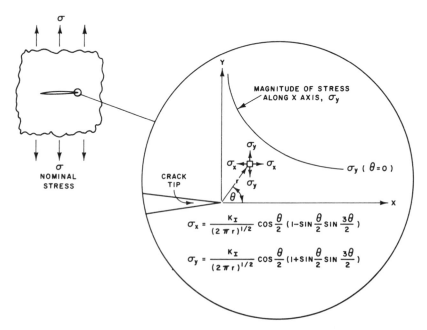

$$\sigma_x = \frac{K_I}{(2\pi r)^{1/2}} \cos\frac{\theta}{2}\left(1-\sin\frac{\theta}{2}\sin\frac{3\theta}{2}\right)$$

$$\sigma_y = \frac{K_I}{(2\pi r)^{1/2}} \cos\frac{\theta}{2}\left(1+\sin\frac{\theta}{2}\sin\frac{3\theta}{2}\right)$$

**Figure 8.2**  Schematic illustration of the elastic stress-field distribution near the tip of a fatigue crack (Mode I deformation).

between the stress-intensity factor and various body configurations, crack sizes, shapes, and orientations, and loading conditions have been established,[5,6] and the more widely used ones were described in Chapter 2.

The critical-stress-intensity factor, $K_c$, $K_{Ic}$, or $K_{Id}$, represents the terminal conditions in the life of a structural component. The total useful life of the component is determined by the time necessary to initiate a crack and to propagate the crack from subcritical dimensions to the critical size, $a_c$. Crack initiation and subcritical crack propagation may be caused by cyclic stresses in the absence of an aggressive environment (fatigue), by an aggressive environment under sustained load (stress-corrosion cracking), or by the combined effects of cyclic stresses and an aggressive environment (corrosion fatigue). All these modes of crack initiation and subcritical crack propagation are localized phenomena that depend on the stress-field intensity at the tip of the notch or crack. Sufficient data are available to show that the rate of subcritical crack growth depends on the stress-intensity factor, $K_I$, which serves as a single-term parameter representative of the stress conditions in the vicinity of the crack tip.[7-12] The use of fracture-mechanics parameters to study the effect of stress concentration on fatigue-crack initiation is presented in this chapter.

## 8.3. Stress Field in the Vicinity of Stress Concentrations

The elastic-stress field in the vicinity of sharp elliptical or hyperbolic notches in a body subjected to tensile stresses normal to the plane of the notch is represented by the following equations:[13]

$$\sigma_x = \frac{K_I}{(2\pi r)^{1/2}} \cos\frac{\theta}{2}\left[1 - \sin\frac{\theta}{2}\sin\frac{3\theta}{2}\right] - \frac{K_I}{(2\pi r)^{1/2}}\frac{\rho}{2r}\cos\frac{3\theta}{2}$$

$$\sigma_y = \frac{K_I}{(2\pi r)^{1/2}} \cos\frac{\theta}{2}\left[1 + \sin\frac{\theta}{2}\sin\frac{3\theta}{2}\right] + \frac{K_I}{(2\pi r)^{1/2}}\frac{\rho}{2r}\cos\frac{3\theta}{2} \qquad (8.1)$$

$$\tau_{xy} = \frac{K_I}{(2\pi r)^{1/2}} \sin\frac{\theta}{2}\cos\frac{\theta}{2}\cos\frac{3\theta}{2} - \frac{K_I}{(2\pi r)^{1/2}}\frac{\rho}{2r}\sin\frac{3\theta}{2}$$

where the coordinates $r$, $\theta$, and $\rho$ are defined in Figure 8.3. The first term in Equations (8.1) defines the magnitude and distribution of the stress field in the vicinity of a fatigue crack. In these equations $K_I$ is the crack-tip stress-intensity factor for a sharp crack of length equal to that of the sharp notch and subjected to the same loading conditions as the notch. The second term in these equations represents the influence of a blunt-tip radius on this stress field. Equations (8.1) also show that on the crack center plane, the stress singularity for narrow elliptical and hyperbolic notches is centered on a line located at $\rho/2$ behind the crack front (Figure 8.3).

Notches in structural components cause stress intensification in the vicinity of the notch tip. The material element at the tip of a notch in a

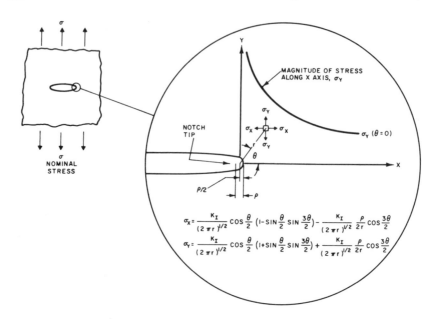

**Figure 8.3**   Schematic illustration of the elastic-stress-field distribution near the tip of an elliptical notch (Mode I deformation).

cyclically loaded structural component is subjected to the maximum stress, $\sigma_{max}$, and to the maximum stress fluctuations, $\Delta\sigma_{max}$. Consequently, this material element is most susceptible to fatigue damage and is, in general, the origin of fatigue-crack initiation. For $r = \rho/2$, the maximum stress on this material element which can be derived from Equations (8.1) is

$$\sigma_{max} = k_t\sigma = \frac{2}{\sqrt{\pi}}\frac{K_I}{\sqrt{\rho}} \tag{8.2}$$

and the maximum stress range is

$$\Delta\sigma_{max} = k_t(\Delta\sigma) = \frac{2}{\sqrt{\pi}}\frac{\Delta K_I}{\sqrt{\rho}} \tag{8.3}$$

where $k_t$ is the stress-concentration factor for the notch. Although these equations are considered exact only when $\rho$ approaches zero, Wilson and Gabrielse[14] showed by using finite element analysis of relatively blunt notches in compact-tension specimens, where the notch length, $a_n$, was much larger than $\rho$, that these relationships are accurate to within 10 percent for notch radii up to 0.18 in.

Equation (8.3) suggests the possible correlation of the fatigue-crack-initiation life for notches with either $\Delta\sigma_{max}$ or $\Delta K_I/\sqrt{\rho}$. This observation is valid within limits of applicability of the equation. Equation 8.3 is accurate for relatively sharp notches. Figure 8.4 shows four geometries that were

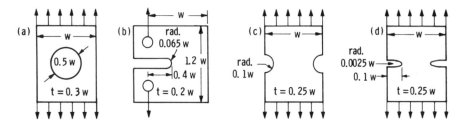

**Figure 8.4**  Notched specimens. (a) Center hole, (b) compact, (c) blunt double-edge notch, and (d) sharp double-edge notch. In each, $w$ = 25.5 mm, $t$ = thickness. (Reprinted with permission. © Society of Automotive Engineers, Inc. SAE Paper No. 820691, 1982)

studied by Dowling.[15]  The corresponding actual stress-concentration factor, $k_t$, and the stress-concentration factor, $k_{fm}$, obtained by using the fracture-mechanics relationship in Equation 8.2, that is, $k_{fm} = (2/\sqrt{\pi})(K_1/\sqrt{\rho})/\sigma$ are presented in Table 8.1.  The data show that the $k_t$ and $k_{fm}$ values differ significantly for the two very blunt notches and are very close for notches with $a_n > \rho$.

TABLE 8.1    Stress-Concentration Factors for the Notches Shown in Figure 8.4.[15]

| Specimen | $k_t$ | $k_{fm}$ |
|---|---|---|
| Center hole | 2.12 | 1.19 |
| Compact | 2.62 | 2.41 |
| Blunt DEN | 2.42 | 1.63 |
| Sharp DEN | 10.7 | 10.3 |

Thorough understanding of fatigue-crack initiation requires the development of accurate predictions of the localized stress and strain behavior in the vicinity of stress concentrations.  Recent developments in elastic-plastic finite-element stress analysis in the vicinity of notches should contribute significantly to the development of quantitative predictions of fatigue-crack-initiation behavior for structural components.[16,17]

## 8.4.  Effect of Stress Concentration on Fatigue-Crack Initiation

The effect of a geometrical discontinuity in a loaded structural component is to intensify the magnitude of the nominal stress in the vicinity of the discontinuity.  The localized stresses may cause the metal in that neighborhood to undergo plastic deformation.  Because the nominal stresses in most structures are elastic, the zone of plastically deformed metal in the vicinity of stress concentrations is surrounded by an elastic-stress field. The deformations (strains) of the plastic zone are governed by the elastic displacements of the surrounding elastic-stress field. In other words, when the structure is stress controlled, the localized plastic zones are strain controlled. Consequently, to predict the effects of stress concentrations on the fatigue behavior of structures, the fatigue behavior of the localized plastic zones has been simulated by testing smooth specimens under strain-controlled conditions, Figure 8.5(a). A better simulation of the effects of stress concentrations on the fatigue behavior of structures is obtained by testing notched specimens under stress-controlled conditions, Figure 8.5(b), because the applied stress can be more directly related to the structural loading.

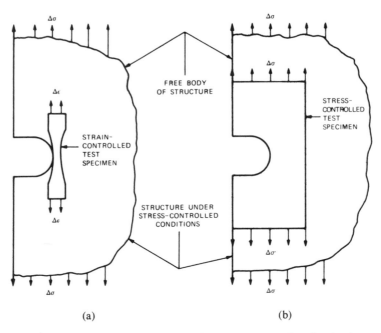

(a)                                              (b)

**Figure 8.5** Test specimen simulation of stress concentrations in structures. (a) Strain-controlled specimen; (b) stress-controlled specimen.

The stress analyses presented in the preceding section facilitated the use of linear-elastic fracture-mechanics parameters in analyzing the fatigue-crack-initiation behavior of notched specimens. The applicability of these parameters to sharp notches has been demonstrated in limited work by Forman,[18] by Constable et al.,[19] and by Jack and Price.[20] Moreover, the applicability of these parameters to sharp and blunt notches has been verified by Barsom and McNicol[21] and by Clark.[22] The fatigue-initiation behavior of HY-130 steel specimens (Figure 8.6) tested by Barsom and McNicol[21] under axial zero-to-tension cyclic loads is presented in Figure 8.7 in terms

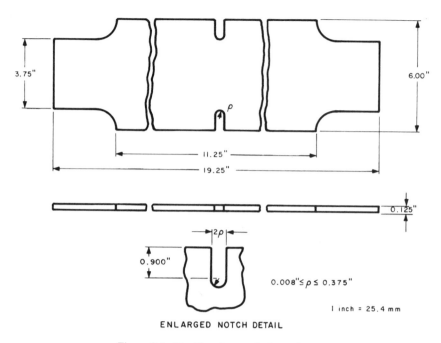

**Figure 8.6**    Double-edge-notched specimens.

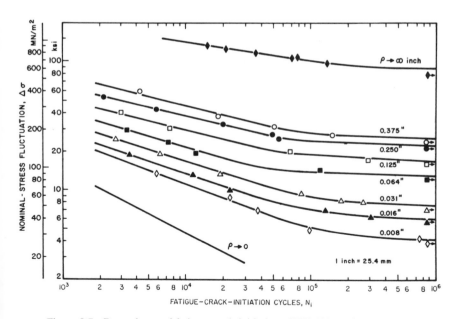

**Figure 8.7**    Dependence of fatigue-crack initiation of HY-130 steel on nominal-stress fluctuations for various notch geometries.

of the number of cycles for fatigue-crack initiation, $N_i$, versus the nominal stress fluctuation, $\Delta\sigma$, and in Figure 8.8 in terms of $N_i$ versus the stress-intensity-factor fluctuation divided by the square root of the notch radius, $\Delta K_I/\sqrt{\rho}$, for all the data presented in Figure 8.7. The stress-intensity factor, $K_I$, for notches was calculated as for a fatigue crack of length equal to the total notch depth. Figure 8.8 shows that a fatigue-crack-initiation threshold, $\Delta K_I/\sqrt{\rho})_{th}$, in HY-130 steel specimens subjected to zero-to-tension axial loads occurs at $K_I/\sqrt{\rho} \approx 85$ ksi (586 MN/m$^2$).

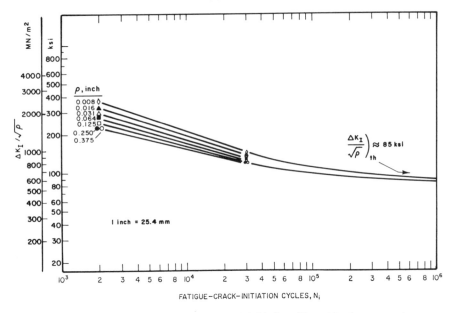

**Figure 8.8** Correlation of fatigue-crack-initiation life with the parameter $\Delta K_I/\sqrt{\rho}$ for HY-130 steel.

The maximum elastic stress at the root of the notch, $\sigma_{max}$, is equal to $2K_I/\sqrt{\pi\rho}$. Despite the fact that the term $\Delta K_I/\sqrt{\rho}$ loses its significance for an unnotched, polished specimen, let us assume that the fatigue-crack-initiation limit in such specimens occurs when $\Delta K_I/\sqrt{\rho} = 85$ ksi. Substituting this into Equation (8.3) shows that the fatigue-crack-initiation threshold of unnotched, polished HY-130 steel specimens tested at a stress ratio, $R$, equal to 0.1 occurs when $\Delta\sigma_{max}$ is equal to 96 ksi (662 MN/m$^2$). This value agrees quite well with the experimental data obtained by testing unnotched, polished specimens (Figure 8.7). The agreement between the calculated value and the experimental test results may be fortuitous, and further work is necessary to establish its significance.

The effect of stress concentration on the fatigue-crack-initiation behavior of type 403 stainless steel having 93.5-ksi yield strength and 110-ksi tensile strength has been studied by Clark.[22] The data (Figure 8.9), which was

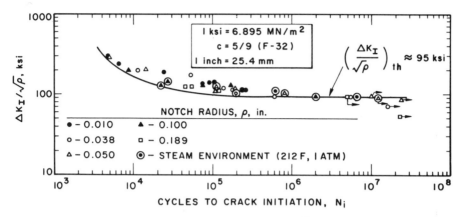

Figure 8.9   Cycles to initiation as a function of $\Delta K/\sqrt{\rho}$ for type 403 stainless steel in air and steam environments (Ref. 22).

obtained by testing notches in compact-tension specimens, show that the number of load fluctuations required to initiate a fatigue crack in the vicinity of a notch tip is related to $\Delta K_I/\sqrt{\rho}$, which in turn is related to the maximum alternating stress in the vicinity of the notch tip. The fatigue-crack-initiation threshold, $\Delta K_I/\sqrt{\rho})_{th}$, in type 403 stainless steel occurred at about 95 ksi.

The fatigue-crack-initiation behavior in various steels was investigated by Barsom and McNicol[23] in three-point bending at a stress ratio, $R$ (ratio of nominal minimum applied stress to nominal maximum applied stress), equal to +0.1. The data were obtained by testing single-edge-notched specimens having a notch that resulted in a stress concentration of about 2.5 (Figure 8.10). The fatigue-crack-initiation behavior of all the specimens tested is presented in Figure 8.11 in terms of the number of cycles for fatigue-crack initiation, $N_i$, versus the ratio of the stress-intensity-factor

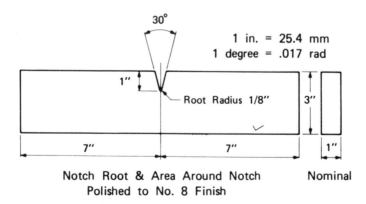

Figure 8.10   Single-edge-notched bend test specimen.

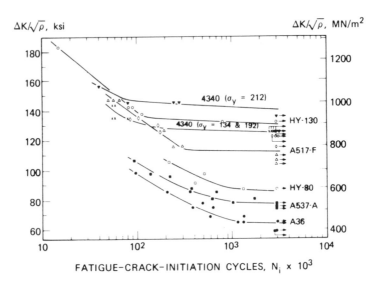

**Figure 8.11**  Fatigue-crack-initiation behavior of various steels.

fluctuation to the square root of the notch-tip radius, $\Delta K_I / \sqrt{\rho}$. Because the same nominal notch length and tip radius were used for all specimens of the steels investigated, the differences in the fatigue-crack-initiation behavior shown in Figure 8.11 are related primarily to inherent differences in the fatigue-crack-initiation characteristics of the steels. The data show that fatigue cracks do not initiate in steel structural components when the body configuration, the notch geometry, and the nominal-stress fluctuations are such that the magnitude of the parameter $\Delta K_I / \sqrt{\rho}$ is less than a given value that is characteristic of the steel. In general, the value of this fatigue-crack-initiation threshold, $\Delta K_I / \sqrt{\rho})_{th}$, increased with increased yield strength or tensile strength of the steel.

## 8.5.  Dependence of Fatigue-Crack-Initiation Threshold on Tensile Properties

As pointed out earlier, high-cycle fatigue behavior, that is, the endurance limit, represents primarily fatigue-crack-initiation behavior (Figure 8.1). Sufficient data are available to correlate endurance limits and tensile strengths of steels.[3] Relations between $\Delta K / \sqrt{\rho}$ and various tensile mechanical properties exist as follows.

### 8.5.1.  Tensile Strength

The steels presented in Figure 8.10 have widely different chemical compositions and mechanical properties. The room temperature tensile

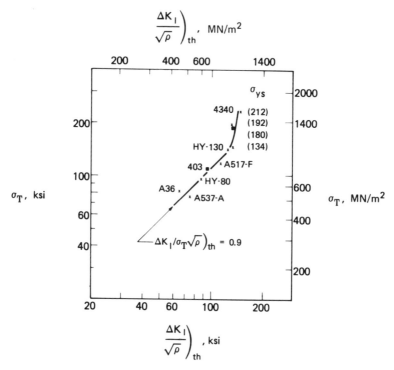

**Figure 8.12**   Dependence of fatigue-crack-initiation threshold on tensile strength.

strengths ranged from 77 to 233 ksi (531 to 1606 MN/m$^2$). These are plotted against the corresponding fatigue-crack-initiation threshold, $\Delta K_I/\sqrt{\rho})_{th}$, in Figure 8.12. The figure also includes data obtained by Clark[24] on type 403 stainless steel and on AISI 4340 steel heat-treated to a yield strength of 180 ksi (1241 MN/m$^2$). The data suggest that for steels having a tensile strength less than about 150 ksi (1034 MN/m$^2$), a given increase in the tensile strength corresponded, approximately, to an equal increase in the magnitude of $\Delta K_I/\sqrt{\rho})_{th}$. On the other hand, for steels having tensile strengths greater than about 150 ksi, an increase of about 100 ksi (689 MN/m$^2$) in the value of the tensile strength resulted in only a 20-ksi (138-MN/m$^2$) increase in the value of $\Delta K_I/\sqrt{\rho})_{th}$. This observation suggests that for high-strength ($\sigma_T > 150$ ksi) steels, the fatigue-crack-initiation threshold is, essentially, independent of tensile strength.

The data presented in Figure 8.12 indicate that, under three-point bending at a stress ratio of 0.1, an approximate value of fatigue-crack-initiation threshold in steels having tensile strengths in the range of about 70 ksi (483 MN/m$^2$) to 150 ksi can be obtained by using

$$\left.\frac{\Delta K_I}{\sigma_T\sqrt{\rho}}\right)_{th} = 0.9 \qquad (8.4)$$

However, for steels having tensile strengths greater than 150 ksi, the fatigue-crack-initiation-threshold values appear to be less than would be predicted by using Equation (8.4). Substituting Equation (8.3) into Equation (8.4) leads to the conclusion that the fatigue-crack-initiation threshold in steels of 150 ksi or less subjected to zero-to-tension bending stresses occurs when the maximum elastic stress calculated at the notch root is equal to the tensile strength of the steel.

Conventional fatigue-life data are usually obtained under fully reversed loading ($R = -1.0$) and show that, in general, the endurance limit occurs at an alternating stress, $\sigma_{alt}$, equal to about one-half of the tensile strength of the steel. The alternating stress is equal to one-half of the total stress range. Consequently, an endurance limit that occurs at an alternating stress of one-half of the tensile strength corresponds to an applied maximum-stress range equal to the tensile strength.

The preceding observations indicate that the fatigue-crack-initiation threshold of various steels is governed, primarily, by the total maximum-stress range.

### 8.5.2. Yield Strength

Barsom and McNicol[21] applied the nondimensional parameter $K_I/\sigma_{ys}\sqrt{\rho}$, where $\sigma_{ys}$ is the 0.2 percent offset yield strength, to normalize the fatigue-crack-initiation threshold behavior of various high-yield-strength martensitic steels. They suggested that the fatigue-crack-initiation threshold in various high-yield-strength martensitic steels is represented by

$$\frac{\Delta K_I}{\sigma_{ys}\sqrt{\rho}} = \text{constant} \tag{8.5}$$

Because the yield-to-tensile ratio in high-yield-strength steels is in the range $0.8 < \sigma_{ys}/\sigma_T < 1.0$, Equations (8.4) and (8.5) are very similar.

The yield strength, $\sigma_{ys}$, and the corresponding fatigue-crack-initiation threshold, $\Delta K_I/\sqrt{\rho}$, of the steels shown in Figure 8.11 are presented in Figure 8.13. The figure also includes data obtained by Clark[24] on type 403 stainless steel and on AISI 4340 steel heat treated to 180-ksi yield strength. The data indicate the existence of an excellent correlation of fatigue-crack-initiation threshold and yield strength in steels having yield strengths in the range of about 40–140 ksi (276–965 MN/m²). The correlation can be represented by

$$\frac{\Delta K_I}{(\sigma_{ys})^{2/3}\sqrt{\rho}} = 5 \tag{8.6}$$

where $\Delta K_I$ is in ksi$\sqrt{\text{in.}}$, $\sigma_{ys}$ is in ksi, and $\rho$ is in in. Tests conducted on steels of yield strengths greater than 140 ksi resulted in fatigue-crack-initiation threshold values less than would be predicted by using Equation (8.6).

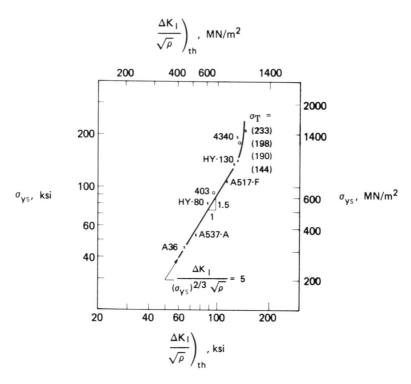

**Figure 8.13**   Dependence of fatigue-crack-initiation threshold on yield strength.

### 8.5.3. Strain-Hardening Exponent

Clausing[25] investigated the tensile properties of various steels and showed the existence of a correlation of strain-hardening exponent with yield strength that was better than a correlation with tensile strength. The relationship between the fatigue-crack-initiation threshold and the strain-hardening exponent of the steels shown in Figure 8.11 was studied. Figure 8.14 shows the plot of the fatigue-crack-initiation threshold and the strain-hardening exponent. Included in this figure are data for type 403 and AISI 4340 steels tested by Clark.[24] The strain-hardening exponent of these two steels was approximated by using the uniform-elongation values. The data show that the fatigue-crack-initiation threshold of the various steels can be predicted by using the strain-hardening exponents and that the relationship between these parameters is given by

$$\frac{\Delta K_1 \sqrt{n}}{\sqrt{\rho}} = 30 \tag{8.7}$$

Figure 8.14 shows that at low values of strain-hardening exponents, that

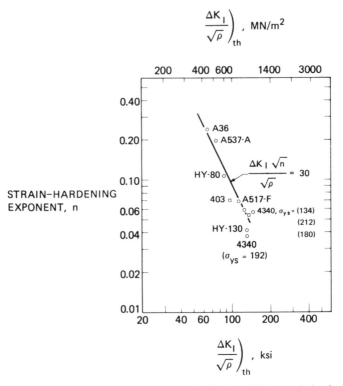

**Figure 8.14** Dependence of fatigue-crack-initiation threshold on strain-hardening exponent.

is, for steels of yield strengths greater than about 140 ksi, the accuracy of Equation (8.7) decreases because of increased data scatter.

### 8.5.4. General Discussion

The correlations represented in the preceding sections show that the fatigue-crack-initiation threshold in various steels is related to the yield strength and to the strain-hardening exponent and, to a lesser extent, to the tensile strength and that the tensile properties of steels are interrelated. The relationships between the tensile properties for various steels were studied by using published data[25-28] and were compared with the relationships suggested by combining Equations (8.4), (8.6), and (8.7). This comparison is shown in Figures 8.15, 8.16, and 8.17. The equations in these figures represent the interrelationships among tensile strength, yield strength, and strain-hardening exponent obtained by combining Equations (8.4), (8.6), and (8.7). These figures show good agreement between the developed relationships and published data of tensile properties for various steels.

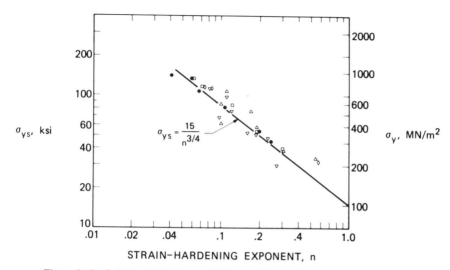

**Figure 8.15**   Relationship between yield strength and strain-hardening exponent.

The interrelationships among tensile properties of various steels were developed from data of martensitic steels and ferrite-pearlite steels. The applicability of these equations to austenitic stainless steels has not been established. However, because of the metastable behavior of some austenitic stainless steels and because stable austenitic stainless steels do not usually conform to the same stress-strain relationship as martensitic and ferrite-pearlite steels, the relationships shown in Figures 8.15, 8.16, and 8.17 may

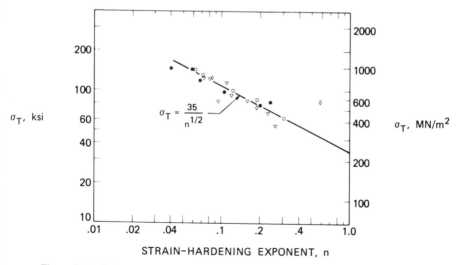

**Figure 8.16**   Relationship between tensile strength and strain-hardening exponent.

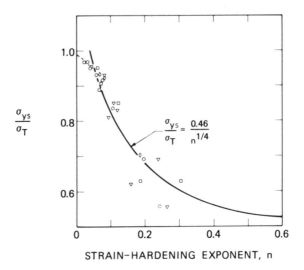

STRAIN-HARDENING EXPONENT, n

**Figure 8.17**  Relationship between yield-to-tensile ratio and strain-hardening exponent.

not be useful to predict the fatigue-crack-initiation-threshold behavior in austenitic stainless steels. Further research is necessary to establish the fatigue-crack-initiation-threshold characteristics of austenitic stainless steels and of other metal alloys.

## 8.6. Generalized Equation for Predicting the Fatigue-Crack-Initiation Threshold for Steels

The discussion in the preceding section showed that, for a stress ratio of 0.1, the fatigue-crack-initiation threshold for notched steel components can be related to the yield strength, tensile strength, or strain-hardening exponent. To develop a generalized equation for predicting the fatigue-crack-initiation threshold for steels, the effect of stress ratio, $R$, on the threshold value must be established. Consequently, the fatigue-crack-initiation behavior for steels having yield strengths between 36 and 110 ksi was studied by Barsom at stress ratios of $-1.0$, 0.1, and 0.5.[29] Analysis of the test results showed that stress ratio had a negligible effect, if any, on the fatigue-crack-initiation threshold from notches when the $\Delta K$ in the $\Delta K/\sqrt{\rho}$ parameter are based on the total stress (tension plus compression) range. A plot of the data in terms of the $\Delta K_{\text{total}}/\sqrt{\rho}$ values at the threshold and $R$ is shown in Figure 8.18[29,30] where $\Delta K_{\text{total}}$ is calculated by using the total stress (tension plus compression) range. The data in Figure 8.18 indicate that the fatigue-

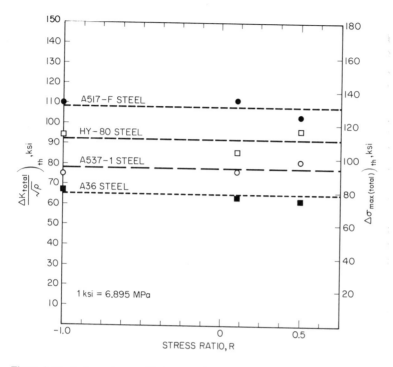

**Figure 8.18**   Independence of fatigue-crack-initiation threshold from stress ratio.

crack-initiation damage that is caused by a compressive-stress range is equal to the damage caused by a tensile-stress range having the same magnitude. Although cracks can initiate under compressive cyclic loading, they would propagate only through the plastically deformed region where inelastic tensile stresses reside in the plastic zone at the tip of the notch and arrest in the vicinity of its boundary. However, under tensile-cyclic loading, once initiated, the crack would propagate beyond the plastically deformed region, causing fracture of the specimen.

Analysis of the available data indicates that the fatigue-crack-initiation threshold for various steels subjected to stress ratios between $-1.0$ and $0.5$ can be estimated best by using the relationship[29,31]

$$\left. \frac{\Delta K_{total}}{\sqrt{\rho}} \right)_{th} = 10\sqrt{\sigma_{ys}} \tag{8.8}$$

where $\Delta K_{total}$ is the stress-intensity factor range, calculated by using the sum of the tensile- and compressive-stress ranges, and $\sigma_{ys}$ is the yield strength of the steel, Figure 8.19.[29,31]

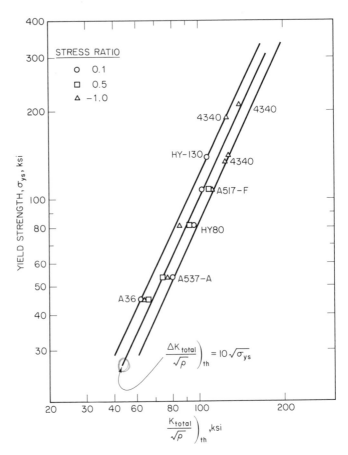

**Figure 8.19** Dependence of fatigue-crack-initiation threshold on yield strength. (From *Metals Handbook* Desk Edition, American Society for Metals, Metals Park, OH 44073, 1985, pp. 32–33. With permission.)

## 8.7. Finite-Initiation-Life Behavior

The fatigue-crack-initiation life of unnotched specimens is strongly dependent on the surface conditions of the specimen. Surface damage and surface irregularities can reduce the initiation life significantly because of the stress concentration. On the other hand, the fatigue-crack-initiation life of un-notched, polished specimens is related to the plastic deformation of the material. The plastic deformation causes the development of slip steps that become the nucleus of the fatigue-crack-initiation site. These observations are equally applicable to notched specimens. The fatigue-crack-initiation life of a notched specimen can be reduced significantly by surface irregularities in the vicinity of the notch tip. In the finite fatigue-crack-initiation-life region, the data presented in Figure 8.8 show that at a constant value of $\Delta K_I/\sqrt{\rho}$, $N_i$ is primarily a function of $1/\rho^2$ (Figure 8.20). This functional relationship indicates that, at a constant value of $\Delta K_I/\sqrt{\rho}$, the number of elapsed cycles required to initiate a fatigue crack at the tip of a notch in a specimen of unit thickness is inversely related to the volume of an element at the notch tip rather than to the surface area of the notch root. In other words, the fatigue-crack-initiation life in the specimens tested by Barsom and McNicol[21] decreased as the volume of the plastically deformed material at the notch tip increased. Because of the high quality of the surface of the notches that were tested, the surface did not affect the number of

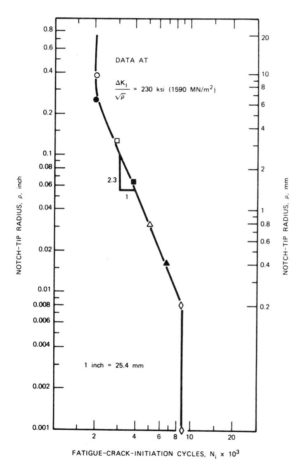

**Figure 8.20**  Dependence of fatigue-crack-initiation life on the notch-tip radius in the region of finite-initiation life of HY-130 steel.

elapsed cycles required to initiate a fatigue crack. The data presented in Figure 8.20 also suggest that the fatigue-crack-initiation life in the finite-life region is independent of notch-tip radius when $\rho$ is large (that is, when the volume of plastically deformed metal is large) and when $\rho$ is small. This behavior occurred in HY-130 steel when $\rho$ was greater than 0.25 in. (6.4 mm) ($k_t < 3$) and when $\rho$ was less than 0.008 in. (0.20 mm) ($k_t > 13$).

The fatigue-crack-initiation life of structural components is influenced by surface irregularities, by internal inclusions and other volumetric stress (strain) raisers, and by the volume of the plastically deformed metal. A mathematical model that depicts the dependence of fatigue-crack-initiation life on surface and volumetric characteristics was presented by Barsom and McNicol.[21] The model was based on a weakest-link theory and Weibull's distribution function.[32–35] It showed that fatigue-crack initiation in structural components can be represented by an areal-risk parameter, a volumetric-risk parameter, a zero-probability stress range, a flaw-density exponent, and a scale parameter.

The data presented in Figure 8.7 were obtained by testing the double-edge-notched specimens shown in Figure 8.6. The notch depth was equal to $a_0 + \rho$, where $a_0$ was a constant equal to 0.90 in. and $\rho$, the root radius, was different for different sets of specimens. Fatigue-crack-initiation life was defined as the number of cycles required to generate an 0.010-in. (0.25-mm) fatigue crack from the root of the notch. The curve for $\rho$ approaching zero (i.e., a fatigue crack) was obtained by calculating the number of cycles required to extend the fatigue crack from an initial length of 0.90 in. to a final length of 0.91 in. (This calculation is described in Chapter 9.) This curve was used by Barsom and McNicol[21] to investigate the transition from the fatigue-crack-initiation threshold, $\Delta K_I / \sqrt{\rho})_{th}$, to the fatigue-crack-propagation threshold, $\Delta K_{th}$, as $\rho$ approaches zero.

The crack-opening displacement, $\delta$, at the tip of a fatigue crack is given by (see Chapter 2)

$$\delta = \frac{K^2}{\sigma_{ys} E} \tag{8.9}$$

where   $K$ = stress-intensity factor, psi$\sqrt{\text{in.}}$
$\sigma_{ys}$ = yield strength, psi.
$E$ = Young's modulus, psi.

The value of $\delta$ at maximum load in a fatigue-cracked HY-130 steel specimen tested under 0–5-ksi$\sqrt{\text{in.}}$ fluctuation in the stress-intensity factor is equal to $0.6 \times 10^{-5}$ in. (152 mm). The value of the tip radius is on the order of $\delta/2$, or $\rho = 0.3 \times 10^{-5}$ in. (76 mm). Substituting this value of $\rho$ in the expression $\Delta K_I / \sqrt{\rho})_{th} = 85$ ksi for HY-130 steel (Figure 8.8) results in a $\Delta K$ value for $R = 0.1$ of 0.15 ksi$\sqrt{\text{in.}}$ (0.16 MN/m$^{3/2}$). Such a small $\Delta K$ value suggests that a preexisting fatigue crack would propagate to failure under the smallest applied stress range, thus negating the existence of a fatigue-crack-propagation threshold, $\Delta K_{th}$, for this steel. The existence of a fatigue-crack-propagation threshold (a stress-intensity-factor fluctuation below which existing fatigue cracks do not propagate under cyclic loading) is presented in Chapter 9, and the data show that, for $R \approx 0$, the $\Delta K_{th}$ for various constructional steels is equal to about 5 ksi$\sqrt{\text{in.}}$

Let us assume that a fatigue-crack-propagation threshold in HY-130 steel does occur under zero-to-tension loads at $\Delta K_{th} = 5$ ksi$\sqrt{\text{in.}}$ Then Equation (8.9) suggests that $\rho = 2.5 \times 10^{-3}$ in. at the fatigue-crack-propagation threshold. Judging by the value of $\rho$, the preceding observations appear to negate the existence of a fatigue threshold in fatigue-cracked HY-130 steel specimens and thus do not agree with available fatigue-threshold data.[36-42] This apparent discrepancy is based on the assumption that the fatigue-crack initiation behavior can be extrapolated to tip radii (i.e., stress concentration factors, $k_t$) well below the 0.008-in. radius tested. The discrepancy may be resolved if the fatigue-crack-initiation behavior of specimens containing root radii smaller than 0.008 in. ($k_t > 13$) is independent of the root radius (stress-concentration factor). Consequently, the fatigue-crack-initiation behavior of specimens containing notches of 0.001-in. tip radius and of the geometry shown in Figure 8.6 was investigated by Barsom and McNicol.[21] The test results are represented in Figure 8.21 superimposed on the data obtained by testing specimens containing notches of 0.008-in. tip radius. The data show, conclusively, that the fatigue-crack-initiation behavior of the specimens tested by Barsom and McNicol is independent of the notch-tip root radius, $\rho$, for $\rho < 0.008$ in. The data presented in Figures 8.7 and 8.8 indicate that the value of the fatigue-threshold stress-intensity factor, $\Delta K_{th}$, for $\rho = 0.008$ in. ($K_t \approx 13$) is about 7.6 ksi$\sqrt{\text{in.}}$, which is in good agreement with the conservative estimate of $K_{th} = 5.5$ ksi$\sqrt{\text{in.}}$ for steels tested at a stress ratio of 0.1 (Chapter 9).

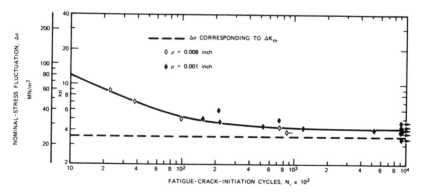

**Figure 8.21**    Fatigue-crack initiation of HY-130 steel for very small root radii.

## 8.8. Generalized Methodology for Predicting Fatigue-Crack Initiation from Notches

Fatigue-crack-initiation behavior from notches for short and long lives can be studied best in terms of the localized strains in the vicinity of the notch. These localized strains can be divided into three regions: (1) elastic-strain field, (2) localized plastic-strain field, and (3) global plastic-strain field. The following is a brief discussion of the fatigue-crack-initiation behavior from notches under the influence of each of these strain fields.

### 8.8.1. Elastic-Strain Field

Fatigue cracks initiate only in regions where plastic deformation occurs. Consequently, fatigue cracks do not initiate in a component whose strain field everywhere including regions of strain concentrations and of residual stresses is elastic.

### 8.8.2. Localized Plastic-Strain Field

Most structural components are subjected to a nominal elastic-stress (-strain) field. However, the localized strains at regions of strain raisers that are embedded in the elastic field can be plastic. Under cyclic loading, these plastically deformed regions can become the nuclei of the fatigue-crack-initiation sites.

The fatigue-crack-initiation behavior for components that contain localized plastic deformation that is surrounded by an elastic-strain field can be analyzed accurately by using the maximum-stress-range parameter, $\Delta\sigma_{max}$. For notches with length larger than $\rho$, $\Delta K_I/\sqrt{\rho}$ is directly related to $\Delta\sigma_{max}$, and the relationship between these two parameters is accurately predicted by Equation (7.3). However, for blunt notches, $\Delta\sigma_{max}$ should be calculated accurately by using theory of elasticity and finite-element analyses when necessary or may be obtained from available handbooks.[43]

The fatigue-crack-initiation behavior for the specimens shown in Figure 8.4 are presented in Figure 8.22[15] as a function of the $\Delta K_I/\sqrt{\rho}$ parameter. The data separation for cycles larger than $10^3$ is caused by the inaccuracy of the relationship between $\Delta K_I/\sqrt{\rho}$ and $\Delta\sigma_{max}$ for very blunt notches. The fatigue-crack-initiation data for lives equal to or less than $10^3$ cycles was obtained under general yielding conditions which caused further separation of the data.

The fatigue-crack-initiation data for blunt and sharp notches that are embedded in an elastic field can be correlated better by using $\Delta\sigma_{max} = k_t(\Delta\sigma)$ rather than $\Delta\sigma_{max} = (2/\sqrt{\pi})(K_I/\sqrt{\rho})$. The use of an accurate value for $k_t$ to calculate $\Delta\sigma_{max}$ results in an excellent correlation for notches whose plastic zone is fully surrounded by an elastic-stress field, Figure 8.23.[15]

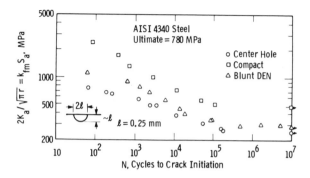

**Figure 8.22**   Correlation of fatigue-crack-initiation-life data for various notched specimens using elastic stress-concentration factors estimated from fracture-mechanics relationships. (Reprinted with permission. © Society of Automotive Engineers, Inc. From SAE Paper No. 820691, 1982.)

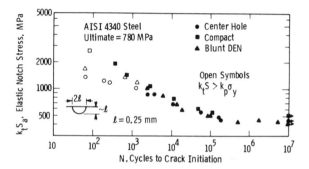

**Figure 8.23**   Correlation of fatigue-crack-initiation-life data for various notched specimens using accurate elastic stress-concentration factors. (Reprinted with permission. © Society of Automotive Engineers, Inc. SAE Paper No. 820691, 1982.)

### 8.8.3. Global Plastic-Strain Field

The transition from a plastic zone that is embedded in an elastic field to general yielding of the specimen is gradual and is strongly influenced by the size of the notch and its plastic zone relative to the size of the specimen. Thus, the data separation in Figure 8.23 was influenced by the large size of the hole relative to the specimen dimensions and would have been different had the ratio of the hole size and specimen width, the specimen geometry, and loading been different.

The most general correlation for fatigue-crack initiation from various notches having plastic zones that are embedded in an elastic field or under general yielding conditions appears to be possible by using the measured stabilized cyclic strain range at the root of the notch, $\varepsilon_a$. Such a correlation is presented in Figure 8.24[15] for the fatigue-crack-initiation data obtained by testing the specimens shown in Figure 8.4. Experimental determination of $\varepsilon_a$ should be conducted for components that would be subjected to general yielding conditions and whose fatigue-crack-initiation lives are very short ($\leqslant 10^3$ cycles). However, most structural components are not subjected to such severe loading conditions, and, therefore, their fatigue-crack-initiation behavior can be determined accurately by using the $\Delta\sigma_{\max}$ parameter.

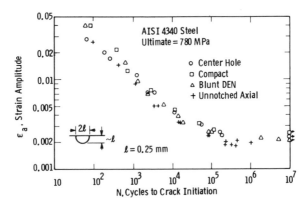

**Figure 8.24** Correlation of fatigue-crack-initiation-life data for various notched specimens using measured notch strains. (Reprinted with permission. © Society of Automotive Engineers, Inc. SAE Paper No. 820691, 1982.)

The analysis and correlations discussed in this chapter can be used to evaluate quantitatively the fatigue-crack-initiation behavior of steels and can be incorporated in design considerations. However, more research is needed to determine the effects of various metallurgical and mechanical parameters on the fatigue-crack-initiation behavior of various metals.

## References

1. S. S. MANSON, *Thermal Stress and Low-Cycle Fatigue*, McGraw-Hill, New York, 1966.
2. B. M. WUNDT, "Effect of Notches on Low-Cycle Fatigue," *ASTM STP 490*, American Society for Testing and Materials, Philadelphia, May 1972.
3. E. G. EELES and R. C. A. THURSTON, "Fatigue Properties of Materials," *Ocean Engineering, 1*, 1968.
4. P. C. PARIS and G. C. SIH, "Stress Analysis of Cracks," *Fracture-Toughness Testing and Its Applications, ASTM STP 381*, American Society for Testing and Materials, Philadelphia, 1965.
5. H. TADA, P. C. PARIS, and G. R. IRWIN, *The Stress Analysis of Cracks Handbook*, Del Research Corporation, Hellertown, Pa., 1973.

6. G. C. SIH, *Handbook of Stress-Intensity Factors*, Institute of Fracture and Solid Mechanics, Lehigh University, Bethlehem, Pa., 1973.

7. P. C. PARIS and F. ERDOGAN, "A Critical Analysis of Crack Propagation Laws," *Transactions of the ASME, Journal of Basic Engineering*, Series D, *85*, No. 3, 1963.

8. J. M. BARSOM, "Fatigue-Crack Propagation in Steels of Various Yield Strengths," *Transactions of the ASME, Journal of Engineering for Industry*, Series B, *93*, No. 4, Nov. 1971.

9. J. M. BARSOM, "Effect of Cyclic-Stress Form on Corrosion-Fatigue Crack Propagation Below $K_{Iscc}$ in a High-Yield-Strength Steel," *Corrosion Fatigue: Chemistry, Mechanics, and Microstructure*, International Corrosion Conference Series, Vol. NACE-2, National Association of Corrosion Engineers, Houston 1972.

10. W. G. CLARK, JR., and H. E. TROUT, JR., "Influence of Temperature and Section Size on Fatigue Crack Growth Behavior in Ni-Mo-V Alloy Steel," *Journal of Engineering Fracture Mechanics, 2*, No. 2, Nov. 1970.

11. H. H. JOHNSON and P. C. PARIS, "Sub-Critical Flaw Growth," *Journal of Engineering Fracture Mechanics, 1*, No. 3, June 1968.

12. R. P. WEI, "Some Aspects of Environment-Enhanced Fatigue-Crack Growth," *Journal of Engineering Fracture Mechanics, 1*, No. 4, April 1970.

13. M. CREAGER, "The Elastic Stress-Field Near the Tip of a Blunt Crack," Master of Science Thesis, Lehigh University, Bethlehem, Pa., 1966.

14. W. K. WILSON and S. E. GABRIELSE, "Elasticity Analysis of Blunt Notched Compact Tension Specimens," *Research Report 71-1E7-LOWFA-R1*, Westinghouse Research Laboratory, Pittsburgh, Feb. 5, 1971.

15. N. E. DOWLING, "A Discussion of Methods for Estimating Fatigue Life," *SAE Technical Paper Series, Paper No. 820691*, Society of Automotive Engineers, 1982.

16. D. F. MOWBRAY and J. E. McCONNELEE, "Applications of Finite Element Stress Analysis and Stress-Strain Properties in Determining Notch Fatigue Specimen Deformation and Life," *Cyclic Stress-Strain Behavior—Analysis, Experimentation, and Failure Prediction, ASTM STP 519*, American Society for Testing and Materials, Philadelphia, May 1973.

17. W. K. WILSON, "Elastic-Plastic Analysis of Blunt Notched CT Specimens and Applications," *Research Report 73-1E7-MAGRR-P1*, Westinghouse Research Laboratories, Pittsburgh, Dec. 31, 1973.

18. R. G. FORMAN, "Study of Fatigue-Crack Initiation from Flaws Using Fracture Mechanics Theory," *Air Force Flight Dynamics Laboratory Technical Report AFFDL-TR-68-100*, Wright-Patterson Air Force Base, Dayton, Ohio, Sept. 1968.

19. I. CONSTABLE, L. E. CULBER, and J. G. WILLIAMS, "Notch-Root-Radii Effects in the Fatigue of Polymers," *International Journal of Fracture Mechanics, 6*, No. 3, Sept. 1970.

20. A. R. JACK and A. T. PRICE, "The Initiation of Fatigue Cracks from Notches in Mild-Steel Plates," *International Journal of Fracture Mechanics, 6*, No. 4, Dec. 1970.

21. J. M. BARSOM and R. C. McNICOL, "Effect of Stress Concentration on Fatigue-Crack Initiation in HY-130 Steel," *ASTM STP 559*, American Society for Testing and Materials, Philadelphia, 1974.

22. W. G. CLARK, JR., "An Evaluation of the Fatigue Crack Initiation Properties

of Type 403 Stainless Steel in Air and Steam Environments," *ASTM STP 559*, American Society for Testing and Materials, Philadelphia, 1974.

23. J. M. Barsom and R. C. McNicol, unpublished data.

24. W. G. Clark, Jr., "How Fatigue Crack Initiation and Growth Properties Affect Material Selection and Design Criteria," *Metals Engineering Quarterly*, Aug. 1974.

25. D. P. Clausing, "Tensile Properties of Eight Constructional Steels Between 70 and $-320°F$," *Journal of Materials, 4*, No. 2, June 1969.

26. J. H. Gross, "PVRC Interpretive Report of Pressure Vessel Research: Section 2—Materials Considerations," *Welding Research Council Bulletin*, No. 101, New York, Nov. 1964.

27. B. F. Langer, "Design-Stress Basis for Pressure Vessels," the William M. Murray Lecture, 1970, *Experimental Mechanics, 11*, No. 1, Jan. 1971.

28. C. P. Royer, S. T. Rolfe, and J. T. Easley, "Effect of Strain Hardening on Bursting Behavior of Pressure Vessels," *Second International Conference on Pressure Vessel Technology: Part II—Materials, Fabrication and Inspection*, American Society of Mechanical Engineers, New York, 1973.

29. R. Roberts, J. M. Barsom, S. T. Rolfe, and J. W. Fisher, *Fracture Mechanics for Bridge Design*, Report No. FHWA-RD-78-69, Federal Highway Administration, Washington, D.C., 1977.

30. M. E. Taylor and J. M. Barsom, "Effect of Cyclic Frequency on the Corrosion-Fatigue Crack-Initiation Behavior of ASTM A517 Grade F Steel," *Fracture Mechanics: Thirteenth Conference, ASTM STP 743*, Richard Roberts, ed., American Society for Testing and Materials, Philadelphia, 1981.

31. J. M. Barsom, "Fracture Mechanics-Fatigue and Fracture," *Metals Handbook—Desk Edition*, American Society for Metals, Metals Park, Ohio, 1985.

32. W. Weibull, "A Statistical Theory of the Strength of Materials," *Ingenieur Vetenskaps Akademie Handlung, 151*, 1939.

33. W. Weibull, "The Phenomenon of Rupture in Solids," *Ingenieur Vetenskaps Akademie Handlung, 153*, 1939.

34. W. Weibull, "A Statistical Distribution Function of Wide Applicability," *Journal of Applied Mechanics, 18*, 1951.

35. K. N. Smith, P. Watson, and T. H. Topper, "A Stress-Strain Function for the Fatigue of Metals," *Journal of Materials, 5*, No. 4, Dec. 1970.

36. R. J. Bucci, W. G. Clark, Jr., and P. C. Paris, "Fatigue-Crack-Propagation Growth Rates Under a Wide Variation of $\Delta K$ for an ASTM A517 Grade F (T-1) Steel," *ASTM STP 513*, American Society for Testing and Materials, Philadelphia, 1972.

37. R. A. Schmidt, "A Threshold in Metal Fatigue," Master of Science Thesis, Lehigh University, Bethlehem, Pa., 1970.

38. P. C. Paris, "Testing for Very Slow Growth of Fatigue Cracks," *MTS Closed Loop Magazine, 2*, No. 5, 1970.

39. J. D. Harrison, "An Analysis of Data on Non-Propagating Fatigue Cracks on a Fracture Mechanics Basis," *British Welding Journal, 2*, No. 3, March 1970.

40. L. P. Pook, "Fatigue Crack Growth Data for Various Materials Deduced from the Fatigue Lives of Precracked Plates," *ASTM STP 513*, American Society for Testing and Materials, Philadelphia, 1972.

41. P. C. PARIS, R. J. BUCCI, E. T. WESSEL, W. G. CLARK, and T. R. MAGER, "Extensive Study of Low Fatigue Crack Growth Rates in A533 and A508 Steels," *ASTM STP 513*, American Society for Testing and Materials, Philadelphia, 1972.

42. J. M. BARSOM, "Fatigue Behavior of Pressure-Vessel Steels," *WRC Bulletin* 194, Welding Research Council, New York, May 1974.

43. R. E. PETERSON, *Stress Concentration Factors*, John Wiley, New York, 1974.

# Fatigue-Crack Propagation under Constant-Amplitude Load Fluctuation

## 9.1. General Background

The life of structural components that contain cracks or that develop cracks early in their lives may be governed by the rate of subcritical crack propagation. Moreover, proof-testing or nondestructive testing procedures or both may provide information regarding the relative size and distribution of possible preexisting cracks prior to service. However, these inspection procedures are usually used to establish upper limits on the size of undetectable discontinuities rather than actual crack size. These upper limits are determined by the maximum resolution of the inspection procedure. Thus, to establish the minimum fatigue life of structural components, it is reasonable to assume that the component contains the largest discontinuity that cannot be detected by the inspection method. The useful life of these structural components is determined by the fatigue-crack-growth behavior of the material. Therefore, to predict the minimum fatigue life of structural components and to establish safe inspection intervals, an understanding of the rate of fatigue-crack propagation is required. The most successful approach to the study of fatigue-crack propagation is based on fracture-mechanics concepts.

Most fatigue-crack-growth tests are conducted by subjecting a fatigue-cracked specimen to constant-amplitude cyclic-load fluctuations. Incremental increase of crack length is measured, and the corresponding number of elapsed load cycles is recorded. The data are presented on a plot of crack length, $a$, versus total number of elapsed load cycles, $N$ (Figure 9.1). An increase in the magnitude of cyclic-load fluctuation results in a decrease of fatigue life of specimens having identical geometry (Figure 9.2). Furthermore, the fatigue life of specimens subjected to a fixed constant-amplitude cyclic-load fluctuation decreases as the length of the initial crack is increased (Figure 9.3). Consequently, under a given constant-amplitude stress fluc-

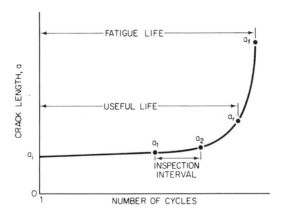

**Figure 9.1**   Schematic representation of fatigue-crack-growth curve under constant-amplitude loading.

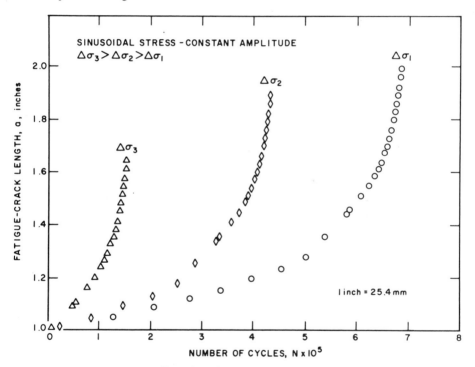

**Figure 9.2**   Effect of cyclic-stress range on crack growth.

tuation, most of the useful cyclic life is expended when the crack length is very small. Various $a$ versus $N$ curves can be generated by varying the magnitude of the cyclic-load fluctuation and/or the size of the initial crack. These curves reduce to a single curve when the data are represented in terms of crack-growth rate per cycle of loading, $da/dN$, and the fluctuation

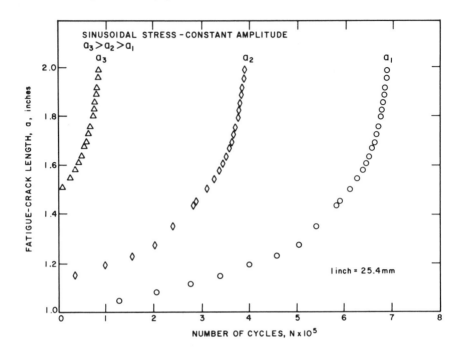

**Figure 9.3**   Effect of initial crack length on crack growth.

of the stress-intensity factor, $\Delta K_I$, because $\Delta K_I$ is a single-term parameter that incorporates the effect of changing crack length and cyclic-load magnitude. The parameter $\Delta K_I$ is representative of the mechanical driving force and is independent of geometry. The most commonly used presentation of fatigue-crack-growth data is a log-log plot of the rate of fatigue-crack growth per cycle of load fluctuation, $da/dN$, and the fluctuation of the stress-intensity factor, $\Delta K_I$.[1]

The fatigue-crack-propagation behavior for metals can be divided into three regions (Figure 9.4). The behavior in region I exhibits a "fatigue-threshold" cyclic stress-intensity-factor fluctuation, $\Delta K_{th}$, below which cracks do not propagate under cyclic-stress fluctuations.[2–16] Region II represents the fatigue-crack-propagation behavior above $\Delta K_{th}$,[17] which can be represented by

$$\frac{da}{dN} = A(\Delta K)^m \qquad (9.1)$$

where   $a$ = crack length.
        $N$ = number of cycles.
        $\Delta K$ = stress-intensity-factor fluctuation.

$A$ and $m$ are constants.

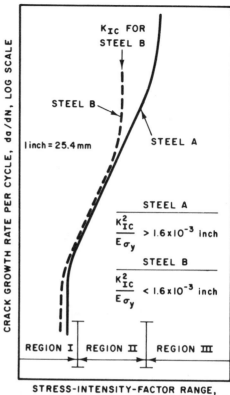

**Figure 9.4**  Schematic representation
of fatigue-crack growth in steel.

STRESS-INTENSITY-FACTOR RANGE,
$\Delta K_I$, LOG SCALE

In region III the fatigue-crack growth per cycle is higher than is that predicted for region II. The data[17-21] show that the rate of fatigue-crack growth increases and that under zero-to-tension loading (that is, $\Delta K = K_{max}$) this increase occurs at a constant value of crack-tip displacement, $\delta_T$, and at a corresponding stress-intensity-factor value, $K_T$, given by

$$\delta_T = \frac{K_T^2}{E\sigma_{ys}} = 1.6 \times 10^{-3} \text{ in. (0.04 mm)} \qquad (9.2)$$

where  $K_T$ = stress-intensity-factor-range value corresponding to onset of
acceleration in fatigue-crack-growth rates.

$E$ = Young's modulus.

$\sigma_{ys}$ = yield strength (0.2 percent offset) (the available data
indicate that the value of $K_T$ can be predicted more closely
by using a flow stress, $\sigma_f$, rather than $\sigma_{ys}$, where $\sigma_f$ is the
average of the yield and tensile strengths).

Acceleration of fatigue-crack-growth rates that determines the transition from region II to region III appears to be caused by the superposition of a ductile tear mechanism onto the mechanism of cyclic subcritical crack extension, which leaves fatigue striations on the fracture surface. Ductile tear occurs when the strain at the tip of the crack reaches a critical value.[22] Thus, the fatigue-rate transition from region II to region III depends on $K_{max}$ and on the stress ratio, $R$.

Equation (9.2) is used to calculate the stress-intensity-factor value corresponding to the onset of fatigue-rate transition, $K_T$ (or $\Delta K_T$ for zero-to-tension loading), which also corresponds to the point of transition from region II to region III in materials that have high fracture toughness, steel A in Figure 9.4—that is, materials for which the critical-stress-intensity factor, $K_{Ic}$ or $K_c$, is higher than the $K_T$ value calculated by using Equation (9.2). Acceleration in the rate of fatigue-crack growth occurs at a stress-intensity-factor value slightly below the critical-stress-intensity factor, $K_{Ic}$, when the $K_{Ic}$ (or $K_c$) of the material is less than $K_T$, steel B in Figure 9.4.[23] Furthermore, acceleration in the rate of fatigue-crack growth in an aggressive environment may occur at the threshold stress-intensity factor, $K_{Iscc}$. The effect of an aggressive environment on the rate of crack growth is discussed in Chapters 11 and 13. Crooker[20] has shown that the $\Delta K_T$ of aluminum and titanium alloys can also be predicted by using Equation (9.2).

## 9.2. Fatigue-Crack-Propagation Threshold

The early investigations of fatigue-crack propagation by Frost and coworkers[2,3] indicated a significant deceleration in the fatigue-crack-growth rates at low stresses. Their data suggested the possible existence of a threshold for fatigue-crack propagation below which fatigue cracks should not propagate. The existence of such a threshold was predicted by an elastic-plastic analysis conducted by McClintock.[4] The work of Lindner[5] and Paris[6] showed that the fatigue-crack-propagation threshold can be established in the context of linear-elastic fracture mechanics and that a threshold stress-intensity-factor range, $\Delta K_{th}$, can be determined below which fatigue cracks should not propagate.

In the later 1960s, Elber[7] noted that fatigue-crack surfaces interfere with each other through a closure mechanism. This plastically induced closure results from the presence of residual plastic deformation left in the wake of the growing fatigue crack. Soon after, Schmidt's[8] experimental studies indicated that crack closure may have a significant effect on the threshold behavior. The presently available data support this observation but indicate that crack closure can be related to four closure mechanisms[9] shown schematically in Figure 9.5: plastically induced closure resulting from the presence of residual plastic deformation left in the wake of the growing fatigue crack, Figure 9.5(c); surface-roughness–induced closure caused by

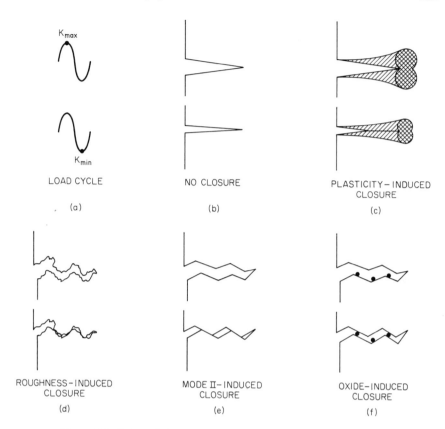

**Figure 9.5**  Schematic illustration of four crack-closure mechanisms.

deviations of the crack trajectory associated with microstructural characteristics of the material (for example, grain size and interlamellar spacing),
Figure 9.5(d); Mode II–induced closure caused by the displacement of the
fatigue crack tip along shear planes, thus preventing a perfect match of the
fatigue-crack surface features left behind the propagating crack, Figure
9.5(e); and environmentally induced closure resulting from corrosion products
within the crack that wedge the crack surfaces, Figure 9.5(f). Several of
these mechanisms can be operative at the same time.

    Several factors may influence the fatigue-crack-propagation threshold,
including[9] yield strength, grain size and other microstructural elements,
mean stress, stress history, residual stress, mode of crack-tip opening,
Young's modulus, temperature, and environment. The effects of most of
these factors on $\Delta K_{th}$ can be explained by their relation to the various
mechanisms of crack closure.

    Sufficient data are available to show the existence of a fatigue-crack-propagation threshold, $\Delta K_{th}$, below which existing fatigue cracks do not propagate under cyclic loading. However, more work is needed to understand better the effects of various factors on the magnitude and the existence of $\Delta K_{th}$ and to use it properly in design. Future work on $\Delta K_{th}$ should better define the effects of microstructure and material properties, various testing procedures, crack and specimen size and geometry, variable-amplitude loading, and various loadings (out-of-plane bending, torsion, and biaxial and triaxial loading). Kitagawa and coworkers[9,15] found a constant $\Delta K_{th}$ value for crack lengths larger than 0.5 mm. However, below this crack size, a transfer occurred where the fatigue limit of an unnotched material, rather than $\Delta K_{th}$, became the threshold condition for propagation. A similar trend for very small cracks was found by Haddad et al.[16] These observations are extremely important for proper use of $\Delta K_{th}$ in structural design. Moreover, preliminary fatigue tests for welded components suggest that the fatigue threshold under variable-amplitude cyclic-loading conditions can change significantly[24] and may be eliminated.

Despite the need for further understanding of $\Delta K_{th}$, the available data for long fatigue cracks subjected to constant-amplitude cyclic loading in various room temperature laboratory atmospheres suggest that the stress ratio is the most important factor affecting the magnitude of $\Delta K_{th}$.

Figure 9.6 presents data published by various investigators[11-13,25] on fatigue-crack-propagation values for steels. The data show that conservative estimates of $\Delta K_{th}$ for martensitic, bainitic, ferrite-pearlite, and austenitic steels subjected to various stress ratios, $R$, larger than $+0.1$ can be predicted from[21]

$$\Delta K_{th} = 6.4(1 - 0.85R) \text{ ksi}\sqrt{\text{in.}}  \qquad (9.3a)$$

or

$$\Delta K_{th} = 7(1 - 0.85R) \text{ MN/m}^{3/2}  \qquad (9.3b)$$

The value of $\Delta K_{th}$ for $R < 0.1$ is a constant equal to 5.5 ksi$\sqrt{\text{in.}}$ (6 MN/m$^{3/2}$).

The data in Figure 9.6 show that the primary factor affecting $\Delta K_{th}$ is the stress ratio and that the mechanical and metallurgical properties of these steels have negligible effect on the fatigue-crack-propagation threshold.

Lindley and Richards[26] have compiled $\Delta K_{th}$ values obtained by various investigators on a wide range of steels, Figure 9.7. These data show that Equations (9.3) can be used to predict lower-bound values of $\Delta K_{th}$ for various steels. Similar data have been published by Sasaki et al.[27] and by Priddle, Figure 9.8,[28] which include weld metal, that support the use of Equations (9.3a) and (9.3b).

Although some of the scatter in the data may be attributed to differences in the materials tested, the most likely source for the scatter was differences

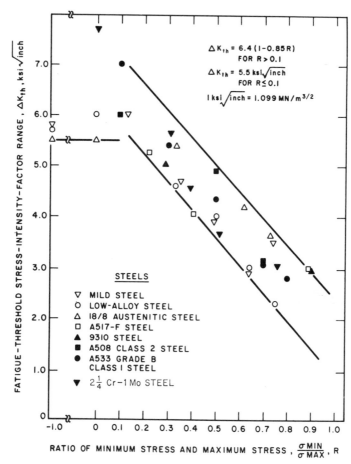

**Figure 9.6**  Dependence of fatigue-threshold stress-intensity-factor range on stress ratio.

in testing procedures used by various investigators because of an absence of a standard test method for determining $\Delta K_{th}$. The American Society for Testing and Materials is in the process of expanding Test Method E-647—81 to include procedures for near-threshold fatigue-crack-growth-rate measurements. A draft of these procedures has been published in Appendix II of Reference 29.

Finally, an analysis of experimental results published in the literature for nonpropagating fatigue cracks in various metals has been conducted by Harrison.[11] His analysis suggests that the fatigue-crack-propagation threshold for a number of materials can be superimposed by using the parameter $\Delta K_{th}/E$, where $E$ is Young's modulus.

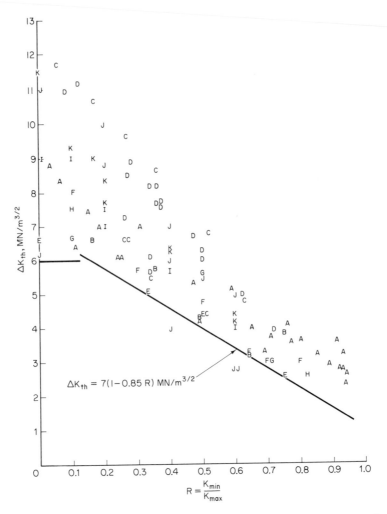

**Figure 9.7**  $\Delta K_{th}$ values for the mild- and low-alloy steels with 0.2 percent proof stress $< 620$ MPa as a function of stress ratio $R$. (From Ref. 26)

## 9.3. Martensitic Steels

Extensive fatigue-crack-growth-rate data for various high-yield-strength ($\sigma_{ys}$ greater than 80 ksi or 552 MN/m$^2$) martensitic steels show that the primary parameter affecting growth rate in region II is the range of fluctuation in the stress-intensity factor and that the mechanical and metallurgical properties of these steels have negligible effects on the fatigue-crack-growth rate in a room temperature air environment.[21,30] The data for these steels fall within a single band, as shown in Figure 9.9, and the upper bound of the scatter

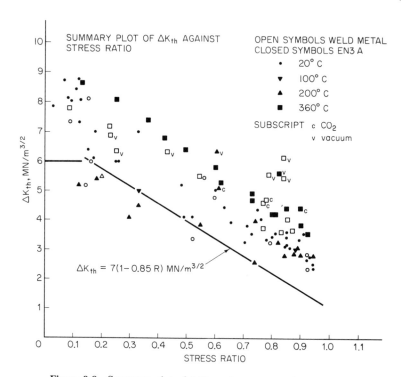

**Figure 9.8**   Summary plot of $\Delta K_{th}$ against stress ratio. (Ref. 28)

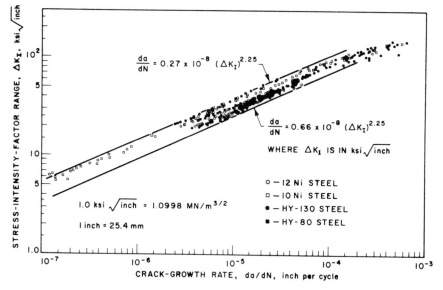

**Figure 9.9**   Summary of fatigue-crack-propagation for martensitic steels.

of the fatigue-crack-propagation-rate data for martensitic steels in an air environment can be obtained from

$$\frac{da}{dN} = 0.66 \times 10^{-8}(\Delta K_I)^{2.25}$$

(9.4)

where $a$ = in.
$K_I$ = ksi$\sqrt{\text{in.}}$

The applicability of Equation (9.4) to martensitic steels ranging in yield strength from 80 to 300 ksi (552 to 2068 MN/m²) has been established.[17,18] The validity of Equation (9.4) has been established further by using data obtained by testing various grades of ASTM A514 steels[30,31] (Figure 9.10);[30] ASTM A517 Grade F steel;[10] ASTM A533 Grade B, Class 1 steel (Figure 9.11);[32] and ASTM A533 Grade A and ASTM A645 steels.[33]

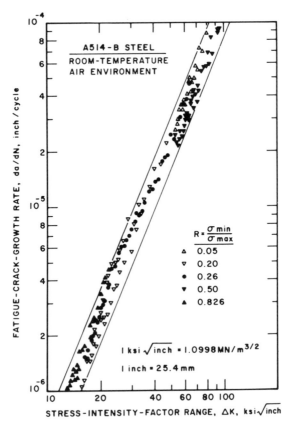

**Figure 9.10**  Crack-growth rate as a function of stress-intensity-factor range for A514 Grade B steel.

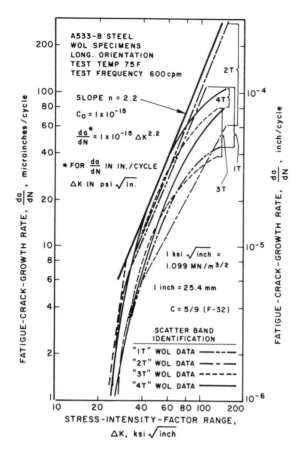

**Figure 9.11** Influence of specimen size on the $da/dN$ versus $\Delta K$ relationship for A533 Grade B, Class 1 steel (Ref. 32).

Equation (9.4) may be used to estimate the fatigue-crack-propagation rate for martensitic steels in a benign environment in region II (Figure 9.4). However, the fatigue-crack-growth rate increases markedly when the fluctuation of the stress-intensity factor, $\Delta K_{\mathrm{I}}$, becomes larger than the value calculated by using Equation (9.2) or when the value of the maximum stress-intensity factor becomes close to the $K_{\mathrm{Ic}}$ (or $K_c$) of the material. Under these conditions the fatigue-crack-growth rate cannot be predicted by Equation (9.4). The useful cyclic life of a structural component subjected to stress fluctuations corresponding to region III–type behavior is of practical interest only in relatively low-life structural applications. Further work is necessary to characterize the fatigue-crack-propagation behavior in region III.

## 9.4. Ferrite-Pearlite Steels

The fatigue-crack-growth-rate behavior in ferrite-pearlite steels[17] prior to the onset of fatigue-rate transition and above $\Delta K_{th}$ is presented in Figure 9.12. The data indicate that realistic estimates of the rate of fatigue-crack growth in these steels can be calculated from

$$\frac{da}{dN} = 3.6 \times 10^{-10}(\Delta K_I)^{3.0} \tag{9.5}$$

where     $a = $ in.
          $\Delta K_I = $ ksi$\sqrt{\text{in.}}$

The data presented in Figures 9.9 and 9.12 and represented by Equations (9.4) and (9.5) show that the rate of fatigue-crack growth at a given $\Delta K_I$ value below the fatigue-rate transition is lower in ferrite-pearlite steels than in martensitic steels. The reason for the differences in the fatigue-crack-growth-rate behavior between martensitic steels and ferrite-pearlite steels is discussed later.

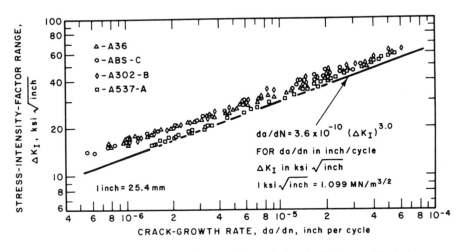

**Figure 9.12**  Summary of fatigue-crack-growth data for ferrite-pearlite steels.

## 9.5. Austenitic Stainless Steels

Extensive fatigue-crack-growth-rate data for austenitic stainless steels in region II have been obtained. Data for solution-annealed and cold-worked type 316 stainless steel and for solution-annealed type 304 stainless steel were obtained by James[34] (Figure 9.13). Data on these types of steels were

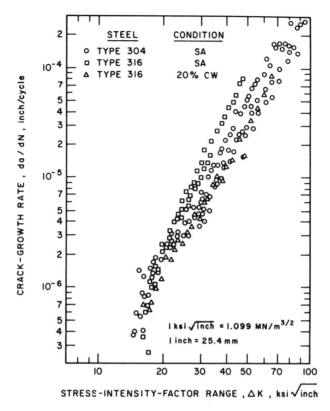

**Figure 9.13**  Fatigue-crack-growth rate of solution-annealed type 304 and 316 stainless steels and 20 percent cold-worked type 316 stainless steel at 75°F.

also obtained by Shahinian et al.[35] (Figure 9.14). Weber and Hertzberg[36] examined the effect of thermomechanical processing on fatigue-crack propagation in type 305 stainless steel. The steel was evaluated in the coarse-grained, cold-worked, and recrystallized conditions. A summary of their data is presented in Figure 9.15 and shows that crack propagation was essentially similar irrespective of thermomechanical processing. The data presented in Figures 9.13 through 9.15 indicate that conservative and realistic estimates of fatigue-crack-propagation rates for austenitic steels in a room temperature air environment can be obtained from

$$\frac{da}{dN} = 3.0 \times 10^{-10}(\Delta K_{\mathrm{I}})^{3.25} \qquad (9.6)$$

where  $a$ = in.
$K_{\mathrm{I}}$ = ksi$\sqrt{\text{in.}}$

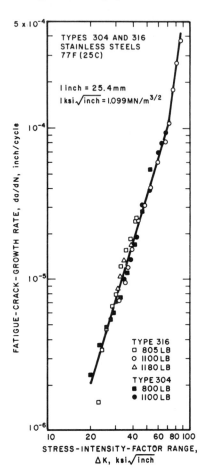

**Figure 9.14** Fatigue-crack-growth rates in type 304 and 316 stainless steels at room temperature as a function of range of stress-intensity factor.

## 9.6. General Discussion of Steels

Engineering estimates of fatigue-crack-growth rates in martensitic, ferrite-pearlite, and austenitic stainless steels can be obtained by using Equations (9.4), (9.5), and (9.6), respectively. These equations are plotted in Figure 9.16 to show the differences in the fatigue-crack-growth-rate behavior of these steels. The fatigue-crack-growth-rate behavior of a given steel may be changed from a behavior that is depicted by Equation (9.5) to that depicted by Equation (9.4). Such changes can be accomplished through thermomechanical processing.

The preceding discussion of fatigue-crack-growth rates in steels was divided into three groups that reflect microstructural differences. These groups also reflect broad variations in mechanical properties. The ranges

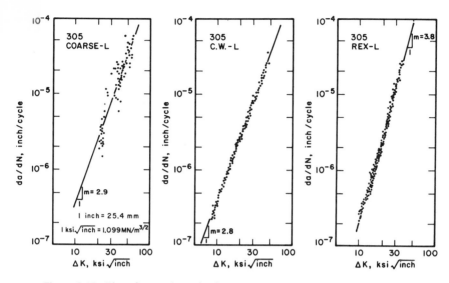

**Figure 9.15**  Plot of stress-intensity-factor range versus crack-propagation rate for "L" orientation of 305 stainless steel in coarse-grained, cold-worked, and recrystallized conditions.

of the tensile strength, yield strength, and strain-hardening exponent of austenitic stainless steels, ferrite-pearlite steels, and martensitic steels for which Equations (9.4), (9.5), and (9.6) are valid are presented in Table 9.1. Considering the wide variation of the chemical compositions and tensile properties of ferrite-pearlite steels, martensitic steels, and austenitic stainless steels, and the relatively small variation in their fatigue-crack-growth rates (Figure 9.16), the use of a single equation to predict fatigue-crack-growth rates for these steels may be justified. Because of the difficulties encountered in the determination of the magnitude of stresses and stress fluctuations in complex structural details, the use of Equation (9.4) to calculate conservative estimates of the rate of fatigue-crack growth in various steels can be justified.

**TABLE 9.1  Ranges in Mechanical Properties of Pressure-Vessel Steels**

| Steel | Yield Strength, $\sigma_{ys}$ (ksi) | Tensile Strength, $\sigma_T$ (ksi) | Strain-Hardening Exponent, $n$ |
|---|---|---|---|
| Austenitic stainless | $30 < \sigma_{ys} < 50$ | $75 < \sigma_T < 95$ | $n > 0.30$ |
| Ferrite-pearlite | $30 < \sigma_{ys} < 80$ | $50 < \sigma_T < 110$ | $0.15 < n < 0.30$ |
| Martensitic | $\sigma_{ys} > 70$ | $\sigma_T > 90$ | $n < 0.15$ |
|  | *Conversion factor* | | |
|  | 1 ksi = $6.895$ MN/m$^2$ | | |

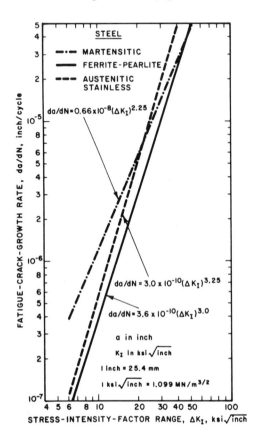

**Figure 9.16** Fatigue-crack-propagation behavior of various steels.

Further justification for the use of Equation (9.4) to calculate conservative estimates of the rate of fatigue-crack growth in various steels was established by Clark.[37] A design example of the use of these equations is presented later in this chapter.

## 9.7. Aluminum and Titanium Alloys

The preceding discussion shows that the region II fatigue-crack-growth-rate behavior of steels in a benign environment is essentially independent of mechanical and metallurgical properties of the material.

Figures 9.17 and 9.18[20] present the fatigue-crack-growth behavior of aluminum and titanium alloys, respectively. The appropriate crack-growth-rate expressions that define the upper-scatter-band behavior of these metal systems are also presented in these figures. The data indicate that, like

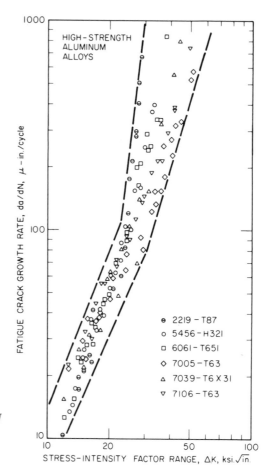

**Figure 9.17**   Summary plot of $da/dN$ versus $\Delta K$ for six aluminum alloys. The yield strengths of these alloys range from 34 to 55 ksi.

steels, the fatigue-crack-growth rates for aluminum and titanium alloys fall within different but distinct scatter bands. However, the scatter bands for aluminum and titanium alloys are larger than for steels. Thus, it is apparent that for a given metal system subjected to a benign environment, the fatigue-crack-growth behavior in region II is not a pertinent factor in material selection considerations within a given metal system.

The fatigue-rate transition from region II to region III for aluminum and titanium alloys has been investigated by Clark and Wessel[38] and by Crooker.[20,39] The data presented in Figures 9.19[38] and 9.20[39] as well as other data obtained by Crooker[20] show conclusively that Equation (9.2) can be used to determine the stress-intensity-factor value corresponding to the transition from region II crack-growth-rate behavior to region III in aluminum and titanium alloys.

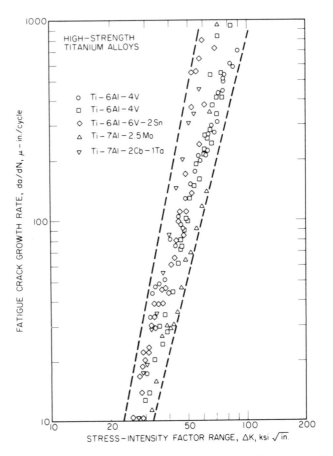

**Figure 9.18**  Summary plot of $da/dN$ versus $\Delta K$ data for five titanium alloys ranging in yield strength from 110 to 150 ksi.

## 9.8. Effect of Mean Stress on Fatigue-Crack Propagation Behavior

The effect of mean load on fatigue-crack-initiation and propagation behavior can be studied by using the load ratio parameter, $R$, where $R$ is equal to $P_{min}/P_{max} = K_{min}/K_{max}$, and $P_{min}$ ($K_{min}$) and $P_{max}$ ($K_{max}$) are the minimum and maximum loads (stress-intensity-factor values) in the cyclic process, respectively.

Several investigations have been conducted to study the effect of mean stress and stress ratio on fatigue-crack-propagation rate.[7,30,40–43] Available experimental data on ASTM A514 Grade B steel show no systematic change

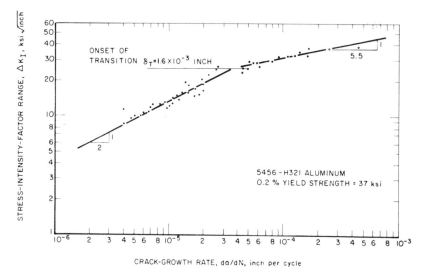

**Figure 9.19**  Fatigue-crack-growth rate as a function of stress intensity-factor range for 5456-H321 aluminum.

in fatigue-crack-growth rate with changes in $R$ value from 0 to 0.82 (Figure 9.10).[30] The data also show that this change in $R$ value has negligible effects on the rate of crack propagation in region II.  On the other hand, Crooker[41] observed that the compression portion of fully reversed tension-compression cycling increased the growth rate in HY-80 steel by approximately 50 percent as compared with zero-tension cycling.  Data on fatigue-crack growth in a 140-ksi (965-MN/m²) yield-strength martensitic steel[42] showed a systematic increase in growth rate with increase in $R$ value from 0 to 0.75 and with decrease in $R$ value from 0 to $-2$, but that maximum nominal stress levels of 0.39–0.94 of the yield strength had no apparent effect (Figure 9.21).  The maximum increase in fatigue-crack-growth rate as a function of variation of $R$ from $-2$ to 0.75 was less than a factor of 2.

Crooker et al.[41,42] studied the effect of stress ratio, $R$, in the range $-2 \leq R \leq 0.75$ by using part-through-crack specimens subjected to axial loads.  Because of the difficulties encountered in measuring crack depth in these specimens, the rate of fatigue-crack growth was calculated by measuring the rate of growth on the surface of the specimen and by assuming that the crack shape did not change (that is, a circular crack front remained circular) as the crack size increased.  Despite reservations relating to the accuracy of this assumption and to the use of part-through-crack specimens for fatigue-crack-growth studies, analysis of the data presented in Figure 9.21 showed that the effect of positive stress ratio, $R \geq 0$, on the rate of

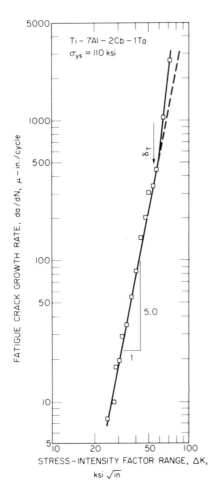

**Figure 9.20**  Fatigue-crack-growth rate in a titanium alloy.

fatigue-crack growth in the 140-ksi yield-strength steel investigated can be predicted by using

$$\frac{da}{dN} = \frac{A(\Delta K)^{2.25}}{(1 - R)^{0.5}} \qquad (9.7)$$

where $A$ is a constant.

The upper bound of the scatter band of martensitic steels that is represented by Equation (9.4) is also presented in Figure 9.21. The data suggest that Equation (9.4) can be used to estimate fatigue-crack-growth rates in martensitic steels subjected to various values of stress ratio.

The preceding discussion indicates that stress ratio has a second-order

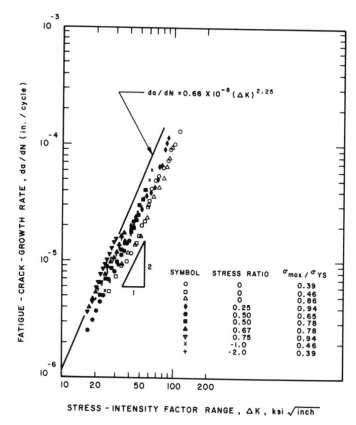

**Figure 9.21**    Effect of stress ratio on the fatigue-crack-propagation rate in a 140-ksi yield-strength martensitic steel.

effect on the rate of crack propagation in region II (Figure 9.4) and that Equations (9.4), (9.5), and (9.6) that represent the upper bound of data scatter can be used to estimate fatigue-crack-growth rates in steels subjected to various values of stress ratio.

The fatigue-crack initiation threshold, $\Delta K_{th}$, has been shown in Section 9.2 to depend on the magnitude of the stress ratio, $R = K_{min}/K_{max}$. Moreover, because $\Delta K_T$ defines the region above which fatigue-crack propagation occurs by striation and ductile dimpling and because ductile dimpling depends on $K_{max}$, then $\Delta K_T$ must depend on $K_{max}$ and on the stress ratio, $R$.

The general effect of stress ratio, $R$, on fatigue-crack-growth rates in regions I, II, and III are presented schematically in Figure 9.22.[44] Data in support of this generalized behavior have been published by Wei.[44]

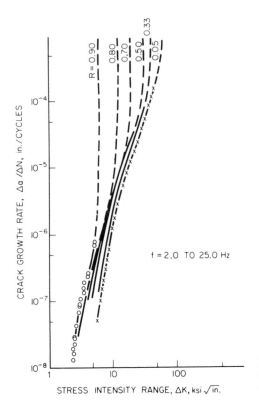

**Figure 9.22** Effect of stress ratio, $R$, on fatigue-crack-growth upper threshold.

## 9.9. Effects of Cyclic Frequency and Waveform

The effect of loading rate on fracture toughness was presented in Chapter 4. The available data show that the rate of loading can affect the fracture toughness significantly and that the magnitude of this effect increases with decrease in yield strength. Because a change in cyclic frequency corresponds to a change in the rate of loading, the effect of cyclic frequency on fatigue-crack-growth rate should be established.

The fatigue-crack-growth rate for A36 steel ($\sigma_{ys} = 36$ ksi) under cyclic frequencies of 6–3000 cycles/min (cpm) is presented in Figure 9.23. Similar data for 12Ni-5Cr-3Mo maraging steel ($\sigma_{ys} = 180$ ksi) were obtained under cyclic frequencies of 6–600 cpm.[22,45,46] The data show that the rates of fatigue-crack growth in a room temperature benign environment were not affected by the frequency of the cyclic-stress fluctuations.

The discussions in the preceding sections show that the stress-intensity-factor fluctuation, $\Delta K_I$, is the primary parameter that governs the rate of crack growth per cycle of loading. Because $\Delta K_I$ is related to the magnitude

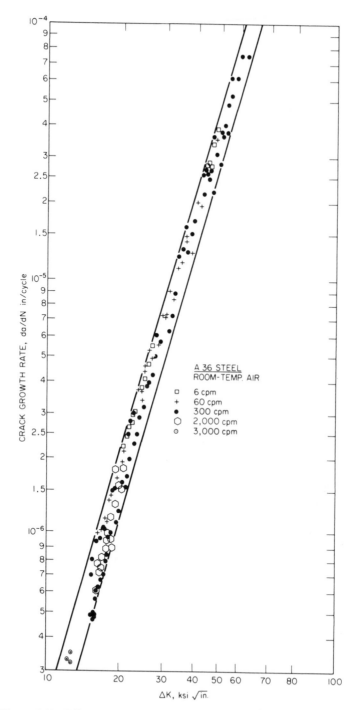

**Figure 9.23** Effect of cyclic frequency on crack-growth rates in a benign environment.

of the stress fluctuation and to the square root of crack length, this parameter does not account for possible differences in growth rates that may exist between various cyclic waveforms such as sine and square waves. Consequently, data on the fatigue-crack-growth rate of 12Ni-5Cr-3Mo maraging steel in a room temperature air environment were obtained under sinusoidal, triangular, square, positive sawtooth ( $\wedge$ ), and negative sawtooth ( $\wedge$ ) cyclic-stress fluctuations. The combined data, presented in Figure 9.24,[45] show that the rates of fatigue-crack growth in a room temperature air environment were not affected by the form of the cyclic-stress fluctuations.

The effects of cyclic frequency and waveform on corrosion-fatigue crack-growth rates are presented in Chapter 13.

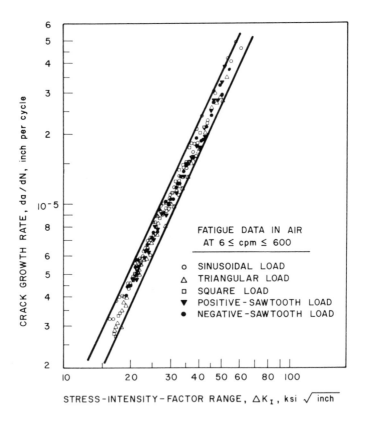

**Figure 9.24** Fatigue-crack-growth rates in 12Ni-5Cr-3Mo steel under various cyclic-stress fluctuations with different stress-time profiles.

## 9.10. Effects of Stress Concentration on Fatigue-Crack Growth

The effect of the stress-concentration factor on fatigue-crack initiation was discussed in Chapter 8. Unfortunately, the effect of the stress-concentration factor on the rate of crack propagation is yet to be investigated. Because the fluctuation of the stress-intensity factor is the primary fatigue-crack-propagation force, it can be surmised that the rate of fatigue-crack propagation is governed by the local stress-intensity-factor fluctuation. That is, the fatigue-crack-propagation rate per cycle, $da/dN$, in the shadow of a notch must be governed by the relationship

$$\frac{da}{dN} = A(\Delta K_{\text{eff}})^m \tag{9.8}$$

where $A$ and $m$ = constants.

$\Delta K_{\text{eff}} \cong k_t(a) \, \Delta\sigma\sqrt{a}$.

$\Delta\sigma$ = nominal stress fluctuation.

$a$ = crack length.

$k_t(a)$ = stress-concentration factor; this factor is a function of crack length such that as the crack propagates outside the field of influence of the notch $k_t(a)$ approaches unity (see Chapter 2 for cracks emanating from notches).

Equation (9.8) could be used to analyze the fatigue-crack-growth behavior of cracks emanating from holes, nozzle corners, or other regions of stress concentrations.

Novak and Barsom[47] summarized the state of the art of all available theoretical $K_I$ analyses for cracks in the vicinity of stress concentrations. The accuracy of some of these analyses was verified by their experimental results on the brittle-fracture behavior for cracks emanating from notches. These analyses can be used in conjunction with Equation (9.8) to predict the rate of growth of cracks in the shadow of stress concentrations.

## 9.11. Fatigue-Crack Propagation in Steel Weldments

Fatigue-crack-growth rates in weldments of various steels are available in the literature.[33,48–51] Fatigue-crack-growth rates in the weld metal and the heat-affected zone of submerged arc weldments of ASTM A533 Grade B, Class 1 steel have been obtained by Clark.[48] In general, fatigue cracks initiated in the heat-affected zone grew into the adjacent weld metal. Figures 9.25 and 9.26 present the fatigue-crack-growth behavior in the weld metal and in the heat-affected zone, respectively. The data show that the rate

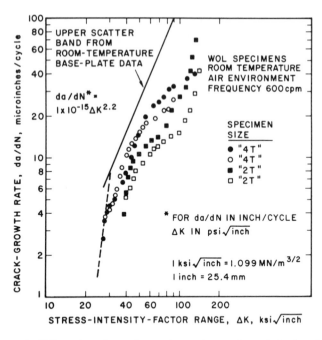

**Figure 9.25**   Effect of specimen size on the crack-growth-rate behavior of A533 Grade B, Class 1 steel weld metal.

of fatigue-crack growth in the weld metal and in the heat-affected zone was equal to or less than that in the base metal and that the upper bound of the scatter band established by testing the base metal at room temperature may be used as a conservative estimate of fatigue properties of the weld metal and the heat-affected zone. This conclusion is confirmed by data obtained on HY-140 steel weldments,[49] on a 5 percent nickel steel (ASTM A645) weldment (Figure 9.27),[33] on type 308 weld metal (Figure 9.28),[50] and on type 316 weld metal.[50]

Extensive fatigue-crack-growth-rate data on various weldments have been published by Maddox.[51] Figure 9.29[51] presents data for structural C-Mn steel weld metals, heat-affected zones, and base metals. A comparison of the data in Figure 9.29 and the equations in Figure 9.16 shows that the upper-scatter-band equations for base metals can be used to obtain conservative engineering estimates of the fatigue-crack-growth rates in base metals, weld metals, and heat-affected zones. Moreover, the data show that the rates of fatigue-crack growth in weld metals and in heat-affected zones are equal to or less than that in the base metal.

As-welded structural components contain residual stresses often of yield stress magnitude.[52] Consequently, fatigue cracks in regions of tensile-residual stresses propagate under high stress ratios. Preceding discussions

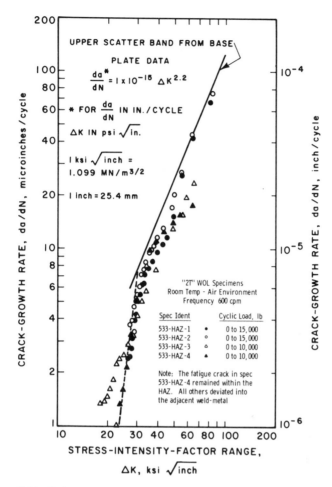

**Figure 9.26**   Fatigue-crack-growth-rate behavior observed in the vicinity of the heat-affected zone of an ASTM A533 Grade B, Class 1 weldment.

showed that high stress ratios affect the magnitude of $\Delta K_{th}$ and the magnitude of $K_T$ but have negligible effect on the fatigue-crack-growth-rate behavior in region II. This observation is supported further by Maddox (Figure 9.30).[51]

## 9.12. Effects of Inhomogeneities on Fatigue-Crack Growth

As pointed out in a previous section, the fatigue-crack-growth rates prior to the fatigue-rate transition are slower in ferrite-pearlite steels than in martensitic steels. Moreover, unlike the fatigue-rate transitions in martensitic

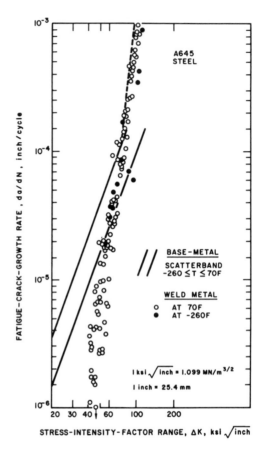

**Figure 9.27**    Fatigue-crack-growth-rate data for A645 steel base metal and weld metal in room air at 70°F and −260°F.

steels, the fatigue-rate transitions in ferrite-pearlite steels occur at slightly higher $\Delta K_I$ values than predicted by Equation (9.2). The use of a flow stress, $\sigma_f$, rather than the yield strength, $\sigma_{ys}$, in Equation (9.2) is suggested to rectify this difference. Both the decrease in fatigue-crack-growth rates and the increase in the $\Delta K_I$ value at the onset of the fatigue-rate transition appear to be related to the composite character of the microstructure of ferrite-pearlite steel or to crack branching or to both.

### 9.12.1. Composite Behavior

A composite material is a material system made up of a mixture of two or more distinct constituents. Most composite constituents can be classified as matrix formers or matrix modifiers. The matrix formers are those constituents which give the composite its bulk form and which hold

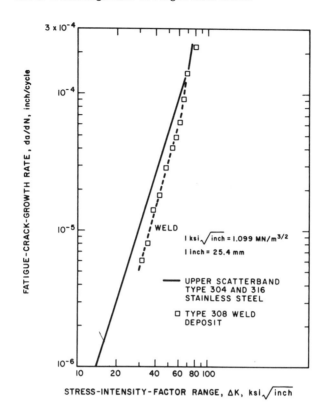

**Figure 9.28**   Comparison of fatigue-crack-growth behavior of type 308 weld deposit with that of type 304 and 316 plate at room temperature.

the matrix modifiers in place. The matrix modifiers determine the character of the internal structure of the composite. In this context, ferrite-pearlite steels can be visualized as particulate composites in which the ferrite is the matrix former and the pearlite is the matrix modifier that is dispersed in the matrix as particles (that is, colonies or patches of irregular shapes).

Fatigue-crack-growth rates in particulate-composite materials might depend on the strength of the particles relative to the strength of the matrix, on the strength of the interface bond between the particles and the matrix, on the size and orientation of the particles, and on the distribution density of these particles. The last determines the mean-free path of the fatigue-crack front during its propagation. The larger the distribution density of the particles, the smaller is the mean-free path, and therefore the larger is the effect of the particles on crack growth. The strengths of the particles and of the interface bond relative to the strength of the matrix determine whether the fatigue-crack-growth rates in the composite will be higher or lower than the rates in the absence of these matrix modifiers.

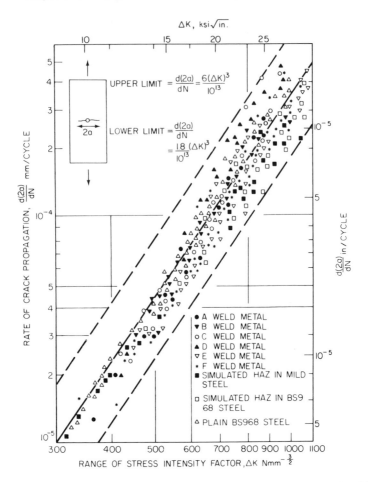

**Figure 9.29**  Plane-strain fatigue-crack-propagation data for C-Mn steel weld metals, heat-affected zones, and base metals.

In ferrite-pearlite-steel composites, the interface bond between the constituents is quite strong. Furthermore, because of the iron carbides in the pearlite, the strength of these matrix modifiers is higher than the strength of the matrix, and the ductility of the matrix is greater than the ductility of the matrix modifier. These properties of the constituents and of the interface tend to improve the strength and toughness of the matrix because the ductile ferrite can distribute the stresses to the stronger pearlite colonies.

The effect of pearlite colonies on the rate of fatigue-crack growth in ferrite-pearlite steels was investigated by using scanning electron micrographs of the fracture profile.[17] The micrographs showed that the plastic deformation under cyclic loading is much more extensive in the ferrite matrix than in the pearlite colonies. The micrographs also showed secondary fatigue cracks

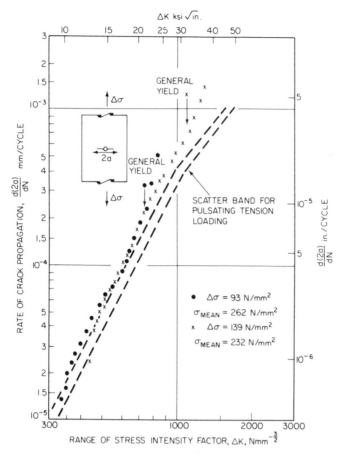

**Figure 9.30**  Effect of high maximum stress (corresponding to $0.8 \times$ yield) on fatigue-crack propagation in C-Mn steel to BS 4360 Grade 50B.

preferentially seeking to propagate around a pearlite colony rather than through it. These observations suggest that the path of least resistance to fatigue-crack growth is through the ferrite matrix and that the pearlite colonies tend to retard crack growth.

Because the properties of the constituents in a composite may affect the rate of fatigue-crack growth, acceleration in the fatigue-crack-growth rates caused by undesirable properties of the matrix modifiers or of the interface or of both would be expected in some composite materials. This behavior has been observed[53] in two 7178 aluminum alloys that differed markedly only in impurity contents.

The preferential propagation of fatigue cracks around a pearlite colony rather than through it causes crack branching. Crack branching was observed in ferrite-pearlite steels but not in martensitic steels.[17] Crack branching in ferrite-pearlite steels is also caused by cracks initiating ahead of the fatigue-crack tip in the large plastically deformed zone whose large size is determined by the low yield strength of the ferrite matrix. These secondary cracks tend to share the crack-opening displacement at the crack tip with the main crack and thereby cause a reduction in the stress-intensity factor. Consequently, the actual fatigue-crack-growth rate is less for a branched crack than the rate estimated for a single crack front. This observation appears to explain the differences in fatigue-crack-growth rates between martensitic and ferrite-pearlite steels; see Equations (9.4) and (9.5). Similarly, at the same crack-opening displacement, the strain at the tip of a branched crack is less than that at the tip of a single crack. Therefore, the value of the crack-opening displacement at the onset of fatigue-rate transition, Equation (9.2), for a branched crack would be expected to be slightly higher than $1.6 \times 10^{-3}$ in., which was shown to be applicable to martensitic steels where severe crack branching was not observed. Examination of the data[17,21] indicated that onset of fatigue-rate transition in martensitic steels and in ferrite-pearlite steels can be predicted more accurately by using the flow stress, $\sigma_f$, in Equation (9.2) rather than the yield strength, $\sigma_{ys}$.

## 9.13. Significance of Fatigue-Rate Transition

The fatigue-rate transition from region II behavior to region III behavior can be predicted by using Equation (9.2). The stress-intensity-factor value, $K_T$, that defines this region of transition under zero-to-tension loads can be calculated by rewriting Equation (9.2) in the form

$$K_T = 0.04\sqrt{E\sigma_{ys}} \tag{9.9}$$

where $K_T$ is in ksi$\sqrt{\text{in.}}$ and $E$ and $\sigma_{ys}$ are in ksi.

Acceleration in the rate of fatigue-crack growth occurs at a stress-intensity-factor value slightly below the critical stress-intensity factor, $K_{Ic}$ (or $K_c$), when the $K_{Ic}$ of the material is less than $K_T$ (steel B in Figure 9.4). Equation (9.9) is used to calculate the stress-intensity-factor value corresponding to onset of fatigue-rate transition, $K_T$, which also corresponds to the point of transition from region II to region III in materials that have high fracture toughness (steel A in Figure 9.4), that is, materials for which the critical stress-intensity factor, $K_{Ic}$ (or $K_c$), is higher than the $K_T$ value calculated by using Equation (9.9). This equation shows that acceleration in fatigue-crack-growth rate in materials having high fracture toughness, $K_{Ic}$ (or $K_c$) > $K_T$, is governed by Young's modulus, $E$, and by the yield strength, $\sigma_{ys}$ (or the flow stress, $\sigma_f$) of the material and is independent of the fracture toughness. Consequently, a significant increase in the fracture toughness of a metal above $K_T$ may have a negligible effect on the fatigue life of a structural component. Moreover, extrapolation of the behavior in region II to $K_{Ic}$ (or $K_c$) values that are higher than $K_T$ could seriously overestimate the useful fatigue life of a structural component.[54] Accurate predictions of finite fatigue lives of structural components subjected to high stress fluctuations require an exact characterization of the fatigue-crack-growth behavior above and below the fatigue-rate transition region and cannot be based on extrapolations.

## 9.14.  Design Example

For most structural materials, the tolerable flaw sizes are much larger than any initial undetected flaws.  However, for structures subjected to fatigue loading (or stress-corrosion cracking), these initial cracks can grow throughout the life of the structure.  Thus, an overall approach to preventing fracture or fatigue failures in large welded structures assumes that a small flaw of certain geometry exists after fabrication and that this flaw can either cause brittle fracture or grow by fatigue to the critical size.  To ensure that the structure does not fail by fracture, the calculated critical crack size, $a_{cr}$, at design load must be sufficiently large, and the number of cycles of loading required to grow a small crack to a critical crack must be greater than the design life of the structure.

Thus, although $S-N$ curves have been widely used to analyze the fatigue behavior of steels and weldments, closer inspection of the overall fatigue process in complex welded structures indicates that a more rational analysis of fatigue behavior is possible by using concepts of fracture mechanics. Specifically, small (possibly large) fabrication flaws may be present in welded structures, even though the structure has been inspected.  Accordingly, a conservative approach to designing to prevent fatigue failure would be to *assume the presence of an initial flaw* and analyze the fatigue-crack-growth behavior of the structural member.  The size of the initial flaw is obviously dependent on the detail geometry, quality of fabrication, and inspection.

A schematic diagram of the general relation between fatigue-crack initiation and propagation is shown in Figure 9.31.  The question of when does a crack "initiate" to become a "propagating" crack is somewhat philosophical and depends on the level of observation of a crack, that is, crystal imperfection, dislocation, microcrack, lack of penetration, and so on.  One approach to fatigue would be to assume an initial flaw size on the basis of the quality of inspection used and then to calculate the number of cycles it would take for this crack to grow to a size critical for brittle fracture.

The procedure to analyze the crack-growth behavior in steels and weld metals using fracture-mechanics concepts is as follows:

1. On the basis of quality of inspection, estimate the maximum initial flaw size, $a_0$, present in the structure and the associated $K_I$ relation (Chapter 2) for the member being analyzed.

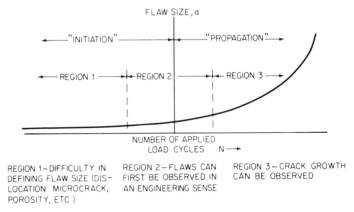

FLAW SIZE, a

← "INITIATION" → ← "PROPAGATION" →

← REGION 1 → ← REGION 2 → ← REGION 3 →

NUMBER OF APPLIED LOAD CYCLES   N →

REGION 1 – DIFFICULTY IN DEFINING FLAW SIZE (DIS- LOCATION MICROCRACK, POROSITY, ETC.)     REGION 2 – FLAWS CAN FIRST BE OBSERVED IN AN ENGINEERING SENSE     REGION 3 – CRACK GROWTH CAN BE OBSERVED

**Figure 9.31** Schematic showing relation between "initiation" life and "propagation" life.

2. Knowing $K_c$ or $K_{Ic}$ and the nominal maximum design stress, calculate the critical flaw size, $a_{cr}$, that would cause failure by brittle fracture.

3. Obtain an expression relating the fatigue-crack-growth rate of the steel or weld metal being analyzed. The following conservative estimates of the fatigue-crack growth per cycle of loading, $da/dN$, have been determined for martensitic steels (for example, A514/517) as well as ferrite-pearlite steels (for example, A36) in a room temperature air environment and were discussed previously,

$$martensitic\ steels: \frac{da}{dN} = 0.66 \times 10^{-8}(\Delta K_I)^{2.25} \qquad (9.4)$$

$$ferrite\text{-}pearlite\ steels: \frac{da}{dN} = 3.6 \times 10^{-10}(\Delta K_I)^3 \qquad (9.5)$$

where $da/dN$ = fatigue-crack growth per cycle of loading, in./cycle.

$\Delta K_I$ = stress-intensity-factor range, ksi$\sqrt{\text{in.}}$

4. Determine $\Delta K_I$ using the appropriate expression for $K_I$, the estimated initial flaw size, $a_0$, and the range of live-load stress (cyclic-stress range).

5. Integrate the crack-growth-rate expression between the limits of $a_0$ (at the initial $K_I$) and $a_{cr}$ (at $K_{Ic}$) to obtain the life of the structure prior to failure.

A numerical example of this procedure is as follows:

1. Assume the following conditions:

a. A514 steel, $\sigma_{ys} = 100$ ksi (689 MN/m$^2$).
b. $K_{Ic} = 150$ ksi$\sqrt{\text{in.}}$ (165 MN/mn$^{3/2}$).
c. $a_0 = 0.3$ in. (7.6 mm), edge crack in tension (Chapter 2).
d. $\sigma_{max} = 45$ ksi (310 MN/m$^2$);
   $\sigma_{min} = 25$ ksi (172 MN/m$^2$);
   $\Delta\sigma = 20$ ksi (138 MN/m$^2$) (live-load stress range).
e. $K_I = 1.12\sqrt{\pi\sigma}\sqrt{a}$ edge crack in tension (Chapter 2).

2. Calculate $a_{cr}$ at $\sigma = 45$ ksi (310 MN/m$^2$):

$$a_{cr} = \left(\frac{K_{Ic}}{1.12\sqrt{\pi\sigma_{max}}}\right)^2 = \left(\frac{150}{1.12(1.77)(45)}\right)^2$$

$$= 2.8 \text{ in. (71.1 mm)}$$

3. Assume an increment of crack growth, $\Delta a$. In this case assume that $\Delta a = 0.1$ in. (2.5 mm). If smaller increments of crack growth were assumed, the accuracy would be increased slightly.

4. Determine the expression for $\Delta K_I$, where $a_{avg}$ represents the average crack size between the two crack increments $a_f$ and $a_j$:

$$\Delta K_I = 1.12\sqrt{\pi}\Delta\sigma\sqrt{a_{avg}}$$

$$= 1.98(20)\sqrt{a_{avg}}$$

5. Using the appropriate expression for crack-growth rate,

$$da/dN = 0.66 \times 10^{-8}(\Delta K_I)^{2.25}$$

solve for $\Delta N$ for each increment of crack growth, replacing $da/dN$ by $\Delta a/\Delta N$:

$$\Delta N = \frac{\Delta a}{0.66 \times 10^{-8}[1.98(20)\sqrt{a_{avg}}]^{2.25}}$$

$$= 12,500 \text{ cycles}$$

6. Repeat for $a = 0.4$–$0.5$ in. (10.2 to 12.7 mm), and so on, by numerical integration, as shown in Table 9.2. The flaw size–life results for this example are presented in Figure 9.32. If only the desired total life is required, the expression for $\Delta N$ can be integrated directly. In this example, direct integration yielded a life of 87,600 cycles, while the numerical technique gave a life of 86,700 cycles.

Note that the total life to propagate a crack from 0.3 to 2.8 in. (7.6 to 71.1 mm) in this example is 86,700 cycles. If the required life were 100,000 cycles, then this design would be inadequate, and one or more of the following changes should be made:

1. Increase the critical crack size at failure [$a_{cr} = 2.8$ in. (71.1 mm)] by using a material with a higher $K_{Ic}$ value.

TABLE 9.2    Fatigue-Crack-Growth Calculations

$$\Delta N = \frac{\Delta a}{0.66 \times 10^{-8}[1.98(\Delta\sigma)\sqrt{a_{avg}}]^{2.25}}$$

where $\Delta a = 0.10$ in. (2.54 mm)

$\Delta\sigma = 20$ ksi (138 MN/m$^2$)

| $a_0$ (in.) | $a_f$ (in.) | $a_{avg}$ (in.) | $\Delta K$ (ksi$\sqrt{\text{in.}}$) | $\Delta N$ (cycles) | $\Sigma N$ (cycles) |
|---|---|---|---|---|---|
| 0.3 | 0.4 | 0.35 | 23.5 | 12,500 | 12,500 |
| 0.4 | 0.5 | 0.45 | 26.7 | 9,750 | 22,250 |
| 0.5 | 0.6 | 0.55 | 29.4 | 7,550 | 29,800 |
| 0.6 | 0.7 | 0.65 | 32.2 | 6,150 | 35,950 |
| 0.7 | 0.8 | 0.75 | 34.6 | 5,200 | 41,150 |
| 0.8 | 0.9 | 0.85 | 36.6 | 4,600 | 45,750 |
| 0.9 | 1.0 | 0.95 | 38.8 | 4,100 | 49,850 |
| 1.0 | 1.1 | 1.05 | 40.5 | 3,700 | 53,550 |
| 1.1 | 1.2 | 1.15 | 42.5 | 3,300 | 56,850 |
| 1.2 | 1.3 | 1.25 | 44.5 | 2,950 | 59,800 |
| 1.3 | 1.4 | 1.35 | 46.1 | 2,700 | 62,500 |
| 1.4 | 1.5 | 1.45 | 47.7 | 2,550 | 65,050 |
| 1.5 | 1.6 | 1.55 | 49.3 | 2,350 | 67,400 |
| 1.6 | 1.7 | 1.65 | 51.0 | 2,200 | 69,600 |
| 1.7 | 1.8 | 1.75 | 52.5 | 2,050 | 71,650 |
| 1.8 | 1.9 | 1.85 | 54.0 | 1,900 | 73,550 |
| 1.9 | 2.0 | 1.95 | 55.6 | 1,800 | 75,350 |
| 2.0 | 2.1 | 2.05 | 56.8 | 1,700 | 77,050 |
| 2.1 | 2.2 | 2.15 | 58.5 | 1,600 | 78,650 |
| 2.2 | 2.3 | 2.25 | 59.6 | 1,500 | 80,150 |
| 2.3 | 2.4 | 2.35 | 60.8 | 1,450 | 81,600 |
| 2.4 | 2.5 | 2.45 | 62.5 | 1,400 | 83,000 |
| 2.5 | 2.6 | 2.55 | 63.5 | 1,350 | 84,350 |
| 2.6 | 2.7 | 2.65 | 64.8 | 1,200 | 85,550 |
| 2.7 | 2.8 | 2.75 | 66.0 | 1,150 | 86,700 |

1 in. = 25.4 mm

1 ksi$\sqrt{\text{in.}}$ = 1.1 MN/m$^{3/2}$

2. Lower the design stress, $\sigma_{max}$, to increase the critical crack size at failure.

3. Lower the stress range ($\Delta\sigma$) to decrease the *rate* of crack growth, thereby increasing the number of cycles required for the crack to grow to the critical size. Note that because the rate of crack growth is a power function of $\Delta\sigma$, or actually $\Delta K$, lowering the stress range slightly has a significant effect on the life.

4. Improve the fabrication quality and inspection capability so that the initial flaw size ($a_0$) is reduced. It is clear from Table 9.2 and Figure 9.32 that most of the life is taken up in the early stages of crack

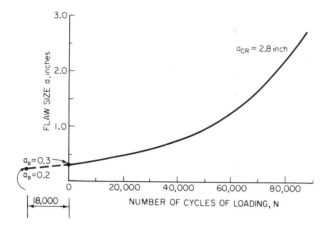

**Figure 9.32**  Fatigue-crack-growth curve.

propagation. In fact, to double the initial crack size during the early stages of propagation requires almost half the total number of cycles. Therefore, any decrease in initial flaw size has a very significant effect on the fatigue life of a structural member.

In this example, if $a_0$ were only 0.2 in. (5.1 mm), the design would be satisfactory. That is, the number of cycles to grow a crack 0.2–0.3 in. (5.1–7.6 mm) is about 18,000 cycles, as indicated in Figure 9.32, which [added to the 86,700 cycles required to grow the crack from 0.3 to 2.8 in. (7.6 to 71.1 mm)] would make the total life equal to 104,700 cycles. It should be noted that for steels with high toughness levels, the state of stress ahead of large cracks may be plane stress, and thus larger cracks could be tolerated than are calculated on the basis of plane-strain behavior. However, because the crack-growth rate increases rapidly for large cracks, the life may not be increased significantly. Moreover, because the crack-growth rate is increasing rapidly for large cracks, a significant increase in the fracture toughness of the material (that is, in the size of the tolerable crack at failure) may result in a negligible increase in the fatigue life of the structural member.

# References

1. P. C. PARIS and F. ERDOGAN, "A Critical Analysis of Crack Propagation Laws," *Transactions of the ASME, Journal of Basic Engineering*, Series D, *85*, No. 3, 1963.
2. N. FROST, "Notch Effects and the Critical Alternating Stress Required to Propagate a Crack in an Aluminium Alloy Subjected to Fatigue Loading," *Journal of the Mechanical Engineering Society, 2*, No. 2, 1960, pp. 109–119.

3. N. E. FROST, "The Growth of Fatigue Cracks," *Proceedings of the First International Conference on Fracture*, Sendai, Japan, 1966, p. 1433.

4. F. A. McCLINTOCK, "On the Plasticity of the Growth of Fatigue Cracks," *Fracture of Solids*, Wiley-Interscience, New York, 1963, p. 65.

5. B. M. LINDNER, "Extremely Slow Crack Growth Rates in Aluminium Alloy 7075-T6," M.S. Thesis, Lehigh University, Bethlehem, Pa., 1965.

6. P. C. PARIS, "Testing for Very Slow Growth of Fatigue Cracks," *MTS Closed Loop Magazine, 2*, No. 5, 1970.

7. W. ELBER, "The Significance of Crack Closure," *Damage Tolerance in Aircraft Structures, ASTM STP 486*, AMERICAN SOCIETY FOR TESTING AND MATERIALS, Philadelphia, 1971, pp. 230–242.

8. R. A. SCHMIDT and P. C. PARIS, "Threshold for Fatigue Crack Propagation and Effects of Load Ratio and Frequency," *Progress in Flaw Growth and Fracture, ASTM STP 536*, AMERICAN SOCIETY FOR TESTING AND MATERIALS, Philadelphia, 1973, p 79.

9. *Fatigue Thresholds—Fundamentals and Engineering Applications*, edited by J. Bäcklund, A. F. Blom, and C. J. Beevers, Vols. I and II, Engineering Materials Advisory Services Ltd., United Kingdom, 1982.

10. R. J. BUCCI, W. G. CLARK, JR., and P. C. PARIS, "Fatigue-Crack-Propagation Growth Rates Under a Wide Variation of $\Delta K$ for an ASTM A517 Grade F (T-1) Steel," *ASTM STP 513*, American Society for Testing and Materials, Philadelphia, 1972.

11. J. D. HARRISON, "An Analysis of Data on Non-Propagating Fatigue Cracks on a Fracture Mechanics Basis," *British Welding Journal, 2*, No. 3, Mar. 1970.

12. L. P. POOK, "Fatigue Crack Growth Data for Various Materials Deduced from the Fatigue Lives of Precracked Plates," *ASTM STP 513*, American Society for Testing and Materials, Philadelphia, 1972.

13. P. C. PARIS, R. J. BUCCI, E. T. WESSEL, W. G. CLARK, and T. R. MAGER, "Extensive Study of Low Fatigue Crack Growth Rates in A533 and A508 Steels," *ASTM STP 513*, American Society for Testing and Materals, Philadelphia, 1972.

14. G. R. YODER, L. A. COOLEY, and T. W. CROOKER, "A Critical Analysis of Grain Size and Yield Strength Dependence of Near-Threshold Fatigue-Crack Growth in Steels," *Naval Research Laboratory Report*, Naval Research Laboratory, Washington, D.C., July 15, 1981.

15. H. KITAGAWA and S. TAKAHASHI, "Applicability of Fracture Mechanics to Very Small Cracks or the Cracks in the Early Stages." *Proceedings of the Second International Conference on Mechanical Behavior of Materials*, American Society for Metals, Metals Park, Ohio, 1976, p. 627.

16. M. H. EL HADDAD, N. E. DOWLING, T. H. TOPPER, and K. N. SMITH, "J Integral Applications for Short Fatigue Cracks at Notches," *International Journal of Fracture*, 16, 1980, p. 15.

17. J. M. BARSOM, "Fatigue-Crack Propagation in Steels of Various Yield Strengths," *Transactions of the ASME, Journal of Engineering for Industry*, Series B, *93*, No. 4, Nov. 1971.

18. J. M. BARSOM, E. J. IMHOF, JR., and S. T. ROLFE, "Fatigue-Crack Propagation in High-Yield-Strength Steels," *Engineering Fracture Mechanics, 2*, No. 4, June 1971.

19. J. M. Barsom, "The Dependence of Fatigue Crack Propagation on Strain Energy Release Rate and Crack Opening Displacement," *ASTM STP 486*, American Society for Testing and Materials, Philadelphia, 1971.

20. T. W. Crooker, "Crack Propagation in Aluminum Alloys Under High-Amplitude Cyclic Load," *Naval Research Laboratory Report 7286*, Washington, D.C., July 12, 1971.

21. J. M. Barsom, "Fatigue Behavior of Pressure-Vessel Steels," *WRC Bulletin*, No. 194, Welding Research Council, New York, May 1974.

22. J. M. Barsom, "Investigation of Subcritical Crack Propagation," Doctor of Philosophy Dissertation, University of Pittsburgh, Pittsburgh, 1969.

23. E. J. Imhof and J. M. Barsom, "Fatigue and Corrosion-Fatigue Crack Growth of 4340 Steel at Various Yield Strengths," *ASTM STP 536*, American Society for Testing and Materials, Philadelphia, 1973.

24. C. G. Schilling and K. H. Klippstein, "Fatigue of Steel Beams by Simulated Bridge Traffic," *Journal of the Structural Division, 103*, No. ST8, August 1977, pp. 1561–1575.

25. S. Suresh and R. O. Ritchie, "Closure Mechanisms for the Influence of Load Ratio on Fatigue Crack Propagation in Steels," U.S. Department of Energy, Contract No. DE-ACO3-76SF00098, April 1982.

26. T. C. Lindley and C. E. Richards, "Near-Threshold Fatigue Crack Growth in Materials Used in the Electricity Supply Industry," *Fatigue Thresholds— Fundamentals and Engineering Applications*, Engineering Materials Advisory Services Ltd., United Kingdom, 1982, pp. 1087–1113.

27. E. Sasaki, A. Ohta, and M. Kosuge, "Fatigue Crack Propagation Rate and Stress Intensity Threshold Level of Several Structural Materials at Varying Stress Ratios $(-1 \sim 0.8)$," *Transactions of the National Research Institute for Metals* (Japan), *19*, No. 4, 1977, pp. 183–199.

28. E. K. Priddle, "The Threshold Stress Intensity Factor for Fatigue Crack Growth in Mild Steel Plate and Weld Metal: Some Effects of Temperature and Environment," *Fatigue Thresholds—Fundamentals and Engineering Applications*, Engineering Materials Advisory Services Ltd., United Kingdom, 1982, pp. 581–600.

29. S. J. Hudak, Jr., and R. J. Bucci, eds., "Fatigue Crack Growth Measurement and Data Analysis," *ASTM STP 738*, American Society for Testing and Materials, 1981, pp. 340–356.

30. J. M. Barsom, "Fatigue-Crack Growth Under Variable-Amplitude Loading in ASTM A514 Grade B Steel," *ASTM STP 536*, American Society for Testing and Materials, Philadelphia, 1973.

31. M. Parry, H. Nordberg, and R. W. Hertzberg, "Fatigue Crack Propagation in A514 Base Plate and Welded Joints," *Welding Journal, 51*, No. 10, Oct. 1972.

32. W. G. Clark, Jr., "Effect of Temperature and Section Size on Fatigue Crack Growth in A533 Grade B, Class 1 Pressure Vessel Steel," *Journal of Materials, 6*, No. 1, March 1971.

33. R. J. Bucci, B. N. Greene, and P. C. Paris, "Fatigue Crack Propagation and Fracture Toughness of 5Ni and 9Ni Steel at Cryogenic Temperatures," *ASTM STP 536*, American Society for Testing and Materials, Philadelphia, 1973.

34. L. A. James, "The Effect of Elevated Temperature upon the Fatigue-Crack

Propagation Behavior of Two Austenitic Stainless Steels," *Mechanical Behavior of Materials*, Vol. III, The Society of Materials Science, Tokyo, 1972.

35. P. Shahinian, H. E. Watson, and H. H. Smith, "Fatigue Crack Growth in Selected Alloys for Reactor Applications," *Journal of Materials, 7*, No. 4, 1972.

36. J. H. Weber and R. W. Hertzberg, "Effect of Thermomechanical Processing on Fatigue Crack Propagation," *Metallurgical Transactions, 4*, Feb. 1973.

37. W. G. Clark, Jr., "How Fatigue Crack Initiation and Growth Properties Affect Material Selection and Design Criteria," *Metals Engineering Quarterly*, Aug. 1974.

38. W. G. Clark, Jr., and E. T. Wessel, "Interpretation of the Fracture Behavior of 5456-H321 Aluminum with WOL Toughness Specimens," *Scientific Paper 67-1D6-BTLFR-P4*, Westinghouse Research Laboratory, Pittsburgh, Sept. 1967.

39. T. W. Crooker, "Factors Determining the Performances of High Strength Structural Metals (Slope Transition Behavior of Fatigue Crack Growth Rate Curves)," *NRL Report of Progress*, Naval Research Laboratory, Washington, D.C., Dec. 1970.

40. K. Walker, "The Effect of Stress Ratio During Crack Propagation and Fatigue for 2024-T3 and 7075-T6 Aluminum," *ASTM STP 462*, American Society for Testing and Materials, Philadelphia, 1970.

41. T. W. Crooker, "Effect of Tension-Compression Cycling on Fatigue Crack Growth in High-Strength Alloys," *Naval Research Laboratory Report 7220*, Naval Research Laboratory, Washington, D.C., Jan. 1971.

42. T. W. Crooker and D. J. Krause, "The Influence of Stress Ratio and Stress Level on Fatigue Crack Growth Rates in 140-ksi YS Steel," *Report of NRL Progress*, Naval Research Laboratory, Washington, D.C., Dec. 1972.

43. D. J. Krause and T. W. Crooker, "Effect of Constant-Amplitude Loading Parameters on Low-Cycle Fatigue-Crack Growth in a 140- to 150-ksi Yield Strength Steel," *Report of NRL Progress*, Naval Research Laboratory, Washington, D.C., Mar. 1973.

44. R. P. Wei, "Fracture Mechanics Approach to Fatigue Analysis in Design," *ASME Paper No. 73-DE-22*, New York, April 1973.

45. J. M. Barsom, "Effect of Cyclic-Stress Form on Corrosion-Fatigue Crack Propagation Below $K_{Iscc}$ in a High-Yield-Strength Steel," in *Corrosion Fatigue: Chemistry, Mechanics, and Micro-Structure*, International Corrosion Conference Series, Vol. NACE-2, National Association of Corrosion Engineers, Houston, 1972.

46. J. M. Barsom, "Corrosion Fatigue Crack Propagation Below $K_{Iscc}$," *Journal of Engineering Fracture Mechanics, 3*, No. 1, July 1971.

47. S. R. Novak and J. M. Barsom, "Brittle Fracture ($K_{Ic}$) Behavior for Cracks Emanating from Notches," *Cracks and Fracture, ASTM STP 601*, American Society for Testing and Materials, Philadelphia, 1976.

48. W. G. Clark, Jr., "Fatigue Crack Growth Characteristics of Heavy Section ASTM A533 Grade B Class 1 Steel Weldments," *ASME Paper No. 70-PVP-24*, American Society of Mechanical Engineers, New York, 1970.

49. W. G. Clark, Jr., and D. S. Kim, "Effect of Synthetic Seawater on the Crack Growth Properties of HY-140 Steel Weldments," *Engineering Fracture Mechanics, 4*, 1972.

50. P. SHAHINIAN, H. H. SMITH, and J. R. HAWTHORNE, "Fatigue Crack Propagation in Stainless Steel Weldments at High Temperature," *Welding Journal, 51*, No. 11, Nov. 1972.

51. S. J. MADDOX, "Assessing the Significance of Flaws in Welds Subject to Fatigue," *Welding Journal, 53*, No. 9, Sept. 1974.

52. T. R. GURNEY, *Fatigue of Welded Structures*, Cambridge University Press, New York, 1968.

53. R. M. N. PELLOUX, "Fractographic Analysis of the Influence of Constituent Particles on Fatigue-Crack Propagation in Aluminum Alloys," *ASME Transactions Quarterly, 57*, No. 2, June 1964.

54. C. M. CARMAN and J. M. KATLIN, "Low Cycle Fatigue Crack Propagation of High-Strength Steels," *ASME Paper No.* 66-MET-3, New York, April 1966.

# Fatigue-Crack Propagation under Variable-Amplitude Load Fluctuation

## 10.1. Introduction

Many engineering structures are subjected to complex fluctuating-load environments. Research conducted since the early 1960s has shown that the variation in the magnitude of the load cycles may affect the fatigue life of components significantly. A thorough understanding of the effect of various variables on the fatigue-crack initiation and propagation under variable-amplitude load fluctuations is essential to the development of accurate prediction methods of fatigue lives for engineering structures. Unfortunately, the effects of variable-amplitude loading on fatigue life are presently not well established. The results of the traditional $S–N$ approach for variable-amplitude loading, like constant-amplitude loading, combine fatigue-crack initiation and propagation in different proportions.[1-5] The fracture-mechanics investigations of fatigue lives under variable-amplitude loading have concentrated exclusively on the fatigue-crack-propagation behavior of metals.[6-11] Some significant results of these investigations are presented in this chapter.

## 10.2. Stress Histories

The development of accurate fatigue-life prediction procedures for variable-amplitude cyclic-load fluctuations requires an understanding and characterizing of stress environments.

The simplest cyclic-stress history is the constant-amplitude sinusoidal-stress fluctuation shown in Figure 10.1. Such a stress history can be represented by a steady mean stress, $\sigma_{mean}$, and a fluctuating stress range, $\Delta\sigma$. The stress range is the algebraic difference between the maximum stress, $\sigma_{max}$, and the minimum stress, $\sigma_{min}$, in the cycle

$$\Delta\sigma = \sigma_{max} - \sigma_{min} \tag{10.1}$$

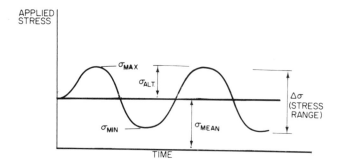

**Figure 10.1**    Terminology for constant-amplitude cyclic-stress loading.

The mean stress is the algebraic mean of $\sigma_{max}$ and $\sigma_{min}$ in the cycle

$$\sigma_{mean} = \frac{\sigma_{max} + \sigma_{min}}{2} \tag{10.2}$$

The alternating stress or stress amplitude is half of the stress range in the cycle

$$\sigma_{alt} = \frac{\Delta\sigma}{2} = \frac{\sigma_{max} - \sigma_{min}}{2} \tag{10.3}$$

A constant-amplitude sinusoidal-stress history can be represented by an analytical function and can be described by various parameters. Unfortunately, many structural components are subjected to random-stress histories that cannot be represented by an analytical function and that lack a describable pattern, as shown in Figure 10.2.

Between the extremes of a simple, constant-amplitude sinusoidal-stress history and a complex, random-stress history, there is a multitude of stress patterns of varying degrees of complexity which can be described by analytic functions and represented by various parameters.[12] Some simple variable-amplitude stress histories are those corresponding to a single or multiple

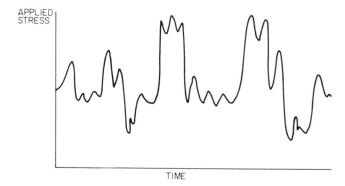

**Figure 10.2**    Random-stress loading.

high-tensile-load cycles superimposed upon constant-amplitude cyclic-load fluctuations (Figure 10.3).

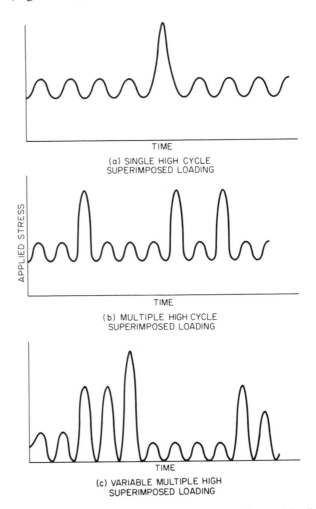

(a) SINGLE HIGH CYCLE
SUPERIMPOSED LOADING

(b) MULTIPLE HIGH CYCLE
SUPERIMPOSED LOADING

(c) VARIABLE MULTIPLE HIGH
SUPERIMPOSED LOADING

**Figure 10.3**  Typical single or multiple high-cycle superimposed loading. (a) Single high-cycle superimposed loading, (b) multiple high-cycle superimposed loading, and (c) variable multiple high superimposed loading.

## 10.3. Probability-Density Distribution

Many engineering structures such as bridges, ships, and others are subjected to variable-amplitude random-sequence load fluctuations. The probability of occurrence of the same sequence of stress fluctuations for a given detail in such structures obtained during a given time interval is very small.

Consequently, the magnitude of stress fluctuations must be characterized to study the fatigue behavior of components subjected to variable-amplitude random-sequence stress fluctuations. The magnitude of the stress fluctuations should be characterized and described by analytic functions. The use of probability-density curves to characterize variable-amplitude cyclic-stress fluctuations appears to be very useful.[10,13–15]

Stress history, or stress spectrum, for a particular location in a structure subjected to variable-amplitude stress fluctuation can be defined in terms of the frequency of occurrence of maximum (peak) stresses. Usually, frequency-of-occurrence data are presented as a histogram, or bar graph (Figure 10.4), in which the height of the bar represents the percentage of recorded maximum stresses that fall within a certain stress interval represented by

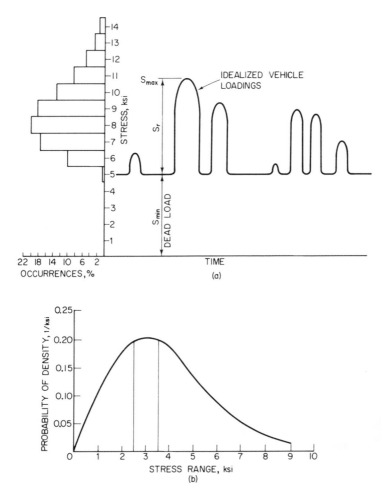

**Figure 10.4**   Frequency-of-occurrence data.

the width of the bar. For example, 20.2 percent of the maximum stresses in Figure 10.4(a) fall within the interval between 7.5 and 8.5 ksi (52 to 59 MN/m$^2$). The frequency of occurrence of stress ranges can be represented by similar plots with the vertical scale changed according to the relationship between $\sigma_{max}$, $\sigma_{min}$, and stress range, $\sigma_r$ or $\Delta\sigma$. Since stress range is the most important stress parameter controlling the fatigue life of structural components, stress range is used to define the major stress cycles in the following discussion.

The frequency-of-occurrence data can be presented in a more general form by dividing the percentage of occurrence for each interval, that is, the height of each bar, in Figure 10.4(a) by the interval width to obtain a probability-density curve such as shown in Figure 10.4(b). Thus, data from sources that use different stress-range intervals can be compared by using the probability-density curve. The area under the curve between any two values of $\Delta\sigma$ represents the percentage of occurrence within that interval.

A single nondimensional mathematical expression can be used to define the probability-density curves for different sets of data. For example, Klippstein and Schilling[14] showed that the following nondimensional mathematical expression, which defines a family of skewed probability-density curves referred to as Rayleigh curves or distribution functions, can be used to fit a probability-density curve accurately to each available set of field data for bridges:

$$p' = 1.011x'e^{-1/2(x')^2} \qquad (10.4)$$

where $x' = (\sigma_r - \sigma_{r\min})/\sigma_{rd}$, $\sigma_r$ (i.e., $\Delta\sigma$) is the stress range, and $\sigma_{r\min}$ (i.e., $\Delta\sigma_{\min}$) and $\sigma_{rd}$ (that is, $\Delta\sigma_d$) are constant parameters that define any particular probability-density curve from the family of curves represented by Equation (10.4). Equation (10.4) is plotted in Figure 10.5(a). As illustrated in Figure 10.5(b), a particular curve from the family is defined by two parameters: (1) the modal stress range, $\sigma_{rm}$, which corresponds to the peak of the curve; and (2) the parameter $\sigma_{rd}$, which is a measure of the width of the curve or the dispersion of the data. The curve could be shifted sideways by changing $\sigma_{rm}$, and the width of the curve could be modified by changing $\sigma_{rd}$. Mathematical expressions for the modal, median, mean, and root-mean-square values of the spectrum are given in Figure 10.5. The root-mean-square (rms) value is defined as the square root of the mean of the squares of the individual values of $x'$ or $\sigma_r$. Stress ($\sigma$) is represented as $S$ in Figure 10.5.

## 10.4. Fatigue-Crack Growth under Variable-Amplitude Loading

Many attempts have been made to predict fatigue-crack-growth behavior under variable-amplitude loading. The following sections present the behavior

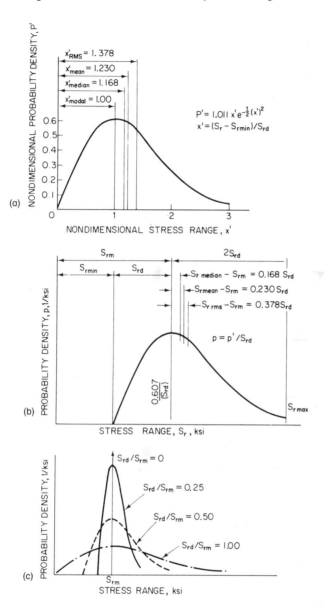

**Figure 10.5**   Characteristics of Rayleigh probability curves.

under simple and complex variable-amplitude loading. The simple loadings correspond to a single or multiple high-tensile-load fluctuations superimposed upon constant-amplitude cyclic-load fluctuations (Figure 10.3). Complex variable-amplitude loadings correspond to multivalue cyclic-load fluctuations.

## 10.5. Single and Multiple High-Load Fluctuations

Several investigators[6-9,11,16-18] observed that changes in cyclic-load magnitude (Figure 10.3) may result in retarded or accelerated fatigue-crack-growth rate. Extensive published data show that the rate of fatigue-crack growth under constant-amplitude cyclic-load fluctuation can be retarded significantly as a result of application of a single or multiple tensile-load cycle having a peak load greater than that of the constant-amplitude cycles (Figure 10.6). Von Euw[19] observed that the minimum value of fatigue-crack-growth rate did not occur immediately following the high-tensile-load cycle but that the rate of growth decelerated to a minimum value. This deceleration region has been termed "delayed retardation," Figure 10.6(b).

Several models have been advanced to explain the phenomenon of crack-growth delay. In general, these models attribute the delayed behavior to crack-tip blunting, residual stresses,[20,21] crack closure,[22] or a combination

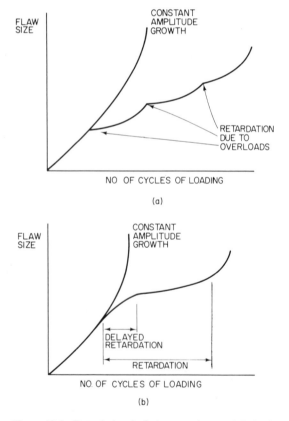

Figure 10.6    Retardation in fatigue-crack-growth behavior.

of these mechanisms.  A crack-tip-blunting model advocates that high-tensile-load cycles cause crack-tip blunting, which in turn causes retardation in fatigue-crack growth at the lower cyclic-load fluctuations until the crack is resharpened.  The residual-stress model suggests that the application of a high-tensile-load cycle forms residual compressive stresses in the vicinity of the crack tip that reduce the rate of fatigue-crack growth.  Finally, the crack-closure model postulates that the delay in fatigue-crack growth is caused by the formation of a zone of residual tensile deformation left in the wake of a propagating crack that causes the crack to remain closed during a portion of the applied tensile-load cycle.  Consequently, fatigue-crack-growth delay occurs because only the portion of the tensile-load cycles that is above the crack-opening level is effective in extending the crack.  These models are useful to describe trends in fatigue-crack-growth-rate behavior caused by single or multiple high-tensile-load cycles but are of little value to predict fatigue lives under these conditions.  Fatigue-crack-growth delay has been shown to be strongly dependent on all the loading variables, such as the stress-intensity-factor fluctuation, the $\Delta K_I$ of the high-tensile-load cycle, the $\Delta K_I$ of the constant-amplitude cycles (Figure 10.7)[11], the stress ratios of these $\Delta K_I$ values, and the number of constant-amplitude cycles between the high-tensile-load cycles.[8,11,23,24]  Extensive research is necessary to further our understanding of the significance of these variables in order to develop equations that can be used to predict accurately the fatigue life of components subjected to single or multiple high-tensile-load cycles.

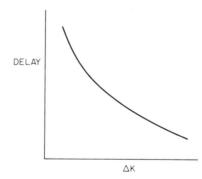

**Figure 10.7**   Schematic showing effect of $\Delta K$ on fatigue-crack-growth delay.

## 10.6. Variable-Amplitude Load Fluctuations

Extensive investigations are currently being conducted to develop methods to predict fatigue lives under variable-amplitude load fluctuations.  Some of these investigations resulted in models that can be used to describe trends in fatigue-crack-propagation rates but have varying degrees of success in predicting fatigue lives under variable-amplitude loads.[10,11,20,21] Presently, the models presented by Barsom[10,13,25] and by Wei and Shih[11] appear to be promising. The model advocated by Wei and Shih is a superposition model where the delay cycles caused by a change in stress magnitude are superimposed on the cycles obtained under constant-amplitude loading, assuming no load interactions. The delay cycles in this model must be estimated from experimental data.

The model advanced by Barsom[10] relates fatigue-crack-growth rate per cycle to an effective stress-intensity factor that is characteristic of the probability-density curve. The development of this model, which was designated the root-mean-square (rms) model, and the supporting experimental data are presented in the following sections.

### 10.6.1. The Root-Mean-Square (RMS) Model

Incremental increase of crack length and the corresponding number of elapsed load cycles can be measured under variable-amplitude random-sequence load spectra. However, unlike constant-amplitude cyclic-load data, the magnitude of $\Delta K_I$ changes for each cycle. Reduction of data in terms of fracture-mechanics concepts requires the establishment of a correlation parameter that incorporates the effects of crack length, cyclic-load amplitude, and cyclic-load sequence.

Barsom[10] attempted to determine the magnitude of constant-amplitude cyclic-load fluctuation that results in the same $a$ versus $N$ curve obtained under variable-amplitude cyclic-load fluctuation when both spectra are applied to identical specimens (including initial crack length). In other words, one of the objectives of his investigation was to find a single stress-intensity parameter, such as mean, modal, or root mean square, that can be used to define the crack-growth rate under both constant- and variable-amplitude loadings. The selected parameter *must* characterize the probability-density curve.

This methodology for analyzing the fatigue behavior under variable-amplitude loading should be applicable to loading conditions that result in relatively smooth, continuous crack-length ($a$) versus number-of-cycles ($N$) curves, as shown in Figure 10.8. Such curves can be obtained from many variable-amplitude loadings and from frequently applied overloads but not from single or infrequent overloads.

Variable-amplitude random-sequence load spectra having Rayleigh probability-density curves of $P_{rd}/P_{rm}$ (or $\sigma_{rd}/\sigma_{rm}$) values equal to 0.5 and 1.0 were investigated as part of Project 12–12 of the National Cooperative Highway Research Program (NCHRP) of the National Academy of Sciences.[10,13,15] A typical portion of the 500-cycle loading block for each is shown in Figure 10.9. Data for crack length versus the corresponding

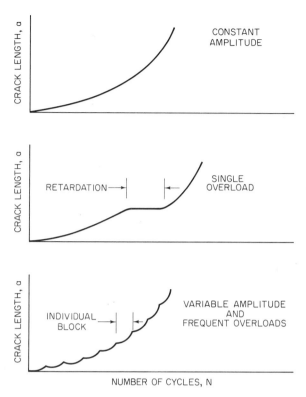

**Figure 10.8** Schematic representation of crack extension under different loading conditions.

number of elapsed load cycles obtained by subjecting identical specimens to these random-sequence load spectra are presented in Figure 10.10. Figure 10.10 also includes data obtained under constant-amplitude cyclic-load fluctuation ($P_{rd}/P_{rm} = 0$). The load range, $P_r$, for every cycle in the constant-amplitude tests was equal to $P_{rm}$. The data show that the fatigue life under constant-amplitude cyclic-load fluctuation is longer than the life obtained under random-sequence load spectra having the same value of $P_{rm}$. The data are re-presented in Figure 10.11 in terms of crack-growth rate and the modal stress-intensity factor, $K_{rm}$.

A better correlation between data obtained under constant-amplitude and variable-amplitude random-sequence load spectra was obtained on the basis of the root mean square of the load distribution, where the root mean square is the square root of the mean of the squares of the individual load cycles in a spectrum. The combined crack-growth-rate data are presented in Figure 10.12 as a function of $\Delta K_{rms}$. The data show that, within the limits of the experimental work, the average fatigue-crack-growth rates per

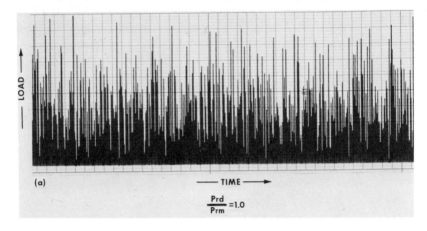

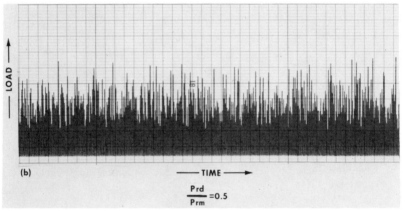

**Figure 10.9** Two variable-amplitude random-sequence load fluctuations investigated.

cycle, $da/dN$, under variable-amplitude random-sequence stress spectra can be represented by

$$\frac{da}{dN} = A(\Delta K_{\text{rms}})^m \tag{10.5}$$

where $A$ and $m$ are constants and

$$\Delta K_{\text{rms}} = \sqrt{\frac{\sum\limits_{i=1}^{k} \Delta K_i^2}{n}}$$

The root-mean-square value of the stress-intensity factor under constant-

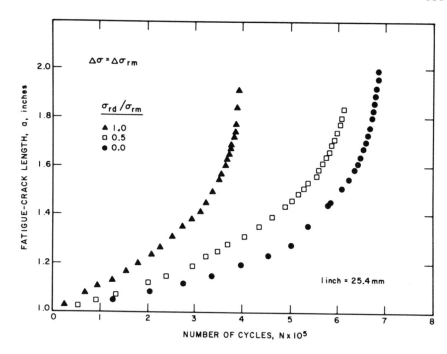

**Figure 10.10**    Crack growth under spectra loads.

amplitude cyclic-load fluctuation is equal to the stress-intensity-factor fluctuation. Consequently, the average fatigue-crack-growth rate can be predicted from constant-amplitude data by using Equation (10.5).

### 10.6.2.  Fatigue-Crack Growth Under Variable-Amplitude Ordered-Sequence Cyclic Load

The root-mean-square stress-intensity factor, $\Delta K_{rms}$, is characteristic of the load-distribution curve and is independent of the order of the cyclic-load fluctuations. To determine whether the order of load fluctuations affects the average rate of crack growth, fatigue tests were performed on identical specimens under random and ordered variable-amplitude cyclic-load fluctuations that represent the same load-distribution curve with $P_{rd}/P_{rm} = 1.0$.[10,13] The tests were conducted at a constant minimum load, $P_{min}$, with $P_{min}/P_{rm} = 0.25$.

Fatigue-crack-growth-rate tests were conducted by using the variable-amplitude random-sequence load fluctuations shown in Figure 10.13(a). In other tests these same load fluctuations were arranged in descending magnitudes, Figure 10.13(b), ascending magnitudes, Figure 10.13(c), and combined

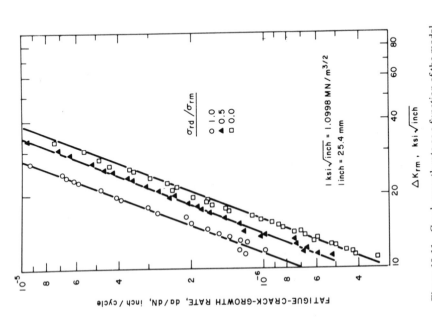

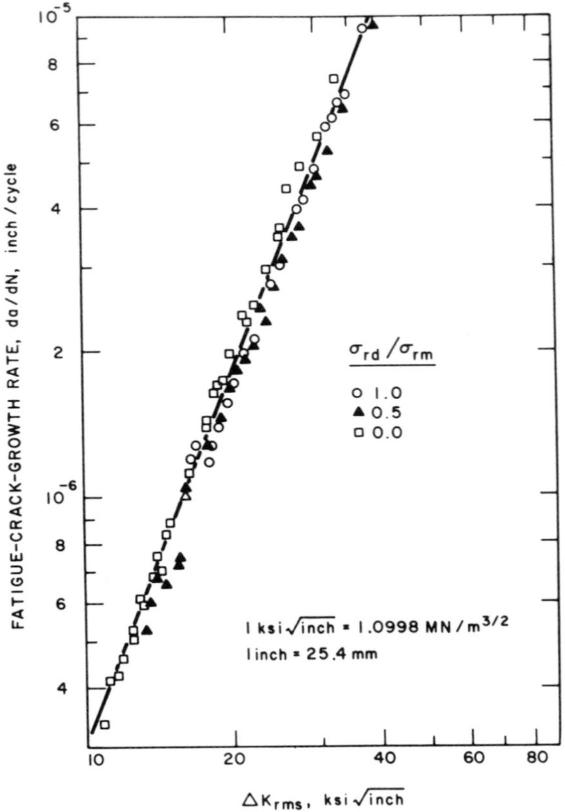

**Figure 10.12**  Crack-growth rate as a function of the root-means-square stress-intensity factor.

**Figure 10.11**  Crack-growth rate as a function of the modal stress-intensity factor.

332

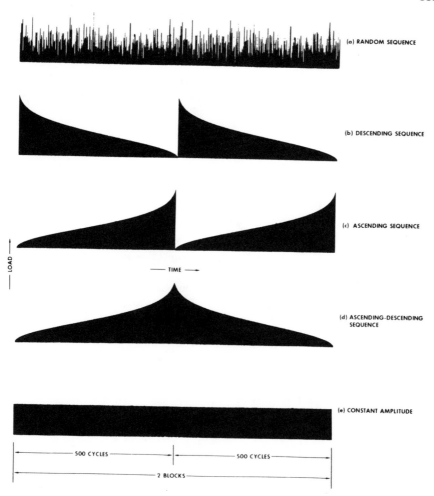

**Figure 10.13** Various random-sequence and ordered-sequence load fluctuations studied to establish the $\Delta K_{rms}$ analysis.

ascending-descending magnitudes, Figure 10.13(d). The fatigue-crack-growth data obtained under these various conditions and under a constant-amplitude cyclic-load fluctuation of $P_r = \Delta P_{rms}$, Figure 10.13(e), are presented in Figure 10.14. The data show that the average rate of fatigue-crack growth is represented accurately by Equation (10.5) regardless of the order of occurrence of the cyclic-load fluctuations.

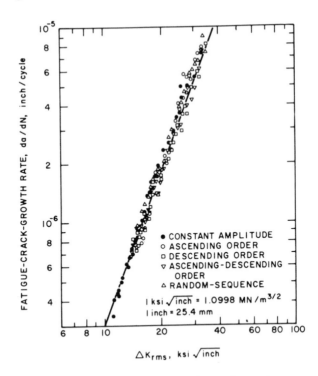

**Figure 10.14**  Summary of crack-growth-rate data under random-sequence and ordered-sequence load fluctuations.

## 10.7. Fatigue-Crack Growth in Various Steels

The preceding results were obtained by testing A514 Grade B steel under variable-amplitude random- and ordered-sequence cyclic-load fluctuations. Because several investigators[6-9,11,16-18] have noted that changes in cyclic-load magnitude can lead to accelerated or retarded rates of fatigue-crack growth in various metals, the applicability of the RMS model to steels of various yield strengths was investigated under NCHRP Project 12–14.[25] Fatigue-crack-growth rates under constant-amplitude and variable-amplitude random-sequence load fluctuations were investigated for A36, A588 Grade A, A588 Grade B, A514 Grade E, and A514 Grade F steels. All loadings followed a Rayleigh probability-density curve, with the ratio of the load-range deviation to the model (peak) load ($P_{rd}/P_{rm}$) equal to either 0 or 1.0. The data presented in Figures 10.15 through 10.19 show that, within the limits of the experimental work, the average fatigue-crack-growth rates per

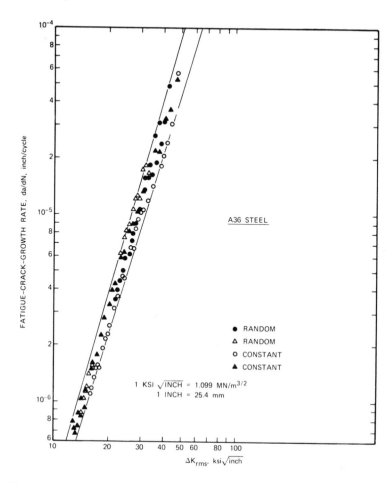

**Figure 10.15** Crack-growth rate as a function of the root-mean-square stress-intensity factor for A36 steel.

cycle, $da/dN$, in various steels subjected to variable-amplitude load spectra can be represented by Equation (10.5). Moreover, the average fatigue-crack-growth rates for the various steels studied under variable-amplitude random-sequence load fluctuations are equal to the average fatigue-crack-growth rates obtained under constant-amplitude load fluctuations when the stress-intensity-factor range, $\Delta K$, under constant-amplitude load fluctuations is equal in magnitude to the $\Delta K_{rms}$ of the variable-amplitude spectra.

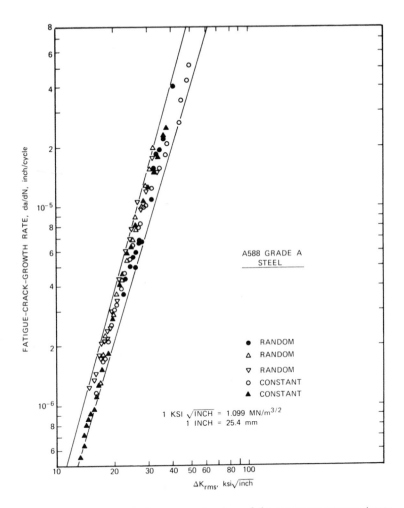

**Figure 10.16**   Crack-growth rate as a function of the root-mean-square stress-intensity factor for A588 Grade A steel.

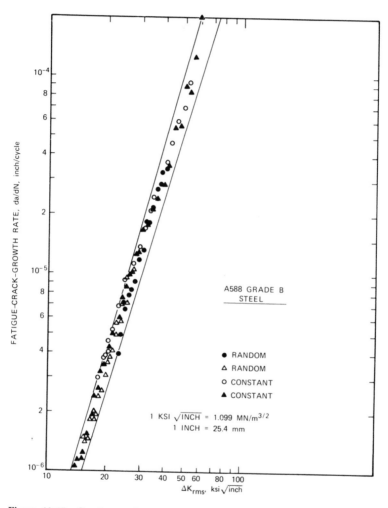

**Figure 10.17**  Crack-growth rate as a function of the root-mean-square stress-intensity factor for A588 Grade B steel.

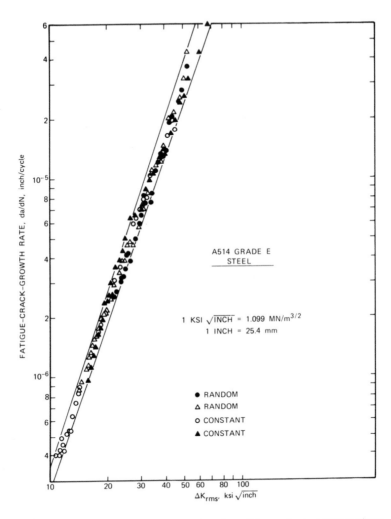

**Figure 10.18** Crack-growth rate as a function of the root-mean-square stress-intensity factor for A514 Grade E steel.

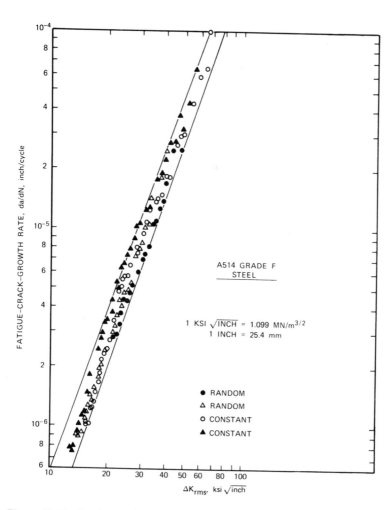

**Figure 10.19**  Crack-growth rate as a function of the root-mean-square stress-intensity factor for A514 Grade F steel.

## 10.8. Fatigue-Crack Growth Under Various Unimodal Distribution Curves

The applicability of the RMS model for correlating crack-growth rates under variable-amplitude random-sequence load fluctuations that follow unimodal distribution curves different from the Rayleigh type have been studied. Fatigue-crack-growth rates under constant-amplitude load fluctuations and

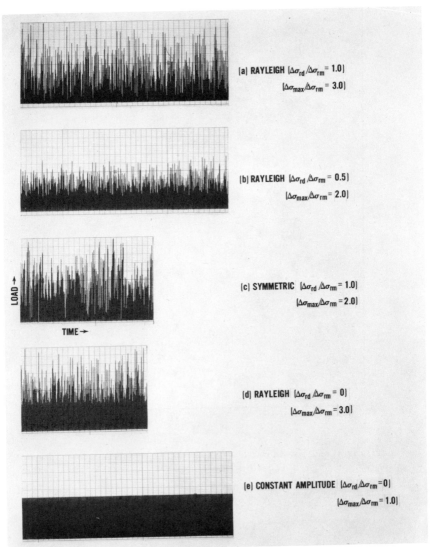

**Figure 10.20**   Single blocks of load fluctuations for various distribution functions.

under four unimodal variable-amplitude random-sequence load fluctuations were investigated by Barsom.[26] A block of variable-amplitude random-sequence load fluctuations of each unimodal distribution curve and of constant-amplitude load fluctuations is presented in Figure 10.20. Each block was applied repeatedly to a single specimen until the test was terminated. The number of cycles in the blocks corresponding to the four distribution curves investigated varied between 302 and 500 cycles. Schematics of the distribution curves corresponding to the various blocks shown in Figure 10.20 are presented in Figure 10.21. These curves cover a wide variation of unimodal distribution curves.

Data of fatigue-crack-growth rate per cycle and the corresponding root-mean-square stress-intensity-factor fluctuation, $\Delta K_{rms}$, obtained by

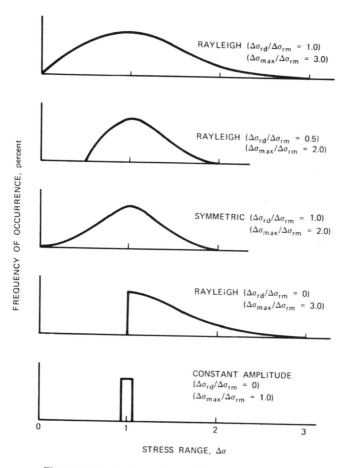

**Figure 10.21** Various unimodal distribution functions.

subjecting identical specimens to the load fluctuations shown in Figure 10.20 are presented in Figure 10.22. The data show that the average fatigue-crack-growth rate, $da/dN$, under constant-amplitude and variable-amplitude random-sequence load fluctuations that follow various unimodal distribution curves can be predicted by using the RMS model, Equation (10.5), of fatigue-crack growth. Further investigations are needed to establish the effects of various parameters, such as variable minimum load, on the rate of fatigue-crack growth under variable-amplitude load fluctuations and the necessary modifications to the RMS model to account for these effects or the development of a better model.

Finally, an example for analysis of the fatigue behavior for a welded structural component subjected to variable-amplitude cyclic loading is presented in Chapter 14.

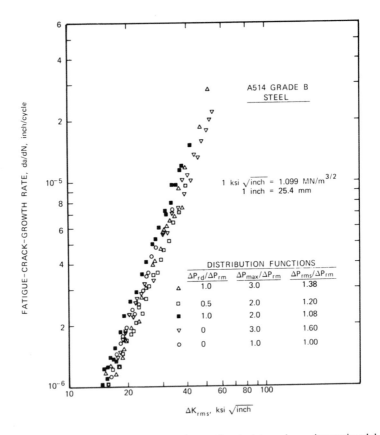

**Figure 10.22**   Summary of fatigue-crack growth-rate data under various unimodal distribution functions.

# References

1. M. A. MINER, "Cumulative Damage in Fatigue," *Journal of Applied Mechanics*, Sept. 1945.

2. A. F. MADAYAG, ED., *Metal Fatigue: Theory and Design*, John Wiley, New York, 1969.

3. T. H. TOPPER, B. I. SANDOR, and JoDEAN MORROW, "Cumulative Fatigue Damage Under Cyclic Strain Control," *Journal of Materials, JMLSA, 4*, No. 1, March 1969.

4. "Effects of Environment and Complex Load History on Fatigue Life," *ASTM STP 462*, American Society for Testing and Materials, Philadelphia, 1970.

5. HIROSHI NAKAMURA and TAKESHI HORIKAWA, "Fatigue Strength of Steel," *Proceedings: The 1974 Symposium on Mechanical Behavior of Materials*, Kyoto, Japan, Aug. 1974. Published by the Society of Materials Science, Japan.

6. J. SCHIJVE, "Significance of Fatigue Cracks in Micro-Range and Macro-Range," *Fatigue Crack Propagation, ASTM STP 415*, American Society for Testing and Materials, Philadelphia, 1967.

7. J. C. MCMILLAN and R. M. N. PELLOUX, "Fatigue Crack Propagation Under Program and Random Loads," *ASTM STP 415*, American Society for Testing and Materials, Philadelphia, 1967.

8. E. F. J. vonEUW, R. W. HERTZBERG, and R. ROBERTS, "Delay Effects in Fatigue Crack Propagation," *ASTM STP 513*, American Society for Testing and Materials, Philadelphia, 1972.

9. R. E. JONES, "Fatigue Crack Growth Retardation After Single-Cycle Peak Overload in Ti-6A1-4V Titanium Alloy," *Engineering Fracture Mechanics, 5*, 1973.

10. J. M. BARSOM, "Fatigue-Crack Growth Under Variable-Amplitude Loading in ASTM A514 Grade B Steel," *ASTM STP 536*, American Society for Testing and Materials, Philadelphia, 1973.

11. R. P. WEI and T. T. SHIH, "Delay in Fatigue Crack Growth," *International Journal of Fracture, 10*, No. 1, Mar. 1974.

12. H. L. LEVE, "Cumulative Damage Theories," *Metal Fatigue: Theory and Design*, edited by A. F. Madayag, John Wiley, New York, 1969.

13. C. G. SCHILLING, K. H. KLIPPSTEIN, J. M. BARSOM, and G. T. BLAKE, "Fatigue of Welded Steel Bridge Members Under Variable-Amplitude Loadings," NCHRP Report 188, Transportation Research Board, Washington, D.C., 1978.

14. K. H. KLIPPSTEIN and C. G. SCHILLING, "Stress Spectrums for Short-Span Steel Bridges," *Fatigue Crack Growth Under Spectrum Loads, ASTM STP 595*, American Society for Testing and Materials, Philadelphia, 1976.

15. C. G. SCHILLING, K. H. KLIPPSTEIN, J. M. BARSOM, and G. T. BLAKE, "Fatigue of Welded Steel Bridge Members Under Variable-Amplitude Loadings," *Research Results Digest, 60*, National Cooperative Highway Research Program, April 1974.

16. H. F. HARDRATH and A. T. MCEVILY, "Engineering Aspects of Fatigue-Crack Propagation," *Proceedings of the Crack Propagation Symposium*, Vol. 1, Cranfield, England, Oct. 1961.

17. J. SCHIJVE, F. A. JACOBS, and P. J. TROMP, "Crack Propagation in Clad 2024-T3A1 Under Flight Simulation Loading. Effect of Truncating High Gust Loads,"

*NLR TR-69050-U*, National Lucht-En Ruimtevaart-laboratorium (National Aerospace Laboratory NLR—The Netherlands), June 1969.

18. C. M. Hudson and H. F. Hardrath, "Effects of Changing Stress Amplitude on the Rate of Fatigue-Crack Propagation of Two Aluminum Alloys," *NASA Technical Note D-960*, NASA, Cleveland, Sept. 1961.

19. E. F. J. vonEuw, "Effect of Single Peak Overloading on Fatigue Crack Propagation," Master's Dissertation, Lehigh University, Bethlehem, Pa., 1968.

20. O. E. Wheeler, "Spectrum Loading and Crack Growth," *General Dynamics Report FZM-5602*, Fort Worth, June 30, 1970.

21. J. Willenborg, R. M. Engle, and H. A. Wood, "A Crack Growth Retardation Model Using an Effective Stress Concept," *Technical Memorandum 71-1-FBR*, Air Force Flight Dynamics Laboratory, Jan. 1971.

22. W. Elber, "The Significance of Fatigue Crack Closure," *ASTM STP 486*, American Society for Testing and Materials, Philadelphia, 1971.

23. F. H. Gardner and R. I. Stephens, "Subcritical Crack Growth Under Single and Multiple Periodic Overloads in Cold-Rolled Steel," *ASTM STP 559*, American Society for Testing and Materials, Philadelphia, 1974.

24. V. W. Trebules, Jr., R. Roberts, and R. W. Hertzberg, "Effect of Multiple Overloads on Fatigue Crack Propagation in 2024-T3 Aluminum Alloy," *ASTM STP 536*, American Society for Testing and Materials, Philadelphia, 1973.

25. J. M. Barsom and S. R. Novak, "Subcritical Crack Growth and Fracture of Bridge Steels," *NCHRP Report 181*, Transportation Research Board, Washington, D.C., 1977.

26. J. M. Barsom, unpublished data.

# 11

## Stress-Corrosion
## Cracking

### 11.1. Introduction

Delayed failure of structural components subjected to an aggressive environment may occur under statically applied stresses well below the yield strength of the material. Failure of structural components under these conditions is caused by stress-corrosion cracking which has long been recognized as an important failure mechanism. Although many tests have been developed to study this mode of failure (Figure 11.1),[1] the underlying mechanisms for stress-corrosion cracking are yet to be resolved, and quantitative design procedures against its occurrence are yet to be established. These difficulties are caused by the complex chemical, mechanical, and metallurgical interactions; the many variables that are known to affect the behavior; the extensive data scatter; and the relatively poor correlation between laboratory test results and service experience.

The traditional approach to studying the stress-corrosion susceptibility of a material in a given environment is based on the time required to cause failure of smooth or mildly notched specimens subjected to different stress levels. This time-to-failure approach, like the traditional $S-N$ approach to fatigue, combines the time required to initiate a crack and the time required to propagate the crack to critical dimensions. The need to separate stress-corrosion cracking into initiation and propagation stages was emphasized by experimental results for titanium alloys.[2] These results showed that some materials that appear to be immune to stress-corrosion in the traditional smooth-specimen tests may be highly susceptible to stress-corrosion cracking when tested under the same conditions using precracked specimens. The behavior of such materials was attributed to their immunity to pitting (crack initiation) and to their high intrinsic susceptibility to stress-corrosion cracking (crack propagation). The following discussion presents the use of fracture-

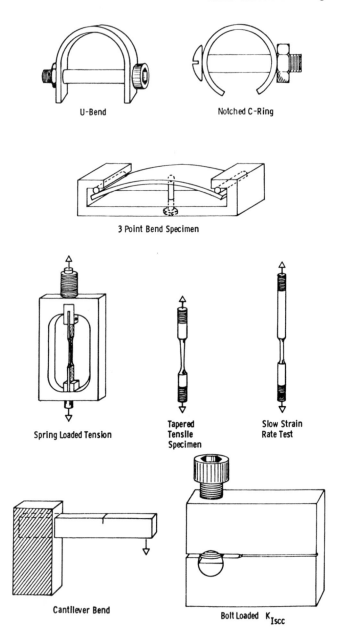

U-Bend

Notched C-Ring

3 Point Bend Specimen

Spring Loaded Tension

Tapered
Tensile
Specimen

Slow Strain
Rate Test

Cantilever Bend

Bolt Loaded $K_{Iscc}$

**Figure 11.1**   Test geometries used to characterize environmentally assisted cracking behavior.

mechanics concepts to study the stress-corrosion cracking of environment-material systems by using precracked specimens.

## 11.2. Fracture-Mechanics Approach

The application of linear-elastic fracture-mechanics concepts to study stress-corrosion cracking has met with considerable success. Because environmentally enhanced crack growth and stress-corrosion attack would be expected to occur in the highly stressed region at the crack tip, it is logical to use the stress-intensity factor, $K_I$, to characterize the mechanical component of the driving force in stress-corrosion cracking. Sufficient data have been published to support this observation.[2-7]

The use of the stress-intensity factor, $K_I$, to study stress-corrosion cracking is based on assumptions and is subject to limitations similar to those encountered in the study of fracture toughness. The primary assumption that must be satisfied when $K_I$ is used to study the stress-corrosion-cracking behavior of materials is the existence of a plane-strain state of stress at the crack tip. This assumption requires small plastic deformation at the crack tip relative to the geometry of the test specimen and leads to size limitations on the geometry of the test specimen. Because these limitations must be established experimentally, and because there does not exist at present a standard test method for stress-corrosion cracking, the limitations established for plane-strain fracture-toughness, $K_{Ic}$, tests (Chapter 3) are usually applied to stress-corrosion-cracking tests.

Various investigators[1-14] have used fracture-mechanics concepts to study the effects of environment on precracked specimens. However, the fracture-mechanics approach to environmental testing did not become widely used until Brown[2,15-16] introduced the $K_{Iscc}$ threshold concept by using pre-cracked cantilever-beam specimens. The $K_{Iscc}$ value at a given temperature for a particular material-environment system represents the stress-intensity-factor value below which subcritical crack extension does not occur under static load in the environment. Since that time, the cantilever-beam test specimen has been used widely to study the stress-corrosion-cracking characteristics of steels,[4,5,10,11,13,15,17] titanium alloys,[14,18-20] and aluminum alloys.[8,21,22]

## 11.3. Experimental Procedures

Experimental procedures for stress-corrosion-cracking tests of precracked specimens may be divided into two general categories. They are time-to-failure tests and crack-growth-rate tests. The time-to-failure tests are similar to the conventional stress-corrosion tests for smooth or notched specimens.[23-26] This type of test using precracked specimens has been widely used since the early work of Brown and Beachem.[2] The crack-growth-rate tests are more complex and require more sophisticated instruments than do the time-to-failure tests. However, data obtained by using crack-growth-rate tests should provide information necessary to enhance the understanding of the kinetics of stress-corrosion cracking and to verify the threshold behavior $K_{Iscc}$, as described shortly.

Various precracked specimens and methods of loading can be used to study the stress-corrosion-cracking behavior of materials in both time-to-failure and crack-growth-rate tests. However, the two most widely used combinations are the cantilever-beam specimen under constant load and the wedge-opening-loading (WOL) specimen under constant displacement conditions (modified WOL specimen) that was developed by Novak and Rolfe.[9]

### 11.3.1. Cantilever-Beam Test Specimen

Figure 11.2 presents a geometry of a cantilever-beam specimen that has been used to study the stress-corrosion-cracking behavior of materials.[4,13,27,28] The specimen is usually face notched 5–10 percent of the thickness, and the notch is extended by fatigue-cracking the specimens at low stress-intensity-factor levels. Then the specimens are tested in a stand similar to that shown in Figure 11.3. Usually, two specimens are monotonically loaded to failure in air to establish the critical stress-intensity

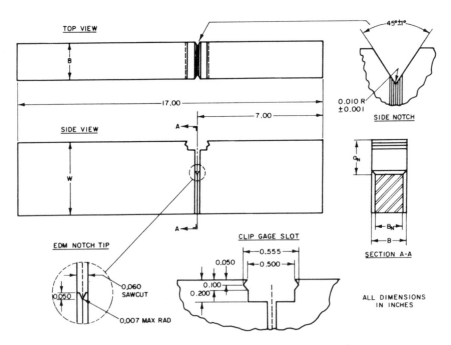

**Figure 11.2**  Fatigue-cracked cantilever-beam test specimen.

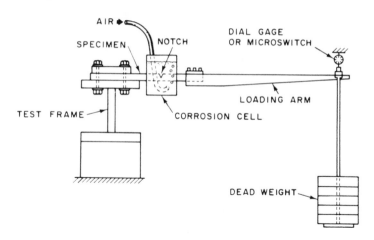

**Figure 11.3**    Schematic drawing of fatigue-cracked cantilever-beam test specimen and fixtures.

factor for failure in the absence of environmental effects ($K_{Ic}$ if ASTM requirements[29] are satisfied and $K_{Ix}$ if they are not satisfied). Subsequently, specimens are immersed in the environment and dead-weight successively loaded to lower initial stress-intensity-factor, $K_{Ii}$, levels. If the material is susceptible to the test environment, the fatigue crack will propagate. As the crack length increases under constant load, the stress-intensity factor at the crack tip increases to the $K_{Ic}$ (or $K_{Ix}$) level, and the specimen fractures. The lower the value of the initial $K_I$, the longer is the time to failure. Specimens that do not fail after a long period of test time, usually 1000 hr for steels, should be fractured and inspected for possible crack extension. The highest plane-strain $K_{Ii}$ level at which crack extension does not occur after a long test time corresponds to the stress-corrosion-cracking threshold, $K_{Iscc}$.

Stress-intensity-factor values for cantilever-beam specimens (Figure 11.2) can be calculated by using the following equation developed by Bueckner[30] for an edge crack in a strip subjected to in-plane bending and modified to account for reduction of thickness due to face notches:

$$K_I = \frac{6M}{(B \cdot B_N)^{1/2}(W - a)^{3/2}} \cdot F\left(\frac{a}{W}\right) \qquad (11.1)$$

where $K_I$ = stress-intensity factor.
   $M$ = bending moment.
   $B$ = gross specimen width.
   $B_N$ = net specimen width.
   $W$ = specimen depth.
   $a$ = total crack length.

$$F\left(\frac{a}{W}\right) = \begin{matrix} 0.36 \\ 0.49 \\ 0.60 \\ 0.66 \\ 0.69 \\ 0.72 \\ 0.73 \end{matrix} \qquad \text{for } \frac{a}{W} = \begin{matrix} 0.05 \\ 0.10 \\ 0.20 \\ 0.30 \\ 0.40 \\ 0.50 \\ 0.60 \text{ and larger} \end{matrix}$$

Figure 11.4 is a schematic representation of test results obtained by using cantilever-beam test specimens. Approximately ten precracked cantilever specimens are needed to establish $K_{Iscc}$ for a particular material and environment.

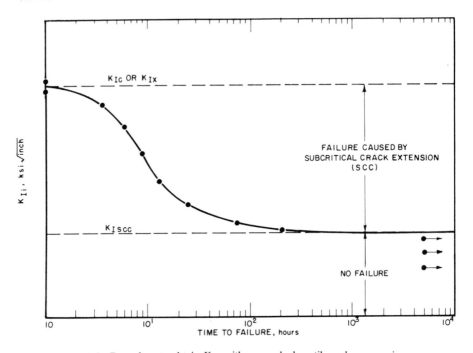

**Figure 11.4**   Procedure to obtain $K_{Iscc}$ with precracked cantilever-beam specimens.

### 11.3.2. Wedge-Opening-Loading Test Specimen

Extensive analytical and experimental investigations have been conducted to study the behavior of wedge-opening-loading (WOL) specimens.[31-33] These specimens have been used to study the fracture toughness,[34] fatigue-crack initiation[35] and propagation,[27] stress-corrosion-cracking,[9] and corrosion-fatigue-crack-growth[27,36] behavior of various materials. The geometry of 1-in.-thick (1-T) WOL specimens is shown in Figure 11.5. The

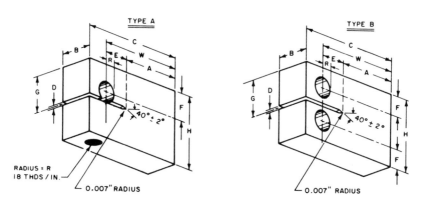

| SPEC | B | W | C | A | E | H | G | D | R | F |
|------|-----|-----|-----|-----|-----|-----|-----|-----|-----|-----|
| IT-A | 1.000 | 2.550 | 3.200 | 1.783 | 0.767 | 2.480 | 1.240 | 0.094 | 0.350 | 1.000 |
| IT-B | 1.000 | 2.550 | 3.200 | 1.783 | 0.767 | 2.480 | 1.240 | 0.094 | 0.250 | 0.650 |

I Inch = 25.4 mm
I degree = 0.017 rad

**Figure 11.5**  Two types of 1-T WOL specimens.

stress-intensity factor, $K_I$, at the crack tip is calculated from[31]

$$K_I = \frac{C_3 P}{B\sqrt{a}} \tag{11.2}$$

where  $P$ = applied load.
  $B$ = specimen thickness.
  $a$ = crack length measured from the loading plane.
  $C_3$ = a function of the dimensionless crack length, $a/W$ (Figure 11.6).
  $W$ = specimen length measured from the loading plane.

Expressed in a polynomial form, $C_3$ for the WOL specimen geometry presented in Figure 11.5 can be represented as

$$30.96\left(\frac{a}{W}\right) - 195.8\left(\frac{a}{W}\right)^2 + 730.6\left(\frac{a}{W}\right)^3 - 1186.3\left(\frac{a}{W}\right)^4 + 754.6\left(\frac{a}{W}\right)^5$$

In the range $0.25 \leq a/W \leq 0.75$, the polynomial is accurate to within 0.5 percent of the experimental compliance.

The WOL specimen was modified by the use of a bolt and loading tup (Figure 11.7) so that it can be self-stressed without using a tensile machine.[9] The crack opening is fixed by the bolt, and the loading is by

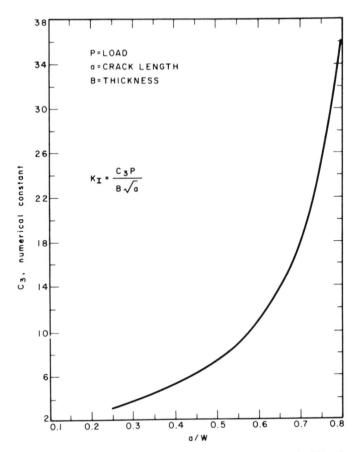

**Figure 11.6**  Relation between $C_3$ used in calculating $K_I$ and $a/W$ ratio.

constant displacement rather than by constant load as in the cantilever-beam specimen. Because a constant crack-opening displacement is maintained throughout the test, the force, $P$, decreases as the crack length increases (Figure 11.7). In cantilever-beam testing, the $K_I$ value increases as the crack length increases under constant load, which leads to fracture for each specimen. In contrast, for the modified WOL specimen, the $K_I$ value decreases as the crack length increases under a decreasing load. The decrease in load more than compensates for the increase in crack length and leads to crack arrest at $K_{Iscc}$. A comparison of these two types of behavior is shown schematically in Figure 11.8. Thus, only a single WOL specimen is required to establish the $K_{Iscc}$ level because $K_I$ approaches $K_{Iscc}$ in the limit. However, duplicate specimens are usually tested to demonstrate reproducibility. Because the bolt-loaded WOL specimen is self-stressed and portable, it can be used to study the stress-corrosion-cracking behavior of materials under actual operating conditions in field environments.

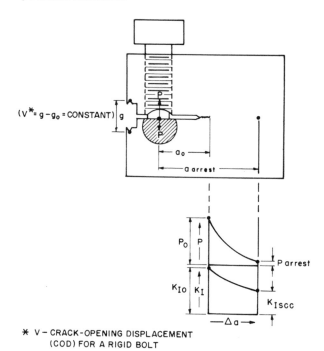

* V – CRACK-OPENING DISPLACEMENT
  (COD) FOR A RIGID BOLT

**Figure 11.7**  Schematic showing basic principle of modified WOL specimen.

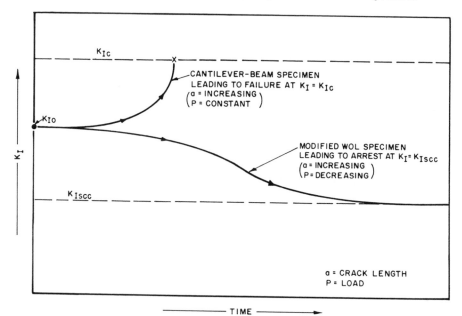

**Figure 11.8**  Difference in behavior of modified WOL and cantilever-beam specimens.

## 11.4. $K_{Iscc}$—A Material Property

Brown and Beachem[37] investigated the $K_{Iscc}$ for environment-material systems by using various specimen geometries. Their results (Figure 11.9) show that identical $K_{Iscc}$ values were obtained for a given environment-material system by using center-cracked specimens, surface-cracked specimens, and cantilever-beam specimens. Smith et al.[18] measured $K_{Iscc}$ values of 10–25 ksi$\sqrt{\text{in.}}$ and 10–22 ksi$\sqrt{\text{in.}}$ by testing center-crack specimens with end loading and wedge-force loading, respectively, for specimens of Ti-8Al-1Mo-1V alloy in 3.5 percent solution of sodium chloride.

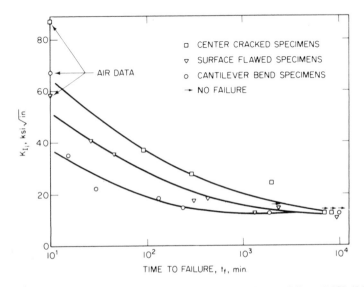

**Figure 11.9**  Influence of specimen geometry on the time to failure (AISI 4340 steel).

$K_{Iscc}$ tests using cantilever-beam specimens and bolt-loaded WOL specimens resulted in identical $K_{Iscc}$ values for each of two 12Ni-5Cr-3Mo maraging steels tested in synthetic sea water.[9] Further test results showed that $K_{Iscc}$ for a specific environment-material system was independent of specimen size above a prescribed minimum geometry limit.[4] On the other hand, the nominal stress corresponding to $K_{Iscc}$, $\sigma_{Nscc}$, was highly dependent on specimen geometry (Figure 11.10), particularly specimen in-plane dimensions such as the height of a cantilever-beam specimen, $W$, and the crack length, $a$. The preceding results indicate that $K_{Iscc}$ for a specific environment-material system is a property of the particular system.

Corrosion products can change the magnitude of the crack-tip driving force by wedging the crack surfaces open. This would be especially applicable for specimens, like the bolt-loaded WOL specimen, that utilize crack arrest

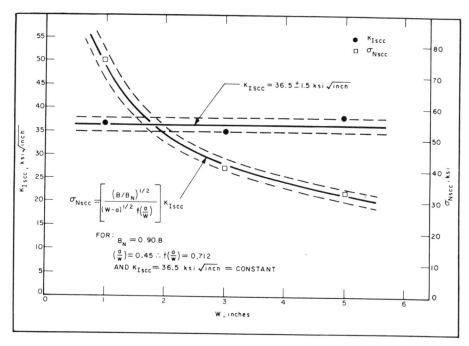

**Figure 11.10**   Effect of $W$ on $K_{Iscc}$ and $\sigma_{Nscc}$ for 18Ni (250) maraging steel.

as a measure of $K_{Iscc}$. Also, because crack-arrest tests appear to be more prone to crack branching, the $K_{Iscc}$ values obtained by increasing-$K_I$ and decreasing-$K_I$ tests can, in some cases, be different.

Imhof and Barsom[27] investigated the effects of thermal treatments on the $K_{Iscc}$ behavior for 4340 steel. Three pieces of 4340 steel were cut from a single plate, and each piece was heat treated to a different strength level. The three pieces were heat treated to a 130-, 180-, and 220-ksi yield strength. The $K_{Iscc}$ for the 130-, 180-, and 220-ksi yield strengths were 111, 26, and 10.5 ksi$\sqrt{\text{in}}$., respectively. These and other results show that thermo-mechanical processing may alter the $K_{Iscc}$ for a given material composition in a specific environment. Available data also show that $K_{Iscc}$ for a given material composition, thermomechanical processing, and environment may be different for different orientations of test specimens. An example of this behavior is observed in aluminum alloys where the susceptibility to stress-corrosion cracking in the short-transverse direction (crack plane parallel to plate surfaces) is greater than in the other directions. Consequently, although $K_{Iscc}$ is a unique property of the tested environment-material system, extreme care should be exercised to ensure the use of the correct $K_{Iscc}$ value for a specific application.

Ideally, $K_{Iscc}$ for a particular material-environment system at a given temperature represents the stress-intensity-factor value below which subcritical crack extension does not occur under static load. However, in practice, $K_{Iscc}$ can be defined as the stress-intensity-factor value corresponding to a low rate of subcritical crack extension that is commensurate with the design service life for the structure.

The mechanisms of stress-corrosion cracking depend on complex chemical, mechanical, and metallurgical interactions that are presently not understood. This lack of understanding and the absence of a standard test method have resulted in significant data scatter. The development of a standard test method is essential for gathering consistent stress-corrosion-cracking test results. However, even under the most ideal test conditions, Clark[38] observed a $\pm 20$ percent scatter in data for a single high-quality AISI 4340 steel bar tested in research-grade hydrogen sulfide gas. Thus, small variations in the chemical composition of the environment or of the material, and in the thermomechanical processing and microstructure of the material, may cause significant differences in the stress-corrosion-cracking behavior.

## 11.5. Validity of $K_{Iscc}$ Data

Standard test methods for stress-corrosion cracking are yet to be established. Discussions of the important parameters that must be considered to ensure valid stress-corrosion cracking have been presented in various publications.[3,4,39] For a specific material-environment system, the two most important parameters are specimen size and duration of the test.

### 11.5.1. Specimen Size

$K_{Iscc}$ is the threshold stress-intensity factor for stress-corrosion cracking under plane-strain conditions. To ensure the existence of plane-strain conditions at the tip of the crack, $K_{Iscc}$ specimens, like $K_{Ic}$ specimens, must satisfy minimum size requirements. However, unlike $K_{Ic}$ specimens, the size requirements for $K_{Iscc}$ specimens are yet to be established. The absence of standard test methods and of requirements on specimen size and test duration has been the cause of significant errors in $K_{Iscc}$ tests. Many of the alleged $K_{Iscc}$ data reported in the literature are not intrinsic properties of the environment-material system tested. A systematic investigation of the influence of state of stress in the neighborhood of the crack tip has not been reported. However, available data[3,16,40] indicate that the apparent $K_{Iscc}$ value for a given environment-material system increases as the deviation from plane-strain conditions increases when this deviation is obtained by changing specimen thickness only and keeping the other in-plane dimensions large relative to the plastic-zone size. Stress-corrosion-cracking data under plane-stress conditions are complicated further because some materials exhibit subcritical crack growth in inert environments under these stress conditions.[3,41]

To circumvent the problems associated with specimen size, Novak[13] suggested a classification to evaluate $K_{Iscc}$ data that is based on the degree of plane-strain conditions existent in geometrically proportionate specimens.

The three suggested classifications of apparent $K_{Iscc}$ behavior are presented in Figure 11.11. The type I, II, and III behaviors correspond to valid $K_{Iscc}$, "partially" valid $K_{Iscc}$, and invalid $K_{Iscc}$, respectively.

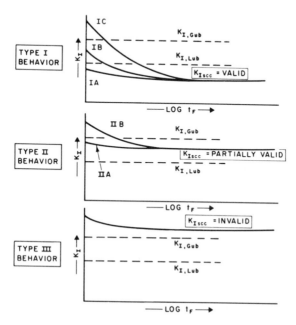

**Figure 11.11**    Three basic types of apparent $K_{Iscc}$ behavior and classification.

The type I or valid $K_{Iscc}$ behavior is that for which the conservative geometrical requirements necessary to ensure plane-strain conditions in $K_{Ic}$ fracture-toughness test specimens are satisfied. Thus, the boundary between type I and type II behaviors, designated $K_{I,Lub}$ in Figure 11.11, corresponds to the highest $K_I$ value that simultaneously satisfies the four equations

$$a_{min} = B_{min} = (W - a)_{min} = \frac{W_{min}}{2} = 2.5 \left( \frac{K_I}{\sigma_{ys}} \right)^2 \qquad (11.3)$$

where  $a_{min}$ = minimum crack length.
$\quad\;\; B_{min}$ = minimum specimen thickness.
$\quad\; W_{min}$ = minimum specimen width.
$\quad\;\; \sigma_{ys}$ = material yield strength.

The boundary between type II and type III behaviors, designated $K_{I,Gub}$ in Figure 11.11, was selected to correspond to the highest $K_I$ value that represents the initial occurrence of a plastic hinge for an elastic–perfectly plastic

material subjected to bending and that simultaneously satisfies the four
equations

$$a_{min} = B_{min} = (W - a)_{min} = \frac{W_{min}}{2} = 1.0\left(\frac{K_I}{\sigma_{ys}}\right)^2 \qquad (11.4)$$

These classifications were proposed to discern the effect of specimen size
on $K_{Iscc}$ behavior and may be used to evaluate published data.

### 11.5.2. Test Duration

Test duration is the second primary parameter that must be understood
to ensure correct test results. The schematic representation for obtaining
$K_{Iscc}$ by using precracked cantilever-beam specimens (Figure 11.4) suggests
that the true $K_{Iscc}$ level was established with test durations greater than
1000 hr. Test durations less than 200 hr (Figure 11.4) would result in
apparent $K_{Iscc}$ values that are greater than the true $K_{Iscc}$ value obtained after
a 1000-hr test duration. The influence of test duration on the apparent $K_{Iscc}$
value obtained by using cantilever-beam test specimens of a 180-ksi yield-
strength, high-alloy steel in room temperature synthetic seawater is shown
in Table 11.1. The data show that an increase of test duration from 100
hr to 10,000 hr decreased the apparent $K_{Iscc}$ value from 170 ksi$\sqrt{\text{in}}$. to 25
ksi$\sqrt{\text{in}}$. Proper test durations depend on specimen configuration, specimen
size, and nature of loading, as well as the environment-material system.
Test durations for bolt-loaded WOL specimens are longer than for cantilever-
beam specimens.[9] In general, test durations for titanium, steel, and aluminum
alloys are on the order of 100, 1000, and 10,000 hr, respectively. The
differences in test duration for different metal alloys are related partly to
the incubation-time behavior in stress-corrosion cracking for the particular
environment-material system. The incubation-time behavior represents the
test time prior to crack extension during which a fatigue crack under sustained
load in an aggressive environment appears to be dormant. The existence
of incubation periods for precracked specimens has been demonstrated by
various investigators.[35-37,39] Benjamin and Steigerwald[42] demonstrated the
dependence of incubation time on prior loading history. Novak demonstrated
the dependence of incubation time on the magnitude of the stress-intensity

**TABLE 11.1   Influence of Cutoff Time on Apparent
$K_{Iscc}$; Constant-Load Cantilever Bend Specimens
(Increasing $K_I$) (Ref. 39)**

| Elapsed Time (hr) | Apparent $K_{Iscc}$ (ksi$\sqrt{\text{in}}$.) |
|---|---|
| 100 | 170 |
| 1,000 | 115 |
| 10,000 | 25 |

factor (Table 11.2).[39] It is apparent that as the applied stress-intensity factor, $K_I$, approaches $K_I$ at fracture, the incubation time must approach zero, and as the applied $K_I$ approaches $K_{Iscc}$, the incubation time approaches infinity. Consequently, specimens subjected to $K_I$ values between $K_{Ic}$ (or $K_I$) and $K_{Iscc}$ can exhibit initiation times that may be of 1000-hr duration or greater.

TABLE 11.2    Influence of $K_I$ on Incubation Time; Constant-Displacement WOL Specimens (Decreasing $K_I$)

| $K_I$ (ksi$\sqrt{\text{in.}}$) | Extent of Crack Growth (in.) | | | | | |
|---|---|---|---|---|---|---|
| | 200 hr | 700 hr | 1400 hr | 2200 hr | 3500 hr | 5000 hr |
| 180 | ND* | 0.35 | 0.76 | 1.00 | 1.12 | — |
| 150 | ND | ND | ND | 0.28 | 0.52 | 0.61 |
| 120 | ND | ND | ND | ND | 0.03 | 0.045 |
| 90 | ND | ND | ND | ND | ND | 0.045 |

*ND: no detectable growth.

The preceding observations show that to evaluate correctly the effect of an environment on a statically loaded structure, the test duration and criterion used to obtain the $K_{Iscc}$ value must be known and evaluated.

## 11.6. $K_{Iscc}$ Data for Some Material-Environment Systems

In general, the higher the yield strength for a given material, the lower the $K_{Iscc}$ value in a given environment.[20,27] The $K_{Iscc}$ for a single plate of 4340 steel tested in 3.5 percent solution of sodium chloride decreased from 111 to 10.5 ksi$\sqrt{\text{in.}}$ as the yield strength was increased from 130 to 220 ksi. In general, the $K_{Iscc}$ in room temperature sodium chloride solutions for steels having a yield strength greater than about 200 ksi is less than 20 ksi$\sqrt{\text{in.}}$. Similar generalizations cannot be made for steels of lower yield strengths. Moreover, stress-corrosion-cracking data for steels having yield strengths less than 130 ksi are very sparse. The results of 5000-hr (30-week) stress-corrosion tests for five steels having yield strengths less than 130 ksi obtained by testing precracked cantilever-beam specimens in room temperature 3 percent sodium chloride solution are presented in Figures 11.12 through 11.16.[43] The results presented in these figures show apparent $K_{Iscc}$ values that ranged from 80 to 106 ksi$\sqrt{\text{in.}}$ (88 to 117 MN/m$^{3/2}$). The apparent $K_{Iscc}$ values measured for most of these steels corresponded to conditions involving substantial crack-tip plasticity. Consequently, linear-elastic fracture-mechanics concepts cannot be used for quantitative analysis of the respective stress-corrosion-cracking behavior.[13] Furthermore, the apparent $K_{Iscc}$ values are suppressed to various degrees below the intrinsic

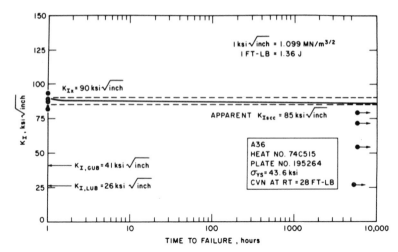

**Figure 11.12** $K_1$ stress-corrosion results for A36 steel in aerated 3 percent NaCl solution of distilled water.

$K_{Iscc}$ values for these steels in a manner similar to the suppression effect for fracture ($K_{Ic}$) behavior.[13]

For such cases, the most important parameter for characterizing the stress-corrosion-cracking behavior is the ratio of the apparent threshold, $K_{Iscc}$, to the value at fracture, $K_{Ix}$, obtained with an identical size specimen (apparent $K_{Iscc}/K_{Ix}$).[13] This ratio, designated the relative index of stress-corrosion-cracking susceptibility, can be used to estimate the relative degradation in fracture behavior of the material as a result of the test solution.

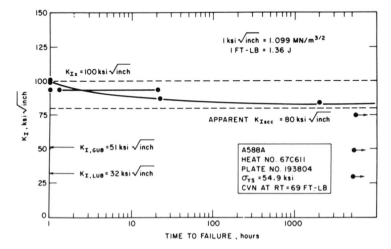

**Figure 11.13** $K_1$ stress-corrosion results for A588 Grade A steel in aerated 3 percent NaCl solution of distilled water.

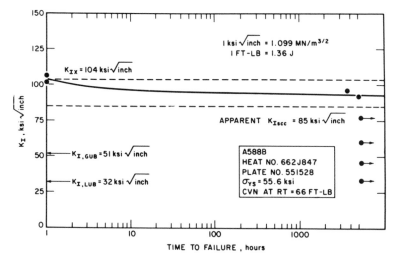

**Figure 11.14**   $K_I$ stress-corrosion results for A588 Grade B steel in aerated 3 percent NaCl solution of distilled water.

General experience with many steels and weldments, some of which are used successfully in long-time environmental service applications, dictates the following broad-based interpretation for assessing the relative stress-corrosion-cracking index: (1) values in the range 0.95–1.00 represent material behaviors that are immune to stress-corrosion cracking, (2) values in the range 0.80–0.95 represent material behaviors that are either moderately

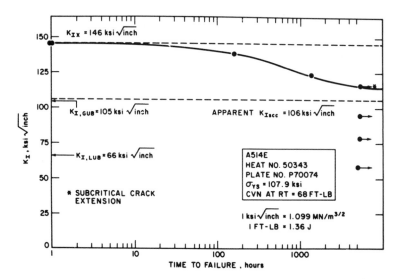

**Figure 11.15**   $K_I$ stress-corrosion results for A514 Grade E steel in aerated 3 percent NaCl solution of distilled water.

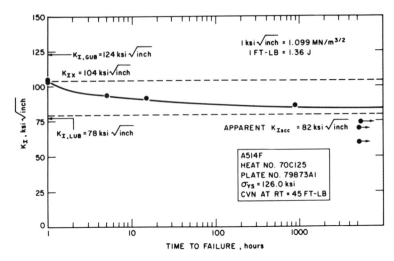

**Figure 11.16** $K_I$ stress-corrosion results for A514 Grade E steel in aerated 3 percent NaCl solution of distilled water.

susceptible to stress-corrosion cracking if the $K_{Iscc}$ value is valid or primarily the result of long-time creep that occurs at the crack tip (under the presence of intensely high stress levels) if the apparent $K_{Iscc}$ value measured is not valid and high levels of crack-tip plasticity are involved, and (3) values less than 0.80 are generally the result of true susceptibility to stress-corrosion cracking for steels regardless of whether the apparent $K_{Iscc}$ value is valid or not.

The results presented in Figures 11.12 through 11.16 show that the five steels yielded relative stress-corrosion-cracking indices of 0.73 or higher for the 5000-hr tests in continuously aerated 3 percent sodium chloride solution. In particular, a value of 0.95, representing essentially immune behavior, was found to occur for the A36 steel. Nearly identical values of 0.80 and 0.82 were measured for the A588A and A588B steels, respectively. These latter values apparently represent primary behavior associated with long-term creep under sustained high-stress-level conditions at the crack tip. The two higher-strength martensitic steels yielded slightly lower ratios, with specific values of 0.73 and 0.79 for the A514E and A514F steels, respectively. That is, although both the fracture-toughness ($K_{Ix}$) and stress-corrosion-cracking threshold (apparent $K_{Iscc}$) values were higher for the A514E steel, the relative stress-corrosion-cracking index was somewhat lower. This difference in behavior of the two A514 steels is somewhat surprising and illustrates the complex nature of the stress-corrosion-cracking mechanism.

Aluminum alloys show high susceptibility to stress corrosion cracking in the short-transverse direction.[8,26] Some aluminum alloys do not exhibit a threshold behavior for stress-corrosion cracking.[21,22]

Some titanium alloys show sustained-load cracking under plane-strain conditions in room temperature air environment. Yoder et al.[44] tested plate samples of eight alloys of the Ti-6Al-4V family. Figure 11.17 summarizes results for one of these alloys tested by using cantilever-bend specimens and part-through crack tension specimens. They concluded that sustained-load cracking in room-temperature air environment is widespread and serious in these alloys and that the resulting degradations ranged from 11 to 35 percent. The magnitude of degradation did not appear to correlate with interstitial contents, processing variables, strength level, or toughness level, but it was orientation dependent.

Novak[13] conducted a systematic study to determine the effect of prior plastic strain on the mechanical and environmental properties of four steels ranging in yield strength from 40 to 200 ksi (550 to 1400 N/mm²). Each steel was evaluated first in the unstrained condition and then after 1 and either 3 or 5 percent plastic strain. The results showed that the value of the stress-intensity factor at fracture decreased with increased magnitude of prior plastic strain. However, the corresponding change in the apparent $K_{Iscc}$ value did not follow any consistent pattern of behavior.

Extensive investigations have been conducted to evaluate the $K_{Iscc}$ behavior of various metals in seawater.[3–7,13,14,20–22,27] The test environments used were synthetic seawater, 3 or 3.5 percent solution of sodium chloride in distilled water, or natural seawater. In general, the results suggest that natural seawater may be more severe than either the synthetic seawater or the solutions of sodium chloride. Tests in natural seawater environments can result in $K_{Iscc}$ values that are 20–30 percent lower than values obtained in synthetic seawater or solutions of sodium chloride.

A systematic investigation of the influence of test temperature on the $K_{Iscc}$ behavior of environment-material systems is yet to be conducted.

$K_{Iscc}$ data for various environment-material systems have been gathered and published.[45–49]

## 11.7. Crack-Growth-Rate Tests

The crack-growth-rate approach to study stress-corrosion-cracking behavior of environment-material systems involves the measurement of the rate of crack growth per unit time, $da/dt$, as a function of the instantaneous stress-intensity factor, $K_I$. Stress-corrosion crack growth has been investigated in various environment-material systems by using different precracked specimens.[3,39] In general, the results suggest that the stress-corrosion crack-growth-rate behavior as a function of the stress-intensity factor can be divided into three regions (Figure 11.18). In region I, the rate of stress-corrosion crack growth is strongly dependent on the magnitude of the stress-intensity factor, $K_I$, such that a small change in the magnitude of $K_I$ results

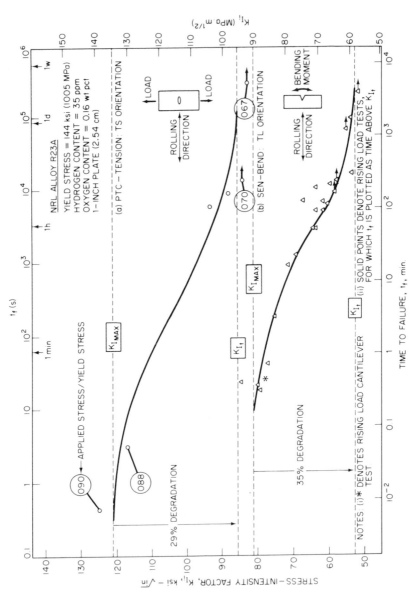

**Figure 11.17** Sustained-load cracking of a commercial-grade Ti-6Al-4V material air environment.

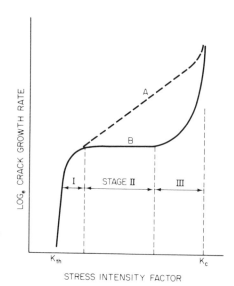

**Figure 11.18**  Schematic illustration of the functional relationship between stress-intensity factor, $K$, and subcritical-crack-growth rate $(da/dt)$.

in a large change in the rate of crack growth. The behavior in region I exhibits a stress-intensity factor value below which cracks do not propagate under sustained loads for a given environment-material system. This threshold stress-intensity factor corresponds to $K_{Iscc}$. Region II represents the stress-corrosion-crack-growth behavior above $K_{Iscc}$. In this region the rate of stress-corrosion cracking for many systems is moderately dependent on the magnitude of $K_I$, type A behavior in Figure 11.18. Crack-growth rates in region II for high-strength steels in gaseous hydrogen as well as other material-environment systems[46-49] appear to be independent of the magnitude of the stress-intensity factor, type B behavior (Figure 11.18). In such cases, the primary driving force for crack growth is not mechanical ($K_I$) in nature but is related to other processes occurring at the crack tip such as chemical, electrochemical, mass-transport, diffusion, and adsorption processes. The crack-growth rate in region III increases rapidly with $K_I$ as the value of $K_I$ approaches $K_{Ic}$ (or $K_c$) for the material.

The characteristic crack-growth behavior for a given environment-material system is determined by the mechanical properties, chemical properties, or both, for the system. The crack-growth approach to study stress-corrosion cracking is of great value in determining the mechanism of crack extension. It can be used to analyze the safety and reliability of structures subjected to stress-intensity-factor values that are above $K_{Iscc}$ and for environment-material systems that do not exhibit a threshold behavior. The bolt-loaded WOL specimen can be used best to determine the stress-corrosion-crack-growth-rate behavior and the $K_{Iscc}$. The rate of growth at $K_{Iscc}$ should be equal to or less than $10^{-5}$ in./hr.

## 11.8. Effect of Composition and Applied Potential

The primary factors that influence $K_{Iscc}$ (aside from strength level, $\sigma_{ys}$) are currently unknown. Less is known about the corresponding factors that influence the kinetics of crack propagation (from a fatigue crack) due to stress-corrosion cracking at $K_I$ levels above $K_{Iscc}$. For example, the very existence of incubation times ($t_{inc}$) that range to 1000 hr or more for precracked steel specimens has only recently been documented.[39] Whereas $t_{inc}$ is known to depend strongly on initial stress-intensity level ($K_{Ii}$), the crack-propagation rate ($da/dt$) appears to be relatively insensitive to stress-intensity level ($K_{Ii}$) for many materials, until $K_{Iscc}$ is approached.[39] However, both $t_{inc}$ and $da/dt$ can vary significantly from one material to another, even at the same strength level ($\sigma_{ys}$) for tests in the same environment. As an example, the kinetics of stress-corrosion cracking for precracked-specimen tests of 18Ni 250-grade maraging steel tested in seawater-type solutions are substantially slower than are those of lean-alloy high-strength steels at the same strength level (D6AC, H-11, AISI 4340). Although a total test time between 10 and 100 hr is generally sufficient to establish $K_{Iscc}$ for such lean-alloy steels,[27,39] corresponding test times of 2000 hr or more are required for 18Ni 250-grade maraging steel.[4,6] Reasons for such differences are currently unknown.

Investigations to study the effect of steel cleanliness and purity on fracture toughness and stress-corrosion cracking have been limited to materials having yield strengths greater than about 180 ksi. As tensile strength increases from 250 to 310 ksi, the improvements in fracture toughness that can be realized by producing ultra-high-purity steel decrease with strength until, at a tensile strength of 310 ksi, steels with ultra-high purity have no higher fracture toughness than steels of normal commercial purity.[50] Studies on stress-corrosion cracking in 18Ni maraging steel indicate that there may be a similar relation between strength and resistance to stress-corrosion cracking as determined by $K_{Iscc}$ measurements.[4-6] For example, at 180-ksi yield strength, increases in the purity of 18Ni maraging steel resulted in increases in $K_{Iscc}$ from 108 ksi$\sqrt{in.}$ to 150 ksi$\sqrt{in.}$, while at a yield-strength level of 250 ksi, a reduction in C, Mn, P, S, and Si contents of an 18Ni maraging steel resulted in an increase of fracture toughness from a $K_{Ic}$ value of 67 ksi$\sqrt{in.}$ to 105 ksi$\sqrt{in.}$, but the resistance to stress-corrosion cracking remained essentially unaffected—exhibiting a $K_{Iscc}$ of 36 ksi$\sqrt{in.}$ in the lower-toughness steel and a value of 25 ksi$\sqrt{in.}$ in the higher-toughness steel. Thus, in maraging steels the improvement in $K_{Iscc}$ that can be realized by improved steel purity appears to be insignificant for the very-high-strength-level steels, for example, 250 ksi.

If the foregoing relations among strength, toughness, and resistance to stress-corrosion were found to hold regardless of steel type, composition, or degree of purity achieved, the use of high-strength steels in aircraft applications could be limited by the relatively low resistance to stress-corrosion cracking even in the useful tensile-strength range of 240–270 ksi, where improvements in steel melting and processing have been found to produce marked improvements in fracture toughness. Insufficient work has been conducted at very-high-purity levels and on steels other than 18Ni maraging steels to determine to what extent the behavior observed to date in the maraging steels is found in other types of steel.

Steels with tensile strengths greater than 240 ksi can have $K_{Ic}$ values greater than 70 ksi$\sqrt{in}$. but rarely have a $K_{Iscc}$ greater than 20 ksi$\sqrt{in}$.[27,51,52] Proctor and Paxton[53] obtained $K_{Iscc}$ values between 8 and 13 ksi$\sqrt{in}$. for 300-grade 18Ni maraging steels of varying purity. Similar data were obtained by Carter,[52] who reported that values of $K_{Ic}$ as high as 71 ksi$\sqrt{in}$. for 300-grade 18Ni maraging steels were obtained by controlling residual elements, but the $K_{Iscc}$ values remained very low at about 7 ksi$\sqrt{in}$. Low-alloy steels such as D6AC, 4340, and 300M of commercial purity also have $K_{Iscc}$ values between 10 and 20 ksi$\sqrt{in}$.

Dautovich and Floreen[54] studied the stress-corrosion cracking behavior of maraging steels and suggested that increases in $K_{Iscc}$ values can be obtained at carbon and sulfur levels below 20 ppm. There is also limited evidence that control of residual elements can lead to improved toughness and stress-corrosion behavior of 9-4-45 steel and 10Ni-Cr-Mo-Co steel.[51,55]

Stress-corrosion cracking tests conducted on steels having yield strengths less than or equal to 180 ksi indicate that the stress-corrosion cracking kinetics ($t_{inc}$ and $da/dt$) are more rapid and $K_{Iscc}$ values less with cathodic potentials compared with identical evaluation under open-circuit conditions.[56,57] In particular, a 180-ksi-strength 17-4 pH steel evaluated in salt water has been observed[57] under cathodic potentials to exhibit a dramatic decrease in $K_{Iscc}$ to a value close to half of that observed under open-circuit conditions. The extent of reduction in $K_{Iscc}$ behavior due to applied electrical potential for higher-strength steels (240–270 ksi) has not yet been investigated under any conditions and may be substantially greater.

## 11.9. General Observations

$K_{Iscc}$ tests, like other linear-elastic fracture-mechanics tests, are conducted on specimens that contain large initial fatigue cracks relative to the other dimensions (thickness and remaining ligament) of the specimen. This requirement is imposed, in part, to ensure that the crack-tip stress field is not affected by the loading arrangement or the front surface of the specimen. Furthermore, linear-elastic fracture mechanics is applicable when the plastically deformed region at the crack tip is small compared to the specimen dimensions and the crack length. Clark[38] investigated the behavior of small cracks and concluded that linear-elastic fracture-mechanics concepts are applicable to small cracks if the crack length is at least 25 time larger than the associated crack-tip plastic-zone size, $r_p$, where $r_p = (1/6\pi)(K/\sigma_{ys})^2$.

The size of initial imperfections in actual engineering structures are usually much smaller than the size of the fatigue crack required for linear-elastic fracture-mechanics tests. Consequently, the stress field at the tip of these imperfections can be affected significantly by the close proximity of surfaces. Moreover, small initial surface imperfections in welded structures usually reside in regions of high residual stress and at geometrical stress concentrations. Under these conditions, the plastic-zone size, if it can be calculated by using linear-elastic fracture-mechanics concepts, would be too large to satisfy Clark's observation, and the initial imperfection may be completely surrounded by plastically deformed material. Consequently, the applicability of linear-elastic fracture-mechanics concepts and test results to such imperfections should be considered very carefully.

$K_{Iscc}$ is the plane-strain stress-intensity-factor threshold for a given material, environment, and temperature below which preexisting fatigue cracks do not propagate under static loads. Because most engineering structures are subjected to load fluctuations, the effect of these fluctuations on the $K_{Iscc}$ value should be known. The very sparse data available[58] suggest that low-magnitude load fluctuations can significantly decrease the $K_{Iscc}$ value. $K_{Iscc}$ can be considered to represent the limit value of $\Delta K_{th}$ in corrosion fatigue as the cyclic frequency for the fluctuating load approaches zero. The magnitude of $\Delta K_{th}$ in corrosion fatigue, as discussed in Chapter 13, appears to increase with decreased cyclic frequency. Consequently, load fluctuations should decrease the magnitude of $K_{Iscc}$. Because most engineering structures are subjected to load fluctuations, the relevance of $K_{Iscc}$ to structural performance for fully immersed cyclically loaded structures must be considered.

Despite the significant progress that has been achieved in stress-corrosion-cracking over the past several years, it can be said that, in general, "there presently is no reliable fundamental theory of stress-corrosion cracking in any alloy-environment system which can be used to predict the performance of equipment even in environments where conditions are readily defined."[59] Thus test results, preferably using the methods described in this chapter, must still be used in design to prevent failures by stress corrosion of statically loaded structures.

# References

1. *Characterization of Environmentally Assisted Cracking for Design—State of the Art*, NMAB-386, National Materials Advisory Board, National Research Council, Washington, D.C., Jan. 1982.
2. B. F. BROWN and C. D. BEACHEM, "A Study of the Stress Factor in Corrosion Cracking by Use of the Precracked Cantilever-Beam Specimen," *Corrosion Science, 5,* 1965.

3.  H. H. Johnson and P. C. Paris, "Subcritical Flaw Growth," *Engineering Fracture Mechanics, 1*, No. 1, 1968.

4.  S. R. Novak and S. T. Rolfe, "Comparison of Fracture Mechanics and Nominal Stress Analysis in Stress Corrosion Cracking," *Corrosion, 26*, No. 4, April 1970.

5.  S. R. Novak and S. T. Rolfe, "$K_{Ic}$ Stress-Corrosion Tests of 12Ni-5Cr-3Mo and 18Ni-8Co-2Mo Maraging Steels and Weldments," *AD482761L*, Defense Documentation Center, Arlington, Va., Jan. 1, 1966.

6.  S. R. Novak, "Comprehensive Investigation of the $K_{Iscc}$ Behavior of Candidate HY-180/210 Steel Weldments," *U.S. Steel Applied Research Laboratory Report No. 89.021-024(1)(B-63105)*, Defense Documentation Center, Arlington, Va., Dec. 31, 1970.

7.  S. R. Novak and S. T. Rolfe, "Fatigue-Cracked Cantilever Beam Stress-Corrosion Tests of HY-80 and 5Ni-Cr-Mo-V Steels," *AD482783L*, Defense Documentation Center, Arlington, Va., Jan. 1, 1966.

8.  D. O. Sprowls, M. B. Shumaker, J. D. Walsh, and J. W. Coursen, "Evaluation of Stress-Corrosion Cracking Using Fracture Mechanics Techniques," *Contract NAS 8-21487, Final Report*, George C. Marshall Space Flight Center, Huntsville, Ala., May 31, 1973.

9.  S. R. Novak and S. T. Rolfe, "Modified WOL Specimen for $K_{Iscc}$ Environmental Testing," *Journal of Materials, JMLSA, ASTM*, Philadelphia, *4*, No. 3, Sept. 1969.

10. E. A. Steigerwald, "Delayed Failure of High-Strength Steel in Liquid Environments," *Proceedings of the American Society for Testing Materials*, Vol. 60, Philadelphia, 1960.

11. H. H. Johnson and A. M. Willner, "Moisture and Stable Crack Growth in a High-Strength Steel," *Applied Materials Research*, Jan. 1965.

12. C. F. Tiffany and J. N. Masters, "Applied Fracture Mechanics," *Fracture Toughness Testing and Its Applications, ASTM STP 381*, American Society for Testing and Materials, Philadelphia, Apr. 1965.

13. S. R. Novak, "Effect of Prior Uniform Plastic Strain on the $K_{Iscc}$ of High-Strength Steels in Sea Water," *Engineering Fracture Mechanics, 5*, No. 3, 1973.

14. R. W. Judy, Jr., and R. J. Goode, "Stress-Corrosion Cracking Characteristics of Alloys of Titanium in Salt Water," *NRL Report 6564*, Naval Research Laboratory, Washington, D.C., July 21, 1967.

15. B. F. Brown, "Stress-Corrosion Cracking and Corrosion Fatigue of High-Strength Steels," *Problems in the Load-Carrying Application of High-Strength Steels, DMIC Report 210*, Defense Metals Information Center, Battelle Memorial Institute, Columbus, Ohio, Oct. 26–28, 1964.

16. B. F. Brown, "A New Stress-Corrosion Cracking Test for High-Strength Alloys," *Materials Research and Standards, 6*, No. 3, March, 1966.

17. H. P. Leckie, "Effect of Environment on Stress Induced Failure of High-Strength Maraging Steels," *Fundamental Aspects of Stress Corrosion Cracking, NACE-1*, National Association of Corrosion Engineers, Houston, 1969.

18. H. R. Smith, D. E. Piper, and F. K. Downey, "A Study of Stress-Corrosion Cracking by Wedge-Force Loading," *Engineering Fracture Mechanics, 1*, No. 1, 1968.

19. R. W. HUBER, R. J. GOODE, and R. W. JUDY, JR., "Fracture Toughness and Stress-Corrosion Cracking of Some Titanium Alloy Weldments," *Welding Journal, 46*, No. 10, Oct. 1967.

20. M. H. PETERSON, B. F. BROWN, R. L. NEWBEGIN, and R. E. GROOVER, "Stress Corrosion Cracking of High Strength Steels and Titanium Alloys in Chloride Solutions at Ambient Temperature," *Corrosion, 23*, 1967.

21. M. O. SPEIDEL, "Stress Corrosion Cracking of Aluminum Alloys," *Metallurgical Transactions, 6A*, No. 4, Apr. 1975.

22. M. O. SPEIDEL and M. V. HYATT, "Stress Corrosion Cracking of High-Strength Aluminum Alloys," *Advances in Corrosion Science and Technology*, Vol. II, edited by M. G. Fontana and R. W. Staehle, Plenum, New York, 1972.

23. H. H. UHLIG, *The Corrosion Handbook*, John Wiley, New York, 1963.

24. H. L. LOGAN, *The Stress Corrosion of Metals*, John Wiley, New York, 1966.

25. A. W. LOGINOW, "Stress Corrosion Testing of Alloys," *Materials Protection, 5*, No. 5, May 1966.

26. D. O. SPROWLS and R. H. BROWN, "What Every Engineer Should Know About Stress Corrosion of Aluminum," *Metals Progress, 81*, Nos. 4 and 5, 1962.

27. E. J. IMHOF and J. M. BARSOM, "Fatigue and Corrosion-Fatigue Crack Growth of 4340 Steel at Various Yield Strengths," *ASTM STP 536*, American Society for Testing and Materials, Philadelphia, 1973.

28. J. M. BARSOM, "Corrosion-Fatigue Crack Propagation Below $K_{Iscc}$," *Engineering Fracture Mechanics, 3*, No. 1, July 1971.

29. "Standard Method of Test for Plane-Strain Fracture Toughness of Metallic Materials," *ASTM E399, Annual Book of ASTM Standards*, American Society for Testing and Materials, Philadelphia, 1974.

30. P. C. PARIS and G. C. SIH, "Stress Analysis of Cracks," *ASTM STP 381*, American Society for Testing and Materials, Philadelphia, 1965.

31. W. K. WILSON, "Review of Analysis and Development of WOL Specimen," *67-7D7-BTLPV-R1*, Westinghouse Research Laboratories, Pittsburgh, Mar. 1967.

32. W. K. WILSON, "Analytical Determination of Stress Intensity Factors for the Manjoine Brittle Fracture Test Specimen," *WERL-0029-3*, Westinghouse Research Laboratories, Pittsburgh, Aug. 1965.

33. M. M. LEVEN, "Stress Distribution in the M4 Biaxial Fracture Specimen," *65-1D7-STRSS-S1*, Westinghouse Research Laboratories, Pittsburgh, Mar. 1965.

34. E. T. WESSEL, "State of the Art of the WOL Specimen for $K_{Ic}$ Fracture Toughness Testing," *Engineering Fracture Mechanics, 1*, No. 1, June 1968.

35. W. G. CLARK, JR., "Evaluation of the Fatigue Crack Initiation Properties of Type 403 Stainless Steel in Air and Steam Environments," *ASTM STP 559*, American Society for Testing and Materials, Philadelphia, 1974.

36. J. M. BARSOM, "Effect of Cyclic-Stress Form on Corrosion-Fatigue Crack Propagation Below $K_{Iscc}$ in a High-Yield-Strength Steel," *Corrosion Fatigue: Chemistry, Mechanics and Microstructure, NACE-2*, National Association of Corrosion Engineers, Houston, 1972.

37. B. F. BROWN and C. D. BEACHEM, "Specimens for Evaluating the Susceptibility of High Strength Steels to Stress Corrosion Cracking," Internal Report, U.S. Naval Research Laboratory, Washington, D.C., 1966.

38. W. G. CLARK, JR., "Applicability of the $K_{Iscc}$ Concept to Very Small Defects," *ASTM STP 601*, American Society for Testing and Materials, Philadelphia, 1976.

39. R. P. WEI, S. R. NOVAK, and D. P. WILLIAMS, "Some Important Considerations in the Development of Stress Corrosion Cracking Test Methods," *Materials Research and Standards, MTRSA, 12*, No. 9, 1972.

40. D. E. PIPER, S. H. SMITH, and R. V. CARTER, "Corrosion Fatigue and Stress-Corrosion Cracking in Aqueous Environments," *A.S.M. National Metal Congress*, Oct. 1966.

41. G. G. HANCOCK and H. H. JOHNSON, "Subcritical Crack Growth in AM350 Steel," *Materials Research and Standards, 6*, 1966.

42. W. D. BENJAMIN and E. A. STEIGERWALD, "An Incubation Time for the Initiation of Stress-Corrosion Cracking in Precracked 4340 Steel," *Transactions of the American Society for Metals, 60*, No. 3, Sept. 1967.

43. J. M. BARSOM and S. R. NOVAK, "Subcritical Crack Growth and Fracture of Bridge Steels," NCHRP Report 181, Transportation Research Board, Washington, D.C., 1977.

44. G. R. YODER, C. A. GRIFFIS, and T. W. CROOKER, "Sustained-Load Cracking of Titanium—A Survey of 6A1-4V Alloys," *NRL Report 7596*, Naval Research Laboratory, Washington, D.C., 1973.

45. "Stress Corrosion Testing," *ASTM STP 425*, American Society for Testing and Materials, Philadelphia, Dec. 1967.

46. R. W. STAEHLE, A. J. FORTY, and D. VAN ROOYEN, eds., *Fundamental Aspects of Stress-Corrosion Cracking, NACE-1*, National Association of Corrosion Engineers, Houston, 1969.

47. B. F. BROWN, ed., *Stress-Corrosion Cracking in High-Strength Steels and in Titanium and Aluminum Alloys*, Naval Research Laboratory, Washington, D.C., 1972.

48. A. AGRAWAL, B. F. BROWN, J. KRUGER, and R. W. STAEHLE, eds., *U.R. Evans Conference on Localized Corrosion, NACE-3*, National Association of Corrosion Engineers, Houston, 1971.

49. M. O. SPEIDEL, M. J. BLACKBURN, T. R. BECK, and J. A. FEENEY, "Corrosion Fatigue and Stress Corrosion Crack Growth in High Strength Aluminum Alloys, Magnesium Alloys, and Titanium Alloys Exposed to Aqueous Solutions," *Corrosion Fatigue: Chemistry, Mechanics and Microstructure, NACE-2*, National Association of Corrosion Engineers, Houston, 1972.

50. L. F. PORTER, "A Discussion of the Paper 'The Role of Inclusions on Mechanical Properties in High-Strength Steels,' " *Journal of Vacuum Science and Technology, 9*, No. 6, Nov.–Dec. 1972.

51. R. T. AULT, C. M. WAID, and R. B. BERTOLE, "Development of an Improved Ultra-High Strength Steel for Forged Aircraft Component," *AFML-TR-71-27*, Air Force, Dayton, Feb. 1971.

52. C. S. CARTER, "Evaluation of a High Purity 18 Percent Ni (300) Maraging Steel Forging," *AFML-TR-70-139*, Air Force, Dayton, June 1970.

53. R. P. M. PROCTOR and H. W. PAXTON, "The Effect of Trace Impurities on the Stress Corrosion Cracking Susceptibility and Fracture Toughness of 18Ni Maraging Steel," *Corrosion Science, 11*, 1971.

54. D. P. DAUTOVICH and S. FLOREEN, "The Stress Corrosion and Hydrogen Embrittlement Behavior of Maraging Steels," presented at NACE Conference, UNIEUX-FIRMINY, France, 1973.

55. B. MRAVIC and J. H. SMITH, "Development of Improved High-Strength Steels

for Aircraft Structural Components,'' *AFML-TR-71-213*, Wright-Patterson Air Force Base, Dayton, Ohio, Oct. 1971.

56. H. P. Leckie and A. W. Loginow, "Stress Corrosion Behavior of High Strength Steels," *Corrosion 24*, No. 9, Sept. 1968, pp. 291–297.

57. R. W. Judy, Jr., C. T. Fujii, and R. J. Goode, "Properties of 17-4 pH Steel," *Naval Research Laboratory (NRL) Report 7639*, Washington, D.C., Dec. 18, 1973.

58. R. R. Fessler and T. J. Barlo, "The Effect of Cyclic Loading on the Threshold Stress for Stress Corrosion Cracking in Mild and HSLA Steels," paper presented at the ASME 3rd National Congress on Pressure Vessel and Piping Technology, San Francisco, 1979.

59. R. W. Staehle, "Evaluation of Current State of Stress Corrosion Cracking," *Fundamental Aspects of Stress Corrosion Cracking, NACE-1*, National Association of Corrosion Engineers, Houston, 1969.

# 12

# Corrosion-Fatigue-Crack Initiation

## 12.1. Introduction

Failure of structural components subjected to fluctuating loads may be caused by the initiation and propagation of cracks. The total useful life of such components is determined by the time (number of cycles) necessary to initiate the crack and to propagate the crack to a critical size. Crack initiation and subcritical crack propagation may be caused by cyclic stresses in the absence of an aggressive environment (fatigue) or by the combined effects of cyclic stresses and an aggressive environment (corrosion fatigue). The relative magnitude of crack-initiation life and crack-propagation life in the total life of a component depends on material properties, structural geometry, applied stresses, and environment.

Corrosion-fatigue behavior of a given material-environment system refers to the characteristics of the material under fluctuating loads in the presence of the particular environment. The corrosion-fatigue behavior of a given material-environment system depends on the metallurgical, mechanical, and electrochemical components of the particular system. Corrosion-fatigue damage occurs more rapidly than would be expected from the individual effects or from the algebraic sum of the individual effects of fatigue, corrosion, or stress-corrosion cracking. The individual effects with the synergism make the corrosion-fatigue mechanism very complex and not well understood with relatively little or no capabilities to predict, a priori, the performance of structural components. Generally, different environments have different effects on the cyclic behavior of a given material. Similarly, the corrosion-fatigue behavior of different materials is usually different in the same environment. The behavior established for a given material-environment system or for a given set of test conditions should not be applied indiscriminately to other systems or conditions.

Significant developments in understanding the corrosion-fatigue crack-propagation behavior of various metal-environment systems has been achieved by using linear-elastic fracture-mechanics methodology. The significant findings are presented in Chapter 13. Unfortunately, very little research has been conducted to study the corrosion-fatigue crack-initiation behavior, especially under low-cyclic-load frequencies and long-life conditions. Recently, linear-elastic fracture-mechanics methodology was used in a systematic study of the corrosion-fatigue crack-initiation behavior for constructional steels in 3.5 percent sodium chloride solution. The most significant findings of this study are presented in this chapter.

## 12.2. Test Specimens and Experimental Procedures

This section presents two specimen geometries and the corresponding test procedures that have been used to study the corrosion-fatigue crack-initiation behavior for constructional steels. However, various other specimens and test procedures can be used to obtain similar test results.

### 12.2.1. Modified Compact-Tension Specimen and Experimental Procedure

Taylor and Barsom[1] used a modified compact-tension (CT) specimen to study the corrosion-fatigue crack-initiation behavior of A517 Grade F steel. The specimens were 9.4 mm thick (0.372 in.) and had in-plane dimensions equal to a 25.4-mm-thick (1.0-in.) CT specimen (Figure 12.1). The specimens contained notches that were milled to a length ($a$) of about 17.0 mm (0.67 in.) or 25.4 mm (1.0 in.) and had tip radii ($\rho$) of 3.3 mm (0.128 in.). The sides of the specimens were ground to No. 6 finish or better, and the surfaces in and around the notch root were then diamond polished to a 6-$\mu$m finish.

The specimens were corrosion-fatigue tested in electrohydraulic testing machines where alignment was obtained by carefully machining the specimens and other auxiliary parts and by using universal joints to load the specimens. The specimens were submerged in an environmental tank (Figure 12.2) made of polymethylmethacrylate (Plexiglas), and all auxiliary parts and the universal joints were made of the same steel investigated (A517 Grade F) to minimize galvanic corrosion. The room temperature 3.5 percent solution of sodium chloride in distilled water environment used by Taylor and Barsom was maintained at a pH of 6.5 ±0.5 and was replaced every 100 hr for all tests that exceeded this period of time. The oxygen content of the solution was not controlled. However, some aeration of the solution occurred as a result of the movement of the specimen and exposure of the surface of the solution to the room air.

The notch-tip surface was inspected periodically by lowering the tank

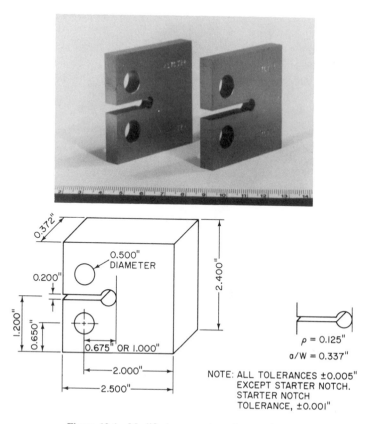

**Figure 12.1**  Modified compact-tension specimens.

(Figure 12.3), drying the notch tip with absorbent cotton swabs, and observing the notch tip at X17 magnification with a microscope. The test was terminated when a crack of 0.25 mm (0.010 in.) or longer was observed at the surface of the notch tip. For high test frequencies (300 cycles per minute), the frequency was temporarily lowered during inspection to facilitate observation of cracking at the notch tip. The test results were presented as the range of cycles for crack initiation bounded by the number of cycles corresponding to the last inspection of the notch tip where no crack was observed and the following inspection where a crack existed.

### 12.2.2. Single-Edge-Notched Specimen
### and Experimental Procedure

Novak[2,3] studied the corrosion-fatigue crack-initiation behavior of constructional steels by using single-edge-notched specimens subjected to cyclic loading under cantilever-bending conditions. All specimens had the geometry

**Figure 12.2** Corrosion tank in operating position.

and dimensions shown in Figure 12.4. The notch was machined to a nominal length of 25.1 mm (0.99 in.); then an additional 0.25 mm (0.01 in.) was removed by polishing the notch tip with a 3- to 6-$\mu$m diamond paste.

All specimens were tested in 3.5 percent solution of NaCl in distilled water that was maintained at a 72 $\pm$ 4°F temperature and a pH of 6.0 $\pm$ 0.8 (Figure 12.5). The solution was changed once a week, and the surface of the solution was exposed to laboratory air.

Crack initiation was detected at X5 to X50 magnification by using a binocular stereomicroscope and a high-intensity light source. Prior to examination of the notch tip, the solution was drained, and loose corrosion products and debris were carefully removed from the notch-tip region. Crack initiation was determined by a range of cycles bounded by the number of cycles corresponding to the last inspection at which no crack was detected and the following inspection where a crack existed.

## 12.3. Corrosion-Fatigue-Crack-Initiation Behavior for Steels

The corrosion-fatigue behavior for metals subjected to load fluctuation in the presence of an environment to which the metal is immune is identical

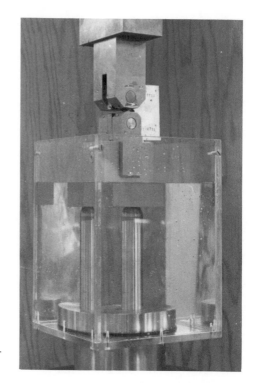

**Figure 12.3**  Corrosion tank in inspection position.

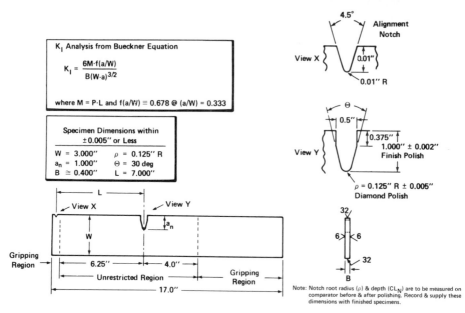

$K_I$ Analysis from Bueckner Equation

$$K_I = \frac{6M \cdot f(a/W)}{B(W \cdot a)^{3/2}}$$

where M = P·L and f(a/W) $\equiv$ 0.678 @ (a/W) = 0.333

Specimen Dimensions within
±0.005″ or Less

| | |
|---|---|
| W = 3.000″ | $\rho$ = 0.125″ R |
| $a_n$ = 1.000″ | $\Theta$ = 30 deg |
| B $\cong$ 0.400″ | L = 7.000″ |

View X

4.5°

Alignment Notch

0.01″

0.01″ R

View Y

$\Theta$

0.5″

0.375″

1.000″ ± 0.002″
Finish Polish

$\rho$ = 0.125″ R ± 0.005″
Diamond Polish

L

View X

View Y

$a_n$

W

Gripping Region

6.25″

4.0″

Unrestricted Region

Gripping Region

17.0″

32

6        6

32

B

Note: Notch root radius ($\rho$) & depth ($CL_N$) are to be measured on comparator before & after polishing. Record & supply these dimensions with finished specimens.

**Figure 12.4**  Corrosion-fatigue crack-initiation specimen.

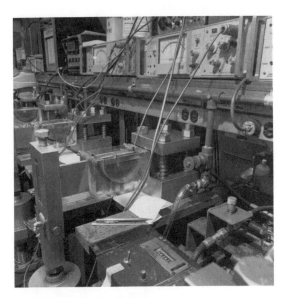

**Figure 12.5** Cantilever-beam specimen in test.

to the fatigue behavior of the metal in the absence of that environment. Consequently, the effect of an environment on the behavior of a material subjected to load fluctuation can be studied by establishing the deviation of the corrosion-fatigue behavior for the environment-material system from the fatigue behavior of the material in a benign environment.

The difference between fatigue and corrosion-fatigue crack-initiation behavior for steel has been studied by Taylor and Barsom[1] and by Novak[2,3] and is presented in the following sections.

### 12.3.1. Fatigue-Crack-Initiation Behavior

The fatigue-crack-initiation behavior for various steels was presented in Chapter 8. The data showed that each steel exhibited a distinct fatigue-crack-initiation threshold at a $\Delta\sigma_{max}$ (or $\Delta K/\sqrt{\rho}$) that can be related to the yield strength of the material.

The fatigue-crack-initiation behavior in air of the ASTM A517 Grade F steel plate whose corrosion-fatigue crack-initiation behavior was investigated by Taylor and Barsom[1] and by Novak[2,3] is presented in Figure 12.6. The test results in this figure include data obtained by Barsom[4] using three-point-bend specimens subjected to stress ratios, $R$, of $-1.0$, $+0.1$, and $+0.5$; data obtained by Novak[2] using cantilever-bend specimens at $R = +0.1$; and data obtained by Taylor and Barsom[1] using compact-tension specimens at $R = +0.1$. The data show that fatigue-crack-initiation life is governed by the total (tension plus compression) maximum-stress range, $\Delta\sigma_{max}$ (or $\Delta K/\sqrt{\rho}$), at the notch tip and that different specimen geometries

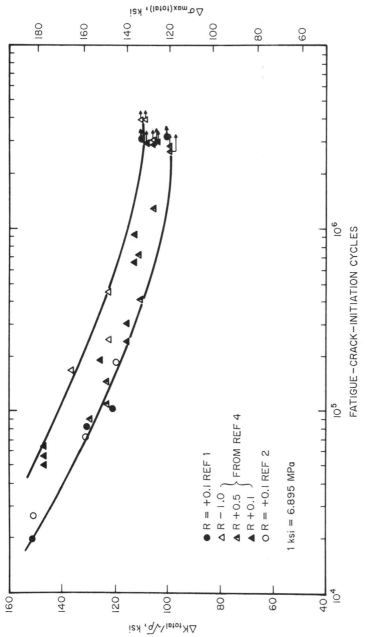

**Figure 12.6** Relationship between fatigue-crack initiation and various stress ratios for A517 Grade F steel.

379

and loadings give similar results. Moreover, the data show a distinct fatigue-crack-initiation threshold that occurred at a $\Delta K/\sqrt{\rho}$ of about 100 ksi (690 MPa), which corresponds to a $\Delta\sigma_{max}$ of 120 ksi (828 MPa). This value is in good agreement with predictions obtained by using the empirical relationships presented in Chapter 8. These relationships indicate that the fatigue-crack-initiation threshold for the A36, A588 Grade A, A517 Grade F, and V150 tested by Novak[2,3] and by Taylor and Barsom[1] occur at a $(\Delta K/\sqrt{\rho})_{th}$ of about 65, 80, 100, and 165 ksi (450, 550, 750, and 1150 MPa), respectively.

### 12.3.2. Corrosion Fatigue Crack-Initiation Behavior

The corrosion-fatigue crack-initiation behavior for A36, A588 Grade A, A517 Grade F, and V150 steels under full immersion conditions in 3.5 percent solution of sodium chloride in distilled water at 12 cycles per minute are presented in Figures 12.7, 12.8, 12.9, and 12.10, respectively. Also shown in these figures is the best-fit equation and the standard deviation for the data and the fatigue-crack-initiation threshold in air for each steel. The data in these figures are presented as a range of cycles bounded by the number of cycles corresponding to the last inspection at which no crack was detected and the following inspection at which a crack existed. The data show that under these test conditions, the environment caused a sub-

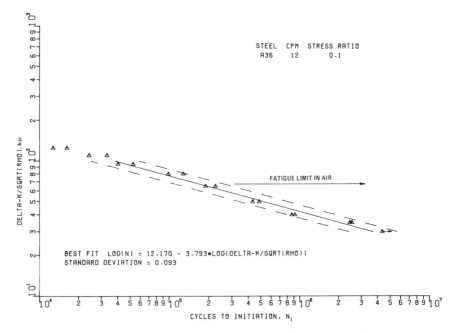

**Figure 12.7**  Corrosion-fatigue crack initiation for A36 steel.

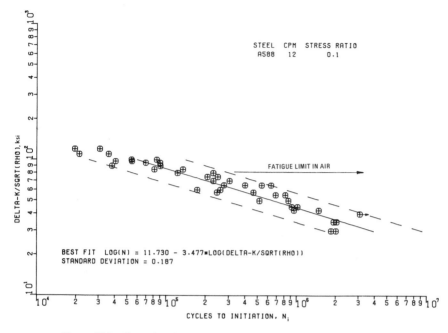

**Figure 12.8**  Corrosion-fatigue crack initiation for A588 Grade A steel.

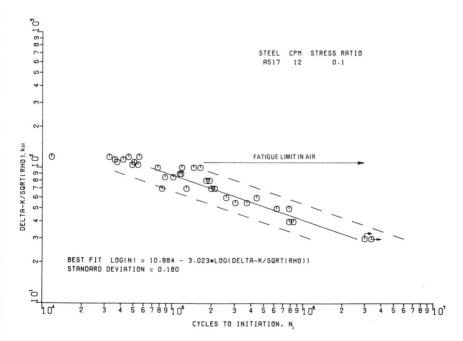

**Figure 12.9**  Corrosion-fatigue crack initiation for A517 Grade F steel.

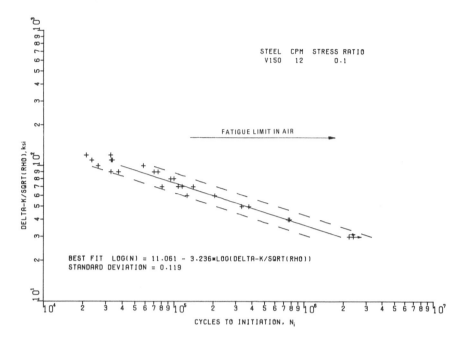

**Figure 12.10**  Corrosion-fatigue-crack initiation for V150 steel.

stantial reduction in the crack-initiation life at $\Delta K/\sqrt{\rho}$ values that are significantly lower than the $\Delta K/\sqrt{\rho}$ value corresponding to the fatigue-crack-initiation threshold in a benign environment. The long-life data points with arrows represent test specimens that were terminated with no indication of corrosion-fatigue crack initiation. These data points may suggest the possible existence of a threshold below which corrosion-fatigue crack-initiation would not occur. Further discussion of corrosion-fatigue crack-initiation threshold is presented in a later section on long-life behavior.

A comparison of the fatigue-crack-initiation behavior in a benign environment and the corrosion-fatigue crack-initiation behavior in 3.5 percent solution of sodium chloride shows that the effect of the environment is small in the region of high-stress range and increases as the stress range decreases. Furthermore, the data show that the corrosion-fatigue crack-initiation life can be represented by a linear relationship of log $\Delta\sigma_{max}$ (or log $\Delta K/\sqrt{\rho}$) and log $N_i$, where $N_i$ is the number of cycles for crack initiation. A generalized relationship for predicting the corrosion-fatigue crack-initiation life for various steels in 3.5 percent solution of sodium chloride is presented in a later section.

The best-fit equations for A36, A588 Grade A, and A517 Grade F suggest the possible dependence of the slope and intercept on the yield strength of the steels such that as the yield strength increases, the slope and intercept values decrease. However, the data for the V150 steel do not support this observation. Further research is needed to establish the relationships, if any, between corrosion-fatigue crack-initiation behavior and properties of the material, the environment, or both.

Figure 12.11 is a superposition of the data presented in Figures 12.7 through 12.10 and shows that, within experimental scatter, the corrosion-fatigue crack-initiation behavior for the four steels was essentially identical. This observation indicates that the corrosion-fatigue crack-initiation behavior for these steels under full immersion conditions in the test environment is independent of chemical composition, microstructure, and mechanical properties (in particular, yield strength) which were significantly different for the steels investigated.

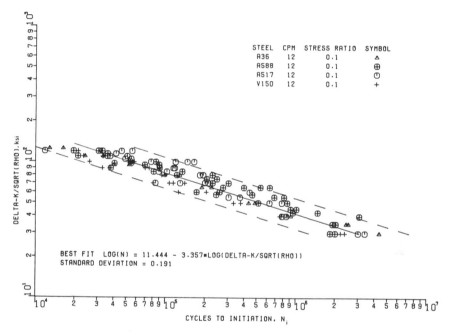

**Figure 12.11**   Corrosion-fatigue-crack initiation for four steels.

### 12.3.3. Effect of Cyclic-Load Frequency

The effects of cyclic-load frequency on the corrosion-fatigue crack-initiation behavior for steels has been investigated in 3.5 percent solution of sodium chloride in distilled water at frequencies equal to 1.2 cycles per minute (cpm) and higher.[1,3,5] The tests were conducted on A588 Grade A and A517 Grade F steels, and the results are presented in Figures 12.12 and 12.13, respectively. The data show a distinct but small increase in the corrosion-fatigue crack-initiation life with increased cyclic-load frequency

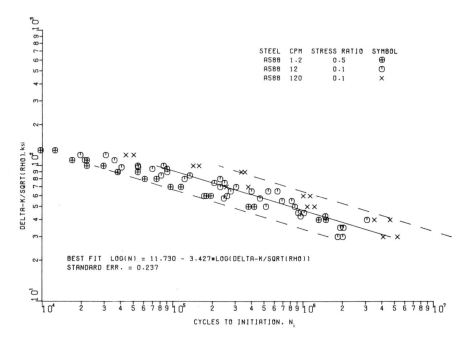

**Figure 12.12**   Corrosion-fatigue-crack initiation for A588 Grade A steel.

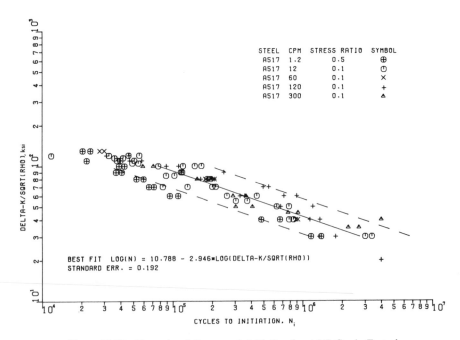

**Figure 12.13**   Corrosion-fatigue-crack initiation for A517 Grade F steel.

from 1.2 cpm to 120 cpm. However, the A517 Grade F corrosion-fatigue crack-initiation life at 300 cpm was less than the life for 120 cpm. This observation may reflect inherent scatter in the data or a possible existence of a unique frequency that results in a minimum environmental and mechanical damage for a given material-environment system.

The data also show that a 100- to 250-fold increase in cyclic-load frequency from 1.2 to 300 cpm resulted in only a 3-fold increase in the mean corrosion-fatigue crack-initiation life. Moreover, the scatter band for the data obtained by testing A588 Grade A and A517 Grade F steels at frequencies of 1.2 to 300 cpm is essentially identical to the scatter band for the various steels tested at 12 cpm (Figure 12.11). Further research is needed to establish the effect, if any, of cyclic-load frequency on the corrosion-fatigue crack-initiation behavior for various material-environment systems. In the meantime, the present data show that $\Delta K / \sqrt{\rho}$ (or $\Delta\sigma_{max}$) is the primary parameter that governs the corrosion-fatigue crack-initiation behavior for structural steels under full-immersion conditions in 3.5 percent solution of sodium chloride in distilled water and that cyclic-load frequency has a secondary effect on the corrosion-fatigue crack-initiation life.

### 12.3.4.  Effects of Stress Ratio

The corrosion-fatigue crack-initiation behavior for A588 Grade A and A517 Grade F steels under full-immersion conditions in room temperature 3.5 percent solution of sodium chloride and at stress ratios, $R$ (ratio of the minimum and maximum nominal stresses, $\sigma_{min}/\sigma_{max}$), of $-1.0$, $+0.1$, and $+0.5$ are presented in Figures 12.14 and 12.15, respectively. The data for A588 Grade A steel (Figure 12.14) show negligible effects of stress ratio on the corrosion-fatigue crack-initiation behavior in this environment, except for the $R = -1.0$ data between $2 \times 10^5$ and $10^6$ cycles to initiation that fell closer to the upper scatter band. Similarly, except for the $R = -1.0$ data between $2 \times 10^5$ and $10^6$ cycles, the data for A517 Grade F steel (Figure 12.15) show negligible effects of stress ratio on corrosion-fatigue crack-initiation behavior. The $R = -1.0$ data for both steels exhibited a noticeable change at about $10^6$ cycles. Further research is needed to establish whether this change represents the true behavior of the environment-material systems tested or an experimental anomaly. The combined data suggest that the corrosion-fatigue crack-initiation behavior for both steels is essentially identical up to $10^6$ cycles and that for $N_i > 10^6$, the A588 Grade A steel exhibited longer initiation life than did A517 Grade F steel.

### 12.3.5.  Long-Life Behavior

Some of the data presented in the preceding sections represent specimens that did not exhibit corrosion-fatigue crack-initiation when the test was terminated at about $2 \times 10^6$ to $4 \times 10^6$ cycles. These data suggest the

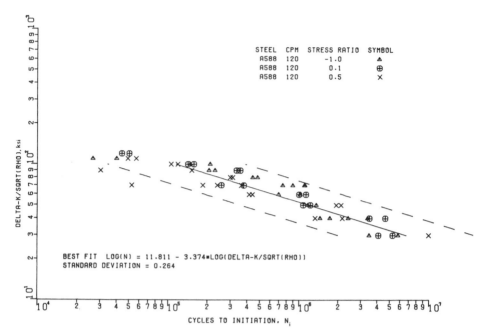

**Figure 12.14**   Corrosion-fatigue-crack initiation for A588 Grade A steel.

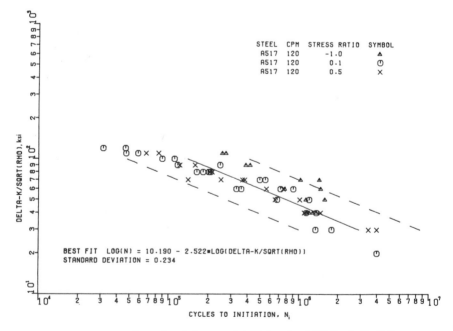

**Figure 12.15**   Corrosion-fatigue-crack-initiation for A517 Grade F steel.

possible existence of a corrosion-fatigue crack-initiation threshold at a $\Delta K/\sqrt{\rho}$ of about 30 ksi. Verification of this behavior required long-term testing at lower $\Delta K/\sqrt{\rho}$ values. Consequently, specimens of A588 Grade A and A517 Grade F were tested at various $\Delta K/\sqrt{\rho}$ values between 30 and 14 ksi. The specimens were subjected to 120 cpm at $R = +0.5$ under full-immersion conditions in room temperature 3.5 percent sodium chloride solution that was changed once a week. The specimens were in test from $4 \times 10^6$ to $7 \times 10^6$ cycles.

The long-life test results are presented in Figure 12.16.[5] The data fall along the linear extension of the log $\Delta K/\sqrt{\rho}$ and log $N_i$ relationship established for the short-life behavior (Figure 12.11) and within the same scatter band.

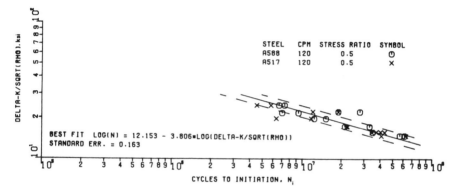

**Figure 12.16**    Long-life corrosion-fatigue-crack initiation.

Each specimen tested exhibited initiation of corrosion-fatigue cracks at the tip of the notch. Consequently, for the conditions used to conduct these tests, no corrosion-fatigue crack-initiation threshold was observed. However, this observation should not be extended indiscriminantly to other material-environment systems or frequencies. For example, preliminary observations suggest the possible existence of thresholds for these steels in this environment with high pH or with cathodic protection.

### 12.3.6. Generalized Equation for Predicting the Corrosion-Fatigue Crack-Initiation Behavior for Steels

Figure 12.17 presents all the corrosion-fatigue crack-initiation data presented in the preceding sections for A36, A588 Grade A, A517 Grade F, and V150 steels under full-immersion conditions in room temperature 3.5 percent solution of sodium chloride in distilled water. These steels represent large variations in chemical composition, thermomechanical processing, microstructure, and mechanical properties (tensile strength, yield strength, elongation, strain hardening, toughness, etc.). The combined data encompass frequencies of 1.2, 12, 60, 120, and 300 cpm and stress ratios

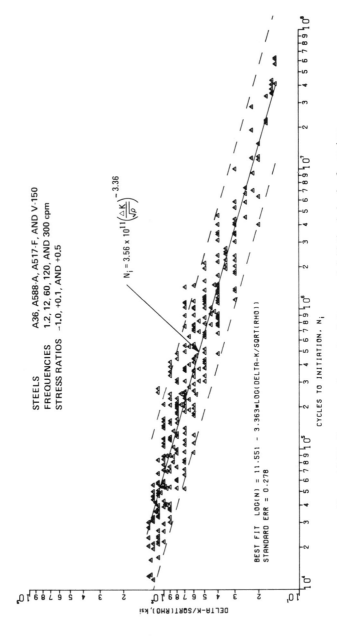

**Figure 12.17** Summary of corrosion-fatigue-crack initiation behavior for various steels.

of $-1.0$, $+0.1$, and $+0.5$ and span four orders of magnitude in corrosion-fatigue crack-initiation life between about $10^4$ and $10^8$ cycles.

Considering the large variation in materials and test conditions, the data fall within a surprisingly narrow scatter band and can be represented by a single linear relationship between $\log \Delta K/\sqrt{\rho}$ and $\log N_i$. The equation for the best-fit line for all the data is

$$N_i = 3.56 \times 10^{11}(\Delta K/\sqrt{\rho})^{-3.36} \tag{12.1}$$

where $\Delta K$ is in ksi and $\rho$ is in inches.

Figure 12.17 includes data that represent the number of cycles corresponding to the last inspection at which no crack was detected and the following inspection at which a crack existed. Thus, equation (12.1) should represent closely the number of cycles for corrosion-fatigue crack initiation. Prediction of corrosion-fatigue crack-initiation life by using the lower bound for the data may be too conservative. Finally, the corrosion-fatigue crack-initiation life for notches having severe stress raisers or severe imperfections on the notch surface or in the immediate vicinity of the notch tip could be significantly less than predicted by using Equation (12.1).

## 12.4. Prevention of Corrosion-Fatigue Failures

Several methods can be used to prevent or circumvent the detrimental effects of corrosion fatigue on structural performance. These methods may be easy to identify, but their implementation may be difficult or very costly. The effectiveness of a given method or a combination of methods depends on the particular material-environment system under consideration. Although a number of parameters are involved in the selection process, several preventive methods can be eliminated by establishing the basic function of the material and of the environment in the system. For example, changing the characteristics of the environment may not be a viable method if the environment is the desired product of the manufacturing operation.

Some methods that are currently in use to prevent or circumvent the detrimental effects of corrosion fatigue on the performance of structural components are presented in this section. A discussion of the advantages and limitations of these methods is beyond the scope of this document.

1. *Isolate the environment and the material.* This can be accomplished by placing a barrier between the environment and the material. Barriers that have been used include metallic coatings (e.g., zinc, chrome), organic coatings (e.g., paint), inorganic coatings (e.g., glass), ceramic and rubber liners, and clading.

2. *Alter the severity of the environment.* This can be accomplished by chemically removing the aggressive constituents in the environment,

by increasing the pH of the environment, or by decreasing the temperature, flow rate, and concentration of the environment.

3. *Apply cathodic protection.* This can be accomplished by externally imposed negative potential or a galvanically generated potential. Sacrificial anodes are frequently used for cathodic protection of structures.

4. *Alter the surface characteristics of the material.* This can be accomplished by inducing favorable compressive stresses on the material surfaces that are exposed to the environment. Compressive surface stress can be induced by using induction hardening (which is used successfully for sucker rods), shot peening, as well as others. These favorable compressive surface stresses may not minimize corrosion-fatigue crack-initiation or general corrosion but could significantly decrease and possibly eliminate crack propagation.

5. *Substitute a more resistant material.* In general, materials that are resistant to a given environment are available. However, care should be exercised in the selection process to ensure (a) that the substitute material possesses all the other properties that are essential for its use in the particular application; (b) that the substitute material will not fail by a mechanism other than corrosion fatigue; (for example, grain-boundary embrittlement, chloride cracking, etc.) to which the other material was immune; and (c) that the material selection is based on its resistance to corrosion fatigue in the environment rather than its resistance to other damage mechanisms such as corrosion or stress-corrosion cracking in the environment.

6. *Design the components to prevent the initiation or the propagation of cracks to a critical size.* For a given material environment system, this can be accomplished by using data similar to that presented in this chapter and in Chapter 13 to design the structural components properly against corrosion-fatigue damage or to establish inspection procedures and inspection intervals that would ensure the safe operation of the structure in the environment of interest.

The preceding presents possible options to prevent or circumvent structural damage caused by corrosion fatigue. The usefulness of any of these methods depends on the particular application. For example, isolating the environment and the material obviates the detrimental effects of the environment but does not minimize damage caused by fatigue. Also, the effectiveness of these methods to prevent corrosion-fatigue crack initiation may be different than for propagation. For example, cathodic protection can have a significant role in preventing or retarding the initiation of corrosion-fatigue cracks but may have little or no effect on the corrosion-fatigue crack-propagation behavior. Because of these and other considerations, extreme care and sound engineering judgment should be exercised in the selection of the appropriate method to prevent or circumvent corrosion-fatigue damage and to ensure the safety, reliability, and cost effectiveness of the structure.

## References

1. M. E. TAYLOR and J. M. BARSOM, "Effect of Cyclic Frequency on the Corrosion-Fatigue Crack-Initiation Behavior of ASTM A517 Grade F Steel," *Fracture Mechanics: Thirteenth Conference, ASTM STP 743*, Richard Roberts, ed., American Society for Testing and Materials, Philadelphia, 1981.
2. S. R. NOVAK, "Corrosion-Fatigue Crack-Initiation Behavior of Four Structural Steels," *Corrosion Fatigue: Mechanics, Metallurgy, Electrochemistry, and Engineering, ASTM STP 801*, American Society for Testing and Materials, Philadelphia, 1983.
3. S. R. NOVAK, "Influence of Cyclic-Stress Frequency and Stress Ratio on the Corrosion-Fatigue Crack-Initiation Behavior of A588-A and A517-F Steels in Salt Water," presented at the Fifteenth National Symposium on Fracture Mechanics, University of Maryland, College Park, Md., 1982.
4. J. M. BARSOM, "Concepts of Fracture Mechanics—Fatigue and Fracture Control," *Fracture Mechanics for Bridge Design*, Federal Highway Administration, *Report No. FHWA-RD-78-69*, U.S. Department of Transportation, Washington, D.C., July 1977.
5. J. M. BARSOM and S. R. NOVAK, "Long-Life Behavior and the Effect of Low Frequency on the Corrosion-Fatigue Crack-Initiation for Steels," unpublished data.

# 13

# Corrosion-Fatigue-Crack Propagation

## 13.1. Introduction

The fatigue-crack-initiation and -propagation behaviors in benign environments were discussed in Chapters 7, 8, 9, and 10. Because structures operate in various environments, the effect of environments on the behavior of structures must be established. The effect of environments on the fracture behavior of statically loaded structural components that contain fatigue cracks was presented in Chapter 11. The effect of environments on the fatigue-crack-initiation behavior was presented in Chapter 12. However, because structural components may contain crack-like imperfections and may be subjected to load fluctuations in aggressive environments, the corrosion-fatigue crack-propagation behavior of metals in various environments is of primary importance.

Several investigators have studied the corrosion-fatigue behavior of various environment-material systems.[1-19] The results of these investigations have helped greatly in the selection of proper materials for a given application. Despite the significant progress that has been achieved to establish the effects of various mechanical parameters on the corrosion-fatigue behavior of environment-material systems, little has been achieved to establish mechanisms of corrosion fatigue in these systems. The available information shows the high complexity of the corrosion-fatigue behavior and suggests that a significant understanding of this behavior can be achieved only by a synthesis of contributions from various fields.

## 13.2. General Behavior

Corrosion-fatigue behavior of a given environment-material system refers to the characteristics of the material under fluctuating loads in the presence of the particular environment. Different environments have different effects on the cyclic behavior of a given material. Similarly, the corrosion-fatigue behavior of different materials is different in the same environment.

The corrosion-fatigue behavior of metals subjected to load fluctuation in the presence of an environment to which the metal is immune is identical to the fatigue behavior of the metal in the absence of that environment. Consequently, the corrosion-fatigue behavior of an environment-material system can be studied by establishing the deviation of the corrosion-fatigue behavior for the environment-material system from the fatigue behavior of the material in a benign environment.

Stress-corrosion crack growth in a statically loaded structure is caused by interactions of chemical and mechanical processes at the crack tip. The highest plane-strain stress-intensity-factor value at which subcritical crack growth does not occur in a material loaded statically in an aggressive environment is designated $K_{Iscc}$. Consequently, to establish the effect of an environment on the fatigue-crack-growth behavior of a material, the fatigue-crack-growth behavior for the material in a benign environment and the $K_{Iscc}$ for the environment-material system should be established first as references.

The generalized fatigue-crack-growth behavior in a benign environment (Figure 13.1), is a special case of the corrosion-fatigue crack propagation behavior for metals. It represents the "corrosion-fatigue" behavior of metals subjected to load fluctuations in the presence of any environment that does not affect the fatigue-crack-growth behavior for the metal. Thus, the corrosion-fatigue behavior for a given environment-material system could be investigated by establishing the base-line fatigue behavior and then by determining the effect of the environment on the fatigue behavior regions I, II, and III (Figure 13.1). However, because $K_{Iscc}$ for an environment-material system defines the plane-strain $K_I$ value above which stress-corrosion crack growth can occur under static loads, the corrosion-fatigue crack-propagation behavior for the environment-material system could be altered when the maximum value of $K_I$, $K_{Imax}$, in a given load cycle becomes greater than $K_{Iscc}$. Consequently, the corrosion-fatigue crack-propagation behavior should be divided into below-$K_{Iscc}$ and above-$K_{Iscc}$ behaviors.

## 13.3. Corrosion-Fatigue Crack-Propagation Threshold

Several recent investigations have significantly increased our understanding of environmental effects on the threshold behavior.[1-4] The fatigue-crack-propagation threshold behavior for steels in room temperature air environment was presented in Chapter 9. The data showed, among other things, that the $\Delta K_{th}$ is strongly dependent on the stress ratio, $R$, and that its value decreases as the value of $R$ increases.

Bucci and Donald[20] observed that the environmental $\Delta K_{th}$ in a 200-

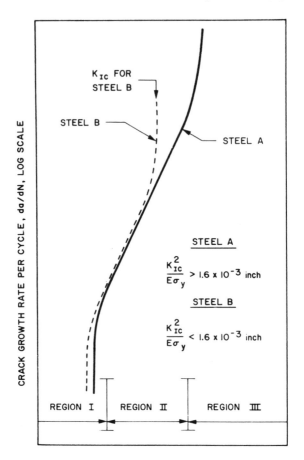

**Figure 13.1** Schematic representation of fatigue-crack growth in steel.

grade maraging steel forging was higher than the $\Delta K_{th}$ in air and that the "salt water appears to produce an inhibitive effect on fatigue cracking at the very low $\Delta K$ levels." Paris et al.[16] also observed that the threshold $\Delta K$ of ASTM A533 Grade B, Class 1 steel in distilled water was greater than that established in room-temperature air. "Since the distilled water retardation of very low crack extension rates was a somewhat surprising result, an additional specimen was tested for which a distilled water environment was provided to the crack tip and its surroundings only after an initial, slow, crack extension rate had been established in room air. Upon application of the distilled water, the rate of crack growth decreased from those initially obtained in air."[16]

  Figure 13.2[2] presents the threshold and near-threshold fatigue-crack-propagation behavior for a $2\frac{1}{4}$Cr-1Mo steel in hydrogen gas and in humid

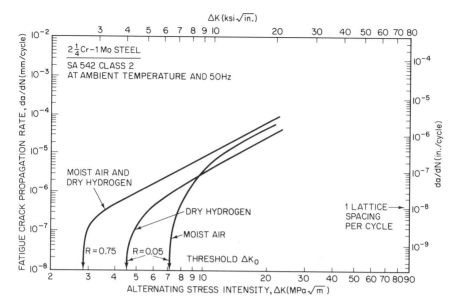

**Figure 13.2**   Fatigue-crack propagation in martensitic $2\frac{1}{4}$Cr-1Mo steel (SA542-2) in moist air and dry hydrogen. (From S. Suresh, G. F. Zamiski, and R. O. Ritchie, "Oxide-Induced Crack Closure: An Explanation of Near-Threshold Corrosion Fatigue Crack Growth Behavior," *Metallurgical Transactions A, 12A,* Aug. 4, 1981, pp. 1435–1443. American Society for Metals, Metals Park, OH 44073.)

air at a frequency of 50 Hz and a stress ratio, $R$, of 0.05 and 0.75. The data show that at $R = 0.05$, the near-threshold fatigue-crack-propagation rate in dry hydrogen was significantly higher than in air and that the $\Delta K_{th}$ for dry hydrogen was lower than the value in air. However, unlike the behavior for $R = 0.05$, the near-threshold fatigue-crack-propagation rate and the threshold for $R = 0.75$ in both environments were identical. Similar trends were observed for other gaseous environments and steels.[2,3] The data in inert gaseous environments indicate that the increase in the near-threshold propagation rate and the decrease in $\Delta K_{th}$ in dry hydrogen from the values in humid air environment are not necessarily related to hydrogen embrittlement. Hydrogen embrittlement and other mechanisms may influence the threshold value or the near-threshold propagation behavior in a given environment-metal system. However, the differences observed between dry gaseous environments and humid air for low-strength steels appear to be related primarily to the formation of corrosion products within the crack in the air environment and their effect on the closure of the crack. (See Chapter 9, Section 9.2.)

Figure 13.3[4] presents threshold and near-threshold propagation behavior for $2\frac{1}{4}$Cr-1Mo steel in moist air (30 percent relative humidity), dry hydrogen,

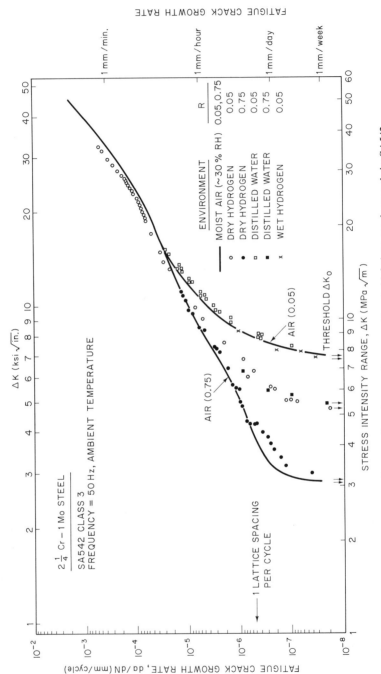

**Figure 13.3** Influence of environment on near-threshold fatigue-crack growth in SA542-3 tested at $R = 0.05$ and 0.75. Room temperature moist-air data are compared with data for wet and dry gaseous hydrogen and distilled water environments.

wet hydrogen, and distilled water environments at $R$ values of 0.05 and 0.75. The data show that, at low stress ratios, humid air, wet hydrogen, and distilled water exhibited essentially identical thresholds and resulted in lower near-threshold crack-propagation rates and higher threshold values than for the dry hydrogen environment. At high $R$ values, the data show a higher $\Delta K_{th}$ in distilled water than in humid air or dry hydrogen.

The concept of reduced cyclic crack-opening displacement caused by corrosion products within the crack was proposed by Paris et al.[16] to explain a higher $\Delta K_{th}$ value in distilled water than in air for pressure-vessel steels tested at an $R$ value of 0.1. This concept suggests that a corrosive environment produces corrosion products which at low stresses wedge the crack open, resulting in a decreased driving force.[1] The concept was also used by Skelton and Haigh[21] to explain the effect of stress ratio on the $\Delta K_{th}$ for CrMoV steels tested at 550°C in vacuum and in oxidizing environments. They found that both inert environments and high stress ratios reduced $\Delta K_{th}$ and that compressive stresses resulted in reduced $\Delta K_{th}$ by compacting the corrosion products which increased the effective cyclic crack-opening displacement.

Skelton and Haigh[21] reported a systematic decrease of $\Delta K_{th}$ as the cyclic frequency was increased from 0.01 to 10 Hz. This finding is very important because many engineering structures are subjected to cyclic frequencies that are less than 1 Hz. Unfortunately, because of the very large number of cycles necessary to establish the $\Delta K_{th}$, most tests are conducted at cyclic frequencies that are higher than 10 Hz.[1] An increase in cyclic frequency to shorten the test time for determining $\Delta K_{th}$ and the use of the results to evaluate the performance of a structure that is subjected to low cyclic frequencies in the service environment can lead to erroneous conclusions.

Limited data related to the effect of cyclic frequency on $\Delta K_{th}$ in aqueous environments has been obtained for various constructional steels.[9] The corrosion-fatigue-crack-growth-rate data for A36 steel tested at 12 cpm indicated that the rate of crack growth decreased significantly at $\Delta K_I$ values less than 20 ksi$\sqrt{\text{in.}}$ (22.0 MN/m$^{3/2}$) (Figure 13.4). Similar behavior was observed in the A588 Grade A, A588 Grade B, A514 Grade E, and A514 Grade F steels tested. To verify this observation, one specimen of A514 Grade E and one specimen of A514 Grade F were tested at 12 cpm in 3 percent sodium chloride solution under cyclic-load fluctuations corresponding to $\Delta K_I$ of about 11 ksi$\sqrt{\text{in.}}$ (12.1 MN/m$^{3/2}$). The test results are presented in Figures 13.5 and 13.6. The data show that a corrosion-fatigue crack-growth-rate threshold, $\Delta K_{th}$, does exist in A514 steels at a value of the stress-intensity-factor fluctuation below which corrosion-fatigue cracks do not propagate at 12 cycles per minute (cpm) in the environment-steel system tested. The value of $\Delta K_{th}$ in the A514 steels tested at 12 cpm in 3 percent

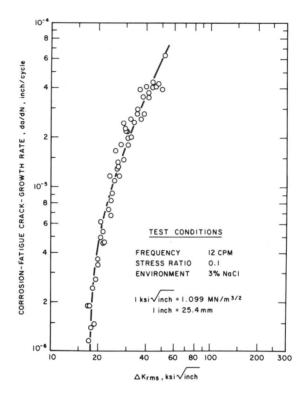

**Figure 13.4**  Corrosion-fatigue-crack-growth rate as a function of the root-mean-square stress-intensity factor for A36 steel.

sodium chloride solution was twice as large as the value of 5.5 ksi$\sqrt{\text{in.}}$ (6.0 MN/m$^{3/2}$) for room temperature air.[22] (For $\Delta K_{\text{th}}$ in air, see Chapter 9.)

## 13.4.  Corrosion-Fatigue-Crack-Propagation Behavior Below $K_{\text{Iscc}}$

The first systematic investigation into the effect of environment and loading variables on the rate of fatigue-crack growth below $K_{\text{Iscc}}$ was conducted on 12Ni-5Cr-3Mo maraging steel (yield strength = 180 ksi) in 3 percent solution of sodium chloride.[8,23,24] The data showed that environmental acceleration of fatigue-crack growth does occur below $K_{\text{Iscc}}$ (Figure 13.7) and that the magnitude of this acceleration is dependent on the frequency of the cyclic-stress-intensity fluctuations. The test results also showed that the fatigue-crack-growth rates in 12Ni-5Cr-3Mo maraging steel tested in a room tem-

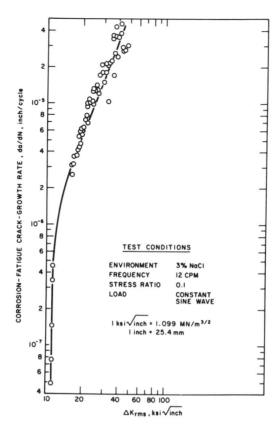

**Figure 13.5** Corrosion-fatigue-crack-growth rate as a function of the root-mean-square stress-intensity factor for A514 Grade E steel.

perature air environment and in a room temperature 3 percent solution of sodium chloride can be represented by

$$\frac{da}{dN} = D(t)(\Delta K)^2 \qquad (13.1)$$

where $D(t)$ depends on the environment-material system and on the sinusoidal cyclic stress-intensity frequencies. That is, the magnitude of the environmental accelerations of fatigue-crack growth can be increased or decreased substantially by changing the environment, the material, and the frequency of loading. In air, $D(t)$ was a constant, independent of frequency. In the sodium chloride solution at high frequencies (cpm > 600), $D(t)$ had essentially the same values as in air, Figure 13.8; thus the environment had negligible effects on the fatigue-crack-growth rate. In the sodium chloride solution

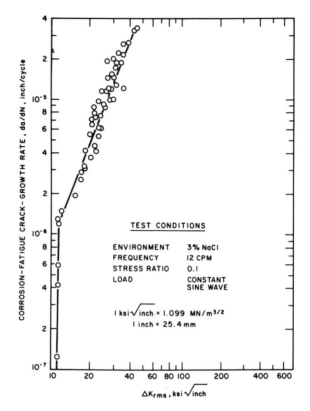

**Figure 13.6**  Corrosion-fatigue-crack-growth rate as a function of the root-mean-square stress-intensity factor for A514 Grade F steel.

at 6 cpm, however, $D(t)$ was three times higher than the value in air, which indicates that the fatigue-crack-growth rate was increased significantly by the environment. The data in Figure 13.8 may be used to predict the value of $D(t)$ for any sinusoidal frequency equal to or greater than about 6 cpm in the environment-material system investigated. Because these results were obtained below $K_{Iscc}$, Barsom[23,24] concluded that the corrosion-fatigue-crack-growth rate for 12Ni-5Cr-3Mo maraging steel in a 3 percent solution of sodium chloride increases to a maximum value and then decreases as the sinusoidal cyclic-stress frequency decreases from 600 cpm to frequencies below 6 cpm. Similar behavior has been established for HY-80 steel[14] and for Ti-8Al-1V-1Mo alloy.[17] The dependence of corrosion-fatigue-crack-growth rate below $K_{Iscc}$ on cyclic frequency has been established for steels,[8,9,14,15,23,24] aluminum alloys,[11,12,18] and titanium alloys.[12,17] However, unlike the data presented in Figure 13.7, curves obtained for different cyclic frequencies

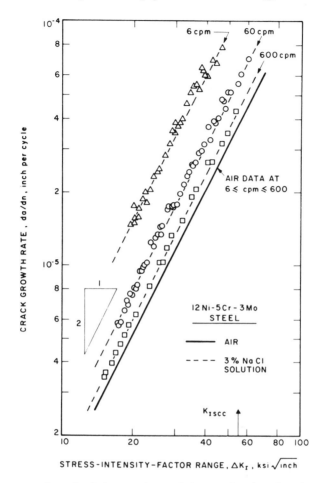

**Figure 13.7**   Corrosion-fatigue-crack-growth data as a function of test frequency.

in various environment-metal systems do not appear to be parallel to each other (Figure 13.9).[25]

The magnitude of the effect of cyclic frequency on the rate of corrosion-fatigue-crack growth depends strongly on the environment-material system.[26,27] The data presented in Figure 13.10[26] show that the 10Ni-Cr-Mo-Co steels tested were highly resistant to the 3 percent solution of sodium chloride and that of the four steels tested, the 12Ni-5Cr-3Mo steel was the least resistant to the 3 percent solution of sodium chloride. Similarly, fatigue-crack-growth rates below $K_{Iscc}$ were accelerated by a factor of 2 when 4340 steel of 130-ksi yield strength was tested at 6 cpm in a sodium chloride solution (Figure 13.11).[27]  Under identical test conditions, the corrosion-

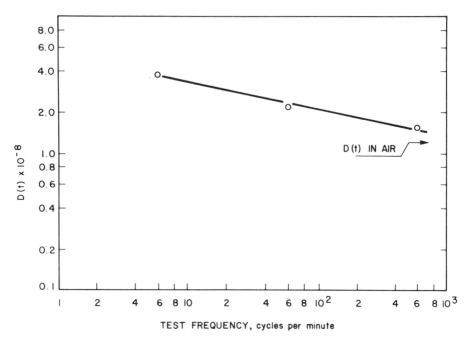

**Figure 13.8** Correlation between time-dependent function $D(t)$ and test frequency for 12Ni-5Cr-3Mo steel tested in sodium chloride solution.

fatigue-crack-growth rates in the same 4340 steel heat treated to 180-ksi yield strength were five to six times higher than the fatigue-crack-growth rates in room temperature air environments (Figure 13.11).

The effect of various additions to aqueous solutions on the stress-corrosion cracking and on the corrosion-fatigue-crack-growth rates at a given cyclic frequency have been investigated for high-strength aluminum alloys (Figure 13.12), magnesium alloys (Figure 13.13), and titanium alloys.[12] A summary of the effects of the additions is presented in Table 13.1.[12] The data suggest that the additions to aqueous solutions that accelerate stress-corrosion crack growth also accelerate the corrosion-fatigue-crack growth and those additions that do not affect stress-corrosion cracking have no effect on corrosion-fatigue-crack growth.

Extensive corrosion-fatigue data have been obtained for various aluminum alloys in water and water vapor environments. The results indicate that these environments have a significant effect on the fatigue-crack-growth rate for aluminum alloys.[11,18,28–32] The effect of water and water vapor on the rate of fatigue-crack growth for a 7075 aluminum alloy is shown in Figure 13.14.

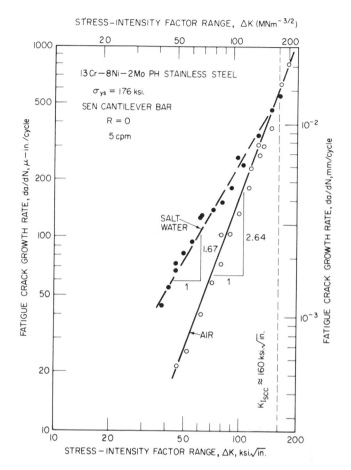

**Figure 13.9**   Air and salt water fatigue-crack-growth-rate behavior of 13Cr-8Ni-2MO PH stainless steel.

Corrosion-fatigue-crack-growth data for various low-yield-strength constructional steels have been investigated below $K_{Iscc}$ to determine the susceptibility of these steels to aqueous environments.[9]

The steels tested were A36, A588 Grade A, A588 Grade B, A514 Grade E, and A514 Grade F steels in distilled water and in 3 percent solution of sodium chloride in distilled water. The tests were conducted under constant-amplitude and variable-amplitude random-sequence sinusoidal load fluctuations at frequencies of 60 and 12 cpm.

The data presented in Figures 13.15, 13.16, and 13.17 show that the addition of 3 percent (by weight) sodium chloride to distilled water had no

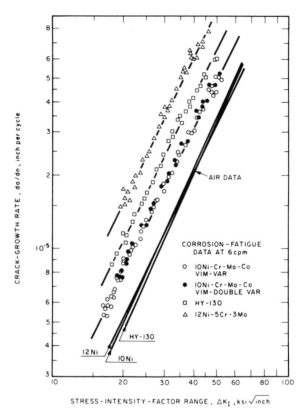

**Figure 13.10** Fatigue-crack-growth rates in air and in 3 percent solution of sodium chloride below $K_{Iscc}$ for various high-yield-strength steel.

effect on the corrosion-fatigue behavior of these steels. Similar data were obtained for the other steels that were tested.[9] The data also show that the corrosion-fatigue-crack-growth-rate behavior at 60 cpm under sinusoidal loads and under square-wave loads is essentially identical. The scatter in test results obtained in a single specimen under corrosion-fatigue conditions was equal to or greater than under fatigue conditions in room temperature air environment. The increase in scatter was caused by the general corrosion of the specimen surfaces, which decreased the accuracy for determining the exact location of the crack tip. Considering the scatter caused by the general corrosion of the specimen surfaces and the inherent scatter observed in fatigue-crack-growth data, the data presented in Figures 13.15 through 13.17 indicate that, at 60 cpm, distilled water and 3 percent solution of sodium chloride in distilled water had negligible effects on the rate of growth of fatigue cracks in the constructional steels investigated. Corrosion-fatigue

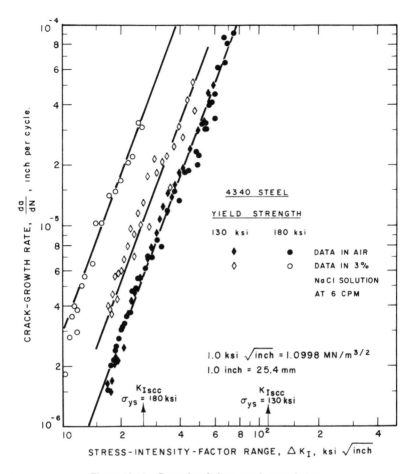

**Figure 13.11**    Corrosion-fatigue-crack-growth data.

data obtained by testing these steels in 3 percent solution of sodium chloride at a stress ratio, $R$, of 0.5 were identical to those obtained at $R = 0.1$, and corrosion-fatigue data obtained for five different heats of A588 steel were also identical.[9]

Corrosion-fatigue data obtained by testing duplicate specimens of A36, A588 Grade A, and A514 Grade F steels in 3 percent solution of sodium chloride at 12 cpm under constant-amplitude sinusoidal loading are presented in Figures 13.18 through 13.20.[9] Similar data were obtained for A588 Grade B and A514 Grade E steels.[9] Superimposed on these figures is the upper bound of data scatter obtained by testing these steels at 60 cpm under constant-amplitude and variable-amplitude random-sequence loading in distilled water and in 3 percent solution of sodium chloride. The long duration

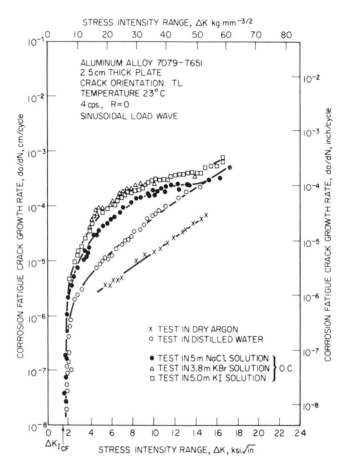

**Figure 13.12** Effect of cyclic stress-intensity range on the growth rate of corrosion-fatigue cracks in a high-strength aluminum alloy exposed to various environments.

of the corrosion-fatigue tests at 12 cpm caused extensive surface corrosion that resulted in greater data scatter than obtained in tests at 60 cpm. The corrosion-fatigue data presented in Figures 13.18 through 13.20 show that the rate of crack growth for the constructional steels tested at 12 cpm and at stress-intensity-factor fluctuations greater than about 15 ksi$\sqrt{\text{in.}}$ (16.5 MN/m$^{3/2}$) was equal to or slightly greater than that observed at 60 cpm.

### 13.4.1. Effect of Cyclic-Stress Waveform

Available data indicate that the environmental effects on the rate of fatigue-crack growth in corrosion fatigue below $K_{\text{Iscc}}$ may be highly dependent on the shape of the cyclic-stress wave.[8] This dependence is illustrated by the difference between the fatigue-crack-growth rate data for 12Ni-5Cr-3Mo

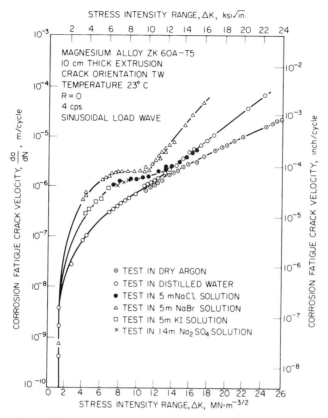

**Figure 13.13** Effect of cyclic stress-intensity range and various environments on the growth rate of corrosion fatigue cracks in a high-strength magnesium alloy.

**TABLE 13.1   Effect of Various Additions to Aqueous Environments on Acceleration of Subcritical Crack Growth in High-Strength Light Metals**

| Alloys | Stress Corrosion | | "True" Corrosion Fatigue (Region II) | |
| | Additions Which *Can* Accelerate Crack Growth | Additions Which *Do Not* Accelerate Crack Growth | Additions Which *Can* Accelerate Crack Growth | Additions Which *Do Not* Accelerate Crack Growth |
|---|---|---|---|---|
| Aluminum base (7079-T651) | $Cl^-$, $Br^-$, $I^-$ | $SO_4^=$ | $Cl^-$, $Br^-$, $I^-$ | $SO_4^=$ |
| Titanium base (Ti-6A1-4V) | $Cl^-$, $Br^-$, $I^-$ | $SO_4^=$ | $Cl^-$, $Br^-$, $I^-$ | $SO_4^=$ |
| Magnesium base (ZK60A-T5) | $Cl^-$, $Br^-$, $I^-$ $SO_4^=$ | | $Cl^-$, $Br^-$, $I^-$ $SO_4^=$ | |

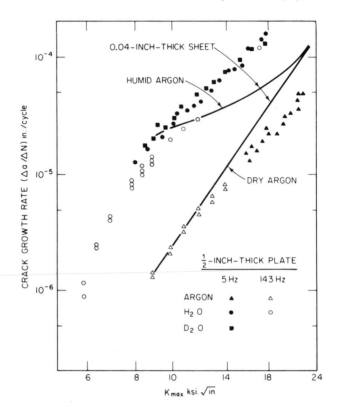

**Figure 13.14** The effect of water and water vapor on the rate of fatigue-crack growth in 7075 aluminum alloy.

steel in a room temperature air environment (Figure 13.21) and in a 3 percent solution of sodium chloride (Figure 13.22) under sinusoidal loading, triangular loading, and square loading at 6 cpm. The tests were conducted on identical specimens in the same bulk environment and at the same maximum and minimum loads. The effect of the cyclic wave on the corrosion-fatigue behavior below $K_{Iscc}$ was obtained from direct comparison between the crack-growth rates per cycle at a constant value of $\Delta K_I$.

The data presented in Figure 13.21 show that the fatigue-crack-growth rates in room temperature air environment are identical under various stress fluctuations and are independent of frequency. The data in Figure 13.22 show that in a sodium chloride solution the crack-growth rates per cycle under sinusoidal and triangular stress fluctuations are almost identical. At a constant frequency, the environment increased the crack-growth rate by the same amount under sinusoidal stress fluctuations as under triangular stress fluctuations. The data also show that environmental effects are negligible when the steel is subjected to a square-wave stress fluctuation. Corrosion-fatigue-crack-growth rates under square-wave loading at 6 cpm

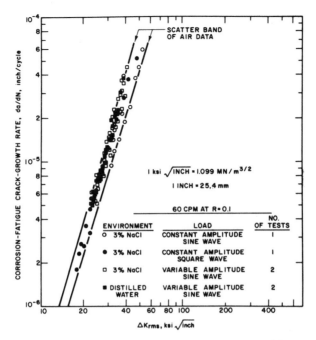

**Figure 13.15**   Corrosion-fatigue-crack-growth rate as a function of the root-mean-square stress-intensity factor for A36 steel.

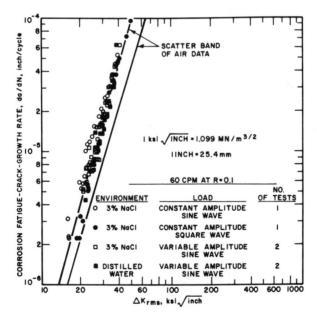

**Figure 13.16**   Corrosion-fatigue-crack-growth rate as a function of the root-mean-square stress-intensity factor for A588 Grade A steel.

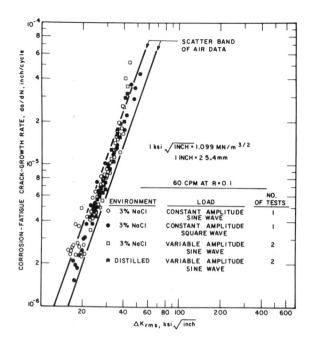

**Figure 13.17**  Corrosion-fatigue-crack-growth rate as a function of the root-mean-square stress-intensity factor for A514 Grade F steel.

for the 12Ni-5Cr-3Mo steel tested in sodium chloride solution were essentially the same as they were in the absence of environmental effects.  By establishing the sinusoidal cyclic frequency that would result in the same environmental effects on the rate of fatigue crack growth, Barsom[8] showed that the environmental damage below $K_{Iscc}$ occurred only during transient loading. Corrosion-fatigue data obtained below $K_{Iscc}$ by using square waves having different dwell times at maximum and minimum loads also showed no environmental effects at constant tensile stresses.

Fatigue-crack-growth data for an aluminum alloy (DTD 5070A) tested in air at 60 cpm showed no difference in the rate of growth under sinusoidal, square, and pulsed waveforms.[33]  On the other hand, fatigue-crack-growth data for 7075-T6 aluminum alloy tested in salt water at 6 cpm under sinusoidal, triangular, and square wave forms showed a behavior very similar to that presented for 12Ni-5Cr-3Mo steel in salt water.[18]  The difference between the results obtained for the DTD 5070A aluminum alloy in air and those obtained for 7075-T6 and for 12Ni-5Cr-3Mo steel in salt water may be due to the relative immunity of the alloy to the air environment, to the high cyclic frequency, or to a difference in the mechanism of corrosion fatigue.

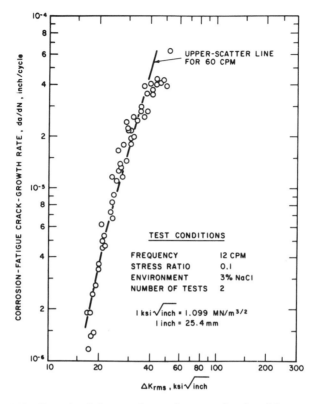

**Figure 13.18**   Corrosion-fatigue-crack-growth rate as a function of the root-mean-square stress-intensity factor for A36 steel.

### 13.4.2. Environmental Effects During Transient Loading

Corrosion-fatigue-crack-growth test results for 12Ni-5Cr-3Mo maraging steel in 3 percent sodium chloride solution under sinusoidal, triangular, and square-wave loading showed that environmental effects in the environment-material investigated are significant only during the transient-loading portion of each cyclic-load fluctuation.[8] The difference between the effects of the sodium chloride solution on the rate of fatigue-crack growth during increasing and decreasing plastic deformation in the vicinity of the crack tip was investigated by using test results obtained for specimens subjected to various triangular cyclic-load fluctuations.

The effects of the environment during increasing plastic deformation in the vicinity of the crack tip were separated from the effects during decreasing plastic deformation by studying the differences in the corrosion-fatigue-crack-growth rate obtained under positive-sawtooth ($\diagup$) and under negative-sawtooth ($\diagdown$) cyclic-load fluctuations.

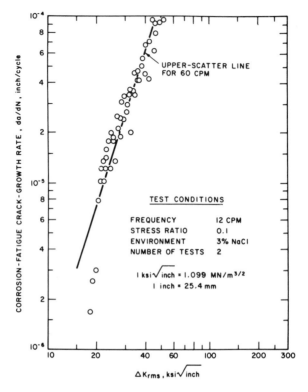

**Figure 13.19**   Corrosion-fatigue-crack-growth rate as a function of the root-mean-square stress-intensity factor for A588 Grade A steel.

The data presented in Figure 13.21[8] show that the rates of fatigue-crack growth in a room temperature air environment under various cyclic-stress fluctuations are not affected by the form of the cyclic-stress fluctuations. Consequently, differences in the rates of corrosion-fatigue-crack growth among triangular waves, positive-sawtooth waves, and negative-sawtooth waves can be attributed primarily to variations in the interaction between plastically deformed metal at the crack tip and the surrounding environment. These variations result from differences in the pattern of stress fluctuations during each cycle.

The corrosion-fatigue-crack-growth-rate data for 12Ni-5Cr-3Mo maraging steel tested in 3 percent sodium chloride solution under various cyclic stress fluctuations are presented in Figure 13.23.[8] The data show that the corrosion-fatigue-crack-growth rates determined with the negative-sawtooth wave and with the square wave are essentially the same as the fatigue-crack-growth rate determined in air. The corrosive effect therefore depends on the wave-

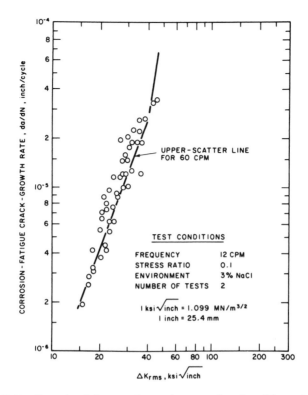

**Figure 13.20**   Corrosion-fatigue-crack-growth rate as a function of the root-mean-square stress-intensity factor for A514 Grade F steel.

form. The negative-sawtooth wave and the square wave show no corrosive effect. The corrosion-fatigue-crack-growth rates determined with sinusoidal, triangular, and positive-sawtooth cyclic-stress fluctuations are identical but are three times higher than the fatigue-crack-growth rate determined in air. Thus, the environment increased the fatigue-crack-growth rate significantly.

Because the corrosive effect increased the rate of fatigue-crack growth below $K_{Iscc}$ by the same amount with the triangular wave as with the positive-sawtooth wave, the corrosive processes in the environment-material system investigated were operative *only* while the tensile stresses in the vicinity of the crack tip were increasing. This conclusion is supported by (1) corrosion-fatigue data obtained with the negative-sawtooth wave, which showed no corrosive effect while the tensile stresses were decreasing, and (2) corrosion-fatigue data obtained with the square wave which showed no corrosive effect at constant tensile stresses.

Fatigue-crack-growth data for 7075-T6 aluminum alloys tested in salt

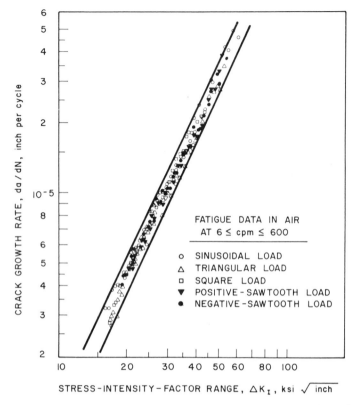

**Figure 13.21**  Fatigue-crack-growth rates in 12Ni-5Cr-3Mo steel under various cyclic-stress fluctuations with different stress-time profiles.

water at 6 cpm under sinusoidal, square, positive-sawtooth, and negative-sawtooth loading showed a behavior very similar to that presented for the 12Ni-5Cr-3Mo maraging steel in salt water.[18]  However, fatigue-crack-growth data for 7075-T651 aluminum alloy tested at 105 cpm in distilled water showed no significant difference in the rate of growth under positive- and negative-sawtooth loadings.[34]  Further work is necessary to resolve the apparent discrepancy in the conclusions relating to the effect of waveform on the corrosion-fatigue behavior for high-strength aluminum alloys in water environments.

### 13.4.3. Effect of Alternate Wet and Dry Conditions

The preceding discussion presents the corrosion-fatigue behavior of steels under full-immersion conditions.  However, many engineering structures

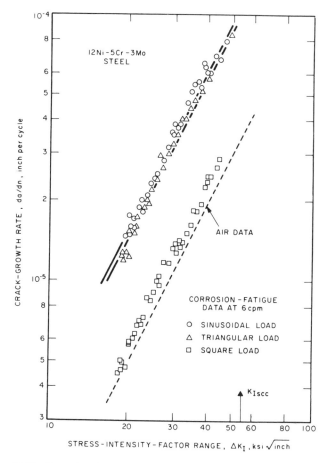

**Figure 13.22**    Corrosion-fatigue-crack-growth rates below $K_{Iscc}$ under sinusoidal, triangular, and square loads.

are subjected to cyclic loading under alternate periods of wet and dry conditions. Therefore, it is necessary to understand the effects of these conditions on the crack-propagation behavior for structural materials to determine more accurately safe service lives and inspection intervals for components that contain fatigue cracks.

Data reported by Barsom and Novak (Figure 13.24)[9] and by Miller et al.[15] indicated a retardation of corrosion-fatigue-crack-propagation rate caused by a dry period that interrupted the corrosion-fatigue test. Because these observations were based on single-specimen tests, a systematic study was conducted by Taylor and Barsom[35] to establish the crack-propagation behavior for steels under wet and dry environment conditions.

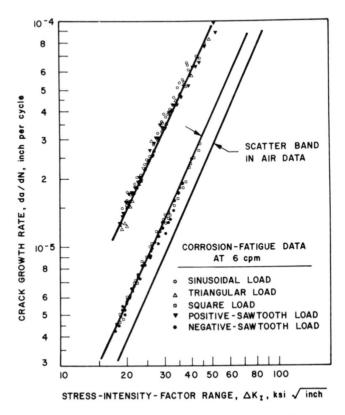

**Figure 13.23** Corrosion-fatigue-crack-growth rates in 12Ni-5Cr-3Mo steel in 3 percent solution of sodium chloride under various cyclic-stress fluctuations with different stress-time profiles.

The study was conducted on A588 Grade A, 4340, and A514 Grade F steels in a room temperature aerated solution of 3 percent sodium chloride in distilled water. The test conditions for each specimen are shown in Table 13.2. All the tests were conducted at a stress ratio, $R$, of 0.1 and the minimum and maximum loads were constant throughout the test. For the corrosion-fatigue tests that were interrupted by one or more drying periods, crack-propagation data were obtained up to a given $\Delta K$ value, and the specimens were then unloaded, removed from the solution, and placed in an air environment for different periods of time as shown in Table 13.2. The specimens were then immersed in the solution to obtain postdrying corrosion-fatigue data under test conditions that were identical to those used prior to drying the specimens.

The crack-length versus number-of-cycles data for all the test specimens are presented in Figures 13.25 and 13.26 and show negligible, if any, retardation

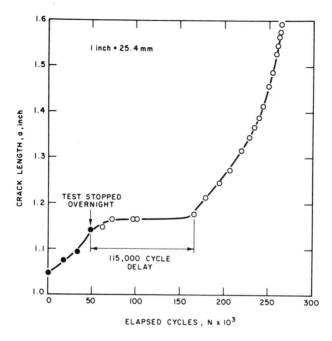

**Figure 13.24**   Retardation of corrosion-fatigue-crack-growth rate under wet and dry environmental conditions for A514 Grade F steel.

TABLE 13.2   **Alternate Wet and Dry Test Conditions**

| Steel | Yield Strength (ksi) | Specimen Number | Environment | Frequency (cpm) | Interruption Condition, $\Delta K$ (ksi$\sqrt{\text{in.}}$) | Duration (hr) |
|---|---|---|---|---|---|---|
| A588 | | | | | | |
| Grade A | 55 | 1-1 | Air | 300 | — | — |
| | | 1-2 | 3% NaCl | 60 | — | — |
| | | 1-3 | 3% NaCl | 60 | 21.1 | 16 |
| | | | | | 37.2 | 68 |
| | | 1-4 | 3% NaCl | 12 | 21.2 | 90 |
| 4340 | 130 | 2-1 | 3% NaCl | 12 | 18.1 | 90 |
| | | | | | 22.9 | 90 |
| | | | | | 31.8 | 20 |
| | | 2-2 | 3% NaCl | 12 | 17.0 | <1 |
| A514 | | | | | | |
| Grade F | 126 | 3-1 | 3% NaCl | 12 | 16.9 | 14 |

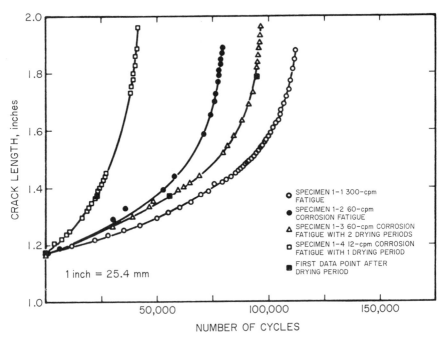

**Figure 13.25** Crack-growth behavior of A588 Grade A steel (tested by using a crack length of 1.180 in. as a common starting point).

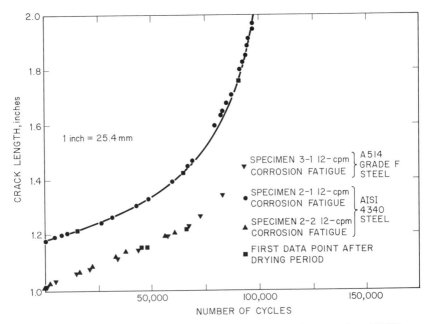

**Figure 13.26** Crack-growth behavior of AISI 4340 steel tempered at 1200°F (650°C) and A514 Grade F steel.

in the corrosion-fatigue-crack-propagation rate in region II, Figure 13.1, caused by drying periods. Consequently, these results suggest that the retardation behavior reported in the literature[9,15] may have been caused by some extraneous circumstance, such as an accidental overload.

## 13.5. Corrosion-Fatigue-Crack Propagation Above $K_{Iscc}$

The rate of corrosion-fatigue crack growth above $K_{Iscc}$ should be greater than the rate below $K_{Iscc}$ because, whenever the magnitude of the maximum stress-intensity-factor value in a given cycle becomes greater than $K_{Iscc}$, the rate of corrosion-fatigue-crack growth should be accelerated by stress-corrosion cracking. Based on the assumption that, above $K_{Iscc}$, the fatigue-crack growth in an aggressive environment is enhanced by the same magnitude as that for crack growth under sustained loads, Wei and Landes[19] hypothesized a linear summation model to predict the corrosion-fatigue behavior above $K_{Iscc}$ for a high-strength steel. The model considers the corrosion-fatigue-crack-growth rate above $K_{Iscc}$ to be the sum of the rate of fatigue-crack growth in an inert reference environment and an environmental component that is computed from the load profile and sustained-load crack-growth data obtained in an identical aggressive environment. Wei and Landes were able to predict satisfactorily the rates of corrosion-fatigue-crack growth for ultra-high-strength steels tested in dehumidified hydrogen (Figure 13.27), distilled water, and water-vapor environments and for a Ti-8Al-1V-1Mo alloy tested in distilled water and sodium chloride solution. This model

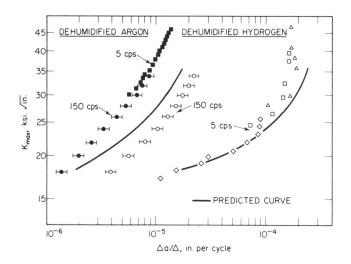

**Figure 13.27**  Fatigue-crack growth in 18Ni (250) maraging steels tested in dehumidified argon and hydrogen (Ref. 19).

can be used to approximate the published rates of corrosion-fatigue-crack growth above $K_{Iscc}$ for ultra-high-strength steels.[14,36–38]

The preceding linear-summation model has limited applicability and should not be used indiscriminately to predict the corrosion-fatigue behavior above $K_{Iscc}$. It can be considered, to a first-order approximation, to be applicable to environment-material systems in which the material is highly susceptible to the environment such as high-strength steels having yield strengths greater than about 200 ksi tested in water. The rates of crack growth under sustained loads in these environment-material systems are usually orders of magnitude greater than under cyclic loads in the absence of the environment. The model neglects the effects of frequency and waveform on the rate of corrosion-fatigue-crack growth below $K_{Iscc}$. Thus, except possibly for square-wave loading, the model is of questionable validity for most environment-material systems. It should be noted that the rate of stress-corrosion cracking for most environment-material systems is very slow when the applied $K_I$ value is only slightly higher than the value for $K_{Iscc}$. Consequently, the corrosion-fatigue behavior below $K_{Iscc}$ for these environment-material systems remains unaltered even when the maximum value for $K_I$ in a given cycle becomes greater than $K_{Iscc}$.

## 13.6. Generalized Corrosion-Fatigue Behavior

Corrosion-fatigue-crack-propagation behavior is a very complex phenomenon. The preceding discussions show that the behavior is strongly dependent on many variables, including frequency, waveform, and stress ratio. Tests at low frequencies are difficult, time consuming, and very costly for region II behavior and are prohibitive for the threshold behavior. Consequently, a clear understanding of the corrosion-fatigue behavior in the various regions does not exist at the present time. However, based on the available data, a simplified schematic characterization of this behavior may be possible.

In an air environment, the fatigue $\Delta K_I$ threshold, $\Delta K_{th}$, in various steels tested at a stress ratio, $R$, of 0.1 is independent of cyclic-load frequency and is equal to about $5.5\,\text{ksi}\sqrt{\text{in}}$. Because hostile environmental effects decrease with increased cyclic-load frequency, the corrosion-fatigue $\Delta K_{th}$ at very high cyclic-load frequencies would have a value close to that of fatigue in air. A $K_{Iscc}$ test can be considered a corrosion-fatigue test at extremely low cyclic-load frequency. In such tests, the rate of crack growth at a stress-intensity-factor fluctuation that is slightly lower than $K_{Iscc}$ is, by definition, equal to zero. Consequently, at extremely low cyclic-load frequencies, $\Delta K_{th}$ should be equal to $K_{Iscc}$. Hence, the value of the environmental $\Delta K_{th}$ at intermediate cyclic-load frequencies must be greater than $5.5\,\text{ksi}\sqrt{\text{in}}$. and less than the value of $K_{Iscc}$ for the environment-material system under consideration. The test results show that, at 12 cpm, the environmental $\Delta K_{th}$ for A514 steels in 3 percent solution of sodium chloride in distilled water was equal to about 11 ksi$\sqrt{\text{in}}$. Based on the preceding observations, a schematic representation of the corrosion-fatigue behavior of steels subjected to different cyclic-load frequencies has been constructed (Figure 13.28). This figure is an oversimplification of a very complex phenomenon.

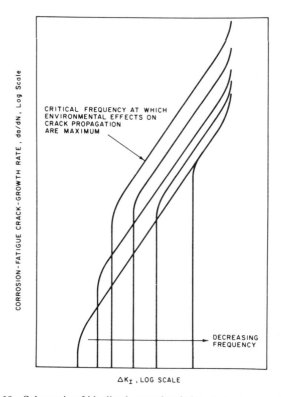

**Figure 13.28**   Schematic of idealized corrosion-fatigue behavior as a function of cyclic-load frequency.

Although significant accomplishments have been made in understanding the corrosion-fatigue behavior for some metal-environment systems, more is needed to improve the procedures for material selection and for structural analysis of engineering structures that are subjected to cyclic loads in aggressive environments.

# References

1. *Fatigue Thresholds—Fundamentals and Engineering Applications*, edited by J. Bäcklund, A. F. Blom, and C. J. Beevers, Engineering Materials Advisory Services Ltd., Vols. I and II, The Chameleon Press Ltd., London, 1982.

2. S. SURESH, G. F. ZAMISKI, and R. O. RITCHIE, "Oxide-Induced Crack Closure: An Explanation of Near-Threshold Corrosion Fatigue Crack Growth Behavior," *Metallurgical Transactions A, 12A*, August 1981, pp. 1435–1443.

3. R. O. RITCHIE, S. SURESH, and C. M. Moss, "Near-Threshold Fatigue Crack Growth in a 2¼Cr-1Mo Pressure Vessel Steel in Air and Hydrogen," *Journal of Engineering Materials and Technology, Transactions of ASME*, Series H, *102*, 1980, p. 293.

4. S. SURESH and R. O. RITCHIE, "Closure Mechanisms for the Influence of Load Ratio on Fatigue Crack Propagation in Steels," Materials Science Division, U.S. Department of Energy, *Report Contract No. DE-AC03-76SF00098*, Washington, D.C., April 1982.

5. *Corrosion Fatigue: Chemistry, Mechanics, and Microstructure*, International Corrosion Conference Series, Vol. NACE-2, National Association of Corrosion Engineers, Houston, 1972.

6. H. H. JOHNSON and P. C. PARIS, "Sub-Critical Flaw Growth," *Journal of Engineering Fracture Mechanics, 1*, No. 3, June 1968.

7. W. G. CLARK, JR., "The Fatigue Crack Growth Rate Properties of Type 403 Stainless Steel in Marine Turbine Environments," *Corrosion Problems in Energy Conversion and Generation*, edited by C. S. Tedman, Jr., Electrochemical Chemical Society, Princeton, N.J., 1974.

8. J. M. BARSOM, "Effect of Cyclic-Stress Form on Corrosion-Fatigue Crack Propagation Below $K_{Iscc}$ in a High-Yield-Strength Steel," *Corrosion Fatigue: Chemistry, Mechanics, and Microstructure*, International Corrosion Conference Series, Vol. NACE-2, National Association of Corrosion Engineers, Houston, 1972.

9. J. M. BARSOM and S. R. NOVAK, "Subcritical Crack Growth and Fracture of Bridge Steels," NCHRP Report 181, Transportation Research Board, Washington, D.C., 1977.

10. T. W. CROOKER and E. A. LANGE, "The Influence of Salt Water on Fatigue Crack Growth in High Strength Structural Steels," *ASTM STP 462*, American Society for Testing and Materials, Philadelphia, 1970.

11. F. J. BRADSHAW and C. WHEELER, "The Influence of Gaseous Environment and Fatigue Frequency on the Growth of Fatigue Cracks in Some Aluminum Alloys," *International Journal of Fracture Mechanics, 5*, No. 4, Dec. 1969.

12. M. O. SPEIDEL, M. J. BLACKBURN, T. R. BECK, and J. A. FEENEY, "Corrosion-Fatigue and Stress-Corrosion Crack Growth in High-Strength Aluminum Alloys, Magnesium Alloys, and Titanium Alloys, Exposed to Aqueous Solutions," *Corrosion Fatigue: Chemistry, Mechanics and Microstructure*, International Corrosion Conference Series, Vol. NACE-2, National Association of Corrosion Engineers, Houston, 1972.

13. D. A. MEYN, "An Analysis of Frequency and Amplitude Effects on Corrosion Fatigue Crack Propagation in Ti-8Al-1Mo-1V," *Metallurgical Transactions, 2*, 1971.

14. J. P. GALLAGHER, "Corrosion Fatigue Crack Growth Behavior Above and Below $K_{Iscc}$," *NRL Report 7064*, Naval Research Laboratory, Washington, D.C., May 28, 1970.

15. G. A. MILLER, S. J. HUDAK, and R. P. WEI, "The Influence of Loading Variables on Environment-Enhanced Fatigue-Crack Growth in High Strength Steels," *Journal of Testing and Evaluation, 1*, No. 6, 1973.

16. P. C. PARIS, R. J. BUCCI, E. T. WESSEL, W. G. CLARK, and T. R. MAGER, "Extensive Study of Low Fatigue-Crack-Growth Rates in A533 and A508 Steels," *ASTM STP 513*, American Society for Testing and Materials, Philadelphia, 1972.

17. R. BUCCI, "Environment Enhanced Fatigue and Stress Corrosion Cracking of a Titanium Alloy Plus a Simple Model for Assessment of Environmental Influence of Fatigue Behavior," Ph.D. Dissertation, Lehigh University, Bethlehem, Pa., 1970.

18. R. J. SELINES and R. M. PELLOUX, *Effect of Cyclic Stress Wave Form on Corrosion Fatigue Crack Propagation in Al-Zn-Mg Alloys*, Department of Metallurgy and Materials Science Report, M.I.T., Cambridge, Mass., 1972.

19. R. P. WEI and J. D. LANDES, "Correlation Between Sustained-Load and Fatigue-Crack Growth in High-Strength Steels," *Materials Research and Standards, MTRSA, 9*, No. 7, July 1969.

20. R. J. BUCCI and J. K. DONALD, *Fatigue and Fracture Investigation of a 200 Grade Maraging Steel Forging*, Del Research Corporation Report, Hellertown, Pa., Oct. 1972.

21. R. P. SKELTON and J. R. HAIGH, "Fatigue Crack Growth Rates and Thresholds in Steels Under Oxidizing Conditions," *Materials Science and Engineering, 36*, 1978, pp. 17–25.

22. J. M. BARSOM, "Fatigue Behavior of Pressure-Vessel Steels," *Welding Research Council (WRC) Bulletin, No. 194*, May 1974.

23. J. M. BARSOM, "Investigation of Subcritical Crack Propagation," Ph.D. Dissertation, University of Pittsburgh, Pittsburgh, 1969.

24. J. M. BARSOM, "Corrosion-Fatigue Crack Propagation Below $K_{Iscc}$," *Journal of Engineering Fracture Mechanics, 3*, No. 1, July 1971.

25. T. W. CROOKER and E. A. LANGE, "Corrosion Fatigue Crack Propagation of Some New High Strength Structural Steels," *Journal of Basic Engineering, Transactions, ASME, 91*, 1969.

26. J. M. BARSOM, J. F. SOVAK, and E. J. IMHOF, JR., "Corrosion-Fatigue Crack Propagation Below $K_{Iscc}$ in Four High-Yield-Strength Steels," *Applied Research Laboratory Report 89.021-024(3)*, U.S. Steel Corporation (available from the Defense Documentation Center), Arlington, Va., Dec. 14, 1970.

27. E. J. IMHOF and J. M. BARSOM, "Fatigue and Corrosion-Fatigue Crack Growth of 4340 Steel at Various Yield Strengths," *ASTM STP 536*, American Society for Testing and Materials, Philadelphia, 1973.

28. J. A. FEENEY, J. C. MCMILLAN, and R. P. WEI, "Environmental Fatigue Crack Propagation of Aluminum Alloys at Low Stress Intensity Levels," *Metallurgical Transactions, 1*, June 1970.

29. A. HARTMAN, "On the Effect of Oxygen and Water Vapor on the Propagation of Fatigue Cracks in 2024-TB3 Alclad Sheet," *International Journal of Fracture Mechanics, 1*, No. 3, Sept. 1965.

30. F. J. BRADSHAW and C. WHEELER, "The Effect of Environment on Fatigue Crack Growth in Aluminum and Some Aluminum Alloys," *Applied Materials Research, 5*, No. 2, 1966.

31. R. P. WEI, "Fatigue-Crack Propagation in a High-Strength Aluminum Alloy," *International Journal of Fracture Mechanics, 4*, No. 2, June 1968.

32. R. P. WEI and J. D. LANDES, "The Effect of $D_2O$ on Fatigue Crack Propagation

in a High Strength Aluminum Alloy," *International Journal of Fracture Mechanics*, 5, 1969.

33. F. J. BRADSHAW, N. J. F. GUNN, and C. WHEELER, "An Experiment on the Effect of Fatigue Wave Form on Crack Propagation in an Aluminum Alloy," *Technical Memorandum MAT93*, Royal Aircraft Establishment, Farnborough, Hants, England, July 1970.

34. S. J. HUDAK and R. P. WEI, "Comments," on paper by J. M. Barsom, "Effect of Cyclic Stress Form on Corrosion Fatigue Crack Propagation Below $K_{Iscc}$ in a High Yield Strength Steel," *Corrosion Fatigue: Chemistry, Mechanics, and Microstructure*, International Corrosion Conference Series, Vol. NACE-2, National Association of Corrosion Engineers, Houston, 1972.

35. M. E. TAYLOR and J. M. BARSOM, unpublished data.

36. E. P. DAHLBERG, "Fatigue Crack Propagation in High Strength 4340 Steel in Humid Air," *Transactions: American Society for Metals*, 58, 1965.

37. W. A. VAN DER SLUYS, "Effect of Repeated Loading and Moisture on the Fracture Toughness of SAE 4340 Steel," *Journal of Basic Engineering, Transactions, ASME*, 87, 1965.

38. W. A. VAN DER SLUYS, "The Effect of Moisture on Slow Crack Growth in Thin Sheets of SAE 4340 Steel Under Static and Repeated Loading," *Journal of Basic Engineering, Transactions, ASME*, 89, 1967.

39. J. P. GALLAGHER and R. P. WEI, "Corrosion Fatigue Crack Propagation Behavior in Steels," *Corrosion Fatigue: Chemistry, Mechanics, and Microstructure*, International Corrosion Conference Series, Vol. NACE-2, National Association of Corrosion Engineers, Houston, 1972.

# Fatigue and Fracture Behavior of Welded Components

## 14.1. Introduction

Welding technology has had a significant impact on industrial developments. Fabrication by welding is an effective method to reduce production and fabrication costs and can be mechanized, computer controlled, and incorporated in assembly lines. Welding fabrication has revolutionized many industries, including shipbuilding and automotive production, and has resulted in the development of various products, such as pressure vessels, that could not otherwise have achieved their present functions.

Welding technology is very complex, and fabrication by welding encompasses characteristics that should be understood to different levels, by the design engineer, the fabricator, and the welder. Some of these characteristics that are pertinent to the present discussion are residual stresses, imperfections, and stress concentrations.

### 14.1.1. Residual Stresses

Fabrication by welding usually induces complex three-dimensional residual stresses that are caused by the heating and cooling effects of welding. The development of these residual stresses in weldments may be demonstrated by considering the following simplified example of a groove weld.

Consider an arc welding process in which a weld groove is filled with molten filler metal, Figure 14.1(a). As this filler metal cools, it contracts along (longitudinal direction) and at right angles (transverse direction) to the groove. This contraction is resisted by the surrounding base metal, resulting in residual tensile ($+$) and compressive ($-$) stresses as indicated in Figures 14.1(b) and 14.1(c). This contraction may also cause distortion

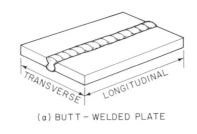

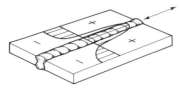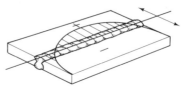

(a) BUTT – WELDED PLATE

(b) LONGITUDINAL RESIDUAL STRESS          (c) TRANSVERSE RESIDUAL STRESS

**Figure 14.1**  Residual stresses for a butt-welded plate. (a) Butt-welded plate, (b) longitudinal residual stress, and (c) transverse residual stress.

of the fabricated component as demonstrated by the exaggerated schematic examples in Figure 14.2. Several practices can be used to minimize these effects, including proper fit-up, proper sequencing and positioning of welds, and preheating and postheating of the welded assembly.

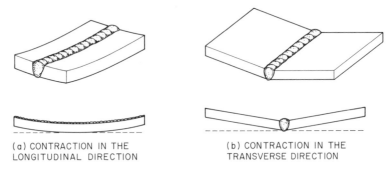

(a) CONTRACTION IN THE                (b) CONTRACTION IN THE
LONGITUDINAL  DIRECTION              TRANSVERSE DIRECTION

**Figure 14.2**  Exaggerated distortions for a single-pass butt-welded plate. (a) Contraction in the longitudinal direction; (b) contraction in the transverse direction.

### 14.1.2. Imperfections

Fabrication by welding may result in various imperfections and cracks in the filler metal or in the base metal.[1,2,3] The various types of imperfections and cracks, their causes, and methods to eliminate them have been the subject of many publications and are beyond the scope of this text. These imperfections and cracks, some of which are shown schematically in Figure 14.3, may be caused by (1) improper design that restricts accessibility and prevents the use of correct electrode angle; (2) incorrect selection of a welding process or welding parameters for the material or joint geometry of interest, such as improper energy input or travel speed; (3) incorrect fabrication procedures, such as incorrect electrode angle or incorrect positioning of the joint; (4) improper care of the electrode or flux, or both, as well as other causes. These generalized observations are presented to emphasize that assurance of good-quality fabrication by welding requires considerations and decisions that start in the design stage and continue throughout the entire fabrication process.

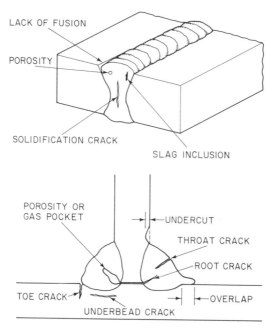

**Figure 14.3**  Imperfections and cracks in welded joints.

### 14.1.3. Stress Concentrations

Fabrication by welding may result in stress-concentration regions in the fabricated joint. Stress concentration may be caused by weld imperfections, or geometrical discontinuities, or both. Examples of stress-concentration regions caused by geometrical discontinuities in fabricated joints are shown in Figure 14.4. These locations usually correspond to the initiation sites for fatigue and fracture of structural components.

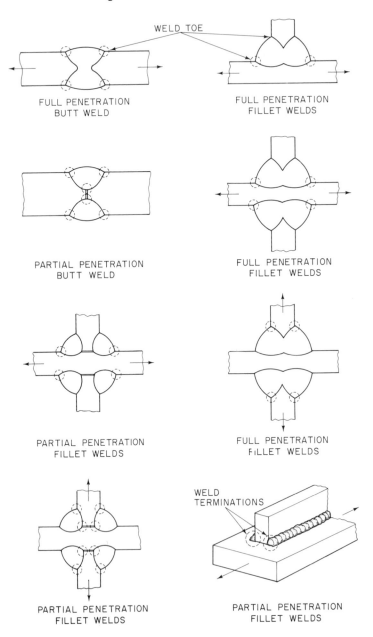

**Figure 14.4** Stress-concentration regions (indicated by dashed circles) for weldments.

## 14.2. Fatigue Behavior of Welded Components

### 14.2.1. General Discussion

Most engineering structures are built in accordance with well-defined specifications and codes that are prepared by technical organizations such as the American Society of Mechanical Engineers (ASME), the American Association of State Highway and Transportation Officials (AASHTO), the American Institute of Steel Construction (AISC), and others. The requirements of these organizations are closely related to other specifications for materials, fabrication, and inspection such as those established by the American Society for Testing and Materials (ASTM) and the American Welding Society (AWS). These codes recognize the possible existence of tolerable imperfections induced by the fabrication process that should not adversely affect the performance of the structure for the intended application. In many instances, attempts to remove allowable imperfections may result in conditions worse than the original condition (Chapter 6 describes such a case). It is also recognized that significant deviations from the appropriate specification may adversely affect the performance of the structure. Consequently, the present discussion relates to the fatigue behavior of welded steel components that have been designed, fabricated, and inspected in accordance with one of these specifications and specifically with the AASHTO specifications.

### 14.2.2. Origins of Fatigue Cracks

The magnitude of the stress intensification at the edge of a given planar imperfection in a unidirectional tensile-stress field depends on its projected size and shape on the plane that is perpendicular to the direction of the primary tensile stress. The maximum stress intensity for a given planar imperfection occurs when the plane of the imperfection is perpendicular to the direction of the primary tensile stress and approaches zero as the plane of the imperfection becomes parallel to the direction of the primary tensile stress. Thus, planar imperfections, like plate laminations, whose plane is parallel to the surfaces of a plate that is subjected to in-plane tensile-stress fluctuations rarely cause a degradation in the fatigue life of the plate. However, these plate laminations can be very detrimental to the fatigue life when the plate is subjected to tensile-stress fluctuations in the through-thickness direction.

The stress intensification caused by a surface imperfection is equal to the stress intensification caused by an embedded imperfection having a size and shape equal to the surface imperfection and its mirror image. Thus, for a given shape, an embedded imperfection must be twice as large as a surface imperfection to cause the same stress intensification as described in Chapter 2. Furthermore, when a component is subjected to bending, the stresses are highest at the surface. Consequently, fatigue cracks initiate more readily from surface imperfections than from embedded ones.

The stress intensification for planar imperfections whose plane is perpendicular to the direction of the primary tensile stress is higher than the stress intensification for a volumetric imperfection having equal planar size and shape projected on the plane perpendicular to the direction of the stress. Consequently, fatigue cracks initiate and propagate more readily from a lack of fusion or cold-crack imperfection than from a gas pocket or porosity having equal projected size and shape on the plane that is perpendicular to the direction of the primary stress.

Fatigue cracks in weldments originate either at internal imperfections, such as porosity, lack of fusion, and trapped slag,[4] or at weld toes and weld terminations, usually from slag intrusion.[4-6] The majority of fatigue cracks in welded structures originate at a weld toe or at a weld termination rather than from internal imperfections. This behavior is attributed to the fact that for a given fatigue life, a much larger embedded imperfection can be tolerated than a surface imperfection. Furthermore, unlike embedded imperfections, surface imperfections that cause fatigue-crack initiation occur in regions of weld toes and weld terminations that are invariably regions of stress concentrations as a result of the geometrical discontinuities of the joint.

The size and frequency of imperfections depend on the welding process, geometry of the weldment including ease of access for welding, and the care exercised in making the weld. Metallographic investigations conducted at Lehigh University and at the Welding Institute on properly welded steel components that were fatigue tested indicate that the maximum depth of the fatigue-crack initiating imperfection at the toe of welds was less than about 0.016 in. (0.4 mm).[4-7] The maximum radius for embedded imperfections, such as gas pockets, in fillet welds was about 0.08 in. (2 mm).[4-7]

Fatigue cracks that initiate at the root of web-to-flange fillet welds are a good example of fatigue cracks that originate from internal imperfections, Figure 14.5.[8] The crack shown in Figure 14.5 initiated at a gas pocket and propagated as an embedded crack that continually changed its shape until it intersected the fillet-weld surface as a penny-shaped crack, Figure 14.5(a). These types of fatigue cracks continue to propagate in all directions in a plane perpendicular to the direction of maximum tensile stress until they become three-corner cracks, Figure 14.5(b). Figure 14.6[8] shows that over 90 percent of the fatigue life of the component was exhausted prior to crack penetration through the back surface of the tension flange.

The majority of fatigue cracks in welded members initiate at a weld toe or at a weld termination near a stiffener, or other attachments such as

TYPE I-FAILURE MODE

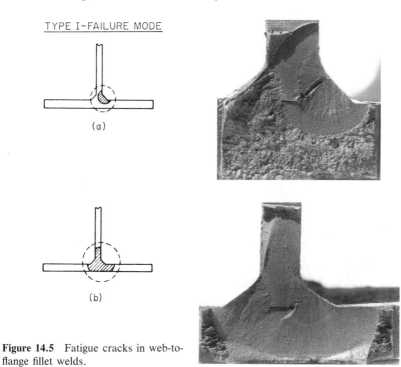

(a)

(b)

**Figure 14.5**   Fatigue cracks in web-to-flange fillet welds.

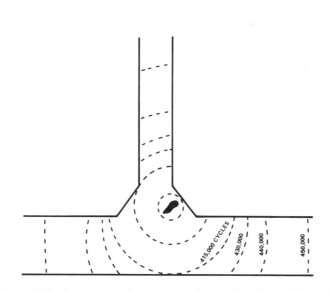

**Figure 14.6**   Stages of crack propagation for a web-to-flange fillet weld.

gusset plate, or end of a cover plate. These are regions of high stress concentration and high residual stresses that may contain small (less than 0.016-in. or 0.4-mm)[4,6] weld imperfections, such as slag intrusion. Moreover, because the surface of the deposited weld metal is invariably rippled, the toe angle between the weld metal and the base metal can vary significantly at neighboring points along the weld toe, resulting in variation in the stress concentration. Figure 14.7[8] shows crack formation at the toe of a longitudinal fillet weld, Figure 14.7(a), and at the toe of a transverse fillet weld, Figure 14.7(b), in cover-plate details. For the cover plate with longitudinal fillet

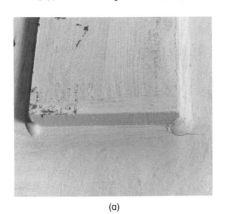
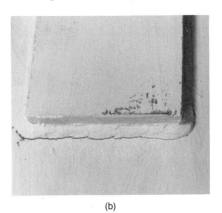

(a)                                          (b)

**Figure 14.7**   Fatigue crack at ends of cover plates. (a) Crack formation at toe of longitudinal fillet weld; (b) crack formation at toe of transverse fillet weld.

welds, the fatigue crack initiates at the termination of the weld and propagates as a part-through crack, Figure 14.8, until it penetrates the opposite surface of the tension flange. The crack then continues to propagate first as a through-thickness crack and then as an edge crack. For the cover plate with transverse fillet weld, Figure 14.7(b), multiple fatigue cracks initiate at the toe of the weld. These cracks propagate as part-through cracks that continually change in shape as their size increases and as they approach adjacent propagating cracks. Subsequently, these cracks link up to form a single part-through crack, Figure 14.9.[5] This part-through crack continues to change its shape as it propagates through the thickness of the flange into the web and then penetrates the back surface of the flange to form a three-corner crack. Over 90 percent of the fatigue life is exhausted prior to the crack penetrating the back surface of the flange. Because the rate of fatigue-crack propagation increases exponentially with increased crack length, most of the fatigue life of welded components is expended when the fatigue crack is small. Consequently, a significant increase in the fracture toughness would have a secondary effect on the fatigue life of the weldment.

Another pertinent observation is that cracks can initiate in compression flanges as well as in tension flanges of welded girders. Fatigue-crack initiation

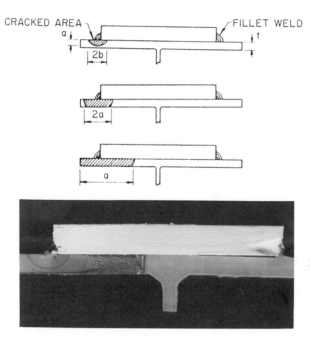

**Figure 14.8**  Stages of crack propagation at the end of a longitudinally welded cover-plate detail.

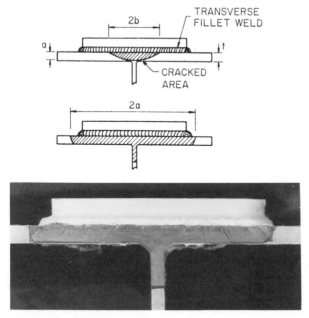

**Figure 14.9**  Stages of crack propagation at the end of a transverse welded cover-plate detail.

and propagation in compression flanges occur in regions of tensile residual stresses. These cracks propagate out of the residual tensile-stress field and stop. The magnitude of the tensile residual stresses depends on welding sequence, weldment geometry, and welding process and can be equal to the yield point of the steel.[5] These residual stresses become redistributed under the influence of cyclic loading.[5]

The preceding brief discussion shows that fatigue cracks in weldments can (1) initiate from small imperfections that are either embedded or on the surface, (2) be located in regions of high stress concentration where the level of stress concentration may vary appreciably in small neighboring locations along the weld toe, and (3) reside in regions of high residual stress that become redistributed under cyclic loading. The imperfections from which fatigue cracks initiate have different characteristics, sizes, and shapes and, in most cases, are very difficult and costly to locate and define nondestructively. Moreover, the fatigue crack changes its shape throughout most of its propagation life. The magnitude of the change depends on the shape and location of the fatigue-crack initiating discontinuity, the stress-field distribution, and the physical shape of the weld and joint configuration.

### 14.2.3. Problems in Application of Fracture Mechanics to Fatigue of Weldments

The usefulness of fracture-mechanics technology for structural design and for failure prevention and analysis depends on the accuracy of the input data for stress, stress range, stress ratio, size and shape of the initial imperfection and critical crack, material properties, stress-intensity-factor solutions, as well as others. Because of inaccuracies in this information and because fracture-mechanics technology is a relatively new and rapidly developing engineering discipline, some problems remain in its application to engineering structures.[9,10] To better predict the fatigue life of welded components by the use of fracture mechanics, further research is necessary (1) to characterize nondestructively the size, shape, and orientation of the fatigue-crack initiating imperfections; (2) to establish the initiation and propagation behavior for very small cracks; and (3) to analyze the interactions between the fatigue cracks emanating from the imperfections and between these cracks and the localized stress field caused by the geometric discontinuities of the component.

**a. Nondestructive characterization of initial imperfections.**    The fatigue life of a component that contains an imperfection can be estimated best by using fracture-mechanics technology. Because the rate of fatigue-crack propagation increases exponentially with increased crack length, most of the fatigue life is expended when the crack is small. Thus, accurate prediction of the characteristics, especially size and shape, of the initial imperfection is of paramount importance for accurate estimate of the fatigue life of a welded component.

The size and frequency of imperfections depend on the welding process, geometry of the weldment including ease of access for welding, and the care exercised in making the weld. The 0.016-in. (0.4-mm) and 0.08-in. (2-mm) imperfections in properly welded components found in the previously mentioned Lehigh University and Welding Institute studies reside in regions of complex geometries that make their detection by nondestructive testing very difficult and extremely costly. A survey of the available literature shows that the probability of nondestructive detection and characterization of such small imperfections is very low, even in demonstration programs which often have simple geometries that can be inspected easily. The probability of detecting these small imperfections in complex weldments is extremely low.

One of the assumptions frequently used in the application of fracture-mechanics technology to predict the fatigue life of components is to set the size of the initial imperfection equal to the largest crack size that can escape detection for the particular nondestructive technique used. Unfortunately, the size of the undetectable crack can be relatively large and depends not only on the flaw detection procedure, Figure 14.10,[10,11] but also on the

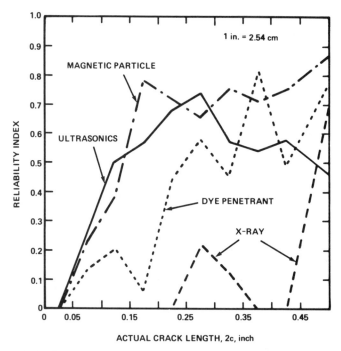

**Figure 14.10**   Comparison of four NDT techniques on reliability of flaw detection in steel cylinders.

operator who uses the instrumentation.[12] The use of this assumption can result in overly conservative estimates of the fatigue life of components and is difficult to justify for building safe and reliable engineering structures that are also economical. Consequently, significant progress in the detection, characterization, and prediction of initial imperfections is required if fracture mechanics concepts are to be used to predict better the fatigue life of complex welded structures (such as bridges, ships, and offshore platforms).

   **b. Analysis of fatigue-crack-propagation behavior for small imperfections.** Extensive research has been conducted to predict the fatigue life of various weldments by using linear-elastic fracture-mechanics concepts.[7,13–18] This effort is based on many assumptions, including the following:

1. Fatigue cracks in welded components initiate from preexisting imperfections that behave like preexisting fatigue cracks. Consequently, the fatigue life of a weldment is governed by the rate of the fatigue-crack propagation, and no credit is given for any initiation life.

2. The stress-intensity factor, $K_I$, and the stress-intensity-factor range, $\Delta K_I$, for the preexisting imperfections can be calculated by using linear-elastic fracture-mechanics concepts.

3. The fatigue-crack-propagation rate for the preexisting imperfections is equal to the rate of propagation obtained by testing specimens that satisfy ASTM requirements and can be represented by the same power-law relationship.

   Fatigue cracks for plane welded beams usually initiate from a gas pocket in the web-to-flange fillet weld. These gas pockets are usually ellipsoidal, with the major axis oriented at about 45° to the plane of the web and the plane of the flange, Figure 14.5. Fracture-mechanics prediction of the fatigue life for these imperfections, which is based on the assumption of a preexisting fatigue crack having the same shape and size as the ellipsoidal imperfection, results in a conservative fatigue life that could be significantly less than the true fatigue life of the weldment because the analysis eliminates the fatigue-crack-initiation life from the total life.

   Fatigue cracks at a weld toe or weld termination initiate from slag intrusions and undercuts. The geometries of these imperfections can be more irregular than those of gas pockets in fillet welds. However, available data indicate that even for cover-plate details, the initiation life can be an important part of the total fatigue life.[13] Because fracture-mechanics methodology is usually used to calculate the fatigue-crack-propagation life only, and because fatigue-crack-initiation life can be a significant portion of the total life as discussed in Chapter 8, fracture-mechanics calculations can result in conservative life predictions for these details.

   Clark[19] investigated the behavior of small cracks and concluded that linear-elastic fracture-mechanics concepts are directly applicable to a small crack-tip plastic-zone size, $r_p$, if the crack length is at least 25 times larger than the associated crack-tip plastic-zone size, where $r_p = (1/6\pi)(K_I/\sigma_{ys})^2$. The imperfections that may exist at the toe of a weld reside in a residual tensile-stress zone that can be equal to the yield strength of the steel. Moreover, based on a maximum design stress of more than one-half the yield strength that is used for bridges and with a stress concentration of about 4 at the end of the cover plate, it is reasonable to assume that the imperfections reside in a plastically deformed region where the local stresses are equal to the yield strength of the steel. Under these conditions, the plastic-zone size, if it can be calculated by using linear-elastic fracture mechanics, is too large to satisfy Clark's conclusion even for the case of cyclic loading in which the plastic zone may be smaller than under equivalent static loading. Consequently, the applicability of linear-elastic fracture mechanics to the small imperfections observed in welded components is questionable. Further research is needed to establish the limits of applicability of fracture-mechanics technology to small imperfections and to develop concepts that could be used to analyze their behavior.

Fracture-mechanics analysis of the fatigue-crack-propagation life of welded components is based on the assumption that the rate of propagation for the small weld imperfections can be represented by a power-law relationship (Chapters 9 and 10). The use of such a relationship implies the applicability of linear-elastic fracture mechanics to small weld imperfections, which, from the preceding discussion, appears to be questionable. Moreover, a power-law relationship is based on experimental data obtained by testing fracture-mechanics–type specimens that contain cracks significantly larger than the weld imperfections under consideration. Thus, the applicability of the relationship between the fatigue-crack-propagation rate, $da/dN$, and the stress-intensity-factor range, $\Delta K_I$, and the correlating constants in the power-law relationship must be established for the small imperfections that are observed in welded components.

Finally, the application of fracture mechanics to predict the fatigue life of welded components is based on a postulation by Maddox[20] that $K_I$ (or $\Delta K_I$) at the toe of a welded joint can be derived from that for the corresponding crack in a flat plate multiplied by a factor $M_k$, which takes into account the stress-concentration effect of the weld shape and joint configuration. Several finite-element analyses have been made to establish the value of $M_k$ for joints involving transverse nonload-carrying fillet welds and for transverse butt welds.[14–17,21] Unfortunately, the results of these analyses differ significantly. Moreover, the available analyses, which are for weldments having idealized geometries, show that the value of $M_k$ depends on joint configuration, weld geometry, weld toe angle, ratio of crack length, and plate thickness, as well as other parameters that are usually very difficult to establish for welded components in actual structures.

These problems indicate that, at the present time, the AASHTO fatigue-design curves described in the following section are an excellent alternative to ensure the structural integrity of welded components subjected to cyclic loadings.

### 14.2.4. AASHTO Fatigue Design Curves for Fabricated Bridge Components

Bridge engineers have recognized, for a long time, the effect of cyclic loading on the structural integrity of welded bridge components. The American Welding Society (AWS) *Specifications for Welded Highway and Railway Bridges*[22] was based on fatigue tests of welded details that were conducted in the 1940s. In the late 1950s, the observation of fatigue cracks at welded details in the American Association of State Highway Officials (AASHO) Road Test[23] bridges indicated the need for further study of the fatigue behavior of welded details and for modifications of the existing specifications. The present AASHTO fatigue design specifications[24,25] are based on extensive fatigue-test results and field experience that have been accumulated since the early 1960s.

The present AASHTO fatigue design specifications are based on experimental curves that relate the fatigue life, $N$, of a welded detail to the total (tension plus compression) applied nominal stress range, $\Delta\sigma$.[5] A large number of tests for a given detail have been conducted to generate a statistically significant stress-range–fatigue-life relationship. The design curves represent the 95 percent confidence limit for 95 percent survival for a given detail.

Figures 14.11 and 14.12 present fatigue-test results for welded beams and cover-plated beams, respectively, fabricated from bridge steels having yield strengths between 36 and 100 ksi (248 and 690 M Pa) and subjected to various minimum loads. Statistical analysis of the available data indicates that the stress range is the primary parameter controlling the fatigue life and that the minimum stress, the maximum stress, and the grade of steel have secondary influence on the fatigue behavior of welded components.[5,13]

Other fatigue tests were conducted on beams and girders with welded attachments and with transverse stiffeners.[7] The available fatigue data for various attachments show that the fatigue strength of a girder with welded attachments is strongly governed by the length of the attachment in the stress direction.[7,25] The longer the attachment, the higher the stress concentration at the toe of the weld and the lower the fatigue strength. Welded attachments longer than 8 in. (203 mm) have fatigue strengths equivalent to welded partial length cover plates. Transverse stiffeners are similar to very short attachments and have fatigue strengths equivalent to those of welded attachments that are 2 in. (50 mm) long or shorter.

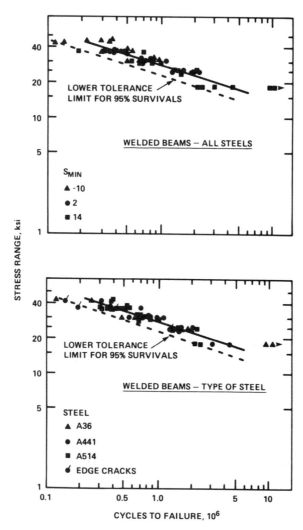

**Figure 14.11**   Effects of minimum stress and steel grade on the fatigue strength of welded beams.

The extensive fatigue data that have been obtained by testing welded bridge details have been used to establish allowable stress ranges for various categories of steel bridge details, Figure 14.13. Each category represents welded-bridge details that have equivalent fatigue strengths. For example, all welded attachments having a length, $L$, in the direction of stress equal to or less than 2 in. (category C) are considered to have equivalent fatigue strength. In reality, under identical fabrication and geometrical conditions,

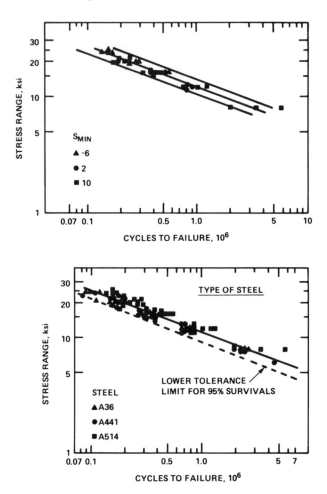

**Figure 14.12**   Effects of minimum stress and steel grade on the fatigue strength of beams with transverse end-welded cover plates.

a 2-in.-long attachment results in a higher stress concentration than does a shorter attachment and, therefore, would have a shorter fatigue life. Because the curve for each category corresponds to the 95 percent confidence limit for 95 percent survival of all the details in a given category, the fatigue-design curves correspond to approximately the shortest lives obtained for details in each category and are, therefore, governed by the details in that category that have the most severe geometrical or weld stress concentration.

The existence of gouges and weld-imperfection stress raisers in a structural detail of a given geometry decrease the fatigue life of the detail. Consequently, significant variability (scatter) in fatigue-life data can be obtained by testing many details of identical geometry but containing different

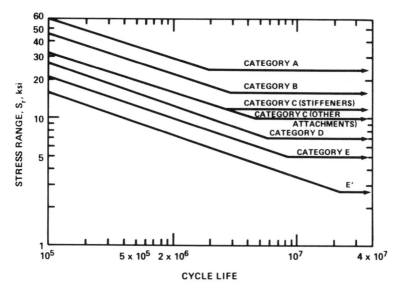

**Figure 14.13**    Design stress range curves for categories A to E′.

size imperfections. This variability in the data is very apparent in the data base used to establish the AASHTO fatigue categories. For example, the longest life obtained for a category C detail (stiffener) that was tested at a stress range, $\Delta\sigma$, of about 25 ksi was about four times longer than the same detail that exhibited the shortest life. The difference in fatigue life for these two specimens was caused primarily by the difference in the size of the initial imperfections that existed in the specimens.

Categories A, B, C, D, and E in Figure 14.13 correspond to plain plate and rolled beams, plain welds and welded beams and plate girders, stiffeners and short attachments (less than 2 in. long), 4-in.-long attachments, and cover-plated beams, respectively. Category E′ corresponds to thick flanges and thick cover plates and suggests that thickness may also affect the fatigue strength of welded girders as has been observed from highway bridges[26] and laboratory tests.[27] The horizontal lines for each category represent the applied nominal stress range corresponding to the fatigue strength (over $2 \times 10^6$ cycles) and are extremely important for highway bridges located on heavily traveled roads. The stress-range threshold corresponding to long life for a given category is related to either the fatigue-crack-initiation threshold or the fatigue-crack-propagation threshold.

The AASHTO fatigue design curves represent the 95 percent confidence limit for 95 percent survival of all the details in a given category and are governed primarily by the details in a given category that have the most severe geometrical discontinuities, imperfections, or both. Because these discontinuities and imperfections minimize or eliminate the fatigue-crack-initiation life, the fatigue life for these details is governed by the fatigue-crack-propagation behavior for the particular geometry and steel. Fatigue-crack-propagation was shown in Chapter 9 to be independent of the strength of the steel. Thus, the AASHTO fatigue design curves should be essentially independent of the strength of the steel. However, it is important to realize that, unlike the fatigue design stress range, the static design stress usually is increased as the strength of the steel is increased.

The fatigue life for a given structural component is determined by the most severe detail in that component. Thus, it is essential to identify that detail and to design the component by using the fatigue category appropriate to the most severe detail it contains.

### 14.2.5. Fatigue Behavior of Welded Components Subjected to Variable-Amplitude Cyclic Loads

Extensive tests were made of simulated-steel highway-bridge members under variable-amplitude random-sequence loading, such as occurs in actual highway bridges.[13]  Welded beams with and without partial-length cover plates and fabricated from both A36 and A514 steels were tested. These details represent the approximate upper and lower bounds, respectively, for fabricated bridge members. The results showed that, like fatigue-crack initiation and fatigue-crack propagation under variable-amplitude loading, Chapters 8 and 10, respectively, the fatigue life for welded components that are subjected to variable-amplitude loading spectra can be predicted by using a single constant-amplitude effective parameter that is a characteristic of the stress-range distribution function.

Figure 14.14[13] compares the cover-plated beam data obtained under variable-amplitude random-sequence cyclic loading and the AASHTO fatigue-design curve for cover-plate ends (category E) on the basis of the root-mean-square stress range. Similarly, Figure 14.15[13] compares the data obtained for welded beams and the AASHTO fatigue-design curve for category B. For both types of details, the appropriate fatigue-design curve closely approximated the lower limit (95 percent tolerance limit) of previous constant-amplitude test results where almost all the data points are above these curves. This shows that the AASHTO fatigue-design curves provide an approximate lower limit for variable-amplitude test results when plotted on the basis of the root-mean-square stress range. The scatter of data in Figures 14.14 and 14.15 is reasonable, considering that data for different steels, minimum stress levels, and welding sequences are included in these plots. The effects of various secondary parameters, such as minimum stress and type of steel, on the fatigue life under variable-amplitude cyclic loading were similar to the effects of these parameters on the fatigue life under constant-amplitude cyclic loading.

Preliminary data suggest that the fatigue limit for over $2 \times 10^6$ cycles

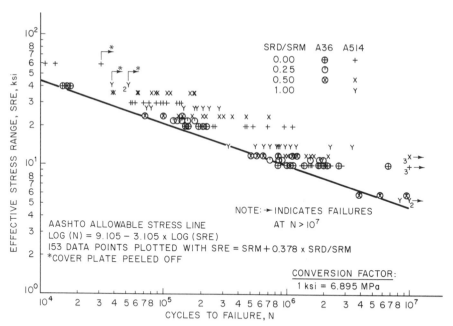

**Figure 14.14** Comparison of cover-plate beam results with AASHTO allowable stress for category E. ($\longrightarrow$ indicates failure at $N > 10^7$.) (Printed with permission. SAE Paper No. 810436, 1981.)

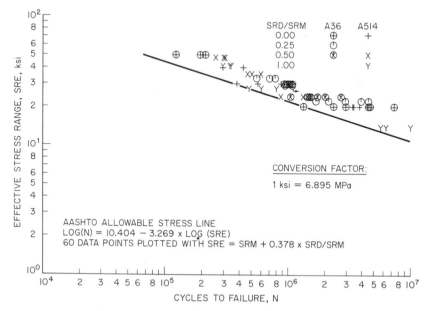

**Figure 14.15** Comparison of welded-beam results with AASHTO allowable stress for category B.

under constant-amplitude loading may not exist under variable amplitude loading if a few (>10 percent) of the stress ranges exceed the stress range corresponding to the fatigue limit under constant-amplitude loading. The data also suggest that, under these conditions, the fatigue behavior for a given structural component can be analyzed by extrapolating the fatigue-design curve for that category to lower stress ranges.

### 14.2.6. Example Problem

The following is a simple example that illustrates the procedure for predicting the fatigue life of a fabricated component subjected to variable-amplitude loading. The calculations are based on the assumption that the fatigue limits for constant-amplitude loading, Figure 14.13, do not exist under variable-amplitude loading and that the long-life behavior is represented by an extrapolation of the finite-life fatigue-design curves.

The fatigue life of a structure is usually governed by the localized behavior of the structural detail which in combination with the applied load fluctuations results in the shortest fatigue life. Thus, a detail with a severe geometrical discontinuity that is subjected to stress fluctuations of small magnitude may exhibit a longer fatigue life than a detail with a moderate geometrical discontinuity that is subjected to large stress fluctuations.

Consider a welded structure that is expected to be subjected each year to the cycles, $N_i$, and their corresponding stress ranges, $\Delta\sigma_i$, that are tabulated in Table 14.1. The fatigue life for this structure can be determined by calculating an effective stress range that represents the various loadings shown in Table 14.1 and the frequency of their occurrence. The effective

TABLE 14.1    Example Problem: Data and Calculations for Fatigue-Life Determination

| $i$ | Number of Cycles per Year ($N_i$) | Stress Range, $\Delta\sigma_i$ (ksi) | $\alpha_i = \dfrac{N_i}{\Sigma_i N_i}$ | $\alpha_i(\Delta\sigma_i)^3$ |
|---|---|---|---|---|
| 1 | 4,500,000 | 0.1 | 0.792 | 0.001 |
| 2 | 800,000 | 0.6 | 0.141 | 0.030 |
| 3 | 140,000 | 5.6 | 0.025 | 4.390 |
| 4 | 232,000 | 7.8 | 0.041 | 19.457 |
| 5 | 3,900 | 10.2 | 0.0007 | 0.743 |
| 6 | 113 | 14.0 | 0.00002 | 0.055 |
| 7 | 2,300 | 15.0 | 0.0004 | 1.350 |

$\Sigma_i N_i = 5,678,313$

$\Sigma_i\, \alpha_i(\Delta\sigma_i)^3 = 25.9$

$\Delta\sigma_{\text{rms}} = \sqrt[3]{\Sigma_i\, \alpha_i\, (\Delta\sigma_i)^3} = 2.96$ ksi

stress range represented by the root-mean-square stress range, $\Delta\sigma_{rms}$, is given by the equation

$$\Delta\sigma_{rms} = \sqrt[3]{\sum_i \alpha_i(\Delta\sigma_i)^3} \tag{14.1}$$

where

$$\alpha_i = \frac{N_i}{\sum_i N_i} \tag{14.2}$$

Table 14.1 presents the details for calculating $\Delta\sigma_{rms}$ which is shown to be equal to 2.96 ksi.

Assume that the structure under consideration is designed to contain structural details that correspond to category E', Figure 14.13, and that these details will be subjected to the loadings presented in Table 14.1. The fatigue lives for category E' details are represented by the E' fatigue-design curve, Figure 14.13, which has the relationship

$$\log N_f = 8.59 - 3 \log \Delta\sigma \tag{14.3}$$

where $N_f$ is the total fatigue life in cycles of loading. Substituting $\Delta\sigma = \Delta\sigma_{rms} = 2.96$ ksi, one obtains $N_f = 15,051,000$ cycles. Thus, the fatigue life in years is given by

$$\frac{\sum_i N_i}{N_f} = 2.65 \text{ yr} \tag{14.4}$$

If this life is too short, the structure should be redesigned to eliminate the category E' details in this section of the structure, or to decrease the stress ranges, or both.

## 14.3. Fracture-Toughness Behavior of Welded Components

### 14.3.1. General Discussion

The fracture toughness and other mechanical properties of steels and weldments depend on several metallurgical factors, including composition, microstructure, and cleanliness. The effects of these factors are complicated by synergisms and, especially for weldments, by a heterogeneous microstructure. A discussion of these factors and their effects is available elsewhere[2,28] and is beyond the scope of this text. However, the following general statements may help to clarify some of the observations related to fracture toughness of weldments.

Chemical elements are added to steel products to obtain certain desired properties such as higher strength, hardenability, toughness, and corrosion resistance. Similarly, chemical elements may be added to filler metals and fluxes to obtain the desired weld metal properties. The effect of a given element on fracture toughness can depend on many factors, including the amount of addition, the interaction of the element with other elements, and the thermomechanical processing of the steel.

The microstructure of a given steel has a significant effect on its fracture-toughness behavior. The microstructural constituents present in structural steels can be ferrite, pearlite, bainite, or martensite. The prominence of each of these constituents depends on the steel composition, processing, and heat treatment. Steels with low hardenability, such as A36, that are subjected to relatively slow cooling rates have ferrite-pearlite microstructures. Quenched and tempered steels, such as A514, have a tempered bainite and tempered martensite microstructure.

Fine-grain microstructures for any of these constituents improve the fracture toughness of the steel. The size of the prior austenite or ferrite grains has been recognized as one of the most important factors that controls the fracture toughness of steels. In general, strength decreases and impact fracture-toughness transition temperature increases as the microstructure changes from tempered martensite, to tempered bainite, to ferrite-pearlite.

The elevated-temperature ($\geq 1650°F$) microstructure of all structural steels is austenite. Rapid cooling rates may transform this microstructure for hardenable steels to as-quenched (untempered) martensite. Untempered martensitic structures contain trapped carbon atoms resulting in high hardness and low fracture toughness. Subsequent heating (tempering) to a temperature where the trapped carbon atoms have mobility defuses the carbon atoms to form carbides. The resulting tempered martensitic structure has lower strength and higher fracture toughness than the untempered martensitic structure. However, tempering followed by slow cooling may embrittle the steel and reduce its fracture toughness, thus offsetting some of the benefits gained by the tempering. This temper embrittlement is one of several embrittling mechanisms that can degrade the fracture toughness of steels and weldments.[2,28,29]

### 14.3.2. Weldments

Many welding processes are available that can produce satisfactory joints in steels. The selection of a particular process is based on many factors that include the thickness and size of the parts to be joined, the position of the weldment, the desired properties and appearance of the finished weldment, the particular application, the cost of fabrication as well as other factors. No single process can be used to produce satisfactory weldments for all steels, thicknesses, and positions. The most suitable process is the one that produces the desired properties in the final product at the lowest possible cost.

In arc welding, which is the most widely used welding process for structural steels, filler metal is melted and used to fill a weld groove. The arc-welding process, welding procedure, and joint geometry influence weld penetration and admixture of the filler metal with the base metal. Because of this admixture, the chemical composition of the base metal can have significant influence on the microstructural and mechanical properties of the weld metal. This influence is significant especially for electroslag and electrogas welds, because they are high-heat-input single-weld-pass processes and, to a lesser extent, for multipass welds by other arc welding processes. The final properties of the weld metal depend on many factors, especially the composition of the weld metal and the conditions governing its solidification and subsequent cooling. Because the heat flow in the weld metal is highly directional toward the adjacent cooler metal, the weld metal develops distinctly columnar grains. Furthermore, the rapid cooling of the weld metal may not allow sufficient time for diffusion of the chemical constituents resulting in microstructural heterogeneities. This segregation and the directional solidification of the weld metal may result in weld-metal properties having pronounced directionality.

For the arc welding processes, the maximum temperature of the weld metal is above that of the base metal joined. This temperature decreases as the distance from the weld increases. Thus, partial melting of the base metal occurs at the weld-metal–base-metal interface, and microstructural changes occur in the base metal in the immediate vicinity of the weld forming a heat-affected zone. The size of this zone is determined by the rate of heating, the volume and temperature of the weld metal, and the rate of cooling of the weld metal and surrounding base metal. These factors as well as the composition and microstructure of the base metal determine the grain size, the grain-size gradient, the microstructure, and therefore, the fracture toughness of the heat-affected zone. Because of the high temperatures and the large variations in temperature gradient and cooling rate, adjacent regions in the heat-affected zone can exhibit large differences in microstructure and properties. In general, for carbon and low-alloy steels, the closer the distance to the weld, the coarser the microstructure. Coarse-grain regions adjacent to the weld interface generally exhibit the poorest toughness.

Grotke[30] divided the heat-affected zone associated with arc welds in plain-carbon and low-alloy steels into five general regions:

1. A partially spheriodized region adjacent to the unaffected base metal where the steel underwent a modest alteration in microstructure . . .

2. A transition region that includes all the microstructures that have undergone partial reaustenitization . . .

3. A grain-refined region where the steel was completely transformed to austenite but at too low a temperature and for too short a time to permit significant grain growth.

4. A grain-coarsened region, resulting from exposure to extremely high austenitizing temperatures.

5. A partially melted region in which incomplete liquidation has occurred, located between the unmelted grain-coarsened region and the entirely fused weld metal.

Illustrations of intermediate heat-affected-zone microstructure associated with bead-on-plate deposits on hot-rolled plain-carbon steel, and on a low-alloy quenched-and-tempered steel are shown in Figures 14.16 and 14.17, respectively. Both welds were made on plates of the same thickness, using identical heat-input conditions; but, the nominal carbon contents were slightly different, 0.20% for the carbon steel and 0.16% for the low-alloy quenched-and-tempered steel.

The heat-affected zone in a single-pass weld forms under the influence of a single thermal cycle. The temperature and temperature distribution from the weld metal into the base metal in a direction perpendicular to the weld is essentially identical at different locations along the weld groove of a simple butt joint for two constant-thickness plates. Consequently, the various microstructural regions in the heat-affected zone can be continuous. However, the weld metal in a multipass weld is built up by the deposition of successive weld beads. The structure and properties of deposited weld beads and existing heat-affected zones are usually altered by the heating effects of subsequent weld-bead deposits. The heat from subsequent weld passes may refine the grain size of the deposited weld metal and existing heat-affected zone, may change the columnar structure of the weld metal to an equiaxed structure, and may temper the microstructure of the existing heat-affected zone. In multipass welds, unlike single-pass welds, the heat-affected zone regions that exhibit relatively low toughness occur intermittently adjacent to the weld interface. In either case, these lower-toughness regions are surrounded by heat-affected zone regions of higher toughness.

### 14.3.3. Fracture-Toughness Tests for Weldments

Weldability is a complex property that is affected by many interrelated factors. Consequently, many fracture-toughness tests have been developed to determine the effects of these factors.[2] Some of these tests are related to the fabrication qualities for the weldment, while others are related to the service performance. The tests for fabrication "must be designed to measure the susceptibility of the weld-metal–base-metal system to such conditions as cracks, porosity or inclusions under realistic and properly controlled conditions of welding."[31] A discussion of these tests is in Reference 2 and is beyond the scope of this text. The service performance tests include yield and tensile strengths, ductility, fracture toughness, stress rupture, stress-corrosion cracking, fatigue, and corrosion fatigue. Some of the fracture-toughness characteristics for weldments are discussed in this section.

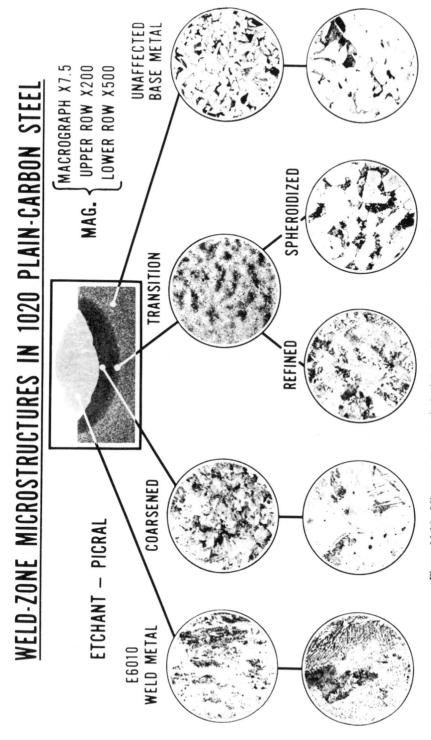

**Figure 14.16** Microstructures typical of the weld metal and the heat-affected base metal in a mild-steel weld.

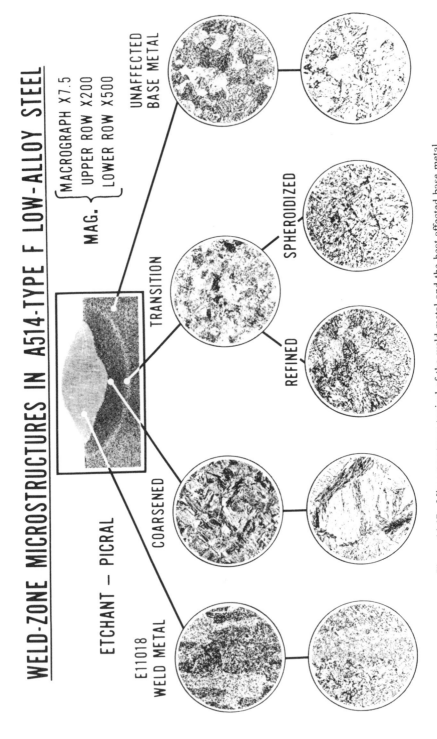

**Figure 14.17** Microstructures typical of the weld metal and the heat-affected base metal in a quenched-and-tempered low-alloy steel weld.

A large number of tests have been developed or adapted to determine the fracture toughness for weldments.[2] These tests include the Charpy V-notch test, various bend tests, the drop weight test, the explosion bulge test, the drop weight tear test, the wide-plate test, and the fracture-mechanics–type tests. Each of these tests has advantages and disadvantages over other tests, and each test measures some parameter that is assumed to represent the fracture toughness for the weldment or some zone within it. Although the subject of fracture-toughness testing of weldments has been studied for many years, there is no general agreement on the best test method to use.

Most fracture-toughness tests have a notch or a fatigue crack. Heterogeneous material properties along the front of the notch or the fatigue crack may cause significant variability in the test results. The magnitude of this variability would depend on many factors, including the rate of change in material properties along and in the vicinity of the notch or crack front, the length and volume of the regions with different properties, the location of the regions along the crack front with respect to each other, and the properties of the surrounding materials.

Weldments exhibit anisotropic heterogeneous material properties. The magnitude and rate of variation in properties can be very large. The volume of a material with essentially uniform properties, especially in the heat-affected zone, can be very small. The low-toughness regions in the heat-affected zone of multipass welds can occur intermittently along the weld interface. These and other factors make the placement of a notch or a fatigue-crack tip in a given microstructural region having uniform properties very difficult. Consequently, significant variability usually is observed in fracture-toughness test results for weldments. Such a variability is observed for Charpy V-notch specimens as well as for $K_{Ic}$ or crack-tip-opening-displacement (CTOD) specimens. Variability in CTOD test results for weld-metal and heat-affected zone specimens is presented in Chapter 17. Furthermore, the properties for a given region may or may not reflect the properties of the welded joint or its performance in an actual structure. Nevertheless, the criteria for acceptance generally are developed in a manner similar to that for base metal. In this sense, fracture-mechanics concepts have been very helpful.

Fracture-toughness test results obtained by testing Charpy V-notch specimens or fracture-mechanics–type specimens ($K_{Ic}$ or CTOD) are influenced by the orientation of the specimen with respect to the weld. Figure 14.18[32] shows the various orientations for fracture-toughness (CVN or fracture-mechanics–type) specimens in a butt welded plate. Orientation 3 is the most commonly used for Charpy V-notch tests of the weld-metal and heat-affected zone.

Despite the problems in determining the fracture toughness for welded joints, the importance of this property for structural performance makes it

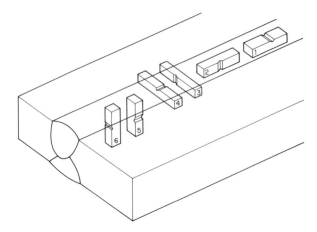

**Figure 14.18**   Possible orientation of fracture-toughness specimens in a butt-welded plate.

necessary to test and attempt to characterize the notch toughness of welded joints. Many codes and standards dictate the specimens and test procedures and establish minimum fracture-toughness values for acceptability. Most of these codes and standards specify the use of the Charpy V-notch specimen.

A book that analyzes the fatigue and fracture behavior of steel-bridge components has been published by Fisher.[33]

# References

1. *Welding Inspection*, American Welding Society, Miami, 1980.
2. R. D. STOUT and W. D. DOTY, *Weldability of Steels*, Welding Research Council, New York, 1978.
3. H. C. CAMPBELL, *Certification Manual for Welding Inspectors*, American Welding Society, Miami, 1977.
4. E. G. SIGNES ET AL., "Factors Affecting the Fatigue Strength of Welded High Strength Steels," *British Welding Journal, 14*, No. 3, 1967.
5. J. W. FISHER ET AL., "Effect of Weldments on the Fatigue Strength of Steel Beams," *NCHRP Report 102*, Transportation Research Board, Washington, D.C., 1970.
6. F. WATKINSON ET AL., "The Fatigue Strength of Welded Joints in High Strength Steels and Methods for Its Improvement," *Proceedings: Conference on Fatigue of Welded Structures*, The Welding Institute, Brighton, England, July 1970.
7. J. W. FISHER ET AL., "Fatigue Strength of Steel Beams with Transverse Stiffness and Attachments," *NCHRP Report 147*, Transportation Research Board, Washington, D.C., 1974.
8. R. ROBERTS, J. M. BARSOM, J. W. FISHER, and S. T. ROLFE, *Fracture Mechanics for Bridge Design* and *Student Workbook—Fracture Mechanics for Bridge*

*Design*, FHWA-RD-78-69, Federal Highway Administration, Office of Research and Development, Washington, D.C., July 1977.

9. J. M. BARSOM, "Fatigue Consideration for Steel Bridges," *Fatigue Crack Growth Measurement and Data Analysis, ASTM STP 738*, edited by S. J. Hudak, Jr., and R. J. Bucci, American Society for Testing and Materials, Philadelphia, 1981.

10. W. C. CLARK, JR., "Some Problems in the Application of Fracture Mechanics," *Scientific Paper 79-1D3-SRIDS-P1*, Westinghouse Research and Development Center, Pittsburgh, 1979.

11. P. F. PACKMAN ET AL., "The Applicability of a Fracture Mechanics Nondestructive Testing Design Criterion," *U.S. Air Force Report AFML-TR-68-32*, Dayton, Ohio, May 1968.

12. W. H. LEWIS ET AL., "Quantitative Measurement of the Reliability of Nondestructive Inspection on Aircraft Structures," *Prevention of Structural Failures*, American Society for Metals, 1978.

13. C. G. SHILLING ET AL., "Fatigue of Welded Steel Bridge Members Under Variable-Amplitude Loadings," *NCHRP Report 188*, Transportation Research Board, Washington, D.C., 1978.

14. T. R. GURNEY, "Theoretical Analysis of the Influence of Toe Defects on the Fatigue Strength of Fillet Welded Joints," *Report No. 32/1977/E*, The Welding Institute, Abington, England, Mar. 1977.

15. T. R. GURNEY and G. O. JOHNSTON, "A Revised Analysis on the Influence of Toe Defects on the Fatigue Strength of Transverse Non-Load-Carrying Fillet Welds," *Report No. 62/1978/E*, The Welding Institute, Abington, England, Apr. 1978.

16. T. R. GURNEY, "Theoretical Analysis of the Influence of Attachment Size on the Fatigue Strength of Transverse Non-Load-Carrying Fillet Welds," *Report No. 91/1979*, The Welding Institute, Abington, England, May 1979.

17. K. H. FRANK, "The Fatigue Strength of Fillet Welded Connections," Ph.D. Thesis, Lehigh University, Bethlehem, Pa., 1971.

18. J. W. FISHER ET AL., "Minimizing Fatigue and Fracture in Steel Bridges," *Structural Integrity Technology*, The American Society of Mechanical Engineers, 1979.

19. W. G. CLARK, JR., "Applicability of the $K_{Iscc}$ Concept to Very Small Defects," *Cracks and Fracture, ASTM STP 601*, American Society for Testing and Materials, Philadelphia, 1976.

20. S. J. MADDOX, "An Analysis of Fatigue Cracks in Fillet Welded Joints," *Research No. E/49/72*, The Welding Institute, Abington, England, 1972.

21. T. R. GURNEY, "Stress Intensity Factors for Cracks at the Toes of Transverse Butt Welds," *Report No. 88/1979*, The Welding Institute, Abington, England, 1979.

22. AMERICAN WELDING SOCIETY, *Specifications for Welded Highway and Railway Bridges*, New York, 1941.

23. J. W. FISHER and I. M. VIEST, "Fatigue Life of Bridge Beams Subjected to Controlled Truck Traffic," Preliminary Publication, 7th Congress, IABSE, Zurich, Switzerland, 1964.

24. AMERICAN ASSOCIATION OF STATE HIGHWAY AND TRANSPORTATION OFFICIALS, *Standard Specifications for Highway Bridges*, Washington, D.C., 1977.

25. J. W. FISHER, *Bridge Fatigue Guide—Design and Details*, American Institute of Steel Construction, Chicago, 1977.

26. D. G. BOWER, "Loading History—Span No. 10, Yellow Mill Pond Bridge I-95, Bridgeport, Conn.," Highway Research Record 428, Transportation Research Board, Washington, D.C., 1973.

27. R. E. SLOCKBOWER and J. W. FISHER, "Fatigue Resistance of Full Scale Cover-Plated Beams," *Fritz Engineering Laboratory Report No. 386-9(78)*, Lehigh University, Bethlehem, Pa., June 1978.

28. A. L. PHILLIPS, ed., *Fundamentals of Welding, Welding Handbook*, Section One, American Welding Society, Miami, 1968.

29. J. A. DAVIDSON, P. J. KONKOL, and J. F. SOVAK, "Assessing Fracture Toughness and Cracking Susceptibility of Steel Weldments—A Review," *Final Report Number FHWA-RD-83*, Federal Highway Administration, Washington, D.C., Dec. 1983.

30. E. GROTKE, "Indirect Tests for Weldability," in *Weldability of Steels*, by R. D. Stout and W. D. Doty, Welding Research Council, New York, 1978.

31. R. D. STOUT, "Direct Weldability Tests for Fabrication," in *Weldability of Steels*, p. 252.

32. D. C. MARTIN, "Direct Weldability Tests for Service," in *Weldability of Steels*, p. 278.

33. J. W. FISHER, *Fatigue and Fracture in Steel Bridges—Case Studies*, John Wiley & Sons, New York, 1984.

# 15

\\ # Fracture Criteria

## 15.1. Introduction

A fracture criterion is a standard against which the expected fracture behavior of a structure can be judged. In general terms, fracture criteria are related to the three levels of fracture performance, namely, plane strain, elastic plastic, or fully plastic, as shown in Figure 15.1. Although it would appear desirable always to specify fully plastic behavior, this is rarely done because it is almost always unnecessary as well as being economically unfeasible in most cases. Furthermore, it is unsound engineering because good design is defined as an optimization of satisfactory structural performance, safety, and economic considerations.

For most structural applications, some level of elastic-plastic behavior at the service temperature and loading rate is a satisfactory fracture criterion. While there may be some cases where fully plastic behavior is necessary (e.g., large dynamic loadings such as submarines being subjected to depth charges) or where plane-strain behavior can be tolerated (e.g., certain short-life aerospace applications where the loading and fabrication can be precisely controlled), for the majority of large complex structures (bridges, ships, pressure vessels, offshore drilling rigs, etc.), some level of elastic-plastic behavior is appropriate. The questions become, "What level of elastic-plastic behavior is required and how can this level of performance be ensured?" The purpose of this chapter is to develop a rational engineering approach to answering these questions.

Unfortunately, the selection of a fracture criterion is often quite arbitrary and is based on service experience for other types of structures that may have no relation to the particular structure an engineer may be designing. An example of the use of a fracture criterion developed for one application and yet which is widely used in many other situations is the 15-ft-lb CVN impact criterion at the minimum service temperature, which was established

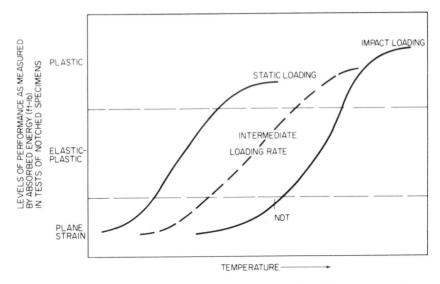

**Figure 15.1** Schematic showing relation between notch-toughness test results and levels of structural performance for various loading rates.

based on the World War II ship failures. This criterion has been widely used for various types of structures, even though the material, service conditions, structural redundancy, and so on may be considerably different from that of the World War II ships for which the criterion was established. Also, because a fracture criterion is only a part of a fracture-control plan, selection of a fracture criterion alone, without considering the other factors involved in fracture control (for example, fatigue) will not necessarily result in a safe structure.

Criteria selection should be based on a very careful study of the particular requirements for a particular structure. In this chapter we shall describe the various factors involved in the development of a criterion, which include

1. The service conditions (loadings, temperature, loading rate, etc.) to which the structure will be subjected.
2. The significance of loading rate on the performance of the structural materials to be used.
3. The desired level of performance in the structure.
4. The consequences of failure.

It should be emphasized that there can be no single *best* criterion for all structures because optimum design should involve economic considerations as well as technical ones. That is, selecting a fracture criteria that is far more conservative than necessary, based on the possible consequences of failure, is unsound engineering. Furthermore, although a study of all aspects of a fracture-control plan, that is, design, fabrication, inspection, operation, and so on, is necessary to ensure the safety and reliability of structures (as will be described in Chapter 16), *the fundamental decision in designing to prevent brittle fracture in any structure is the development of suitable fracture criteria for the structural materials to be used in the structure.* Thus the establishment of the proper fracture criteria for a given structural application should be the basis of subsequent material selection, structural design, fabrication procedures, and inspection requirements.

There are two general parts to a fracture criterion:

*1. The general test specimens to categorize the material behavior.* Throughout the years, various fracture criteria have been specified using notch-toughness tests such as CVN impact, NDT, dynamic tear, and, more recently, the fracture-mechanics test specimens described in Chapter 3 which are used to measure the critical stress-intensity factor. Other fracture mechanics test specimens currently used for elastic-plastic behavior include $J_{Ic}$, CTOD, and $R$-curve specimens, as described in Chapter 17. The test specimen used for a particular application should be that one which most closely models the actual structural behavior. However, selection of the general test specimen to use is often based on past experience, empirical correlations, as well as economics and convenience of testing rather than on the basis of the test specimen that most closely models the actual structure.

*2. The specific notch-toughness value or values.* The second and more difficult part of establishing a fracture criterion is the selection of the specific level of performance in a particular test specimen in terms of measurable values for material selection and quality control.

The specified values in any criterion should consider both safe structural performance and cost and are always subject to considerable differences in opinion between knowledgeable engineers. Accordingly, one of the main objectives of this textbook is to provide some rational guidelines for the engineer to follow in establishing toughness criteria for various structural applications.

## 15.2. General Levels of Performance

The primary design criterion for most large structures such as bridges, pressure vessels, ships, and so on, is still based on strength and stability requirements such that nominal elastic behavior is obtained under conditions of maximum loading. Usually the strength and stability criteria are achieved by limiting the maximum design stress to some percentage of the yield strength. In many cases, *fracture toughness also* is an important design criterion, and yet specifying a notch-toughness criterion is much more difficult, primarily because

1. Establishing the specific level of required notch toughness (i.e., the required CVN or $K_{Ic}$ value at a particular test temperature and rate of loading) is costly and time consuming and is a subject with which design engineers should become familiar.
2. There is no well-recognized single "best" approach. Therefore, different experts will have different opinions as to the "best" approach, although the science of fracture mechanics is helping to overcome this difficulty.
3. The cost of structural materials increases with increasing levels of inherent notch toughness. Thus economic considerations must be included when establishing any toughness criterion.

Materials that have extremely high levels of notch toughness under even the most severe service conditions (earthquakes, ice movement, dynamic loading, etc.) are available, and the designer can always specify that these materials be used in critical locations within a structure. However, because the cost of structural materials generally increases with their ability to perform satisfactorily under more severe operating conditions, a designer generally does not wish to specify arbitrarily more notch toughness than is required for the specific application. In the same sense, a designer does not specify the use of a material with a very high yield strength for a compression member if the design is such that the critical buckling stress is very low. In the former case, the excessive notch toughness is unnecessary, and in the latter case, the excessive yield strength is unnecessary. Both cases are examples of unsound engineering.

The problem of establishing specific fracture-toughness requirements that are not excessive but are still adequate for normal service conditions is a long-standing one for engineers. However, by using concepts of fracture mechanics, rational fracture criteria can be established for fracture control in different types of structures.

Previously, the maximum allowable flaw size in a member has been shown to be related to the notch toughness and yield strength of a structural material as follows:

$$a = C\left(\frac{\text{critical } K}{\text{yield strength}}\right)^2$$

For conditions of maximum constraint (plane strain), such as might occur in thick plates or in regions of high constraint, the flaw size becomes proportional to $(K_{Ic}/\sigma_{ys})^2$, $(K_{Ic}(t)/\sigma_{ys}(t))^2$, or $(K_{Id}/\sigma_{yd})^2$, where both the toughness and yield strength should be measured at the service temperature and loading rate of the structure.

Thus, the ratio of the critical $K$ and the yield strength (either in the valid plane-strain region or extrapolated into the elastic-plastic region) becomes a good index for measuring the relative toughness of structural materials. For most structural applications, it is desirable that the structure tolerate large flaws without fracturing; therefore, the use of materials with high $K_{Ic}/\sigma_{ys}$, $K_{Ic}(t)/\sigma_{ys}(t)$, or $K_{Id}/\sigma_{yd}$ ratios (i.e., elastic-plastic behavior) is a desirable condition.

The basic question in establishing a fracture criterion for large structures becomes "How high must this ratio be for a particular structural material to ensure satisfactory and safe performance in large complex structures?"

No simple answer exists because the engineer must take into account such factors as the design life of the structure, consequences of a failure in a structural member, redundancy of load path, probability of overloads, and fabrication and material cost. However, fracture mechanics can provide a conservative engineering approach to evaluate this question rationally. It is assumed that discontinuities will initiate and propagate in most large complex structures and that local yield-stress loading and plane-strain conditions may exist in parts of the structure (although the use of thin plates tends to minimize the possibility of plane-strain behavior). Therefore, the $K_{Ic}/\sigma_{ys}$ ratio obtained at the appropriate loading rate for materials used in a particular structure—either by direct measurement (Chapters 3 and 4) or by empirical correlation with auxiliary fracture-toughness tests (Chapter 5)—is one of the primary material parameters that defines the relative safety of a structure against brittle fracture.

If a structure is loaded "slowly" ($\sim 10^{-5}$ in./in./sec), the $K_{Ic}/\sigma_{ys}$ ratio is the controlling toughness parameter. If, however, the structure is loaded "dynamically" ($\sim 10^1$ in./in./sec or impact loading), the $K_{Id}/\sigma_{yd}$ ratio is the controlling parameter. At intermediate loading rates, the ratio of $K_{Ic}(t)$ and $\sigma_{ys}(t)$ determined at the appropriate loading rate is the controlling parameter. Definitions and test conditions for each of these ratios are as follows:

1. $K_{Ic}$: critical plane-strain stress-intensity factor under conditions of static loading as described in ASTM Test Method E-399—"Standard Method of Test for Plane Strain Fracture Toughness of Metallic Materials (Chapter 3)."

2. $\sigma_{ys}$: static tensile yield strength obtained in "slow" tension test as described in ASTM Test Method E-8—"Standard Methods of Tension Testing of Metallic Materials."

3. $K_{Ic}(t)$: critical plane-strain stress-intensity factor measured at an intermediate loading rate corresponding to $t$ seconds to maximum load.

4. $\sigma_{ys}(t)$: tensile yield strength measured at an intermediate loading rate corresponding to $t$ seconds to yield load.

5. $K_{Id}$: critical plane-strain stress-intensity factor as measured by "dy-

namic'' or "impact" tests (Chapter 3). The test specimen is similar
to a $K_{Ic}$ test specimen but is loaded rapidly.

6. $\sigma_{yd}$: dynamic tensile yield strength obtained in "rapid" tension test
at loading rates comparable to those obtained in $K_{Id}$ tests. This value
is difficult to obtain experimentally. A good engineering approximation
for $\sigma_{yd}$ under loading rates comparable to impact conditions based on
experimental results of structural steels is

$$\sigma_{yd} = \sigma_{ys} + (20\text{--}30 \text{ ksi})$$

Use the higher number for low-strength steels (40–60 ksi) and the
lower number for high-strength steels (80–120 ksi).

Because high constraint (thick plates—plane-strain conditions) at the
tip of a crack can lead to premature fracture, the engineer should strive
for the lowest possible degree of constraint (thin plates—plane-stress conditions) at the tip of a crack. Generally, this behavior can be accomplished
by specifying the largest possible ratio of the appropriate critical $K$ and
yield strength, consistent with economic considerations.

Structural materials whose toughness and plate thickness are such that
the critical $K$-to-yield-strength ratio for service loading rates is less than
about $\sqrt{t/2.5}$ exhibit elastic plane-strain behavior as described in Chapter
3 and generally fracture in a brittle manner. These materials usually are
not used as primary tensile load-carrying members for most structural applications because of the high level of constraint at the tip of a crack and
the rather small critical crack sizes at design stress levels. Fortunately,
most structural materials have toughness levels such that they do *not* exhibit
plane-strain behavior at service temperatures, service loading rates, and
common structural sizes normally used. However, very thick plates or
plates used to form complex geometries where the constraint can be very
high may be susceptible to brittle fractures even though the inherent notch
toughness as measured by small-scale laboratory tests appears satisfactory.
Still, there are some situations where this level of behavior can be used
successfully (such as for rail steels). In these situations, parameters other
than fracture toughness (such as stress state and fatigue-crack initiation
and propagation) control the behavior of the structure.

Structural materials whose toughness levels are such that they exceed
the plane-strain limits described above and in Chapter 3 exhibit *elastic-
plastic* fractures with varying amounts of yielding prior to fracture. The
tolerable flaw sizes at fracture vary considerably but can be fairly large.
Fracture is usually preceded by the formation of large plastic zones ahead
of the crack. Most structures are built of materials that exhibit some level
of elastic-plastic behavior at service temperatures and loading rates, and
thus the appropriate critical $K$ values are difficult to obtain and cannot be
measured directly. Chapter 17 describes elastic-plastic tests to determine
these levels of fracture toughness.

Many structural materials exhibit ductile plastic fractures preceded by large deformation at service temperature and loading rates. Obviously, $K_{Ic}$, $K_{Ic}(t)$, or $K_{Id}$ values cannot be measured. This type of behavior is very desirable in structures and represents considerable notch toughness. However, this level of toughness is rarely necessary and thus is usually not specified, except for unusual cases such as submarine hulls or nuclear structures.

Photographs comparing the general fracture appearance of these three levels of behavior are presented in Figures 15.2, 15.3, and 15.4. *Plane-strain* behavior refers to fracture under elastic stresses with little or no shear-lip development and is essentially brittle. A fracture surface typical of this behavior is shown in Figure 15.2; note the lack of deformation. *Plastic* behavior refers to ductile failure under general yielding conditions

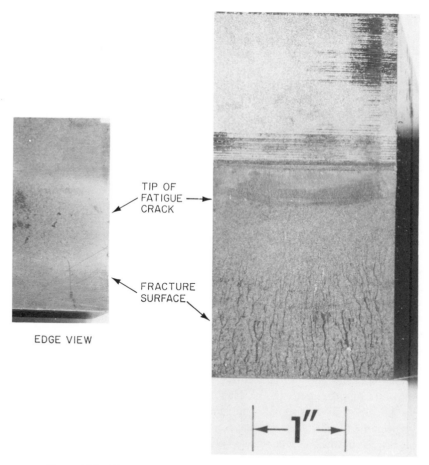

**Figure 15.2**  Photograph of plane-strain fracture surface and edge view.

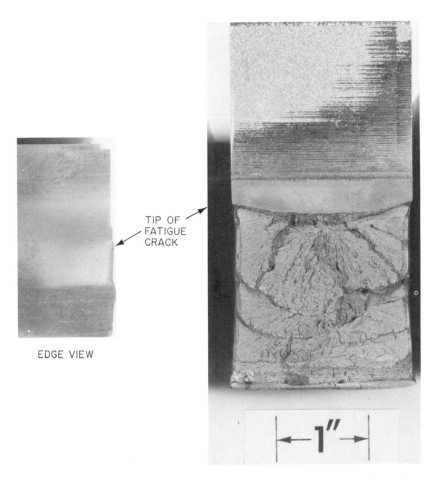

**Figure 15.3** Photograph of elastic-plastic (mixed-mode) fracture surface and edge view.

accompanied, usually, by the development of very large shear-lip development. A fracture surface typical of this behavior is shown in Figure 15.4, along with the side view showing considerable deformation. The transition between these two extremes is the *elastic-plastic* region, which is also referred to as the mixed-mode region. A fracture surface typical of this behavior is shown in Figure 15.3. However, fracture-surface appearance by itself is neither necessary nor sufficient to indicate the mode of fracture. Ductile fractures can propagate in the elastic-plastic or plastic regions with little or no shear-lip development when the geometry of the detail prevents the development of 45° shear planes.

For structural steels, these three levels of performance are usually described in terms of the transition from brittle to ductile behavior as measured by various types of notch-toughness tests. This general transition

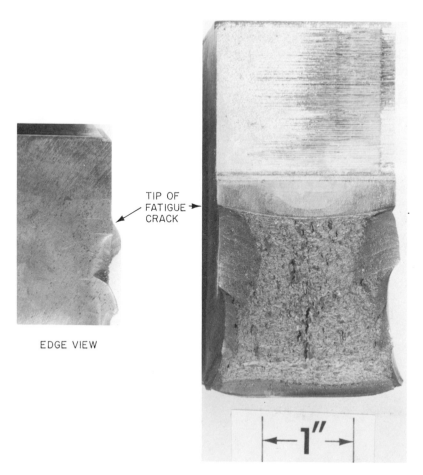

TIP OF
FATIGUE
CRACK

EDGE VIEW

**Figure 15.4**   Photograph of plastic fracture surface and edge view.

in fracture behavior was related to the three levels of behavior schematically in Figure 15.1. For static loading, the transition region occurs at lower temperatures than for impact (or dynamic) loading, for those structural materials that exhibit a transition behavior. (This general behavior was discussed in Chapter 4.) Thus, for those structures that are subjected to static loading, a static transition curve (i.e., $K_{Ic}$ test results) should be used to predict the level of performance at the service temperature. For structures subjected to impact or dynamic loading, the impact transition curve (i.e., $K_{Id}$ test results) should be used to predict the level of performance at the service temperature. For structures subjected to some intermediate loading rate, an intermediate loading-rate transition curve (i.e., $K_{Ic}(t)$ test results) should be used to predict the level of performance at the service temperature. Because the actual loading rates for many structures are not well known, the impact loading curve (Figure 15.1) is often used to predict the service performance of structures. However, this may be unduly conservative and often does not properly model the service behavior for many types of structures that are loaded statically or at intermediate loading rates, such as bridges.

Figures 15.5 and 15.6 show the three general levels of performance for typical structural steels as measured by various types of notch-toughness tests. It should be emphasized that although the upper limit of plane-strain behavior is reasonably well established as $K_{Ic}/\sigma_{ys}$, $K_{Ic}(t)/\sigma_{ys}(t)$, or $K_{Id}/\sigma_{yd}$ = $\sqrt{t/2.5}$, the boundary between elastic-plastic and plastic behavior *is not well established*. The boundaries shown in Figures 15.5 and 15.6 are general guidelines based on the authors' experience in which the beginning of the plastic region is defined as that region where the transition curves begin to level off.

Because most structural materials exhibit nonplane-strain behavior (i.e., either elastic-plastic or plastic behavior as shown in Figure 15.1) at service temperatures and loading rates, critical $K$ values are difficult to obtain for most structural materials at service temperatures and loading rates. This fact represents a difficult situation when material properties must be specified (although it is a highly desirable one from the viewpoint that plane-strain brittle fractures are rare). Therefore, for those situations

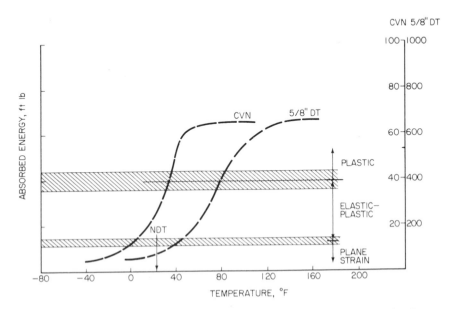

**Figure 15.5** Relation among plane-strain, elastic-plastic, and plastic levels of performance for an ABS-C steel as measured by NDT, CVN, and DT test results. (Ranges are approximate.)

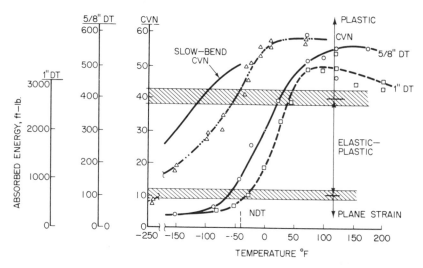

**Figure 15.6**  Relation among plane-strain, elastic-plastic, and plastic levels of performance for an A517 steel as measured by various notch-toughness tests.

where it is desired to specify a minimum level of notch toughness that exceeds plane-strain behavior (which is usually the case), auxiliary fracture toughness test methods (such as described in Chapter 5) or elastic-plastic fracture mechanics tests (such as described in Chapter 17) must be used to ensure that the structural material exhibits nonplane-strain behavior at the service temperature.

## 15.3. Consequences of Failure

Although rarely stated as such, the basis of structural design of large complex structures is an attempt to optimize the desired performance requirements relative to cost considerations (including materials, design, and fabrication) so that the probability of failure (and its economic consequences) is low. Generally, the primary criterion is the requirement that the structure support its own weight plus any applied loads and still have the nominal stresses be less than either the tensile yield strength (to prevent excessive deformation) or the critical buckling stress (to prevent premature buckling).

Although brittle fractures can occur in riveted or bolted structures, the evolution of welded construction with its emphasis on monolithic structural members has led to the desirability of including some kind of fracture criterion for most structures, in addition to the strength and buckling criteria already in existence. That is, if a fracture initiates in a welded structure, there usually is a continuous path for crack extension. However, in riveted or bolted structures, which generally consist of many individual plates or shapes, a continuous path for crack extension rarely exists. Thus any cracks that may extend generally are arrested as soon as they traverse a single plate or shape. Consequently, there can be a large difference in the possible fracture behavior of welded structures, compared with either riveted or bolted structures.

Failure of most engineering structures is caused by the initiation and propagation of cracks to critical dimensions. Because crack initiation and propagation for different structures occur under different stress and environmental conditions, no single fracture criterion or set of criteria should be used for all types of structures. Most criteria are developed for particular structures based on extensive service experience or empirical correlations and thus are valid only for a particular design, fabrication method, and service use.

However, one of the biggest reasons that no single fracture criterion should be applied uniformly to the design of different types of structures is the fact that the *consequences of structural failure* are vastly different for different types of structures. For example, a fracture criterion for steels used in seagoing ship hull structures is that the NDT (nil-ductility) temperature be 30°F for a minimum service temperature of 30°F. This criterion was based on the assumption that ships are subjected to full-impact loading. The excellent service experience of this type of structure is such that even this requirement is considered by some to be too conservative. Nonetheless, in view of the consequences of failure of a ship, that is, either in terms of loss of life or of cargo, this criterion appears reasonable, if the assumption that ships are loaded dynamically is true.

In contrast, the fracture criterion for the steels proposed to be used in the hull structure of stationary but floating nuclear power plants inside a protective breakwater was that the NDT be $-30°F$, for a minimum service temperature of $+30°F$. Thus, for *less severe loading* (because the platforms are stationary), the steels in the floating nuclear power plants were required to have an NDT temperature 60°F *below* their service temperature, compared with seagoing ship steels whose NDT temperature generally is at their service temperatures. However, for the hull structure of floating nuclear power plants, where service experience is nonexistent and the occurrence of a brittle fracture might have resulted in significant environmental damage and the drastic curtailment of an entire industry, the fracture criterion was extremely conservative because of the consequences of failure, even though the design of the stationary floating hull structure was similar to that of seagoing ship hull structures.

Another example of consequences of failure is the *lack* of a necessity for specifying a fracture criterion for a piece of earthmoving equipment where the consequences of failure of a structural member may be loss of the use of the equipment for a short time until the part can be replaced. If the consequences of failure are minor, then specification of a toughness requirement that might increase the cost of each piece of equipment significantly may be unwarranted.

Thus, the consequences of failure should be a major consideration when determining

1. The need for some kind of fracture criteria, and
2. The level of performance (plane strain, elastic plastic, or fully plastic) to be established by the toughness criteria.

Each class or type of structure must be evaluated carefully and the consequences of failure factored into the selected fracture criterion. In fact, in his 1971 AWS Adams Memorial Lecture, Pellini has stated that "one should not use a design criterion in excess of real requirements because this results in specifications of lower NDT and therefore, increased costs." Needless to say, determining "real requirements," that is, balancing safety and reliability against economic considerations, is a difficult task.

## 15.4. Original 15-ft-lb CVN Impact Criterion for Ship Steels

Although occasional brittle fractures were reported in various types of structures (both welded and riveted) prior to the 1940s,[1,2] it was not until the rapid expansion in all-welded ship construction during the early 1940s that brittle fracture became a well-recognized structural problem. During the early 1940s, over 2500 Liberty ships, 500 T-2 tankers, and 400 Victory ships were constructed as a result of World War II. Because the basic designs of each of these three types of ships (Liberty, Victory, and tankers) were similar, it was possible to analyze any structural difficulties on a statistical basis.

The first of the Liberty-type ships were placed in service near the end of 1941, and by January 1943, there were ten major fractures in the hull structures of the ships that were in service at that time. Numerous additional failures throughout the next few years led to the establishment of an investigative board to conduct an investigation into the design and methods of construction of welded steel merchant ships. In 1946, this board made its final report[3] to the Secretary of the Navy.

The role of materials, welding, design, fabrication, and inspection is described in various extensive reviews of this problem,[3-10] and the interested reader is referred to these documents. Of interest in this particular section is the work leading to the development of the 15-ft-lb notch-toughness criterion, which is still widely used for many other types of structures.

In the development of the 15-ft-lb criterion, samples of steel were collected from approximately 100 fractured ships and submitted to the

National Bureau of Standards for examination and tests. This particular study resulted in the collection of an extremely complete body of data relative to the failure of large welded ship hull structures. The plates from fractured ships were divided into three groups:

1. Those plates in which fractures originated, called *source plates.*
2. Those plates through which the crack passed, called *through plates.*
3. Those plates in which a fracture stopped, called *end plates.*

An analysis of the Charpy V-notch impact test results showing the frequency distribution of the 15-ft-lb Charpy V-notch transition temperature of the source, through, and end plates is presented in Figure 15.7. These

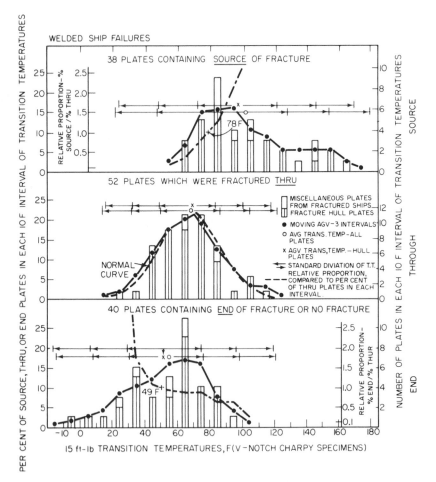

**Figure 15.7** Frequency distribution of 15-ft-lb V-notch Charpy transition temperatures for fractured ship plates (Ref. 10, reproduced with permission of the National Academy of Sciences).

results show that the *source* plates had a high average 15-ft-lb transition temperature, about 95°F. The through plates had a more normal distribution of transition temperatures, and the average 15-ft-lb transition temperature was lower, that is, about 65°F. The *end* plates had the lowest average 15-ft-lb transition temperature (about 50°F) and a distribution with a long tail at lower transition temperatures.

Parker[10] points out that "the character of these distribution curves was not wholly unanticipated; one would suspect that the through plates would be most representative of all ship plates and hence might tend toward a normal distribution, whereas the *source* and *end* plates were *selected* for their role in the fracturing of the ship. A factor in this selection is the notch toughness of the plate as measured in the Charpy test. The overlapping of the *source*- and *end*-plate distributions with the through-plate distribution can be explained by factors involved in the fracturing in addition to the notch toughness of the plate. For example, a plate in the through-fracture category having a transition temperature between about 60°F and 90°F might have been a fracture *source* plate under more severe stress conditions such as those in the region of a notch; under less severe conditions of average stress, it might have been an *end* plate. Even though factors other than notch toughness contributed to the selection of a plate for its role in fracturing, it should be pointed out that statistical analyses have indicated that the differences in transition temperature and energy at failure temperature between the *source* and *through* and *through* and *end* plates are not due to chance."

The extremely low probability of the differences in Charpy properties of plates in the three fracture categories being due to chance permitted the development of several criteria that are very important to engineers. Figure 15.8 shows that only 10 percent of the *source* plates absorbed more than 10-ft-lb in the V-notch Charpy test at the failure temperature; the highest value encountered in this category was 11.4 ft-lb. At the other extreme, 73 percent of the *end* plates absorbed more than 10 ft-lb at the failure temperature. It was therefore concluded, based on these data and for steels of this quality, that in the large ship hull structures brittle fractures are not likely to initiate in a plate that absorbs more than 10 ft-lb in a Charpy V-notch test conducted at the anticipated operating temperature.

Thus, a slightly higher level of performance, namely, the *15-ft-lb* transition temperature as measured with a CVN impact test specimen, was selected as a fracture criterion on the basis of actual service behavior of a large number of similar-type ship hull structures. Since the establishment of this relation between CVN values and service behavior in ship hulls, the 15-ft-lb transition temperature has been a widely used fracture criterion, even though it was developed only for a particular type of steel and a particular class of structures, namely, ship hulls.

Fortunately for the engineering profession and the general public safety, similar statistical correlations between test results and service failures do

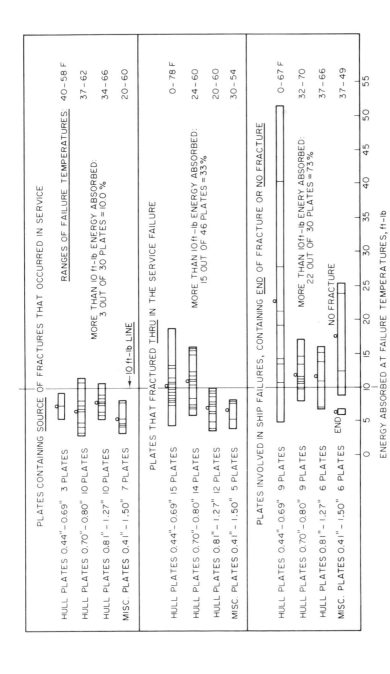

**Figure 15.8** Relation of energy absorbed by V-notch Charpy specimens at the temperature of ship failure to the nature of the fractures in ship plates (Ref. 10, reproduced with permission of the National Academy of Sciences).

not exist for any other class of structures because there have not been such a large number of failures in any other type of structure. However, the difficulty of obtaining service experience creates a problem for the design engineer in establishing toughness criteria for new types of structures.

## 15.5. Transition-Temperature Criteria

Since the time of the World War II ship failures, the fracture characteristics of low- and intermediate-strength steels generally are described in terms of the transition from brittle to ductile behavior as measured by impact tests. This transition in fracture behavior can be related schematically to various fracture states, as was shown in Figure 15.1. Typical fracture surfaces showing each of these three regions were shown in Figures 15.2, 15.3, and 15.4.

For static loading, the transition region occurs at lower temperatures than for impact (dynamic) loading, depending on the yield strength of the steel (Figure 15.9). Thus, for structures subjected to static loading, the static transition curve should be used to predict the level of performance at the service temperature. For structures subjected to impact or dynamic rates of loading, the impact transition curve should be used to predict the level of performance at the service temperature. For structures subjected

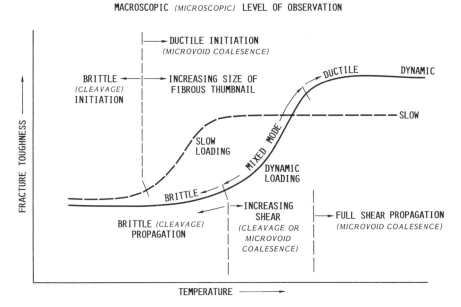

**Figure 15.9**  Schematic showing relation between slow and dynamic fracture toughness and fracture appearance.

to some intermediate loading rate, an intermediate loading-rate transition curve should be used to predict the level of performance at the service temperature. If the loading rate for a particular type of structure is not well defined, and the consequences of failure are such that a fracture will be extremely harmful, a conservative approach is to use the impact loading curve to predict the service performance. As was noted in Figure 15.1, the nil-ductility transition temperature is close to the upper limit of plane-strain conditions under conditions of *impact* loading.[11,12]

After establishing the loading rate for a particular structure, and the corresponding loading rate for the test specimen to be used, the next step in the transition-temperature approach to fracture-resistant design is to establish the *level* of material performance required for satisfactory structural performance. That is, as shown schematically in Figure 15.10 for impact loading of three arbitrary steels 1, 2, and 3, one of the following three general levels of material performance should be established at the service temperature for primary load-carrying members in a structure:

a. Plane-strain behavior.
b. Elastic-plastic (mixed-mode) behavior.
c. Fully plastic behavior.

Using the schematic results shown in Figure 15.10, and an arbitrary minimum service temperature as shown, steel 1 would exhibit plane-strain behavior at the minimum service temperature, whereas steels 2 and 3 would exhibit elastic-plastic and fully plastic behavior, respectively.

As an example of the transition-temperature approach, assume that a 30-ft-lb Charpy V-notch impact test value is required for ship hull steels.

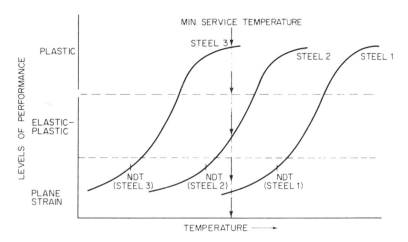

**Figure 15.10**  Schematic showing relation between level of performance as measured by impact tests and NDT for three arbitrary steels.

Figure 15.11 compares the average toughness levels of several grades of ABS ship hull steels and shows that, according to a 30-ft-lb CVN criterion, the CS-grade steel can be used at service temperatures as low as $-90°F$, whereas the CN-grade steel can be used only to about $-30°F$, and the B-grade steel meets this requirement only down to service temperatures of about $+10°F$.

One limitation to the transition-temperature approach sometimes occurs with the use of materials that do not undergo distinct transition-temperature behavior or with materials that exhibit a low-energy shear behavior. Figure 15.12 shows the relationship of low-energy performance compared to normal behavior and to very high-level toughness behavior (such as obtained in an HY-80-type steel used for military applications).

Low-energy shear behavior usually does not occur in low- to intermediate-strength structural steels ($\sigma_{ys} \leq 100$ ksi) but sometimes is found

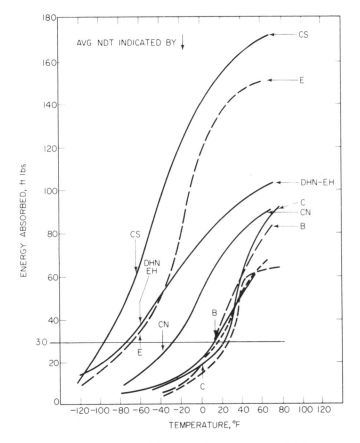

**Figure 15.11**  Comparison of minimum service temperature of ship steels using an arbitrary criterion of 30 ft-lb.

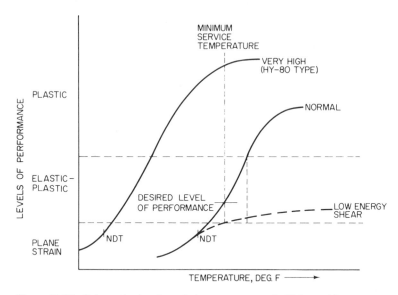

**Figure 15.12**  Schematic showing relation among normal-, high-, and low-energy shear levels of performance as measured by impact tests.

in high-strength steels ($\sigma_{ys} > 100$ ksi). For example, if the desired level of performance is as shown in Figure 15.12, a material exhibiting low-energy shear behavior may never achieve this level of performance at *any* temperature. For these high-strength materials, the through-thickness yielding and the leak-before-burst methods described in the following sections are very useful to establish fracture criteria.

## 15.6. Through-Thickness Yielding Criterion

The through-thickness yielding criterion for structural steels is based qualitatively on two observations.[13] First, increasing the design stress in a particular application (which generally requires a higher-yield-strength material) results in more stored elastic energy in a structure, which means that the fracture toughness of the steel should also be increased to have the same degree of safety against fracture as a structure with a lower working stress. Second, increasing plate thickness promotes a more severe state of stress, namely, plane strain. Thus, a higher level of toughness is required to obtain the same level of performance in thick plates as would be obtained in thin plates.

By using concepts of linear-elastic fracture mechanics,[14,15] a quantitative approach to the development of toughness requirements is based on the requirement that in the presence of a large sharp crack in a large plate, through-thickness yielding should occur before fracture. Specifically, the requirement is based on the ratio of plate thickness to plastic-zone size ahead of a large sharp flaw.

From a qualitative viewpoint, the effect of plate thickness on the fracture toughness of steel plates tested at room temperature has been generally established.[14,15] As the plate thickness is decreased, the state of stress changes from plane strain to plane stress, and ductile fractures generally occur along 45° planes through the thickness. Except for very brittle materials, failure is usually preceded by through-thickness yielding and is not catastrophic. Conversely, in thick plates the state of stress is generally plane strain, and fractures usually occur normal to the direction of loading. Except for ductile materials, through-thickness yielding does not occur prior to fracture, and failure may be unstable. The transition in state of stress from plane strain to plane stress is responsible for a large increase in toughness and is quite desirable. Thus, if plane-stress behavior can be assured, structural members should fail only when plastically overloaded.

The behavior of most structural members is somewhere between the two limiting conditions of plane stress and plane strain. Hahn and Rosenfield[16,17] have shown that in terms of either through-thickness strain or crack-opening displacement, there is a significant increase in the rate at which through-thickness deformation occurs when the following relation exists,

$$\left(\frac{K_{Ic}}{\sigma_{ys}}\right)^2 \frac{1}{t} \geqslant 1 \qquad (15.1)$$

where $K_{Ic}$ = plane-strain stress-intensity factor, ksi$\sqrt{in.}$

$\sigma_{ys}$ = yield strength, ksi.

Brown and Srawley,[15] as well as ASTM Committee E-24 on Fracture Testing of Metals,[18] have indicated that the following relation must be satisfied to ensure plane-strain behavior:

$$\left(\frac{K_{Ic}}{\sigma_{ys}}\right)^2 \frac{1}{t} \leqslant 0.40 \qquad (15.2)$$

Detailed analysis of numerous $K_{Ic}$ test results as well as other experimental evidence[16,17,19,20,21] indicates that plane-strain conditions may exist at somewhat smaller plate thicknesses than required by Equation (15.2). However, assume that Equation (15.1) represents an upper bound for plane-strain behavior and may be used to define the condition at which considerable through-thickness yielding begins to occur. This type of behavior is desirable in structural applications and can be used as a criterion to obtain satisfactory performance in structures where through-thickness yielding can occur, such as in large thin plates that contain through-thickness cracks and in which prevention of fracture is an important consideration.

Equation (15.1), which is based on Hahn and Rosenfield's experimental observation of the plastic-zone size in silicon steel, is similar to Irwin's $\beta_I$

criterion for $\beta_1 = 1.$[22] In this model, Irwin proposed that the minimum $K_{Ic}$ value for a "leak-before-break" criterion be as follows:

$$\beta_1 = \frac{1}{t}\left(\frac{K_{Ic}}{\sigma_{ys}}\right)^2 \simeq 1.5 \qquad (15.3)$$

Therefore,

$$t \simeq \frac{1}{1.5}\left(\frac{K_{Ic}}{\sigma_{ys}}\right)^2 \qquad (15.4)$$

Thus, for through-thickness yielding to occur in the presence of a large sharp crack in a large plate, a fracture-toughness criterion based on Equation (15.1) appears reasonable on the basis of both experimental and theoretical considerations. This condition is conservative and there are many design applications for which this type of performance is not required. However, it is a desirable goal in establishing toughness requirements for large thin plates that may contain through-thickness cracks and are used in critical applications.

Equations 15.1 through 15.4 suggest that, for a given yield strength, $K_{Ic}$ must increase as the thickness, $t$, increases. Such a relationship between $K_{Ic}$ and $t$ is inconsistent with the fracture-toughness transition behavior for steels which indicates that, above a given temperature, $K_{Ic}/\sigma_{ys}$ is essentially independent of plate thickness. Thus, Equations 15.1 through 15.4 become too conservative as the plate thickness increases above about 2 inches.

Rearranging Equation (15.1), a toughness criterion for steels to obtain through-thickness yielding before fracture can be developed in terms of yield strength and plate thickness as follows:

$$K_{Ic} \geq \sigma_{ys}\sqrt{t} \quad \text{for } t \leq 2 \text{ in.} \qquad (15.5)$$

From this relation, the $K_{Ic}$ values required for through-thickness yielding before fracture at any temperature were developed for steels with various yield strength levels and plate thicknesses. These values, shown in Table 15.1 and plotted in Figure 15.13, demonstrate the marked dependence of toughness on yield strength and plate thickness if through-thickness yielding is to precede fracture. That is, the desired toughness increases linearly with yield strength and with the square root of plate thickness.

Service experience suggests that the required toughness on the basis of Equation (15.5) may be excessive for thick plates. Until investigations of very thick plates establish the necessary toughness requirements, the use of Equation (15.5) should be limited to plates less than 2 in. thick.

It is generally not possible to determine $K_{Ic}$ when the plate has sufficient toughness according to the criterion just described. According to the ASTM recommended practice,[18] the plate thickness for valid results should be

$$\left(\frac{K_{Ic}}{\sigma_{ys}}\right)^2 \frac{1}{t} < 0.40 \qquad (15.2)$$

**TABLE 15.1**  $K_{Ic}$ **Values Required for Through-Thickness Yielding Before Fracture**

| Yield Strength, $\sigma_{ys}$ (ksi) | Plate Thickness, $t$ (in.) | $K_{Ic}$ (ksi$\sqrt{\text{in.}}$) |
|:---:|:---:|:---:|
| 40 | $\frac{1}{2}$ | 28 |
| | 1 | 40 |
| | 2 | 57 |
| 60 | $\frac{1}{2}$ | 42 |
| | 1 | 60 |
| | 2 | 85 |
| 80 | $\frac{1}{2}$ | 57 |
| | 1 | 80 |
| | 2 | 113 |
| 100 | $\frac{1}{2}$ | 71 |
| | 1 | 100 |
| | 2 | 141 |
| 120 | $\frac{1}{2}$ | 85 |
| | 1 | 120 |
| | 2 | 170 |
| 140 | $\frac{1}{2}$ | 99 |
| | 1 | 140 |
| | 2 | 198 |
| 160 | $\frac{1}{2}$ | 113 |
| | 1 | 160 |
| | 2 | 226 |
| 180 | $\frac{1}{2}$ | 127 |
| | 1 | 180 |
| | 2 | 255 |
| 200 | $\frac{1}{2}$ | 141 |
| | 1 | 200 |
| | 2 | 283 |

Rearranging terms, the maximum valid $K_{Ic}$ that can be measured by current practice is

$$K_{Ic} = 0.63\sigma_{ys}\sqrt{t}$$

However, the required $K_{Ic}$ according to the through-thickness yielding criterion is

$$K_{Ic} = 1.0\sigma_{ys}\sqrt{t}$$

Having the required toughness greater than the maximum toughness that can currently be measured means that the specimen should behave in

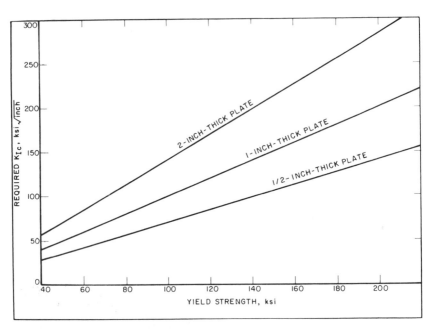

**Figure 15.13**   $K_{Ic}$ required to satisfy through-thickness yielding before fracture criterion.

a somewhat ductile manner in the test and that elastic plane-strain behavior will not be obtained in materials that meet the criterion.

## 15.7. Leak Before Break

The leak-before-break criterion was proposed by Irwin et al.[23] as a means of estimating the necessary toughness of pressure-vessel steels so that a surface crack could grow through the wall and the vessel "leak" before fracturing. That is, the critical crack size at the design stress level of a material meeting this criterion would be greater than the wall thickness of the vessel so that the mode of failure would be leaking (which would be relatively easy to detect and repair) rather than fracture.

    Figure 15.14 shows schematically how such a surface crack might grow through the wall into a through-thickness crack having a length approximately equal to $2B$. Thus the leak-before-break criterion assumes that a crack of twice the wall thickness in length should be stable at a stress equal to the nominal design stress.

    The value of the general stress intensity, $K_I$, for a through crack in a large plate (Figure 15.15) is

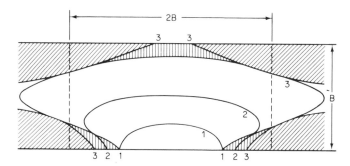

**Figure 15.14**  Spreading of a part-through crack to critical size (line 3) for the short-crack failure models. Lines 1, 2, and 3 are assumed to represent crack edge positions during increase of crack size. Shaded regions are shear lip (vertical shading) or potential shear lip (slant shading).

$$K_I^2 = \frac{\pi \sigma^2 a}{1 - \frac{1}{2}(\sigma/\sigma_{ys})^2} \qquad (15.6)$$

where  $2a$ = crack length.

$\quad\quad \sigma$ = tensile stress normal to the crack.

$\quad\quad \sigma_{ys}$ = yield strength.

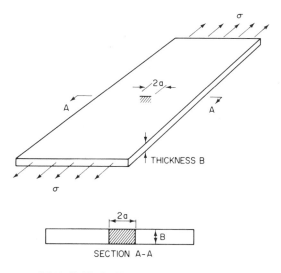

SECTION A-A

FINAL THROUGH-THICKNESS CRACK

$2a \cong 2B$

**Figure 15.15**  Final dimensions of through-thickness crack.

(Note that for low values of design stress, $\sigma$, this expression reduces to $K_I = \sigma\sqrt{\pi a}$.)

At fracture, $K_I = K_c$ (assuming plane-stress behavior) and

$$K_c^2 = \frac{\pi\sigma^2 a}{1 - \frac{1}{2}(\sigma/\sigma_{ys})^2} \tag{15.7}$$

Because standard material properties are usually obtained in terms of $K_{Ic}$, the following relation between $K_c$ and $K_{Ic}$ is used to establish the leak-before-break criterion in terms of $K_{Ic}$,

$$K_c^2 = K_{Ic}^2(1 + 1.4\beta_{Ic}^2) \tag{15.8}$$

where

$$\beta_{Ic} = \frac{1}{B}\left(\frac{K_{Ic}}{\sigma_{ys}}\right)^2, \quad \text{a dimensionless parameter}$$

Thus, substituting for $K_c$ and $\beta_{Ic}$, the following general relation is obtained:

$$\frac{\pi\sigma^2 a}{1 - \frac{1}{2}(\sigma/\sigma_{ys})^2} = K_{Ic}^2\left[1 + 1.4\left(\frac{K_{Ic}^2}{B\sigma_{ys}^2}\right)^2\right] \tag{15.9}$$

In the leak-before-break criterion, the depth of the surface crack, $a$, is set equal to the plate thickness, $B$ (Figure 15.16), and we obtain

$$\frac{\pi\sigma^2 B}{1 - \frac{1}{2}(\sigma/\sigma_{ys})^2} = K_{Ic}^2\left[1 + 1.4\left(\frac{K_{Ic}^4}{B^2\sigma_{ys}^4}\right)\right]$$

or

$$\frac{\pi\sigma^2}{1 - \frac{1}{2}(\sigma/\sigma_{ys})^2} = \frac{K_{Ic}^2}{B}\left[1 + 1.4\left(\frac{K_{Ic}^4}{B^2\sigma_{ys}^4}\right)\right]$$

where  $\sigma$ = nominal design stress, ksi.
$\sigma_{ys}$ = yield strength, ksi.
$B$ = vessel wall thickness, in.
$\sigma$ = maximum permissible design stress level, ksi.
$K_{Ic}$ = plane-strain crack toughness (ksi $\sqrt{\text{in.}}$) required to satisfy the leak-before-break criterion for a material with a particular $\sigma_{ys}$, a vessel with wall thickness $B$, and design stress $\sigma$.

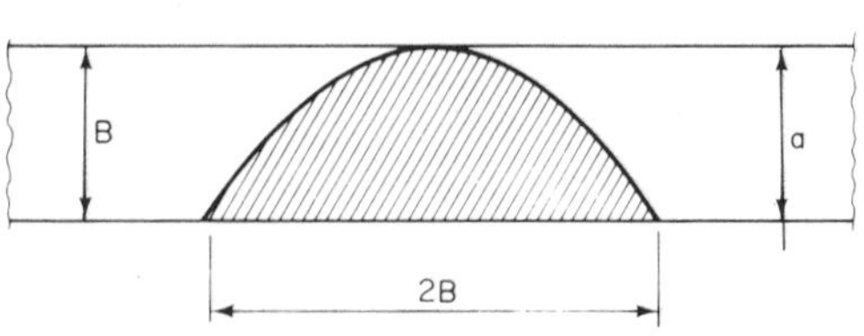

**Figure 15.16**  Assumed flaw geometry for leak-before-burst criterion.

In this expression, $\sigma$, $\sigma_{ys}$, and $K_{Ic}$ are the design stress, yield strength, and material toughness at the particular service temperature and loading rate.

Irwin et al.[23] have calculated various values of the nominal design stress that will satisfy the foregoing leak-before-break criterion for three structural steels, namely, A212B, A302B, and HY-80 steel. In their example, they estimated $K_{Id}$ values based on the NDT temperature plus either 60° or 120°F for each of the three steels. The basic input data are presented in Table 15.2. Using these estimates of material properties, they selected three thickness values (1, 2.5, and 5.0 in.) and calculated the allowable dynamic design stress that would satisfy the leak-before-break criterion for each of the three wall thicknesses chosen for the example. That is, the critical crack size for the 1-in.-thick vessel is 1 in., and for the 5-in.-thick vessel it is 5 in. Obviously, because the toughness levels are the same for all vessel thicknesses studied, the allowable design stress should decrease significantly with increasing vessel thickness.

**TABLE 15.2    Actual and Estimated Material Properties Used by Irwin et al.[23] in Examples of Leak-Before-Break Criterion**

| | Actual Test Values | | | Estimated Material Properties | | | | |
| | | | NDT $+60°F$ | | | NDT $+120°F$ | | |
| Steel | $\sigma_{ys}$ $+70°F$ (ksi) | NDT (°F) | NDT $+60°F$ (°F) | $\sigma_{yd}$ (ksi) | $K_{Id}$ (ksi$\sqrt{\text{in.}}$) | NDT $+120°F$ (°F) | $\sigma_{yd}$ (ksi) | $K_{Id}$ (ksi$\sqrt{\text{in.}}$) |
|---|---|---|---|---|---|---|---|---|
| A212B | 36 | 20 | 80 | 55.9 | 51.1 | 140 | 50.1 | 57.0 |
| A302B | 62 | 0 | 60 | 82.7 | 74.4 | 120 | 77.8 | 79.1 |
| HY-80 | 82 | −150 | −90 | 122 | 116 | −30 | 113 | 126 |

The results of their calculations are presented in Table 15.3 and show the allowable design stress to satisfy the leak-before-break criterion. Note that the general trend in each case is to decrease the allowable design stress by a factor of about 3 as the wall thickness is increased from 1 to 5 in. The same general trend for static design stresses would be expected if $K_{Ic}$ values could be obtained for these steels. $K_{Id}$ values were used because it was easier to approximate these values from the $K_{Id}$ at NDT.

Because of the uncertainties in stress level at intersections and of the effects of residual stress, and as a general conservative approach, the leak-before-break criterion can also be established assuming that the nominal stress, $\sigma$, is equal to the yield stress, $\sigma_{ys}$. Thus, for $\sigma = \sigma_{ys}$, the general

TABLE 15.3    Allowable Design Stresses to Satisfy Leak-Before-Break Criterion
for Three Steels Having Cracks Equal to Wall Thickness

| Steel | Estimated Dynamic Values | B (in.) | $\beta_{Ic}$ | $\dfrac{\sigma}{\sigma_{yd}}$ | Allowable Design Stress, $\sigma_d$, to Satisfy the Leak-Before-Break Criterion (ksi) |
|---|---|---|---|---|---|
| A212 | At 80°F | 1 | 0.84 | 0.65 | 36 |
| | $\sigma_{yd}$ = 55.9 ksi | 2.5 | 0.33 | 0.34 | 19 |
| | $K_{Id}$ = 51.1 ksi$\sqrt{in}$. | 5 | 0.17 | 0.23 | 13 |
| | At 140°F | 1 | 1.29 | 1.10 | 55 (greater than $\sigma_{yd}$) |
| | $\sigma_{yd}$ = 50.1 ksi | 2.5 | 0.52 | 0.45 | 23 |
| | $K_{Id}$ = 57.0 ksi$\sqrt{in}$. | 5 | 0.26 | 0.29 | 15 |
| A302B | At 60°F | 1 | 0.81 | 0.63 | 52 |
| | $\sigma_{yd}$ = 82.7 ksi | 2.5 | 0.32 | 0.33 | 28 |
| | $K_{Id}$ = 74.4 ksi$\sqrt{in}$. | 5 | 0.16 | 0.23 | 19 |
| | At 120°F | 1 | 1.03 | 0.76 | 59 |
| | $\sigma_{yd}$ = 77.8 ksi | 2.5 | 0.41 | 0.39 | 29 |
| | $K_{Id}$ = 79.1 ksi$\sqrt{in}$. | 5 | 0.21 | 0.26 | 20 |
| HY-80 | At −90°F | 1 | 0.912 | 0.69 | 84 |
| | $\sigma_{yd}$ = 122 ksi | 2.5 | 0.364 | 0.36 | 44 |
| | $K_{Id}$ = 116 ksi$\sqrt{in}$. | 5 | 0.182 | 0.24 | 30 |
| | At −30°F | 1 | 1.25 | 0.88 | 99 |
| | $\sigma_{yd}$ = 113 ksi | 2.5 | 0.50 | 0.44 | 50 |
| | $K_{Id}$ = 126 ksi$\sqrt{in}$. | 5 | 0.25 | 0.29 | 33 |

criterion reduces to

$$\frac{\pi\sigma_{ys}^2}{1 - \frac{1}{2}(\sigma_{ys}/\sigma_{ys})^2} = \frac{K_{Ic}^2}{B}\left[1 + 1.4\left(\frac{K_{Ic}^4}{B^2\sigma_{ys}^4}\right)\right]$$

$$2\pi\sigma_{ys}^2 = \frac{K_{Ic}^2}{B} + (1.4)\frac{K_{Ic}^6}{B^3\sigma_{ys}^4}$$

or

$$\frac{1.4K_{Ic}^6}{B^3\sigma_{ys}^4} + \frac{K_{Ic}^2}{B} = 2\pi\sigma_{ys}^2$$

As an example of the use of this criterion, the engineer must first select the nominal yield-strength steel that he or she wishes to use, then determine the wall thicknesses (these two factors might be established on the basis of a general strength criterion to withstand a given internal pressure), and, finally, select the required minimum toughness level necessary to meet the criterion. *Then*, from the steels available, the engineer must select that

**TABLE 15.4  $K_{Ic}$ Values Required to Satisfy Leak-Before-Break Criterion for Yield-Strength Loading**

| Material Yield Strength and Assumed Applied Stress (ksi) | Vessel Thickness, $B$ (in.) | Required $K_{Ic}$ (ksi $\sqrt{\text{in.}}$) |
|---|---|---|
| 40 | $\frac{1}{2}$ | 35 |
| | 1 | 50 |
| | 2 | 70 |
| | 4 | 100 |
| 80 | $\frac{1}{2}$ | 70 |
| | 1 | 100 |
| | 2 | 140 |
| | 4 | 195 |
| 120 | $\frac{1}{2}$ | 105 |
| | 1 | 145 |
| | 2 | 210 |
| | 4 | 295 |
| 160 | $\frac{1}{2}$ | 145 |
| | 1 | 195 |
| | 2 | 280 |
| | 4 | 325 |
| 200 | $\frac{1}{2}$ | 180 |
| | 1 | 245 |
| | 2 | 345 |
| | 4 | 490 |

one or ones which meet the criterion. Final material selection would be on the basis of the foregoing, plus other criteria, such as cost, fabrication, and so on.

The required $K_{Ic}$ values that will satisfy the leak-before-break criterion at yield strength levels ranging from 40 to 200 ksi and for wall thicknesses ranging from $\frac{1}{2}$ to 4 in. are presented in Table 15.4 and Figure 15.17. These results illustrate the significant effect of an increase in thickness or yield strength when selecting materials to satisfy the leak-before-break criterion. Note the similarity of the through-thickness yielding and leak-before-break criterion by comparing the results presented in Table 15.1 and 15.4. Thus this criterion can be used to ensure a high level of notch toughness (elastic-plastic behavior). The required $K_{Ic}$ values for $\sigma_{\text{design}} = \sigma_{ys}/2$ are presented in Table 15.5 and are plotted in Figure 15.18, showing the effect of design stress level on the required $K_{Ic}$ values.

The leak-before-break criterion, like the through-thickness yielding criterion, becomes too conservative as the thickness increases above about 2 inches.

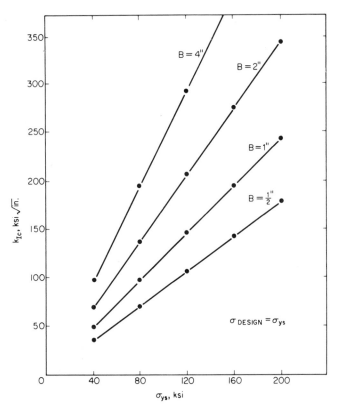

**Figure 15.17**  Effect of yield strength and thickness on $K_{Ic}$ required to satisfy leak-before-burst criterion ($\sigma_{des} = \sigma_{ys}$).

**TABLE 15.5** $K_{\mathrm{Ic}}$ **Values Required to Satisfy Leak-Before-Break Criterion for a Design Stress Equal to One-Half the Yield Strength**

| Material Yield Strength (ksi) | Assumed Applied Stress (ksi) | Vessel Thickness, $B$ (in.) | Required $K_{\mathrm{Ic}}$ (ksi$\sqrt{\text{in.}}$) |
|:---:|:---:|:---:|:---:|
| 40 | 20 | $\frac{1}{2}$ | 23 |
|  |  | 1 | 32 |
|  |  | 2 | 46 |
|  |  | 4 | 65 |
| 80 | 40 | $\frac{1}{2}$ | 46 |
|  |  | 1 | 65 |
|  |  | 2 | 92 |
|  |  | 4 | 130 |
| 120 | 60 | $\frac{1}{2}$ | 69 |
|  |  | 1 | 97 |
|  |  | 2 | 137 |
|  |  | 4 | 194 |
| 160 | 80 | $\frac{1}{2}$ | 92 |
|  |  | 1 | 130 |
|  |  | 2 | 183 |
|  |  | 4 | 259 |
| 200 | 100 | $\frac{1}{2}$ | 114 |
|  |  | 1 | 162 |
|  |  | 2 | 229 |
|  |  | 4 | 324 |

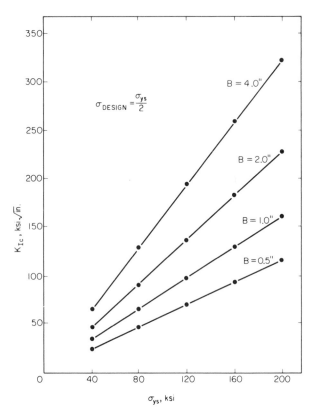

**Figure 15.18** Effect of yield strength and thickness on $K_{Ic}$ required to satisfy leak-before-burst criterion ($\sigma_{des} = \sigma_{ys}/2$).

## 15.8. Comparisons of Various Methods to Establish Fracture Criteria

The preceding sections described various notch-toughness criteria that have been proposed for use in structural applications. Obviously, one of the most desirable criteria is one which is based directly on the appropriate plane-strain critical $K$ values that are used to calculate stress–crack-size trade-offs, as discussed in Chapter 6. However, the *direct* use of linear-elastic fracture mechanics assumes that only materials whose plane-strain fracture toughness values can be determined using the test method described in Chapter 3 are used. As has been discussed before, most structural materials exhibit a *greater* level of performance than plane strain at service temperatures and loading rates. Furthermore, most *criteria* are established to ensure a level of performance greater than plane strain.

    Table 15.6 summarizes the various criteria just discussed as well as their applications. Before selecting a particular criterion, the engineer should

**TABLE 15.6  Comparison of Various Methods to Establish Fracture Criteria**

| Name | General Use | Level of Performance (See Figure 12.1) | Test Specimen Used | Applications |
|---|---|---|---|---|
| $K_{Ic}$, $K_{Ic}(t)$, or $K_{Id}$ | All materials | Plane strain | $K_{Ic}$ standard (ASTM E-399) or $K_{Id}$ nonstandard | Linear-elastic behavior<br>Quantitative stress–flaw-size calculations<br>Aerospace applications<br>Very-high-strength materials<br>Low-temperature applications |
| 15 ft-lb | Low- and medium-strength (30–140 ksi) structural steels | Upper plane strain to lower elastic plastic | ASTM standard (E-23); Charpy V-notch impact | Ship hull steels<br>Bridge steels<br>Applications where moderate level of toughness is required under impact conditions |
| Fracture analysis diagram | All materials exhibiting transition-temperature behavior | All levels; plane strain, elastic plastic, fully plastic | Nil-ductility transition (ASTM E-208); dynamic tear (Mil Standard 1601) | Transition-temperature region<br>Applications where intermediate to high levels (elastic plastic) of dynamic toughness are required; defines NDT, FTE, and FTP criteria |
| Through-thickness yielding | Structural metals | Elastic plastic | $K_{Ic}$ or CVN to estimate $K_{Ic}$ | Accounts for desired increase in toughness with increasing yield strength and plate thickness<br>Large thin plates with edge or through-thickness cracks |

**TABLE 15.6** (*Continued*)

| Name | General Use | Level of Performance (See Figure 12.1) | Test Specimen Used | Applications |
|------|-------------|----------------------------------------|--------------------|--------------|
| Leak-before-break | Primarily pressure vessels to ensure "leaking" before unstable crack growth | Elastic plastic | $K_{Ic}$ or CVN to estimate $K_{Ic}$ or $K_c$ (nonstandard plane stress); $R$-curve (see Chapter 16) | Pressure vessels Design stress–flaw-size calculations possible as a function of vessel thickness |

The following three elastic-plastic methods that can be used to establish criteria are described in Chapter 17; for purposes of comparison, they are included in this table.

| Name | General Use | Level of Performance (See Figure 12.1) | Test Specimen Used | Applications |
|------|-------------|----------------------------------------|--------------------|--------------|
| Crack-tip opening displacement | All structural materials but primarily low- to medium-strength steels | Elastic plastic to fully plastic | Crack-tip opening-displacement bend specimen; British standard—similar to slow-bend $K_{Ic}$ test specimen (E-399) | Entire transition temperature; widely used in United Kingdom and Japan; Alaskan line pipe |
| $R$-curve | All structural materials but primarily high-strength materials | Elastic plastic to fully plastic | Oversize compact tension specimen | Primarily used in studies to measure resistance to stable crack extension; plane-stress behavior |
| $J$-integral | All structural materials | Elastic plastic (directly related to $K_{Ic}$ in elastic region) and upper shelf | Similar to $K_{Ic}$ test specimens | Elastic-plastic behavior; E813 standard requires stable crack growth; quantitative stress–flaw-size calculations possible; nuclear industry |

make sure that the general guidelines discussed in this chapter are thoroughly understood and that the particular criterion used does indeed model the particular structure as close as possible.

Also shown in Table 15.6 are three elastic-plastic test methods that can be used to establish fracture criteria, as described in Chapter 17, as well as the Pellini's Fracture Analysis Diagram (FAD) approach.[11,12]

# References

1. H. G. ACKER, "Review of Welded Ship Failures," *Ship Structure Committee Report, Serial No. SSC-63*, U.S. Coast Guard, Washington, D.C., Dec. 15, 1953.

2. D. P. BROWN, "Observations on Experience with Welded Ships," *Welding Journal*, Sept. 1952, pp. 765–782.

3. *Final Report of a Board of Investigation to Inquire into the Design and Methods of Construction of Welded Steel Merchant Vessels*, GPO, Washington, D.C., 1947.

4. F. JONASSEN, "A Resumé of the Ship Fracture Problem," *Welding Journal, Research Supplement*, June 1952, pp. 316-S–318-S.

5. M. L. WILLIAMS and G. A. ELLINGER, "Investigation of Fractured Steel Plates Removed from Welded Ships," *Ship Structure Committee Report, Serial No. NBS-1*, U.S. Coast Guard, Washington, D.C., Feb. 25, 1949.

6. M. L. WILLIAMS, M. R. MEYERSON, G. L. KLUGE, and L. R. DALE, "Investigation of Fractured Steel Plates Removed from Welded Ships," *Ship Structure Committee Report, Serial No. NBS-3*, U.S. Coast Guard, Washington, D.C., June 1, 1951.

7. M. L. WILLIAMS, "Examination and Tests of Fractured Steel Plates Removed from Welded Ships," *Ship Structure Committee Report, Serial No. NBS-4*, U.S. Coast Guard, Washington, D.C., April 2, 1953.

8. M. L. WILLIAMS and G. A. ELLINGER, "Investigation of Structural Failures of Welded Ships," *Welding Journal, Research Supplement*, Oct. 1953, pp. 498-S–527-S.

9. M. L. WILLIAMS, "Analysis of Brittle Behavior in Ship Plates," *Ship Structure Committee Report, Serial No. NBS-5*, U.S. Coast Guard, Washington, D.C., Feb. 7, 1955. (Also presented in *ASTM STP 158*, American Society for Testing and Materials, Philadelphia, 1954.)

10. E. R. PARKER, *Brittle Behavior of Engineering Structures*, prepared for the Ship Structure Committee, John Wiley, New York, 1957.

11. W. S. PELLINI, 1971 AWS Adams Lecture, *Principles of Fracture-Safe Design, Part I, Welding Journal Research Supplement*, Mar. 1971, pp. 91-S–109-S.

12. W. S. PELLINI, 1971 AWS Adams Lecture, *Principles of Fracture-Safe Design, Part II, Welding Journal Research Supplement*, Apr. 1971, pp. 147-S–162-S.

13. S. T. ROLFE, J. M. BARSOM, and MAXWELL GENSAMER, "Fracture-Toughness Requirements for Steels" presented at the Offshore Technology Conference, Houston, May 18–21, 1969.

14. "Fracture Toughness Testing and Its Applications," *ASTM STP 381*, American Society for Testing and Materials, Philadelphia, 1965.

15. "Plane Strain Crack Toughness Testing," *ASTM STP 410*, American Society for Testing and Materials, Philadelphia, 1967.

16. G. T. HAHN and A. R. ROSENFIELD, "Sources of Fracture Toughness: The Relation Between $K_{Ic}$ and the Ordinary Tensile Properties of Metals," *ASTM STP 432*, American Society for Testing and Materials, Philadelphia, 1968, pp. 5–32.

17. G. T. HAHN and A. R. ROSENFIELD, "Plastic Flow in the Locale of Notches and Cracks in Fe-3Si Steel Under Conditions Approaching Plane Strain," *Ship Structure Committee Report SSC-191*, U.S. Coast Guard, Washington, D.C., Nov. 1968.

18. "Standard Method of Testing for Plane-Strain Fracture Toughness of Metals," *ASTM Standards*, Vol. 03.01, *Metals—Mechanical Testing*, 1985.

19. A. R. ROSENFIELD, P. K. DAI, and G. T. HAHN, *Proceedings of the International Conference on Fracture*, Sendai, Japan, 1965.

20. S. T. ROLFE and S. R. NOVAK, discussion of "What Does an Engineer Need to Know About Measurement of Fracture Toughness When Using High-Strength Structural Steels," ASME Metals Engineering Conference, Houston, April 2–5, 1967.

21. S. T. ROLFE and S. R. NOVAK, "Slow-Bend $K_{Ic}$ Testing of Medium-Strength High-Toughness Steels," *ASTM STP 463*, American Society for Testing and Materials, Philadelphia, 1970.

22. G. R. IRWIN, "Relation of Crack-Toughness Measurements to Practical Applications," *ASME Paper No. 62-MET-15*, presented at Metals Engineering Conference, Cleveland, April 9–13, 1962.

23. G. R. IRWIN, J. M. KRAFFT, P. C. PARIS, and A. A. WELLS, "Basic Aspects of Crack Growth and Fracture," *NRL Report 6598*, Naval Research Laboratory, Washington, D.C., Nov. 21, 1967.

# 16

# Fracture-Control Plans

## 16.1. Introduction

The objective in structural design of large complex structures, such as bridges, ships, pressure vessels, aircraft, and so on, is to optimize the desired performance and safety requirements and cost (i.e., the overall cost of materials, design, fabrication, operation, and maintenance). In other words, the purpose of engineering design is to produce a structure that will perform the operating function efficiently, safely, and economically. To achieve these objectives, engineers make predictions of service loads and conditions, calculate stresses in various structural members resulting from these loads and service conditions, and compare these stresses with the critical stresses in the particular failure modes that may lead to failure of the structure. Members are then proportioned and materials specified so that failure does not occur by any of the pertinent failure modes. Because the response to loading can be a function of the member geometry, an iterative process may be necessary.

Possible failure modes that usually are considered are

1. General yielding or excessive plastic deformation.
2. Buckling or general instability, either elastic or plastic.
3. Subcritical crack growth (fatigue, stress-corrosion, or corrosion fatigue) leading to loss of section or unstable crack growth.
4. Unstable crack extension, either ductile or brittle, leading to either partial or complete failure of a member.

Although other failure modes exist, such as corrosion or creep, the above-mentioned failure modes are the ones that usually receive the greatest attention. Furthermore, of these four failure modes, engineers usually concentrate on only the first two and assume that proper selection of materials and design stress levels will prevent the other two failure modes from occurring. This reasoning is not always true and has led to several catastrophic structural failures. In good structural design, *all* possible failure modes should be considered. In this particular textbook, failure by fracture or subcritical crack growth by fatigue, stress corrosion, or corrosion fatigue are the dominant failure modes to be considered. However, it is assumed that the engineer looks at all possible failure modes and designs structures to prevent failure by any of the possible failure modes.

In the case of brittle fracture or fatigue, many of the fracture-control guidelines that have been followed to minimize the possibility of brittle fractures in structures are familiar to structural engineers. These guidelines include the use of structural materials with good notch toughness, elimination or minimization of stress raisers, control of welding procedures, proper inspection, and the like. When these general guidelines are integrated into specific requirements for a particular structure, they become part of a fracture-control plan. A fracture-control plan is therefore a specific set of recommendations developed for a particular structure and should not be indiscriminantly applied to other structures.

The four basic elements of a fracture-control plan are as follows:

1. *Identification* of the factors that may contribute to the fracture of a structural member or to the failure of an entire structure. Description of service conditions and loadings.

2. *Establishment* of the relative contribution of each of these factors to a possible fracture in a member or to the failure of the structure.

3. *Determination* of the relative efficiency and trade-offs of various design methods to minimize the possibility of either fracture in a member or failure of the structure.

4. *Recommendation* of specific design considerations to ensure the safety and reliability of the structure against fracture. This would include recommendations for desired levels of material performance as well as material selection, design stress levels, fabrication, and inspection.

The development of a fracture-control plan for large structures, such as bridges, airplanes, ships, and so on, is complex. Despite the difficulties, attempts to formulate a fracture-control plan for a given application, even if only partly successful, should result in a better understanding of the possible fracture characteristics of the structure under consideration.

The total useful life of a structural component is generally determined by the time necessary to initiate a crack and to propagate the crack from subcritical dimensions to a critical size. The life of the component can be prolonged by extending the crack-initiation life and/or the subcritical-crack-propagation life. Consequently, the crack-initiation, subcritical-crack-propagation, and unstable-crack-propagation (fracture) characteristics of structural materials, as well as their fracture behavior, are primary considerations in the formulation of fracture-control guidelines for structures.

Unstable crack propagation is the final stage in the useful life of a structural component subject to failure by the fracture mode. This stage is governed by the material toughness, the crack size, and the stress level. Consequently, unstable crack propagation cannot be attributed *only* to material toughness, or *only* to stress conditions, or *only* to poor fabrication, but rather to particular combinations of the foregoing factors. However, if any of these factors is significantly different from what is usually obtained in a particular type of structure, the possibility of failure may increase markedly.

Structural materials that have adequate notch toughness to prevent brittle fractures at service temperatures and loading rates are available. However, when these structural materials are used in conjunction with inadequate design or poor fabrication, or both, the safety and reliability of a structural component cannot be guaranteed because the useful life of a structural component is governed by the time required to initiate a crack and to propagate the crack to terminal conditions (for example, leaks or unstable crack propagation). Thus, the useful life of a structural component depends on the magnitude and fluctuation of the applied stress, the magnitude of the stress-concentration factors and the size, the shape and orientation of the initial discontinuity, the stress-corrosion susceptibility, the fatigue characteristics, and the corrosion-fatigue behavior of the structural material in the environment of interest. Because most of the useful life is expended in initiating and propagating cracks at low values of the stress-intensity factor (Chapters 7–10), an increase in the fracture toughness of the steel may have a small effect on the useful life of a structural component whose primary mode of failure is fatigue. Despite these facts, oversimplification of failure analyses has led some to advocate the philosophy that structures should be fabricated of "forgiving materials." Such materials are characterized as having sufficient notch toughness to fracture in a ductile manner at operating temperatures and under impact loading even though the structure may be subjected to slow or intermediate loading rates. The use of these materials is advocated to "forgive" any mistakes that may be committed by the fabricator, inspector, designer, or user. While this approach to ensure the safety and reliability of structures does work most of the time, it perpetuates a false sense of security, places an unjustifiable burden on the structural material, and unnecessarily increases the cost of the structure.

Another prevalent yet unfounded philosophy advocates that the primary cause of fractures in welded structures is the inherent inferior characteristics of the weldments. These characteristics include yield-strength residual stresses and weld discontinuities. Advocates of this philosophy often neglect to consider environmental effects, cyclic history, and stress redistribution caused by load fluctuation or proof tests, and when an obvious weld discontinuity does not exist in the vicinity of the fracture origin, the cause of failure is attributed to "microweld" defects. Although residual stress and weld discontinuities can contribute to failure, the foregoing oversimplification can lead to an incorrect fracture analysis of a structural failure. The use of oversimplification or gross exaggerations in the analysis of failures, exemplified by these philosophies, can lead to erroneous conclusions. Correct diagnosis and preventive action can be established only after a thorough study of *all* the pertinent parameters related to the specific problem under consideration. An integrated look at all these parameters and their synergistic effect on the safety and reliability of a structure is necessary to develop a fracture-control plan.

As pointed out earlier, the recent development of fracture mechanics has been extremely helpful in synthesizing the various elements of fracture-control plans into more unified quantitative plans than was possible previously. Specifically, fracture mechanics has shown that although numerous factors (e.g., service temperature, material toughness, design, welding, residual stresses) can contribute to brittle fractures in large welded structures, there are three primary factors that control the susceptibility of a structure to brittle fracture, namely,

1. Notch toughness of a material at a particular service temperature, loading rate, and plate thickness.
2. Size of crack or discontinuity at possible locations of fracture initiation.
3. Tensile stress level, including residual stress.

All three factors can be interrelated using concepts of fracture mechanics to predict the susceptibility of a structure to brittle fracture. When the particular combination of stress and crack size in a structure reaches the critical stress-intensity factor for a particular specimen thickness and loading rate, fracture can occur. It is the specific intent of any fracture-control plan to establish the possible ranges of $K_I$ that might be present throughout the lifetime of a structure because of the various service loads and to ensure that the critical stress-intensity factor ($K_c$, $K_{Ic}$, $K_{Ic}(t)$, $K_{Id}$, or other critical values such as $K_{Iscc}$) for the materials used is sufficiently large so that the structure will have a safe life.

One of the key questions in developing a fracture-control plan for any particular structure is how large must the degree of safety and reliability be for the particular structure in question. The degree of safety and reliability needed, sometimes referred to as the factor of safety, is often specified by a code. However, the degree of safety depends on many additional factors such as consequences of failure or redundancy and thus varies even within a generic class of structures.

Accordingly, a fracture-control plan is developed only for the specific structure under consideration and can vary from one which must, in essence, provide assurance of very low probability of service failures to one which may allow for occasional failures during manufacturing or service. An example of the former situation would be a nuclear power plant structure where the consequences of a structural failure are such that even a minor failure may not be tolerable. In the latter case, the consequences of failure might be minimal, and it would be more efficient and economical periodically to maintain and replace parts rather than design them so that no failures occur. An example of a situation where a service failure could be tolerated might be that of the loading bed of a dump truck where periodic inspection would indicate when plates would need to be repaired or replaced because the consequences of failure would be minor.

In commenting on fracture-control plans, Irwin[1] has noted that

> For certain structures, which are similar in terms of design, fabrication method, and size, a relatively simple fracture control plan may be possible, based upon extensive past experience and a minimum adequate toughness criterion. It is to be noted that fracture control never depends solely upon maintaining a certain average toughness of the material. With the development of service experience, adjustments are usually made in the design, fabrication, inspection, and operating conditions. These adjustments tend to establish adequate fracture safety with a material quality which can be obtained reliably and without excessive cost.

## 16.2. Historical Background

Prior to about 1940, fracture-control plans or fracture-safety guidelines did not really exist in a formalized sense because the number of brittle fractures was small. Most large structures were built from low-strength materials using thin, riveted plates with the structural members arranged so that in the event of failure of one plate the fracture was usually arrested at the riveted connections. Thus although there were some exceptions as noted in Chapter 1, most failures were not catastrophic. While it is true that the number of failures in particular structural situations was reduced by various design modifications based on experience from any service failures, the *first general overall fracture-control guideline* was merely to use lower design stress levels. One of the early fracture-control applications was in the boiler and pressure-vessel industry where the allowable stress was decreased continually as a certain percentage of the maximum tensile stress, thereby decreasing the number of service failures in succeeding years.

A *second general guideline to fracture control* was to eliminate stress concentrations as much as possible. In the early 1940s brittle fracture as a potentially serious problem for large-scale structures was brought to the designer's attention by the large number of World War II ship failures. The majority of fractures in the Liberty ships started at the square hatch

.corners of square cutouts at the top of the shear strake. The design changes involving rounding and strengthening of the hatch corners, removing square cutouts of the shear strake, and adding rivets and crack arresters in various locations led to an immediate reduction in the incidence of failures. The use of crack arresters and improved work quality in the tankers reduced the incidence of failures in these vessels. Thus the second general type of fracture-control guideline was that of design improvement to minimize stress concentrations.

The ship-failure problem called attention to general inadequacies of designing to prevent fracture primarily by just limiting stresses or minimizing stress concentrations. The World War II shipbuilding program was the first large-scale use of welding to produce monolithic structures where cracks could propagate continuously from one plate to another, in contrast to the multiple-plate usage for riveted structures. While problems of fracture were apparent prior to the large-scale use of welding, there was evidence that the nature of welded structures provided for continued extension of a fracture, which can result in total failure of a structure. Previous experience with riveted structures indicated that brittle fractures were usually limited to a single plate.

The *third general type of fracture control* was to improve the notch toughness of materials. The first material-control guideline occurred about the late 1940s and early 1950s following extensive research on the cause of the ship failures. The particular fracture-control guideline was the observation that plates having a CVN impact energy value greater than 10–15 ft-lb at the service temperature did not exhibit complete fracture but rather arrested any propagating cracks. Studies indicated that generally the ship fractures experienced during World War II did not occur at temperatures above the 15-ft-lb Charpy V-notch impact energy transition temperature. This observation led to the development of the 15-ft-lb transition-temperature criterion as a means of fracture control. Early in the 1950s development of the nil-ductility transition-temperature test led to another general fracture-control guideline for structural materials, namely, that structural steels have an NDT temperature below their service temperature when loaded dynamically.

Since that time, and particularly during the 1960s, the development of general criteria for prevention of brittle fractures, or at least analyses of methods for designing to prevent brittle fractures, have been numerous. These included the development of the fracture analysis diagram (FAD) by the Naval Research Laboratory, the early use of fracture-mechanics concepts in the Polaris missile motor case failures, and the turbine-rotor generator spin test analyses. In the 1960s and 1970s, rapid development of fracture mechanics as a research tool and eventually as an engineering tool led to the use of fracture-mechanics concepts in the development of fracture criteria for several structures such as nuclear pressure vessels, various aircraft structures, and bridges. In the 1980s fracture-control plans have become required as structures become more complex, for example, the space shuttle and offshore drilling rigs.

However, the three basic elements of fracture control—that is, (1) use a lower design stress, (2) minimize stress concentrations, and (3) use materials with improved notch toughness—have long been known to engineers. Fracture mechanics' basic contribution is to make these guidelines *quantitative* and to show the relative importance of each of these elements.

Although the literature has been dominated with the use of fracture mechanics as a research tool, Irwin[1] has stated that

> the practical objective of fracture mechanics is a continuing increase in the efficiency with which undesired fractures of structural components are prevented. The fracture control plans most commonly used in the past have not possessed a high degree of efficiency. Generally the visible portions of such plans consist in statements of minimum required toughness and in statements of inspection standards. No quantitative connection between these two fracture control elements is employed in establishing these statements. The unseen parts consist mainly in adjustments of design and fabrication. After enough years of experience so that further adjustments of design and fabrication are unnecessary, the fracture control plan is complete. The plan then consists of certain minimum toughness requirements and adherence to certain inspection requirements plus the state-of-art methods of design and fabrication which fracture failure experience showed to be desirable or necessary. Proof testing has the great advantage, where employed, because much of the fracture failure experience necessary for development of the plan tends to occur in the proof test rather than in service. The use of transition temperature based measurements rather than fracture mechanics based measurements of fracture toughness is not a significant disadvantage to the reliability of such plans, once established. However, the efficiency may be low and use of fracture mechanics methods would be helpful in designing modifications toward improved efficiency.

## 16.3. Development of a Fracture-Control Plan

Section 16.1 presented the four general elements of a fracture-control plan to prevent failure by the fracture mode. It should be reemphasized that *all* possible failure modes, for example, buckling, yielding, and corrosion, should be considered during a structural design. Textbooks on design of particular types of structures (bridges, buildings, pressure vessels, aircraft, etc.) should be used to design to prevent failures by buckling, yielding, and so on, and appropriate textbooks should be consulted to design to prevent chemical failures such as by corrosion. The purpose of this textbook is to provide technical information and design guidelines that can be used to prevent failure by fracture or subcritical crack growth leading to fracture.

The four general elements of a fracture-control plan are described in as much detail as possible, realizing that the specific details of any fracture-control plan depend on the particular structure being analyzed.

### 16.3.1. Identification of the Factors That May Contribute to the Fracture of a Structural Member or to the Failure of an Entire Structure. Description of Service Conditions and Loadings

The first step in all structural design is to establish the probable loads and service conditions throughout the design life of a structure. Usually the live loads are specified by codes, for example, American Association of State Highway and Transportation Officials (AASHTO) for bridges, or performance criteria such as operating pressures in pressure vessels or payload requirements in aircraft structures. These types of loadings usually are reasonably well defined, although there are certain types of structures such as ships where the loading is not well defined. Nonetheless, the first step is to establish the probable loads.

From a fracture-toughness viewpoint, a major decision is determining the *rate* at which these loads are applied, since this establishes whether $K_{Ic}$ for slow loading, or a $K_{Ic}(t)$ value for an intermediate loading rate, or $K_{Id}$ for impact loading should be the controlling toughness parameter.

Wind loads, hurricane loadings, sea-wave loadings, and so on are established by field measurements in the particular location of the structure in the world. Also field measurements on similar types of structures are used to estimate the effect of these loads on structural response and behavior. Earthquake loadings for particular regions are based on experience plus various code requirements, as well as a judgment factor related to the degree of conservatism desired for a particular structure.

Repeated or fatigue loading is usually established by the particular design function of the structure. That is, bridges are subjected to fatigue by the movement of vehicles, and thus fatigue must be considered in the design of bridges, whereas the loads on buildings are primarily dead loads so fatigue is usually neglected. Fatigue loadings can be constant amplitude, such as in rotating machinery, or variable amplitude, such as in bridges or aircraft structures. Regardless of the type of loading, fatigue can result in subcritical crack growth by various means, as described in Chapters 7–14. Thus even though the initial flaw size may be small (based on quality of fabrication), the possibility of larger cracks is present when the structure is subjected to repeated loading.

Chemical or environmental factors such as corrosion, stress corrosion, cavitation, and so on must be considered for various structures, depending on the particular design function of the structure. Crack growth by stress-corrosion and corrosion fatigue were described in Chapters 11–13.

From a fracture viewpoint, temperature can have a significant effect on the service behavior of structural members. For those structures fabricated from materials that exhibit a brittle to ductile transition in behavior (primarily the low- to medium-strength structural steels), the minimum service temperature must be established.

Quality of fabrication, which generally controls the initial crack or defect size, is another factor that should be established so that some estimate of possible initial flaw sizes can be made.

Obviously, the inherent notch toughness ($K_{Ic}$, $K_{Ic}(t)$, or $K_{Id}$) of a structural material based on the particular chemistry, heat treatment, and so on is a primary factor that may contribute to the fracture behavior of the material and is mentioned last only because the other factors are often slighted.

The point is that for each structure for which a fracture-control plan is desired, the designer should consider all possible loadings and service conditions *before* the selection of materials or allowable stresses. In this sense, the basis of structural design in all large complex structures should be an attempt to optimize the *desired performance requirements* in terms of the service loadings relative to *cost considerations* so that the probability of failure is low. Accordingly, if the failure mode is brittle fracture, then the critical-$K$-to-yield ratio (at the appropriate loading rate, temperature, and plate thickness) should be selected to minimize the probability of fracture. However, if desired performance requirements are such that the overall weight of the structure must be minimized, then *another* possible "failure mode" is *nonperformance* because of excessive weight, and therefore a higher allowable design stress must be used. Because the allowable design stress is usually some percentage of the yield strength, a higher-yield-strength material usually is specified.

These two requirements of (1) a high critical-$K$-to-yield-strength ratio and (2) a high yield strength ($\sigma_{ys}$) are conflicting, and a compromise often must be reached. However, as long as this analysis is made *prior to* material selection and establishment of design or allowable stresses, the designer has a good chance of achieving a "balanced design" in which the *performance requirements* as well as the *prevention of failure requirements* both are met as economically as possible. A design example for selection of a material for a specific pressure-vessel application based both on performance and minimum weight was presented in Chapter 6 and showed the advantages of the fracture-mechanics approach to design.

Often it is not possible to change materials for a particular structure because of existing codes, past practices, or economics. In these cases, a limited fracture-control plan can still be effective by reducing the design stress levels, restricting the range of operating temperatures, improving fabrication and inspection, and so on. Thus, even though it is desirable to develop complete fracture-control plans during the early design stages, it is not always possible to do so. However, considerable benefits can still be realized by limited fracture-control plans developed at later stages in the design of a structure. In all cases, the designer should identify as closely as possible all service conditions and loadings to which the structure will be subjected.

### 16.3.2. Establishment of the Relative Contribution of Each of These Factors to a Possible Fracture in a Member or to the Failure of the Structure

Fracture mechanics technology has shown that the many factors that can contribute to fracture in large welded structures (for example, service loadings, fatigue, quality of fabrication, loading rate, material toughness), can be incorporated into the three primary factors that control the susceptibility of a structure to brittle fracture, namely, (1) tensile stress level ($\sigma$) and stress range ($\Delta\sigma$), (2) material toughness (preferably measured in terms of $K_c$, $K_{Ic}$, or $K_{Id}$), and (3) crack size, shape, and orientation ($a$). As shown schematically in Figure 16.1, these factors can be related quantitatively using the $K_I$ relationships developed in Chapter 2.

The contribution of the various types of loadings to the possibility of a brittle fracture occurs primarily in the calculation of the maximum value of stress which can occur in the vicinity of a crack. The calculation of these stresses ranges from simple calculations of $P/A$ or $M \cdot y/I$ to extremely complex elasticity solutions for various structural shapes such as plates,

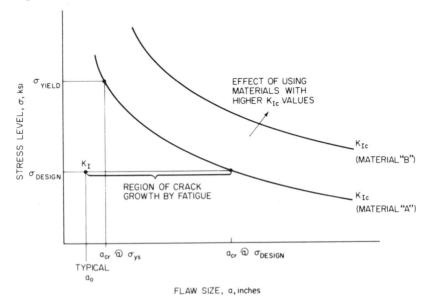

**Figure 16.1** Schematic showing relation among stress, critical flaw sizes, and material toughness.

shells, and box girders. The task of calculating nominal stresses has been made easier by digital computers and new methods of stress analysis, such as the finite element analysis method.

For welded construction, the possibility of residual stresses exists, and the local stress level can be of yield-strength magnitude. In fact, it is often assumed that local yielding exists in the vicinity of stress concentrations. The ductility of structural materials is relied upon to redistribute these stresses so that premature failure does not occur. However, if the possibility of brittle fracture exists, the designer may want to assume that localized yield stresses are present in portions of the structure where cracks can be present and to calculate the resistance to fracture accordingly. That is, determine the critical crack size at yield stress loading (Figure 16.1) and compare it with the maximum possible flaw size based on fabrication and inspection capabilities. However, local yielding can occur in the vicinity of stress concentrations and in regions of tensile residual stresses such as for weldments. This local yielding causes a redistribution of stresses such that during subsequent loadings those regions may exhibit a purely elastic-stress behavior.

Figure 16.2 is a schematic showing the effect of local residual stresses on crack growth as well as the effects of plane-strain or plane-stress conditions on subsequent fatigue-crack growth. Figure 16.3 is a schematic showing the effect of service temperature on critical flaw size, and Figure 16.4 is a schematic showing the design use of $K_{Iscc}$. Referring to Figures 16.1–16.4,

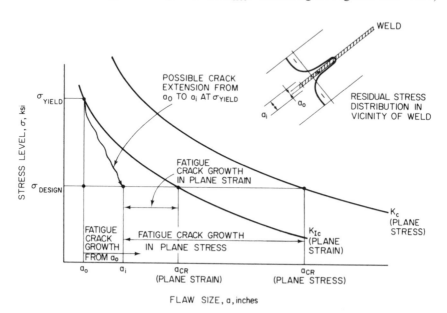

**Figure 16.2** Schematic showing effect of local residual stresses and plane-strain to plane-stress transition (loss of constraint) on fatigue-crack growth.

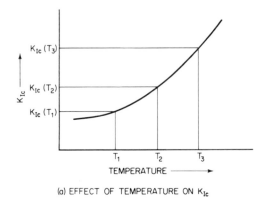

(a) EFFECT OF TEMPERATURE ON $K_{Ic}$

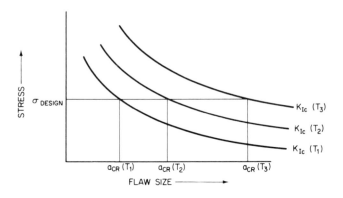

(b) EFFECT OF $K_{Ic}$ ON CRITICAL FLAW SIZE

**Figure 16.3** Schematic showing effect of temperature on critical flaw size. (a) Effect of temperature on $K_{Ic}$; (b) effect of $K_{Ic}$ on critical flaw size.

the following general conclusions can be made with respect to fracture control regarding the relative contribution of stress level, material toughness, and crack size:

1. In regions of high residual stress, where the actual stress can equal the yield stress over a small region, the critical crack size is computed for $\sigma_{yield}$ instead of the design stress, $\sigma_{design}$. If, for the particular structural material being used, both the base metal and the weld metal are sufficiently tough (e.g., material B in Figure 16.1), the critical crack size for full yield stress loading should be satisfactory. Under fatigue loading, a crack can grow out of the residual stress zone, and the critical crack size becomes the value at the design stress level. Note that the "critical crack size" in a particular material is dependent on the particular design stress level and therefore is *not* a material property. However, this conservative approach does not account for

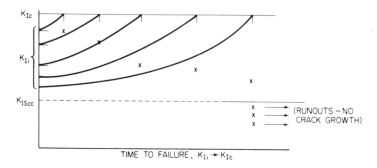

(a) RELATION BETWEEN $K_{Ii}$, $K_{Ic}$, $K_{IScc}$

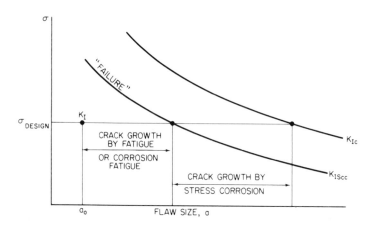

(b) RELATION BETWEEN CRACK GROWTH BY FATIGUE OR
CORROSION FATIGUE AND BY STRESS CORROSION

**Figure 16.4**    Schematic showing design use of $K_{Iscc}$. (a) Relation among $K_{Ii}$, $K_{Ic}$, and $K_{Iscc}$; (b) relation between crack growth by fatigue or corrosion fatigue and by stress corrosion.

erection stresses, or initial loading or for crack blunting caused by high residual stresses.

2. Depending on the level of toughness of the material, any crack which does initiate in the presence of residual stresses ($a_0$, Figure 16.2) could arrest quickly as soon as the crack propagates out of the region of high residual stress. However, the initial crack size for any subsequent fatigue-crack growth will be fairly large ($a_i$, Figure 16.2).

3. For the design stress level, determine the calculated critical crack size. If it is large, compared, for example, to the plate thickness,

subcritical crack growth (by fatigue) should lead to relaxation of the stress ahead of the crack, resulting in plane-stress or elastic-plastic behavior. For this case the critical stress-intensity factor for plane stress, $K_c$, will be greater than $K_{Ic}$ or $K_{Id}$, which is an additional degree of conservatism (Figure 16.2).

4. Materials with low toughness values can still be used if the applied stresses under tensile loads are reduced, or if residual compressive stresses applied to the component by shot-peening or induction case hardening are used to inhibit crack initiation such as the case for gear teeth or landing gears, or if the distribution of stresses causes the initiation and propagation of cracks in a decreasing stress field, or if the crack orientation is not in a critical plane to cause unstable crack extension such as the case for "shelling" in rails.

5. The relative contribution of temperature is, of course, to establish the particular level of toughness to be used in calculating the stress–crack-size trade-offs. That is, as shown in Figure 16.3(a), $K_{Ic}$ (or $K_{Id}$) can vary with temperature for certain structural materials. Thus the $K_{Ic}$ values at temperatures $T_1$, $T_2$, and $T_3$ are different and lead to different values of critical flaw size even though the design stress remains constant, Figure 16.3(b).

6. The relative effect of fatigue loading is to grow a crack of initial size $a_0$ to $a_{cr}$, (Figure 16.1). The number of cycles (or time in the case of stress corrosion) required to grow an existing crack to critical size was discussed in Chapters 7–14.

7. Subcritical crack growth by stress corrosion was discussed in Chapter 11. For materials susceptible to stress corrosion, a good design practice for sustained-load applications is to use $K_{Iscc}$ as a limiting design curve (rather than $K_{Ic}$) so that "failure" is defined as the start of stress-corrosion crack growth, as shown in Figure 16.4(b). This conservative practice is followed because once stress-corrosion crack growth is started, it is only a matter of time until the $K_I$ level reaches $K_{Ic}$ and complete failure occurs. That is, fatigue-crack growth is usually considered to be deterministic because the number of cycles of loading can be estimated. However, the rate of stress-corrosion crack growth is extremely difficult to predict because of the large number of variables such as chemistry of the corrodent, concentration of corrodent at the crack tip, temperature, and so on. Thus, once the $K_{Iscc}$ level is reached, a realistic design approach for sustained-load applications is to consider failure to be imminent.

Using these guidelines, the relative contribution of material properties ($K_{Ic}$, $K_c$, $K_{Ic}(t)$, $K_{Id}$, $K_{Iscc}$, $da/dN$), stress level ($\sigma$), and crack size ($a$) can be established for a given set of service conditions for a particular structure.

### 16.3.3.  Determination of the Relative Efficiency
and Trade-offs of Various Design Methods
to Minimize the Possibility of Either Fracture
in a Member or Failure of the Structure

Previously, it has been established that the three primary factors that control the susceptibility of a structure to brittle fracture are

1. Tensile stress level.
2. Size of crack or severity of discontinuity.
3. Material toughness at the particular temperature, loading rate, and plate thickness.

Thus there are three general design approaches to minimizing the possibility of brittle fracture in a structural member, and each of these is directly related to the preceding factors:

1. Decrease design stress.
2. Minimize initial discontinuities.
3. Use materials with improved notch toughness.

Each of these design approaches has been used by engineers in various types of structures for many years.  The technology of fracture mechanics

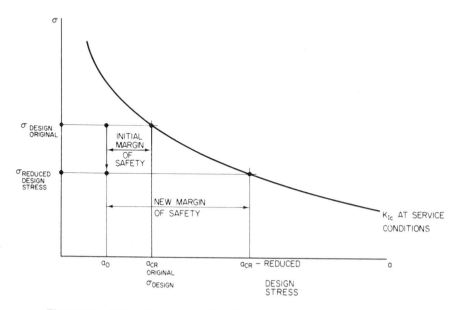

**Figure 16.5**  Schematic showing effect of lowering the design stress on fracture control.

merely makes the process more quantitative. Using fracture-mechanics terminology, fracture control is simply making sure that $K_I < K_c$ (or $K_{Ic}$, $K_{Id}$, etc.) at all times, much like keeping $\sigma < \sigma_{ys}$ to prevent yielding.

For example, Figure 16.5 shows that for the same quality of fabrication and inspection as well as the same critical material toughness, $K_{Ic}$ or $K_c$, reducing the design stress or stress fluctuation leads to a new margin of safety. This new margin of safety can be either a larger margin of safety against fracture or an increased fatigue life because of the possibility of larger subcritical crack growth before failure. Specific examples of this fact were presented in Chapter 5.

Figure 16.6 shows the general effect of improving the quality of fabrication and inspection while using the same design stress level and material. Finally Figure 16.7 shows the general effect of using a structural material with improved notch toughness and the same design stress and quality of fabrication. These examples lead to the general conclusion that because $K_{Ic} \simeq C\sigma\sqrt{a}$, it would be expected that increasing $K_{Ic}$ or decreasing $\sigma$ would have a greater effect on the resistance to fracture than would reducing the initial crack size, $a_0$. Also, it is usually easier to control $\sigma$ or $K_{Ic}$ than $a_0$. However, because the primary parameter that governs the rate of subcritical crack growth is the stress-intensity factor, $K_I$, or the stress-intensity factor range, $\Delta K_I$, raised to a power of 2 or greater, decreasing $\sigma$ or $\Delta\sigma$ results in a much more significant increase in the useful fatigue life of most structural components than does increasing $K_{Ic}$.

The specific fracture-control approach followed for a particular type of structure obviously depends on the service conditions applicable to the

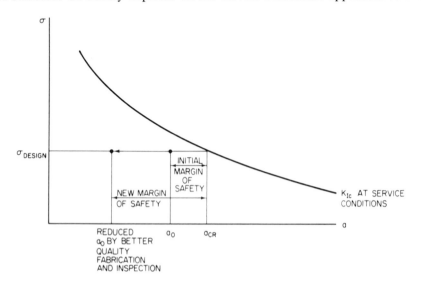

**Figure 16.6**  Schematic showing effect of reducing the initial flaw size on fracture control.

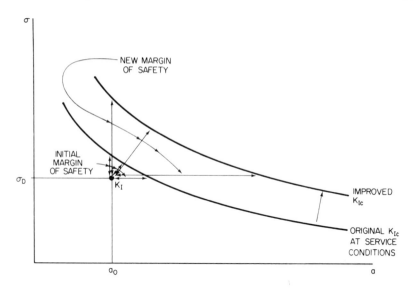

**Figure 16.7**  Schematic showing effect of using material with better toughness (improved $K_{Ic}$) on fracture control.

particular type of structure.  However, several general guidelines are as follows:

1. For cases where the structural loadings are well known, or can be controlled through the use of pressure relief valves such as in pressure vessels, moderate changes in design stress can be relied upon as a primary basis of fracture control.  Reducing the design stress range level has an added benefit in that the fatigue-crack growth is proportional to a power function of the stress-range level and thus the fatigue-crack growth is reduced significantly.

2. Conversely, for structures where the loadings are not well defined, using structural materials with good notch toughness is desirable.

3. For complex welded structures, such as ships or bridges, where the quality of fabrication and inspection techniques are not as well defined as in other types of structures such as aircraft structures, the use of structural materials with some known minimum level of notch toughness is very desirable.

4. If the design stress levels are well known and the quality of fabrication and inspection can be controlled quite closely, such as in the manufacture of small pressure vessels, then materials with relatively low notch toughness can be used safely.

5. For structures where the consequences of failure are such that failures cannot be tolerated, such as the pressure vessel in a nuclear reactor,

or the hull structure of a floating nuclear power plant, or the main structure in an airplane, then *all three* factors should be controlled.

For those cases where fatigue-crack growth is a consideration, such as in bridges, the total useful design life of a structural component can be estimated from the time necessary to initiate a crack and to propagate the crack from subcritical dimensions to the critical size. The life of the component can be prolonged by extending the crack-initiation life and/or the subcritical-crack-propagation life. In an engineering sense, the initiation stage is that region in which a very small initial discontinuity or crack grows to become a measurable propagating crack in fatigue. The subcritical-crack-growth stage is that region in which a propagating fatigue crack follows one of the existing crack-growth "laws," for example, $da/dN = A \cdot \Delta K^m$ as described in Chapters 9 and 10. The unstable crack growth stage is that region in which fatigue-crack growth is *very* rapid, or a brittle fracture occurs, or ductile tearing occurs, resulting in loss of section and failure.

The effect of each of the three primary factors that control the susceptibility of a structure to failure—(tensile stress level ($\sigma$), flaw size ($a$), and material toughness ($K_{Ic}$, $K_{Id}$, CVN, etc.)—on the total life of a structure subjected to fatigue loading may be summarized as follows:

*Tensile stress*   Large effect on life (region I—Figure 16.8) because the rate of fatigue-crack growth is decreased significantly as the applied stress range is decreased ($\sigma_1$ compared with $\sigma_2$, Figure 16.8). Design stress range ($\sigma_{max} - \sigma_{min}$) is the primary factor to control.

*Flaw size*   Large effect on life (region II—Figure 16.8) because the rate of fatigue crack growth for *small* flaws is very low. Quality of fabrication and inspection is the primary factor to control.

*Material toughness*   (i) Large effect on life in moving from plane-strain behavior to elastic-plastic behavior (region III—Figure 16.8). The AASHTO Material Toughness Requirements ensure this level of performance under intermediate rates of loading. The three levels of performance, namely, plane-strain, elastic-plastic, and plastic behavior, are shown schematically in Figure 16.9. For most structural applications, some moderate level of elastic-plastic behavior at the service temperature and loading rate constitutes a satisfactory performance criterion.

(ii) Small effect on life in moving from elastic-plastic behavior to plastic behavior (region IV—Figure 16.8) because the rate of fatigue-crack growth be-

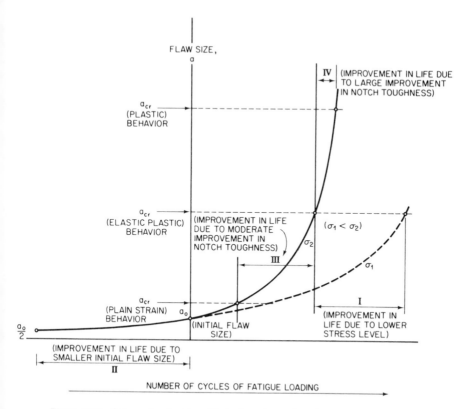

**Figure 16.8**   Schematic showing effect of notch toughness, stress, and flaw size on improvement of life of a structure subjected to fatigue loading.

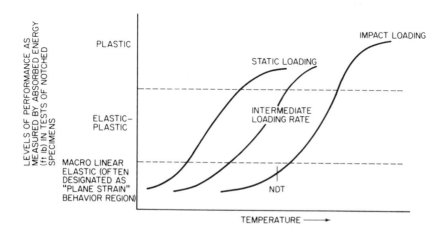

**Figure 16.9**   Schematic showing relation between notch-toughness test results and levels of structural performance for various loading rates.

comes so large that even if the critical crack size ($a_{cr}$) is doubled or even *tripled,* the effect on the remaining fatigue life is small. However, the mode of failure may change.

In summary, use of stress–crack-size–toughness relations such as presented in Figures 16.1–16.8 should be used to determine the general design approach to minimize the possibility of brittle fracture in the particular type of structure being considered. It should be noted that in most structures, all three methods of fracture control are usually used to some degree.

Although the preceding three methods of minimizing the possibility of brittle fracture (*use lower design stress, better fabrication, or tougher materials*) are the basic design approaches, other design methods can be used which are also very effective. These methods include the following:

1. Use *structural materials* whose notch toughness is such that the material does not fail by brittle fracture even under the most severe operating conditions to which the structure may be subjected. The use of HY-80 steel for submarine hull structures is an example of this method. This method is an extreme use of the method just described of using materials with improved notch toughness. However, as shown in Figure 16.8, this method is not that effective for structures subjected to fatigue loading.

2. Provide *multiple-load* fracture paths so that a single fracture cannot lead to complete failure. Multiple-load-path structures should not be confused with redundant structures.

A redundant structure is one in which the laws of statics are insufficient to solve for the loads and stresses, for example, a beam with fixed ends. In contrast, a simply supported single-span bridge structure is nonredundant because the reactions can be determined by the laws of statics. If the geometry of this single-span determinant structure is a single box girder such that failure of the single tension flange leads to collapse of the bridge, then the structure is also a *single-load-path* structure. However, if the geometry consists of eight independent structural shapes with a concrete deck, then the structure is a *multiple-load-path* structure and is much more resistant to complete fracture than the single box girder.

Another example is a truss member composed of one structural shape (e.g., a wide-flange shape in tension) or multiple shapes (e.g., four to ten eye-bar members parallel to each other). The former is a single-load-path structure, while the latter is a *multiple-load-path structure.*

The distinguishing feature is whether or not, in the event of fracture of a primary structural member, the load can be transferred to and carried by other members. If so, the structure is a multiple-load-path one; if not, the structure is a single-load-path one. In this sense, multiple-load-path structures are usually more resistant to failure than single-load-path structures. For example, if a single member fails in a single-load-path structure, the entire structure may collapse, as occurred with the Silver Bridge at Point Pleasant, West Virginia.[2] Conversely, if a single member fails in a multiple-load-path structure, the entire structure may not collapse. This type of behavior was demonstrated in the failure of the Kings Bridge in Australia.[3] That is, at the instant of failure, the failed span in the Kings Bridge contained three cracked girders. One member had cracked while the girders were still in the fabrication shop. A second girder failed during the first winter the bridge was opened to traffic, a full 12 months before the failure of the bridge. Failure of a third girder led to final failure, although architectural concrete sidewalls (which added to the multiple-load paths of the overall structure) prevented complete collapse. Similar examples of the importance of multiple-load paths (as well as adequate inspection) can be shown by analysis of individual failures of other structures during the past several years.

Therefore, assuming that the load can be transferred and carried by other members, the factor of safety applicable to toughness and applied stress should be greater for single-load-path structures than for multiple-load-path structures. If redistribution of load occurs suddenly or if the crack-driving force is not restricted, the fracture behavior of the structure will be governed, primarily, by the dynamic stability of the structure. Under these conditions, fracture toughness above a given value and the design stresses may have negligible effect on the fracture behavior of the structure. The arrest of unstable crack propagation in these cases must be based on a fully dynamic theory that considers the dynamic behavior of the structure and of the propagating crack. Unfortunately, fully dynamic theories that can be applied to engineering structures, such as bridges, ships, offshore rigs and so on, are yet to be developed. The best method to minimize the consequences of failure under these conditions is by proper selection of crack arresters and by design methods that reduce the driving force.

Fatigue-crack propagation in multiple-load-path structures may occur essentially under constant deflection, which corresponds to a decreasing stress-field intensity, rather than under constant load. Thus, cracks propagating in multiple-load-path structures may eventually arrest and, although individual structural components will have to be replaced or repaired, complete failure of the structure is not expected to occur so long as redistribution of load can occur.

3. Provide *crack arresters* so that in the event that a crack should initiate, it will be arrested before complete failure occurs. Crack arresters or a fail-safe philosophy (i.e., in the event of "failure" of a member, the structure is still "safe") have been used extensively in the aircraft industry as well as the shipbuilding industry. A properly designed crack-arrest system must satisfy four basic requirements, namely,

  a. Be fabricated from structural materials with a high level of notch toughness.

    b. Have an effective local geometry, such as is shown in Figure 16.9.

    c. Be located properly within the overall geometry of the structure.

    d. Be able to act as an energy-absorbing system or deformation-restricting system in cases such as gas-pressurized line pipes where the high rate of loading continually deforms the structure.

    4. Control fatigue-crack growth. If fatigue-crack growth is possible, the same general design methods described for fracture control also apply because of the correspondence between a fracture-control plan based on restriction of fatigue-crack initiation, fatigue-crack propagation, and fracture toughness of materials and a fracture-control plan based on fabrication, inspection, design, and fracture toughness of materials. The fact that fatigue-crack initiation, fatigue-crack propagation, and fracture toughness are functions of the stress-intensity fluctuation, $\Delta K$ (as described in Chapters 7–14), and of the critical stress-intensity factor, $K_{Ic}$—which are in turn related to the applied nominal stress (or stress fluctuation), the crack size, and the structural configuration—demonstrates that a fracture-control plan for various structural applications in which fatigue is a consideration depends on the same factors in which just fracture is a consideration. This type of fracture control was described in Figure 16.8.

    5. Reduce the rate of load application. Although not really a design method to prevent brittle fracture, the fact that many structures are loaded at slow to intermediate loading rates where their notch toughness is quite satisfactory even though their notch toughness is considered to be very low on the basis of *impact* loading-rate tests leads to an understanding of why there are so few brittle fractures in older structures (Figure 16.10).

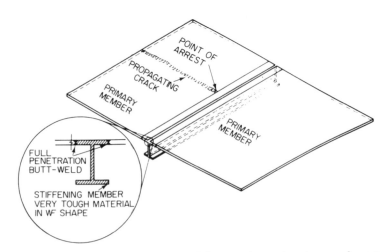

**Figure 16.10** Schematic showing out-of-plane crack arrester where crack must grow through entire wide-flange($_{WF}$) shape before it can continue.

The recent understanding of the *loading-rate* shift (Chapters 4–6) leads to this conclusion. Thus, if the structure can be designed such that it is loaded *slowly,* so that the controlling toughness parameter is $K_{Ic}$ rather than $K_{Id}$, the possibility of fracture is reduced considerably. Thus, control of the loading rate is a very effective method of fracture control.

### 16.3.4. Recommendation of Specific Design Considerations to Ensure the Safety and Reliability Against Fracture. This Would Include Recommendations for Desired Levels of Material Performance as Well as Material Selection, Design Stress Levels, Fabrication, and Inspection

This step in the development of a fracture control plan should be the easiest if the preceding steps have all been followed and all technical and performance factors have been properly considered. However, it is usually the most difficult because it involves making the difficult decision of just how much fracture control can be justified *economically.*

Although not widely stated as such, the general nature of structural design of all types of large complex structures is such that the possibility of failure does exist, particularly when the possibility of overloads or natural loads such as earthquakes must be considered. The goal of good sound engineering design is to optimize structural performance consistent with economic considerations such that the probability of structural failure is minimized. To eliminate *completely* the possibility of structural failure (where failure can be yielding, instability, brittle fracture, etc.) is essentially impossible if structures are to perform their design function economically. For the yielding and general buckling modes of failure, sufficient experience has been acquired so that the designer can usually prevent these types of failure quite economically, and the various codes and specifications reflect this fact. However, this is not always true for brittle fractures. Thus the very real question of just how much notch toughness is necessary *economically* to perform satisfactorily *functionally* is very difficult to answer.

All factors related to toughness criteria must be considered, and an economic decision must be made, based on technical input obtained regarding the level of performance to be specified. Once this level of performance (plane strain, elastic plastic, fully plastic) has been established for the service loadings and conditions, then some material-toughness property must be specified. Even if this toughness level can be specified directly in terms of $K_{Ic}$, or $K_{Id}$ (which is usually not the case unless plane-strain behavior is specified), setting material specifications on the basis of $K_{Ic}$ or $K_{Id}$ values is essentially prohibitive, both from an economic and a technical viewpoint. The $K_{Ic}$ test is just too complex and too expensive to conduct on a routine quality-control basis, and there is no standardized $K_{Id}$ test. Also, as is often the case, the desired toughness level is above that which can be measured using linear-elastic fracture-mechanics test methods. Hence some auxiliary test specimen must be used, based on the various correlations described in Chapter 5, or elastic-plastic fracture mechanics must be used, Chapter 17.

In fact, most fracture criteria and material specifications for various types of structures are based on concepts of fracture mechanics. However, the actual material requirements (in addition to strength, ductility, etc.) generally are specified in terms of CVN impact, or NDT temperature, requirements rather than $K_{Ic}$ or $K_{Id}$ requirements. Two such examples of the development of fracture-control plans and material specifications are the ASME Code Section III requirements for nuclear pressure vessels[3] and the AASHTO Guide Specification for Steel Bridge Members.[4] In the first case, the material-toughness requirements are specified using NDT and CVN impact specimens, and in the second case the toughness requirements are specified using CVN impact specimens. These two fracture-control plans are described in the following two sections.

## 16.4. Fracture-Control Plan for Nuclear Pressure Vessels

### 16.4.1. General Background

In January 1971, the Pressure Vessel Research Committee formed a task group to review[4] criteria for ferritic-material-toughness requirements for pressure-retaining components of the reactor coolant pressure boundary operating below 700°F. These criteria, when used in addition to the stress limits allowed by the ASME Code, should permit the establishment of safe procedures for operating nuclear reactor components under normal, upset, and testing conditions; emergency and faulted conditions should be considered on a case-by-case basis.

It is almost self-evident that the presence of flaws, defects, and other sources of high stress concentration in structural members can decrease the load-carrying capability. However, for engineering design analysis purposes, this general observation must be made quantitative and must include the following:

1. The relation of the strength reduction to the size and geometry of the flaw.
2. The important material properties which determine the magnitude of the strength reduction.
3. The influence of the geometry of the structural member and the kind and the rate of applied loading.
4. The effect of nonapplied loads such as residual stresses from welding.

For applications involving structural grades of carbon and alloy steels, an engineering design procedure based on the "transition-temperature" concept, largely developed in the 1940s, has been widely adopted and used with very good success in preventing the occurrence of unexpected failures of the brittle-fracture type. In essence, this procedure is based on the marked change in fracture behavior and characteristics from "low" to "high" values with increasing temperature when certain varieties of test specimens are used. Specifically, this transition behavior is best exhibited when the specimen contains a sharp notch or a crack. The Charpy V-notch and the drop weight tests are among the most commonly used specimens for this purpose. The resulting transition temperatures can be defined for the Charpy specimen in terms of an energy level, the amount of deformation, or the fracture appearance. For the drop weight test, the nil-ductility temperature is defined in ASTM Specification E-208 and is the temperature above which the specimen will sustain a specific amount of deformation without cracking to either edge of the specimen.

The transition-temperature procedure is applied in design by permitting loading on the structure only at temperatures higher than the transition temperature by a specified temperature increment. The increment to be used is determined by one or a combination of several methods: (1) correlation with service experience, (2) correlation with model tests, and (3) engineering judgment.

Although the transition-temperature procedure has the virtues of simplicity and successful experience, it does not inherently have the capability of quantitatively treating some of the items mentioned earlier. The linear-elastic fracture-mechanics concept has these capabilities. This concept is based on an elastic analysis of the stresses in the neighborhood of the tip of a sharp crack in solids. Thus, it is essentially a stress-analysis–based approach, as described in Chapter 2.

The basic analysis assumes elastic behavior of the stresses in the body, including the region around the crack. It is found that the stress distribution near the crack tip is always the same and that the stress magnitudes all depend on a single quantity termed the *stress-intensity factor,* generally designated as $K$. For loadings which produce an opening mode of displacement between the crack surfaces, the stress-intensity factor is further designated as $K_I$.

Equations are available for the calculation of the value of $K_I$ in terms of the applied load and the crack size for various combinations of crack geometries, structural member dimensions and shapes, and type of applied loading. All these equations have an identical general form, namely,

$$K_I = C\sigma\sqrt{\pi a} \qquad (16.1)$$

where $\sigma$ = nominal (gross section) tensile stress perpendicular to the plane of the crack at the location of the crack, psi.

$a$ = characteristic crack dimension, such as crack depth for surface cracks, in.

$C$ = nondimensional constant whose value depends on the crack geometry, the ratio of the crack size to the size of the structural member, and the type of loading (tension, bending, etc.).

The basic premise of linear-elastic fracture mechanics is that unstable propagation of an existing flaw will occur when the value of $K_I$ attains a critical value designated as $K_{Ic}$ (Chapter 6). The $K_{Ic}$, generally called the fracture toughness of the material, is a temperature-dependent material property (Chapter 4). The implementation of the fracture-mechanics concept as a fracture-control design procedure then consists of two essential steps:

1. Determine the $K_{Ic}$ properties of the material using suitable test specimens and conditions.

2. Determine the actual or anticipated flaw size in the structural component and calculate the limiting value of stress which will keep the value of $K_I$ in the component less than $K_{Ic}$. A safety factor may be applied to the stress, and a safety margin may also be incorporated in the flaw size by choosing a reference flaw size considerably larger than the actual or anticipated maximum flaw size.

In the case of structural carbon and alloy steel materials, it has been found that fracture-toughness properties are dependent on temperature and on the loading rate imposed on the flaw. The loading-rate effect leads to several categories of fracture toughness values as follows:

$K_{Ic}$    = static initiation fracture toughness obtained under slow loading conditions.

$K_{Ic}(t)$ = intermediate initiation fracture toughness obtained under monotonically increasing load that causes fracture in $t$ seconds.

$K_{Id}$    = dynamic initiation fracture toughness obtained under impact loading rates.

$K_{Ia}$    = crack-arrest fracture toughness obtained from the value of $K_I$ under conditions where a rapidly propagating fracture is arrested within a test specimen.

In structural steels, experimental evidence shows that $K_{Id}$ and $K_{Ia}$ generally are less than $K_{Ic}$ or $K_{Ic}(t)$ at a given temperature. Later in this section, a quantity designed as $K_{IR}$, the reference value of the fracture toughness, is used; it is the lower bound of the available $K_{Ic}$, $K_{Ic}(t)$, $K_{Id}$, and $K_{Ia}$ values for the pertinent materials.

Experimental investigations of the temperature effect on the fracture toughness of carbon and alloy steels of the lower- and intermediate-yield-strength grades show that $K_{Ic}$, $K_{Ic}(t)$, $K_{Id}$, and $K_{Ia}$ all exhibit a sharp increase with temperature over a relatively small temperature range. Further, analytical studies and correlations have indicated that this temperature range can be related to the transition temperature of each material item as determined by the Charpy and/or drop weight tests. A specific example of this kind of correlation is the indexing of the temperature dependence of $K_{IR}$ to the drop weight nil-ductility transition temperature. Other empirical correlations between fracture toughness ($K_{Ic}$, $K_{Ic}(t)$, $K_{Id}$) and Charpy test values (ft-lb) have also been developed.

The foregoing discussion has been a capsule summary of the transition-temperature and fracture-mechanics concepts which form useful bases for a fracture-control design procedure for nuclear power plant components. In summary, the toughness requirements and design criteria developed in the PVRC document make use of both concepts. Fracture mechanics provides the formulas and equations for defining a relation between the type and magnitude of the applied stresses, the crack-tip stress-intensity factor, and the size and geometry of the postulated flaw. The transition-temperature approach provides methods for estimating the material's fracture toughness and for verifying the temperature dependence of the fracture toughness.

There are *four major regions of reactor coolant boundary requiring criteria,* as follows:

1. Vessel shell and head regions remote from geometric discontinuities. These may be subjected to sufficient radiation to affect the mechanical and toughness properties.
2. Pumps and valve bodies and vessel regions at or near nozzles and flanges not subjected to significant radiation.
3. Piping.
4. Bolting.

In this condensation, only region 1 will be considered. For the other regions, reference is made to the original fracture-control plan, *Welding Research Council Bulletin No. 175.*

### 16.4.2. Allowable $K_{IR}$ Values

Figure 16.11 shows the relationship that can be conservatively expected among the critical (or reference, $R$) stress-intensity factor, $K_{IR}$, and a temperature which is related to the reference nil-ductility temperature, $RT_{NDT}$. This curve is based on the lower bound of dynamic and crack-arrest $K_{Id}$ and $K_{Ia}$ values measured as a function of temperature on specimens of A533 B-1, A508-2, and A508-3 steels that have yield strengths of 50 ksi or less. The data used for the derivation of Figure 16.11 are described in a later section on derivation of the $K_{IR}$ curve. No available data points for static,

$$K_{IR} = 1.223 \exp\left\{0.0145\,(T-(RT_{NDT}-160))\right\} + 26.777$$

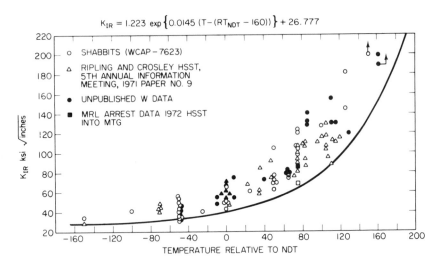

**Figure 16.11**  Derivation of curve of reference stress-intensity factor ($K_{IR}$).

dynamic, or arrest tests fall below the curve. An analytical approximation to the curve in Figure 16.11 is

$$K_{IR} = 26.777 + 1.223 \exp\{0.0145[T - (RT_{NDT} - 160)]\} \qquad (16.2)$$

### 16.4.3. Testing Requirements and Radiation-Induced Changes

The following acceptance tests for materials in shell or head regions remote from discontinuities are recommended:

1. Charpy V-notch tests are required for materials having a nominal section thickness greater than $\frac{1}{2}$ in. Drop weight tests are required when geometrically possible.

2. Properties are to be determined in the direction of the maximum general primary membrane stress. This is the hoop direction in a cylindrical shell, but in a spherical shell or head, the specified properties are required in both tangential-longitudinal and tangential-transverse directions. Therefore, for spherical shells or heads, test specimens should be oriented in the direction normal to the principal direction in which the metal was worked (other than thickness direction), so that the specimens represent the generally lower toughness of that orientation.

3. The NDT temperature, $T_{NDT}$, must be determined by drop weight tests (ASTM E-208—69) using P-1–, P-2–, or P-3–type specimens from the quarter-thickness location. At the temperature $T_{NDT} + 60°F$, conduct

three CVN tests using specimens from the quarter-thickness location. If all CVN values are $\geq 40$ mils of lateral expansion, then $T_{NDT}$ is the reference temperature, $RT_{NDT}$. If any value is less than 40 mils, conduct CVN tests to determine the temperature, $T_{40}$, at which all CVN values are $\geq 40$ mils. Then the reference temperature $RT_{NDT} = T_{40} - 60°F$. Thus $RT_{NDT}$ is the higher of $T_{NDT}$ or $T_{40} - 60°F$. An alternative requirement of 35 mils of lateral expansion, but not less than 50 ft-lb absorbed energy, would also seem reasonable. If the material is to be subjected to significant radiation, data from surveillance specimens must be obtained and used as required by specifications of ASTM E-185—70—"Recommended Practice for Surveillance Tests on Structural Materials in Nuclear Reactors."

4. The procedure must be applied to specimens representing base material, heat-affected zone, and weld metal.

5. The impact properties described in item 3 must be met throughout the life of the component. Since neutron radiation increases the transition temperature and reduces the fracture toughness of ferritic steels, determination must be made of the margin between initial and end-of-life properties to provide assurance that the required properties will be obtained at the quarter-thickness location throughout life. Further, since the degree of radiation-induced change varies widely depending on the neutron fluence, the temperature at the point of irradiation, and the relative sensitivity to radiation of the particular steel component, a systematic surveillance program as required by ASTM E-185—70 is essential. Maintenance of the required properties in the materials of lowest (limiting) end-of-life toughness, from those in test 4, and with due consideration of the variable factors noted must be verified by surveillance specimens. The radiation-induced temperature change $\Delta T_{NDT}$ and the maintenance of the required impact properties at $RT_{NDT} + 60°F$ at the quarter-thickness location is to be established by tests of surveillance specimens conducted in accordance with the methods of ASTM. It is not feasible to name a specific fluence above which a surveillance program is mandatory. Available data show, however, that if the fluence at the inner wall is less than $10^{18}_{nvt}$ no significant radiation damage is to be expected. For higher values of end-of-life fluence, the omission of a surveillance program should be justified by showing that for the particular lots of base and weld metal being used the reactor vessel shell will not become more limiting than other parts of the vessel.

The recommended tests and acceptance standards given were based on the following considerations:

1. The use of both drop weight and CVN tests gives protection against

the possibility of errors in the conducting of tests or the reporting of test results.

2. The CVN requirements are given in terms of *lateral expansion* rather than absorbed energy because this provides protection from variation in yield strength from initial heat treatment and the change in yield strength produced by irradiation.

3. The requirement of 40 mils lateral expansion or, alternatively, 35 mils and 50 ft-lb at $RT_{NDT}$ + 60°F throughout the life of the component provides assurance of adequate fracture toughness at upper shelf temperatures.

4. The CVN test at $T_{NDT}$ + 60°F serves to weed out nontypical materials such as those which might have a low transition temperature but an abnormally low energy absorption on the upper shelf.

### 16.4.4. Reference Flaw Size

After the $K_{IR}$ that the material is capable of providing has been established, the next step in the procedure for obtaining allowable loading is the choice of a postulated defect size. For normal and upset conditions, as defined in the ASME code (Section III, NB-3113) in shell or head regions remote from discontinuities, a very conservative defect size is recommended. It consists of a sharp surface flaw, perpendicular to the direction of maximum stress, having a depth of one-fourth of the section thickness over most of the thickness range of interest. Since the nondestructive test procedures used are not strongly dependent on thickness above 12 in. and below 4 in., the postulated depth is assumed to be constant beyond those values, as shown in Figure 16.12. The assumed shape of the defect is semielliptic, with a length six times its depth. Consideration was also given in some

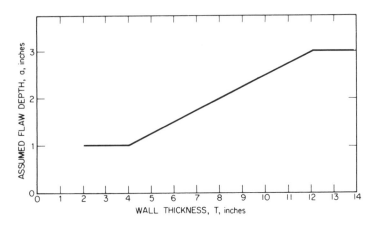

**Figure 16.12**   Reference flaw depth.

cases to the more conservative crack of infinite length. This assumed quarter-thickness defect provided the rationale for using the quarter-thickness location for test specimens and the same location for the radiation damage calculations.

If the ASME code's 2 percent radiography standard is taken at face value, it may be noted that the postulated defect is 12.5 times as large in linear dimensions as the code-allowable dimensions and thus has over 150 times the area. It is not safe to assume that no defects larger than those that the code allows will ever occur, but it does seem reasonable to assume that with the combination of examinations (generally radiography) required by Section III of the ASME code and the volumetric examination (generally ultrasonic mapping) required by Section XI, there is a very low probability that a defect larger than about four times that allowed by the code will escape detection. The postulated defect is about three times as large in linear dimensions and has about ten times the area of even that conservative value.

Other defect shapes and locations might have been chosen to develop the recommended criteria. Buried defects are more apt to escape detection than surface defects, but when not too near the surface, they produce only about half the $K_I$ value for a given size, so the use of a buried defect would not be as conservative. Surface defects longer than six times their depth would produce higher $K_I$ values, but even if an infinitely long defect had been postulated, the recommended properties and calculation methods would still provide protection against a defect having a depth one-sixth instead of one-fourth of the section thickness.

*In summary, the postulation of the large defect size, the use of dynamic test data to give the lower-bound toughness value, and the safety factors (discussed shortly) on the allowable stresses combine to provide a large degree of conservatism in the safety and reliability of nuclear vessels.*

### 16.4.5. Allowable Loading

The final step in the calculation of allowable loading is the calculation of the applied $K_I$ and comparison with the $K_{IR}$ value at the appropriate temperature. In a vessel shell or head remote from discontinuities, the significant loadings are (1) general primary membrane stress due to pressure and (2) thermal stress due to thermal gradient through the thickness during startup and shutdown. Effects of residual stress are not included in the recommended procedure because

1. Peak values in a postweld heat-treated component usually are less than 20 percent of the yield strength.
2. Service stresses and radiation effects both tend to reduce residual stresses during the life of the component.
3. Conservatisms throughout the whole recommended procedure and the

safety factors applied appear to be ample to cover any incalculable adverse effects.

Therefore, the procedure for calculating the allowable system pressure during startup consists of adding the $K_I$ corresponding to the membrane tension produced by pressure to the $K_I$ produced by thermal gradient at a desired startup and shutdown rate and requiring that the sum of these $K_I$ values not exceed the available $K_{IR}$ of Figure 16.11 at each temperature. Methods for calculating these two $K_I$ values for the assumed maximum credible defect are described in the PVRC document. These methods result in the curves shown in Figures 16.13 and 16.14. The $K_I$ produced by pressure is

$$K_{I \text{ pressure}} = [M_t] \times [\text{general primary membrane stress}]$$

Figure 16.13 shows $M_t$ as a function of wall thickness. The four lines shown for the $M$ values are to be used for stress values 0.1, 0.5, 0.7, and 1.0 times the material yield strength. The $M$ values for other stress levels can be interpolated linearly with sufficient accuracy since they are not strongly affected by the stress level. For calculated stresses above yield level, linear-elastic fracture mechanics is not applicable.

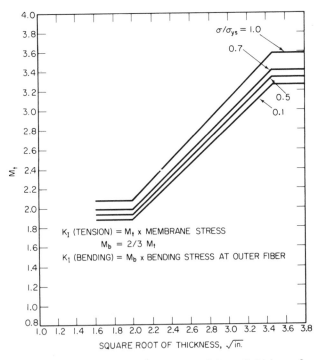

**Figure 16.13** $M_t$ and $M_b$ versus the square root of the wall thickness for semi-elliptical surface flaw $\frac{1}{4}$T deep and $1\frac{1}{2}$T long.

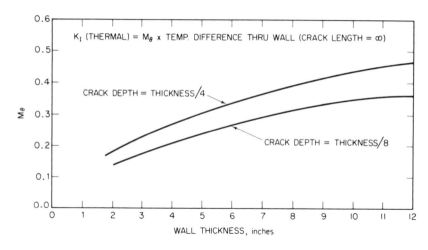

**Figure 16.14**    Stress-intensity factor caused by thermal stress.

The $K_I$ produced by thermal gradient is

$$K_{I\ therm.} = [M_\theta] \times [\text{temperature difference through wall}]$$

Figure 16.14 shows $M_\theta$ as a function of wall thickness, based on the following assumptions:

1. An assumed shape of the temperature gradient.
2. The shutdown starts after a steady-state temperature has been attained.
3. The rate of change of temperature is of the magnitude associated with startup and shutdown, that is, less than about 100°F/hr. The result would be overly conservative if applied to rapid temperature changes.
4. The postulated defect has a depth one-fourth of the wall thickness and infinite length. (The values of $M_\theta$ for ⅛-thickness crack depth are also shown in Figure 16.14.)

For some purposes, Figure 16.15 may be found more convenient to use than Figure 16.14. It is based on a radius-to-thickness ratio of 10 and uses cooling rate directly, thus avoiding the need to calculate the temperature difference across the wall.

The governing operating condition is most apt to be *shutdown* because decreasing temperature produces tensile stress in the portion of the reactor vessel wall which is subjected to the greatest radiation damage.

The requirement for startup, shutdown, and normal and upset operating conditions is that, throughout the life of the component at each temperature, the following inequality holds:

$$K_{I\ pressure} + K_{I\ therm.} \leq K_{IR}    \text{from Figure 16.11}$$

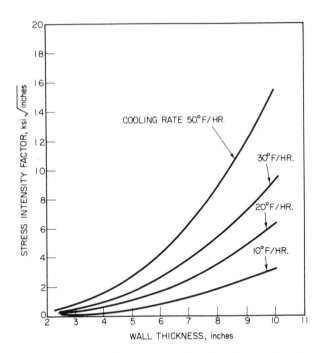

**Figure 16.15**  Stress-intensity factor caused by thermal stress for cylinders with radius/thickness = 10.

### 16.4.6. Safety Factors

The methods described here for the calculation of allowable loadings do not include any safety factors beyond the use of conservative assumptions for the available $K_{IR}$ and the defect size. The application of additional "factors of safety" is a function of the code-writing organizations rather than the research organizations such as PVRC, since the choice of safety factors depends on the chosen relationship between economics and conservatism and on the consequences of failure. If additional conservatism is desired, it is suggested that in the equation for normal operating conditions $K_{I\,pressure}$ be multiplied by a factor in the range 1.5–2.0. Due to its secondary and self-relieving nature, the factor for $K_{I\,therm.}$ can be smaller. In addition, the values given here for $K_{I\,pressure}$ are based on a crack length six times its depth. Therefore, it should be adequate to use a safety factor of 1.0–1.3 on $K_{I\,therm.}$.

During hydrotest of the vessel and/or system, there is no thermal stress, and there is less uncertainty regarding the stress values. Also, the consequences of failure are less severe. Therefore, less conservatism is justifiable, and the suggested safety factor for the system hydrotest after installation should be in the range 1.25–1.5.

The suggested requirement for a shop hydrotest of an individual component is that the test be made at a temperature 60°F higher than the reference temperature $RT_{NDT}$.

### 16.4.7. Derivation of $K_{IR}$ Curve

In general, tests to determine $K_{Ic}$, $K_{Ic}(t)$, $K_{Id}$, and $K_{Ia}$ are expensive and time consuming, but the advantages of using linear-elastic fracture mechanics for fracture analysis are so strong that many ways have been developed to estimate critical stress-intensity factors from data from other tests. Enough data exist on A533, Grade B, Class 1, and A508 steels to relate actual measured $K_{Ic}$ values to results from other tests with a very high degree of confidence.

One of the more commonly used fracture tests is the drop weight test. The NDT determined from the drop weight test represents the upper temperature where brittle fracture occurs under defined conditions of strain and flaw size. Because conditions and geometry are defined, the critical stress-intensity factor can be calculated, and this has been done by Irwin and others. The relationship between dynamic yield strength ($\sigma_{yd}$) and $K_{Id}$ resulting from this work is

$$K_{Id} = (\text{factor}) \times \sigma_{yd}$$

where the factor is between 0.5 and 0.78 (Chapter 5) and a value of 0.6 is commonly used.

The development of the $K_{IR}$ versus temperature curve (where temperature is relative to the NDT) was done by plotting all known data— $K_{Id}$, $K_{Ia}$, and low-temperature $K_{Ic}$ versus the temperature relative to the drop weight NDT of the same plate or forging. A lower-bound curve was drawn and the values at NDT (i.e., $T - \text{NDT} = 0$) checked against the analytical relationships discussed earlier. The value of $K_{IR}$ at NDT on this curve is equal to $0.55\sigma_{yd}$ of the steels tested. The resulting curve, therefore, represents a very conservative assumption as to the critical stress-intensity versus temperature properties of materials similar to those tested, as related to the measured NDT.

Figure 16.11 shows the $K_{IR}$ versus NDT-adjusted temperature curve, on which are plotted all pertinent data available. Only $K_{Id}$ and $K_{Ia}$ data are shown, because $K_{Ic}$ values are invariably much higher, so they do not affect the lower bound.

To facilitate computer calculations, the equation representing this curve is also shown, and actual values for reference are as follows:

$$K_{IR} = 26.777 + 1.223 \exp\{0.0145[T - (\text{NDT} - 160)]\}$$

| $RT_{NDT}$ | $K_{IR}(ksi\sqrt{in.})$ |
|---|---|
| NDT − 160 | 28 |
| NDT − 140 | 28 |
| NDT − 120 | 29 |
| NDT − 100 | 30 |
| NDT − 80 | 31 |
| NDT − 60 | 32 |
| NDT − 40 | 34 |
| NDT − 20 | 36 |
| NDT | 39 |
| NDT + 20 | 43 |
| NDT + 40 | 49 |
| NDT + 60 | 56 |
| NDT + 80 | 66 |
| NDT + 100 | 80 |
| NDT + 120 | 98 |
| NDT + 140 | 121 |
| NDT + 160 | 153 |
| NDT + 180 | 196 |

## 16.5. Fracture-Control Plan for Steel Highway Bridges

### 16.5.1. Introduction

Steel bridges have an excellent service record extending over millions of operational service years. However, the collapse of the Point Pleasant Bridge[2] in the United States as well as other localized fractures in bridges has led to an increasing concern about the possibility of catastrophic fractures in steel bridges. This concern is accentuated by a growing awareness by material scientists and design engineers of the use of fracture-mechanics methodology in other applications and the increasing emphases on safety and reliability of structures.

Most bridge structures in existence perform safely and reliably. The safety and reliability of these structures have been achieved by improving the weak links that were observed during the life of each component in the system. Present specifications on material, design, and fabrication generally are based on correlations with service experience. The comparatively few service failures in steel bridges indicate that steel properties, design, and fabrication procedures used for bridges basically are satisfactory.

Notch-toughness requirements often are developed to be used in conjunction with good design, fabrication, and inspection procedures, without being specific as to how "good" procedures are defined. The American Association of State Highway and Transportation Officials approach to fracture control has been to specify the materials, design, fabrication (welding), construction, inspection, and maintenance in four separate specifications as follows:

1. Standard specifications for highway bridges.
2. Standard specifications for transportation materials and methods of sampling and testing.
3. Standard specifications for welding of structural steel highway bridges.
4. Manual for maintenance inspection of bridges.

The controls described in these specifications have worked very well, and the incidence of bridge failures has been extremely low throughout the history of the AASHTO organization. When failures have occurred, they have usually been attributed to the fact that the AASHTO specifications were not followed.

In the 1970s, AASHTO recognized the need for a more specific fracture-control plan for nonredundant steel bridges and developed the "Guide Specifications for Fracture Critical Non-Redundant Steel Bridge Members."[5] Specifically this guide requires additional controls on the material notch toughness, welding, and inspection of fracture-critical bridge members compared with redundant bridge members. Fracture-critical members (FCMs) or member components are tension members or tension components of members whose failure would be expected to result in collapse of the bridge. The guide specifications are to be used in conjunction with all existing AASHTO requirements and are based on numerous research studies designed to translate research into engineering practice.

Based on considerable research and study, including a steel-industry sponsored research program,[6,7] the Committee on Bridges and Structures of the American Association of State Highway and Transportation Officials adopted the Charpy V-notch (CVN) impact-toughness requirements for bridge steels in fracture-critical members as shown in Table 16.1. These requirements for primary tension members of bridge steels were based primarily on the information presented in the subsequent sections.

### 16.5.2. Fracture-Control Plan for Fracture-Critical Members

A well-conceived fracture-control plan recognizes that the performance of a structure or a structural component is governed not only by material properties but also by the design, fabrication, erection, inspection, and final use of the structure. The objective in developing fracture-toughness specifications for a structure should be to establish the necessary and sufficient toughness values that ensure the adequacy of the material for the intended application. In that sense, the AASHTO fracture-control plan for steel highway bridges really consists of the four separate specifications for design,

**TABLE 16.1  Base Metal Charpy V-Notch Requirements for Fracture-Critical Members in Steel Highway Bridges**

| AASHTO | ASTM Designation | Thickness (in.) | Zone 1[b] | Zone 2[c] | Zone 3[d] |
|---|---|---|---|---|---|
| M 183 | A 36 | Up to 1½ in. inclusive | 25 @ 70°F | 25 @ 40°F | 25 @ 10°F |
| " | " | Over 1½ in. up to 4 in. | 25 @ 70°F | 25 @ 40°F | 25 @ −10°F |
| M 223 & | A 572* & | Up to 1½ in. inclusive | 25 @ 70°F | 25 @ 40°F | 25 @ 10°F |
| M 222 | A 588* | Over 1½ in. to 2 in. inclusive mechanically fastened or welded | 25 @ 70°F | 25 @ 40°F | 25 @ −10°F |
| | | Over 2 in. to 4 in. inclusive mechanically fastened or welded | 25 @ 70°F | 25 @ 40°F | 25 @ −10°F |
| | | Over 2 in. to 4 in. mechanically fastened | 25 @ 70°F | 25 @ 40°F | 25 @ −10°F |
| | | Over 2 in. to 4 in. inclusive welded | 30 @ 70°F | 30 @ 40°F | 30 @ −10°F |
| M 244 | A 514 & | Up to 1½ in. inclusive mechanically fastened or welded | 35 @ 0°F | 35 @ 0°F | 35 @ −30°F |
| | A 517 | Over 1½ in. to 2½ in. mechanically fastened or welded | 35 @ 0°F | 35 @ 0°F | 35 @ −30°F |
| | | Over 2½ in. to 4 in. mechanically fastened | 35 @ 0°F | 35 @ 0°F | 35 @ −30°F |
| | | Over 2½ in. to 4 in. welded | 45 @ 0°F | 45 @ 0°F | Not permitted |

[a]The CVN-impact testing shall be "P" plate frequency testing in accordance with AASHTO T-243 (ASTM A673). For Zone 3 requirements only, Charpy impact tests are required on each plate at each end. The Charpy test pieces shall be coded with respect to heat/plate number and that code shall be recorded on the mill-test report of the steel supplier with the test result. If requested by the Engineer, the broken pieces from each test (three specimens, six halves) shall be packaged and forwarded to the Quality Assurance organization of the state. Use the average of three (3) tests. If the energy value for more than one of three test specimens is below the minimum average requirements, or if the energy value for one of the three specimens is less than two-thirds (2/3) of the specified minimum average requirements, a retest shall be made and the energy value obtained from each of the three retest specimens shall equal or exceed the specified minimum average requirements.

[b]Zone 1: Minimum Service Temperature 0°F and above.

[c]Zone 2: Minimum Service Temperature from −1°F to −30°F.

[d]Zone 3: Minimum Service Temperature from −31°F to −60°F.

*If the yield strength of the material exceeds 65 ksi, the temperature for the CVN value for acceptability shall be reduced by 15°F for each increment of 10 ksi above 65 ksi. The yield strength is the value given in the certified "Mill Test Report."

materials, welding, and maintenance currently used by AASHTO designers in addition to the recently developed "Guide Specifications for Fracture Critical Non-Redundant Steel Bridge Members."

Analysis of bridge test results[8-10] indicates that the maximum loading rates observed in bridges are closer to slow-bend loading rates than to impact loading rates. The loading times in bridges are greater than 1 sec, which corresponds to a strain rate of less than $10^{-3}$ sec$^{-1}$ on the elastic-plastic boundary in the vicinity of the crack tip. Thus, a strain rate of $10^{-3}$ sec$^{-1}$ can be used as a conservative measure of the maximum strain rate for bridges.

The $K_{Ic}$ data obtained by testing A36 and A572 Grade 50 steels at a strain rate of $10^{-3}$ sec$^{-1}$ indicate that fracture does not occur under plane-strain conditions when the test temperature is greater than about $-80°F$ ($-62°C$) (Figures 16.16 and 16.17). The fracture-toughness transition region

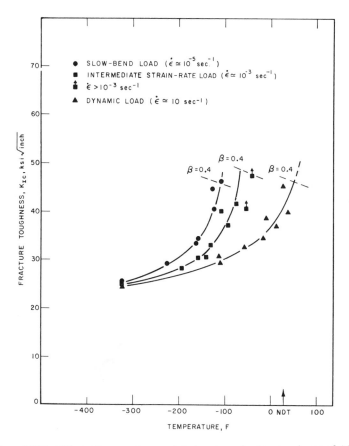

**Figure 16.16**    Effect of temperature and strain rate on fracture toughness of A36 steel.

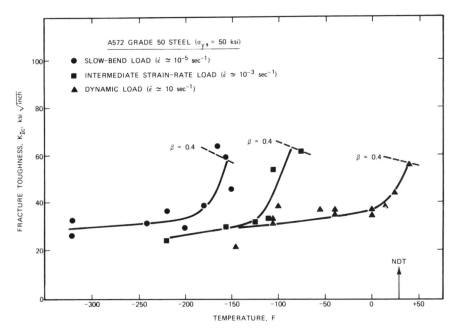

**Figure 16.17**   Effect of temperature and strain rate on fracture toughness of A572 Grade 50 steel ($\sigma_{ys}$ = 50 ksi).

(the temperature region bounded by the lower end of the fracture-toughness transition temperature and the highest temperature at which fracture does occur under plane-strain conditions) for the steels investigated is on the order of 50°F (28°C). A fracture-toughness transition temperature of $-130$°F ($-90$°C) corresponded to a nonplane-strain fracture behavior at $-80$°F ($-62$°C) for these steels tested at the intermediate strain rate of $10^{-3}$ sec$^{-1}$.

One of the requirements used in the development of the AASHTO toughness requirements was that fracture of a 50-ksi (345-MN/m$^2$) yield-strength steel subjected to the intermediate strain rate does not occur under plane-strain conditions. This requirement should be satisfied when the fracture-toughness transition temperature of the steel under the intermediate strain rate occurs at about $-80$°F. Because $K_{Ic}$ tests are expensive and difficult to conduct, and because of the apparent correspondence between $K_{Ic}$ test results and CVN test results, the CVN test was selected as the reference test for the AASHTO fracture-toughness requirements. The fracture-toughness transition temperature is the temperature at which the fracture toughness of the steel begins to increase rapidly from plane-strain behavior to fully ductile behavior. The CVN test results (Figure 16.18) show that this transition behavior at 15 ft-lb under the intermediate rate of loading at $-80$°F should ensure a nonplane-strain fracture behavior at the minimum operating temperature of $-30$°F ($-34$°C) for a 50-ksi yield-strength steel.

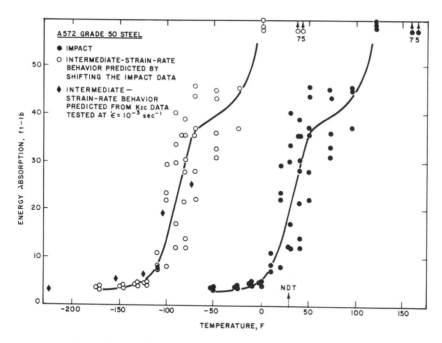

**Figure 16.18**   Charpy V-notch energy-absorption behavior for impact loading and intermediate strain-rate loading of standard CVN specimens.

The use of 15 ft-lb as a reference at $-80°F$ to ensure elastic-plastic behavior at $-30°F$ corresponds to a critical $K$ value equal to about 50 ksi $\sqrt{\text{in}}$. This value of $K$ is about twice the lowest value obtained at the beginning of the fracture-toughness transition and, therefore, is well into rather than at the beginning of the fracture-toughness transition.

Although the intermediate-loading-rate test is the test that more properly describes the expected service performance of bridge steels, the impact-loading-rate test is much easier to conduct and analyze and is less expensive than an intermediate-loading-rate CVN test. Consequently, the difference in fracture-toughness behavior at the two strain rates was used to develop the toughness values in terms of the impact test rather than an intermediate-loading-rate test. The temperature shift between the CVN and $K_{Ic}$ curves of a 50-ksi yield-strength steel tested at a strain rate of $10^{-3}$ sec$^{-1}$ and at an impact strain rate of 10 sec$^{-1}$ was on the order of 120°F (67°C) (Figure 16.18). Consequently, the requirement of a 15-ft-lb CVN impact value at $+40°F$ (4°C) is equivalent to a 15-ft-lb CVN value under intermediate strain rate at $-80°F$ ($-62°C$), which in turn corresponds to a nonplane-strain fracture behavior at the assumed minimum operating temperature of $-30°F$ ($-34°C$). Thus, a CVN fracture-toughness requirement of 15 ft-lb at $+40°F$ was imposed on all primary tension members of 50-ksi yield-strength steels for bridge applications. This same requirement was also imposed on all primary tension members of bridge steels having yield strengths less than 50 ksi, which is a conservative requirement for these steels. To account for possible scatter in test results and to obtain an increased level of notch toughness for fracture-critical members, a 25-ft-lb requirement was established for FCMs, as shown in Table 16.1.

The 25-ft-lb CVN impact-toughness requirement at $+40°F$ for steels of 50-ksi yield strength or less was based on a $-30°F$ minimum operating temperature. The preceding procedure can be used to develop toughness requirements for any minimum operating temperature. The resulting toughness requirement for 50-ksi yield-strength steels is 25 ft-lb at a test temperature that is $70°F$ ($39°C$) higher than the specified minimum operating temperature. Thus the CVN test temperatures and the minimum operating temperatures are linearly related. To minimize the proliferation of a variety of testing temperatures, and the resulting problems in the design and fabrication of steel bridges, the variable testing temperatures were comprehended by establishing three zones of service temperatures and providing temperatures and CVN values for each zone. The three zones of service temperatures and the corresponding test temperatures for bridge steels for 50-ksi yield strength or less are presented in Table 16.1.

The general relations between service temperatures and test temperatures for A36 steel satisfying the requirements of each of the three service-temperature zones are shown in Figure 16.19. These results show that, because of the loading-rate shift, CVN-toughness levels greater than 25 ft-lb are expected at intermediate loading rates approximately $70°F$ below the impact testing temperature. In terms of the NDT temperature measured using drop weight test specimens under impact loading, the minimum service temperature is approximately $70°F$ below NDT.

The specifications of the American Society for Testing and Materials for A572 Grade 50 and A588 steels require a minimum yield-strength value of 50 ksi. Consequently, these steels may have yield strengths that are significantly higher than 50 ksi. The data in Figure 16.20 show that the magnitude of the temperature shift between static and impact loading rates decreased with increased yield strength. The magnitude of the decrease in the temperature shift is about $15°F$ ($8°C$) for every 10-ksi ($69\text{-MN}/\text{m}^2$) increase in yield strength. To ensure the same fracture behavior for A572 Grade 50 and A588 steels having yield strengths significantly greater than 50 ksi, the CVN test temperature was decreased incrementally as the yield strength increased. The CVN requirements for zones 1, 2, and 3 were restricted to A572 Grade 50 and A588 steels having yield points between 50 and 65 ksi, inclusive. When the yield strength of these steels exceeds 65 ksi ($448 \text{ MN}/\text{m}^2$), the temperature for the CVN value for acceptability was reduced by $15°F$ for each increment of 10 ksi above 65 ksi.

The foregoing philosophy, which is based on fracture-mechanics con-

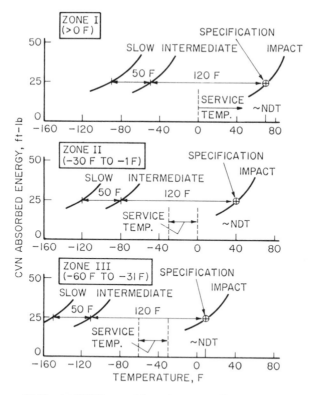

**Figure 16.19**    AASHTO material-toughness specifications for A36 steel.

cepts, was used to develop toughness requirements for bridge steels of 100-ksi (689-MN/m$^2$) yield strength (ASTM A514 and A517). These steels show a temperature shift of 60°F (33°C) between static and impact loading rates (Figures 16.21 and 16.22). Moreover, increasing the design stress (which generally requires a higher-yield-strength steel) results in more stored elastic energy in a structure. Thus, the fracture toughness of the steel should also be increased to ensure the same degree of safety against fracture as the structure with the lower design stress. The resulting fracture-toughness requirements for high-strength bridge steels range from 30 to 45 ft-lb for increased strength level and plate thickness as presented in Table 16.1.

As shown in Table 16.1, the magnitude of the CVN impact requirements and the difference between minimum service temperature and test temperature vary depending upon yield strength and plate thickness, consistent with the discussions in Chapter 4. For the highest-strength steel, almost no credit is given for the temperature shift, resulting in an added degree of conservatism.

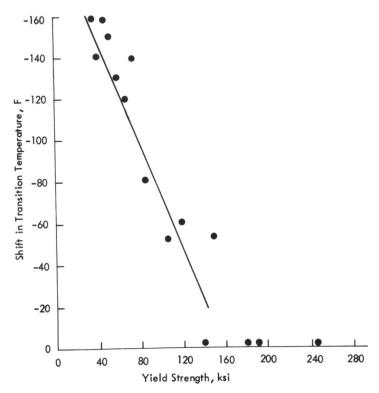

**Figure 16.20** Effect of yield strength on shift in transition temperature between impact and slow-bend $K_{Ic}$ or CVN tests.

In summary, the basis for the AASHTO material-toughness specification is fracture mechanics. However, because the desired level of performance is outside the range of linear-elastic fracture-mechanics behavior and because toughness values cannot be currently measured by existing fracture-mechanics tests, correlations between $K_{Ic}$ and CVN test results were used to establish the material-toughness requirements shown in Table 16.1. These toughness requirements depend on the existence of a strain-rate shift to obtain the desired toughness levels at a service (intermediate) loading rate 70°F (21°C) below the actual specification loading rate (impact), which has been verified by several investigators, as described in Chapter 4.

### 16.5.3. Fracture Tests of Welded Beams

To verify the adequacy of the toughness requirements presented in Table 16.1, beam specimens of A36 steel and A572 Grade 50 steel were used to study the fracture behavior of simulated bridge members under extreme service conditions. The specimens were designed to include two common structural details that adversely affect fatigue and fracture strength: (1) a cover-plate end and (2) a transverse stiffener. The cover-plate end is

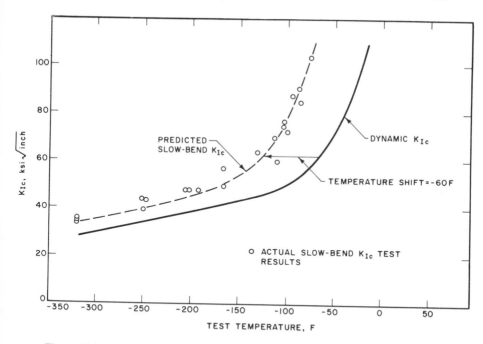

**Figure 16.21**    Use of CVN test results to predict the effect of loading rate on $K_{Ic}$ for A517 Grade F steel.

the most severe common structural detail with respect to fatigue[11] and probably also fracture.[12]

Beam specimens of A36 and A572 Grade 50 steels were subjected to a cyclic stress range of 21 ksi (145 MG/m$^2$) for 100,000 cycles or more. This loading corresponds to the maximum allowable fatigue loading specified by AASHTO[9] for cover-plate ends in either steel but is much more severe than the cyclic loadings measured in actual bridges.[11]

After the specimens had been subjected to cyclic loading, some were cooled to $-30°F$ ($-34°C$) and then loaded under an impulse load to the maximum allowable AASHTO bending stress, which was 20 ksi (138 MN/m$^2$) for A36 steel and 27 ksi (186 MN/m$^2$) for A572 Grade 50 steel. If the specimen did not fail, it was then subjected to impulses of 36 ksi (248 MN/m$^2$), the specified minimum yield point of A36 steel, and 50 ksi (345 MN/m$^2$), the specified minimum yield point of A572 Grade 50 steel. The total time for the 20-ksi impulse was approximately 1 sec—about the same time as for the impulses observed in field measurements of truck loadings.[8] Since the stress levels in the field measurements were generally below 6 ksi, the strain rates in the test impulse were well above the strain rates observed in the field. In short, the test temperature was below the minimum temperature expected to occur in actual highway bridges in the continental United States, the loading rates were well above the loading rates observed

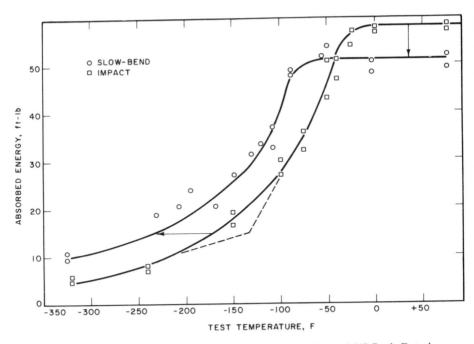

**Figure 16.22**   Slow-bend and impact CVN test results for an A517 Grade F steel.

in field measurements, and the beams were subjected to the highest allowable cyclic stresses and the maximum number of cycles specified by AASHTO for cover-plate ends at that stress level. Thus, the fracture stress was determined for nonredundant (single-load-path) bridge members containing the *most severe* common structural detail and for the *most severe combination* of temperature, strain rate, and prior fatigue loading that could reasonably be expected to occur in actual highway bridges in the continental United States.

All the simulated bridge members tested contained fatigue cracks at the end of the cover plate which had initiated and propagated when the specimens were subjected to the full allowable fatigue loadings permitted by AASHTO. However, these simulated highway-bridge members had sufficient fracture toughness under the most severe test temperatures and strain rates to prevent premature fracture when the members were subjected to the maximum design stress permitted by AASHTO at that time—20 ksi for A36 steel and 27 ksi for A572 Grade 50 steel. Specifically, the nominal stresses at actual fracture (which exclude residual stresses and stress in-tensification caused by stress concentrations) were equal to about 34 ksi (234 MN/m$^2$) and 46 ksi (317 MN/m$^2$) for the A36 and the A572 Grade 50 steels, respectively; these values are close to the specified minimum yield point of the respective steels. Thus, the simulated highway bridge members fractured at stress levels about 70 percent higher than the maximum design stress permitted by AASHTO at that time.

Under the extreme prior fatigue-loading conditions, which far exceeded those usually encountered in actual bridges, cracks developed at the cover-plate ends and ranged in depth from 0.03 to 0.375 in. (0.76 to 9.5 mm). Fatigue-life estimates suggest that the useful fatigue life of the beams had been almost exhausted. In service, members with cracks about 0.4 in. (10.2 mm) deep would soon fail by fatigue regardless of the level of fracture toughness of the steel. Thus, an increase in fracture toughness above the values measured in the steels investigated would not significantly improve the service performance of such bridge members as discussed previously.

The preceding discussion should not imply that crack sizes significantly larger than 0.4 in. cannot exist in actual bridge components. That such large cracks can be tolerated is explained by the redundancy built into the design of many bridges, by the propagation of fatigue cracks in a decreasing stress field, by the small stresses applied to actual bridges compared with the maximum design stresses, and by the extremely low probability that the severest detail, lowest temperature, and highest strain rate will occur simultaneously.

In summary, the test results suggest that a beam with an end-welded cover plate subjected to 100,000 cycles of an applied 21-ksi (145-MN/m²) stress range would be expected to contain a crack about 0.4 in. (10.2 mm) deep and with a length equal to the width of the cover plate. The results of the fracture tests showed that these cracks would not result in fracture of the beams under the maximum allowable design stress, even when the beams are subjected to the maximum loading rate that occurs in bridges and the minimum operating temperature. The test results also indicated that a substantial increase in the fracture toughness of the steel would have resulted in a negligible increase in the useful life of the tested beams.

## 16.6. Comprehensive Fracture-Control Plans— George R. Irwin

Dr. George Irwin has prepared the following general comments on fracture-control plans,[1] which are reprinted in their entirety.

For certain structures, which are similar in terms of design, fabrication method, and size, a relatively simple fracture control plan may be possible, based upon extensive past experience and a minimum adequate toughness criterion. It is to be noted that fracture control never depends solely upon maintaining a certain average toughness of the material. With the development of service experience, adjustments are usually made in the design, fabrication, inspection, and operating conditions. These adjustments tend to establish adequate fracture safety with a material quality which can be obtained reliably and without excessive cost. A fair statement of the basic philosophy of fracture control for such structures might be as follows. Given that the material possesses strength properties within the specified limits, and given that the fracture toughness lies above a certain minimum requirement (Nil-Ductility Temperature, Fracture Appearance Transition Temperature, Plane-Strain or Plane-Stress Crack Toughness), then it is assumed that past experience indicates well enough how to manage design, fabrication, and inspection so that fracture failures in service occur only in small tolerable numbers.

With the currently increasing use of new materials, new fabrication techniques, and novel designs of increased efficiency, the preceding simple fracture control philosophy has tended to become increasingly inadequate. The primary reason is the lack of suitable past experience and the increased cost of paying for this experience in terms of service fracture failures. Indeed, modern technology is beginning to exhibit situations with space vehicles, jumbo-jet commercial airplanes, and nuclear power plants for which not even *one* service fracture failure would be regarded as acceptable without consequences of disaster proportions. Consideration must be given, therefore, to comprehensive plans for fracture control such as one might need in order to provide assurance of zero service fracture failures. A review of the fracture control aspects of a comprehensive plan may be advantageous even for applications such that the required degree of fracture control is moderate. One reason for this would be that an understanding of how to minimize manufacturing costs in a rational way is assisted when we assemble all of the elements which contribute to product quality and examine their relative effectiveness and cost. In the present case, the quality aspect of interest is the degree of safety from service fractures.

After these introductory comments, it is necessary to point out that the concept termed comprehensive fracture control plan is quite recent and cannot yet be supported with completely developed illustrations. We know in a general way how to establish plans for fracture control in advance of extensive service trials. However, until a number of comprehensive fracture control plans have been formulated and are available for study, detailed recommendations to guide the development of such plans for selected critical structures cannot be given.

The available illustrative examples of fracture control planning which might be helpful are those for which a large number of the elements contributing to fracture control are known. At least in terms of openly available information, these examples are incomplete in the sense that the fracture control elements require collection, re-examination with regard to relative efficiency, and careful study with regard to adequacy and optimization. Substantial amounts of information relative to fracture control are available in the case of heavy rotating components for large steam turbine generators, components of commercial jet airplanes (fuselage, wings, landing gear, certain control devices), thick-walled containment vessels for BW and PW cooled nuclear reactors, large diameter underground gas transmission pipelines, and pressure vessel components carried in space vehicles. Certain critical fracture control aspects of these illustrative examples are as follows.

A. Large Steam Turbine Generator Rotors and Turbine Fans
   1. Vacuum de-gassing in the ladle to reduce and scatter inclusions and to eliminate hydrogen.
   2. Careful ultra-sonic inspection of regions closest to center of rotation.
   3. Enhanced plane-strain fracture toughness.
B. Jet Airplanes
   1. Crack arrest design features of the fuselage.
   2. Fracture toughness of metals used for beams and skin surfaces subjected to tension.
   3. Strength tests of models.
   4. Periodic re-inspection.
C. Nuclear Reactor Containment Vessels
   1. Quality uniformity of vacuum degassed steel.
   2. Careful inspection and control of welding.
   3. Uniformity of stainless cladding.
   4. Proof testing.
   5. Investigations of low cycle fatigue crack growth at nozzle corners and of cracking hazard from thermal shock.
D. Gas Transmission Pipelines
   1. Adequate toughness to prevent long running cracks.
   2. High-stress-level, in-place, hydrotesting.
   3. Corrosion protection.
E. Spacecraft Pressure Vessels
   1. Surface finishing of welds so as to enhance visibility of flaws.
   2. Heat treatment to remove residual stress and produce adequate toughness within given limits of strength.
   3. Adjustment of hydrotesting to assure adequate life relative to stable crack growth in service.

In a large manufacturing facility, the inter-group cooperation necessary to achieve successful fracture control on the basis of a comprehensive fracture control plan may require special attention. In general, the comprehensive plan will contain various elements pertaining to Design, Materials, Fabrication, Inspection, and Service Operation. These elements should be directly or indirectly related to fracture testing information. However, the coordination of the entire plan to ensure its effectiveness is not a priori a simple task. The following outline lists certain fracture control tasks under functional headings which might, in some organizations, imply separate divisions or departments.

I. Design
   A. Stress distribution information.
   B. Flaw tolerance of regions of largest fracture hazard due to stress.
   C. Estimates of stable crack growth for typical periods of service.
   D. Recommendation of safe operating conditions for specified intervals between inspection.
II. Materials
   A. Strength properties and fracture properties.

$\sigma_{YS}$, $\sigma_{UTS}$, $K_{Ic}$, $K_c$.

$K_{Iscc}$ for selected environments.

$da/dN$ for selected levels of $\Delta K$ and environments.

   B. Recommended heat treatments.

   C. Recommended welding methods.

  III. Fabrication

    A. Inspections prior to final fabrication.

    B. Inspections based upon fabrication control.

    C. Control of residual stress, grain coarsening, grain direction.

    D. Development or protection of suitable strength and fracture properties.

    E. Maintain fabrication records.

  IV. Inspection

    A. Inspections prior to final fabrication.

    B. Inspections based upon fabrication control.

    C. Direct inspection for defects using appropriate non-destructive evaluation (NDE) techniques.

    D. Proof testing.

    E. Estimates of largest crack-like defect sizes.

  V. Operations

    A. Control of stress level and stress fluctuations in service.

    B. Maintain corrosion protection.

    C. Periodic in-service inspections.

From the above outline, one can see that efficient operation of a comprehensive fracture control plan requires a large amount of inter-group coordination. If a complete avoidance of fracture failure is the goal of the plan, this goal cannot be assured if the elements of the fracture control plan are supplied by different divisions or groups in a voluntary or independent way. It would appear suitable to establish a special fracture control group for coordination purposes. Such a group might be expected to develop and operate checking procedures for the purpose of assuring that all elements of the plan are conducted in a way suitable for their purpose. Other tasks might be to study and improve the fracture control plan and to supply suitable justifications, where necessary, of the adequacy of the plan.

## References

1. GEORGE R. IRWIN, private communication.
2. Point Pleasant Bridge, National Transportation Safety Board, "Collapse of U.S. 35 Highway Bridge, Point Pleasant, West Virginia, December 15, 1967," *Report No. NTSB-HAR-71-1*, Washington, D.C., 1971.
3. R. B. MADISON and G. R. IRWIN, "Fracture Analysis of Kings Bridge, Melbourne," *Journal of the Structural Division, ASCE, 97*, No. ST9, Sept. 1971.
4. PVRC Ad Hoc Task Group on Toughness Requirements, "PVRC Recommendations on Toughness Requirements for Ferritic Materials," *WRC Bulletin No. 175*, Aug. 1972.
5. AASHTO, *Guide Specifications for Fracture Critical Non-Redundant Steel Bridge Members*, Sept. 1978.

6. J. M. Barsom, J. F. Sovak, and S. R. Novak, "AISI Project 168-Toughness Criteria for Structural Steels: Fracture Toughness of A36 Steels," *U.S. Steel Corporation Research Laboratory Report 97.021-001(1)*, May 1, 1972 (American Iron And Steel Institute, Washington, D.C.).

7. J. M. Barsom, J. F. Sovak, and S. R. Novak, "AISI Project 168-Toughness Criteria for Structural Steels: Fracture Toughness of A572 Steels," *U.S. Steel Corporation Research Laboratory Report 97.021-001(2)*, Dec. 29, 1972 (American Iron And Steel Institute, Washington, D.C.).

8. G. R. Cudney, "Stress Histories of Highway Bridges," *Journal of the Structural Division, ASCE, 94*, No. ST12, Dec. 1968, pp. 2725–2737.

9. R. B. Madison, "Application of Fracture Mechanics to Bridges," *Fritz Engineering Laboratory Report No. 335.2*, Lehigh University Institute of Research, Bethlehem, Pa., June 1969.

10. Highway Research Board of the NAS-NRC Division of Engineering and Industrial Research, the AASHTO Road Test, Report 4, Bridge Research, Special Report CID, Publication No. 953, National Academy of Science–National Research Council, Washington, D.C., 1962.

11. J. W. Fisher, K. H. Frank, M. A. Hirt, and B. M. McNamee, "Effect of Weldments on the Fatigue Strength of Steel Beams," *NCHRP Report 102*, Washington, D.C., 1970.

12. C. G. Schilling, K. H. Klippstein, J. M. Barsom, S. R. Novak, and G. T. Blake, "Low-Temperature Tests of Simulated Bridge Members," *Journal of the Structural Division, ASCE, 101*, No. ST1, January 1975.

# 17

# Elastic-Plastic Fracture Mechanics

## 17.1. Introduction

Almost all low- to medium-strength structural steels that are used in the section sizes of interest for large complex structures such as bridges, ships, pressure vessels, and so on, are of insufficient thickness to maintain plane-strain conditions under slow-loading conditions at normal service temperatures. Thus for many structural applications, the linear-elastic analysis used to calculate $K_{Ic}$ values is invalidated by the formation of large plastic zones and elastic-plastic behavior. One approach to the fracture analysis of these materials is to use empirical correlations to approximate $K_{Ic}$, $K_{Ic}(t)$, or $K_{Id}$ values, as described in Chapter 5. In addition, considerable effort is being devoted to the development of elastic-plastic fracture-mechanics analyses as an extension of the linear-elastic analyses.

The primary extensions of linear-elastic fracture mechanics into the elastic-plastic region are the following:

1. $R$-curve analysis.
2. $J$-integral.
3. Crack-tip opening displacement (CTOD).

The test method for $R$-curve determination is described in ASTM E-561—81—"Standard Practice for $R$-Curve Determination." ASTM E813—81—"Standard Test for $J_{Ic}$, A Measure of Fracture Toughness"—describes the $J$-integral Test Method. ASTM is preparing a test method for CTOD testing, although the British have had a standard for CTOD testing for years, the most recent one being British Standard 5762—"Method for Crack-Tip Opening Displacement (CTOD) Testing."

Although all three methods can be used to test specimens at any temperature throughout the transition-temperature region, there are some limitations imposed by the ASTM Test Methods. A schematic representation of fracture behavior throughout the transition-temperature region (as measured by different test methods) is shown in Figure 17.1. Current test methods are such that $K_{Ic}$ behavior is restricted to the lower shelf, where plane-strain linear-elastic conditions exist. On the other hand, $J_{Ic}$ behavior is restricted to the upper transition and upper-shelf regions where fibrous tearing occurs. Because the $R$-curve analysis uses an elastic $K_R$ analysis (corrected for plastic-zone adjustment), it generally is restricted to the lower transition region, although $R$-curves are used to analyze the behavior of structural materials in the upper transition region where stable crack growth occurs using either the $J$-integral analysis or a CTOD analysis.

Because the CTOD test method is based on the determination of a critical strain at fracture from a load-displacement record which does not require a stress analysis, the CTOD test method can be used throughout the entire transition-temperature region. In practice, it is used primarily in the lower transition and upper transition regions, which are the regions of interest for most structural steels. Thus, with the restrictions imposed in the current ASTM Test Methods as well as the economics of fracture-toughness testing, the CTOD test method is the fracture-mechanics test method most applicable over the entire transition-temperature region.

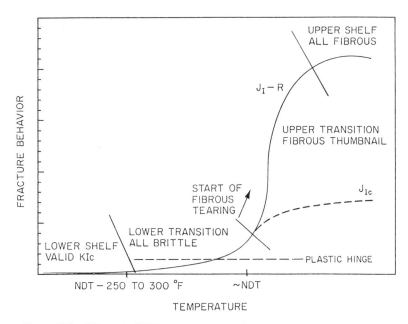

**Figure 17.1**   Schematic CTOD-temperature-transition curve showing four regions.

## 17.2. Crack-Tip Opening Displacement

### 17.2.1. Analysis

In 1961, Wells[1] proposed that the fracture behavior in the vicinity of
a sharp crack could be characterized by the opening of the notch faces,
namely, the crack-tip opening displacement, as shown in Figure 17.2. Furthermore he showed that the concept of crack-opening displacement was
analogous to the concept of critical crack extension force ($G_c$ as described
in Chapter 2), and thus the CTOD values could be related to the plane-
strain fracture toughness, $K_{Ic}$. Because CTOD measurements can be made
even when there is considerable plastic flow ahead of a crack, such as
would be expected for elastic-plastic or fully plastic behavior, this technique
might still be used to establish critical design stresses or crack sizes in a
quantitative manner similar to that of linear-elastic fracture mechanics.

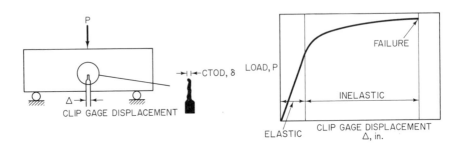

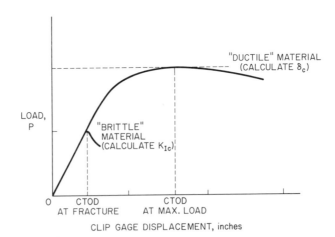

**Figure 17.2** Relation between $K_{Ic}$ and CTOD test behavior; $K_{Ic}$/CTOD test
setup and load displacement records from $K_{Ic}$/CTOD tests.

Using a crack-tip plasticity model proposed by Dugdale,[2] referred to as the strip-yield-model analysis, it is possible to relate the CTOD to the applied stress and crack length. The strip yield model consists of a through-thickness crack in an infinite plate that is subjected to a tensile stress normal to the plane of the crack (Figure 17.3). The crack is considered to have a length equal to $2a + 2r_y$. At each end of the crack there is a length $r_y$ that is subjected to yield-point stresses that tend to close the crack or, in reality, to prevent it from opening. Thus the length of the "real" crack would be $2a$. Another way of looking at the behavior of this model is to assume that yield zones of length $r_y$ spread out from the tip of the real crack, $a$, as the loading is increased. Thus the displacement at the original crack tip, $\delta$, which is the CTOD, increases as the real crack length increases or as the applied loading increases. The basic relationship developed by Dugdale is

$$\delta = 8\frac{\sigma_{ys}a}{\pi E} \ln \sec\left(\frac{\pi}{2}\frac{\sigma}{\sigma_{ys}}\right) \tag{17.1}$$

where $\sigma_{ys}$ = yield strength of the material, ksi.
   $a = \frac{1}{2}$ real crack length, in.
   $\sigma$ = nominal stress, ksi.
   $E$ = modulus of elasticity of the material, ksi.

Using a series expansion for $\ln \sec [(\pi/2)(\sigma/\sigma_{ys})]$, this expression

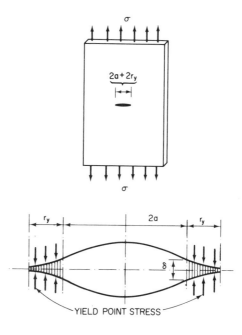

**Figure 17.3**  Dugdale strip-yield model.

becomes

$$\delta = \frac{8\sigma_{ys}a}{\pi E}\left[\frac{1}{2}\left(\frac{\pi}{2}\frac{\sigma}{\sigma_{ys}}\right)^2 + \frac{1}{12}\left(\frac{\pi}{2}\frac{\sigma}{\sigma_{ys}}\right)^4 + \frac{1}{45}\left(\frac{\pi}{2}\frac{\sigma}{\sigma_{ys}}\right)^6 + \cdots\right] \quad (17.2)$$

For nominal stress values less than $\frac{3}{4}\sigma_{ys}$, a reasonable approximation for $\delta$, using only the first term of this series, is

$$\delta = \frac{\pi\sigma^2 a}{E\sigma_{ys}} \quad (17.3)$$

In Chapter 2, it was shown that, for a through-thickness crack of length $2a$,

$$K_I = \sigma\sqrt{\pi a} \quad (17.4)$$

or

$$K_I^2 = \sigma^2\pi a$$

Thus $\delta E\sigma_{ys} = K_I^2$, and since $E = \sigma_{ys}/\varepsilon_{ys}$, the following relation exists:

$$\frac{\delta}{\varepsilon_{ys}} = \left(\frac{K_I}{\sigma_{ys}}\right)^2 \quad (17.5)$$

Also, the strain energy release rate, $G$, is equal to

$$G = \frac{\pi\sigma^2 a}{E} \quad (17.6)$$

$$G = \delta \cdot \sigma_{ys}$$

At the onset of crack instability under plane-strain conditions, where $K_I$ reaches $K_{Ic}$ and CTOD reaches a critical value, $\delta_c$,

$$\frac{\delta_c}{\varepsilon_{ys}} = \left(\frac{K_{Ic}}{\sigma_{ys}}\right)^2 \quad (17.7)$$

Because $(K_{Ic}/\sigma_{ys})^2$ can be related to the critical crack size in a particular structure, it is reasonable to assume that the parameter $\delta_c/\varepsilon_{ys}$ can likewise be related to the critical crack size in a particular structure. The advantage of the CTOD approach is that CTOD values can be measured throughout the entire plane-strain, elastic-plastic, and fully plastic behavior regions, whereas $K_{Ic}$ values can be measured only in the plane-strain region or approximated in the early portions of the elastic-plastic region.

As with the $K_I$ analysis, the application of the CTOD approach to engineering structures requires the measurement of a fracture-toughness parameter, $\delta_c$, which is a material property that is a function of temperature, loading rate, specimen thickness, and possibly specimen geometry, that is, notch acuity, crack length, and overall specimen size.

After considerable study of various methods to measure $\delta_c$, the British Standards Institution has published a "Standard Methods for Crack-Tip Opening Displacement (CTOD) Testing."[3] Basically the $\delta_c$ specimen is a

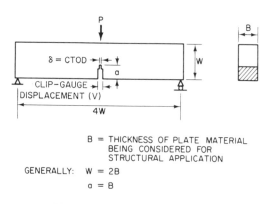

Figure 17.4   CTOD test specimen.

slow-bend test specimen similar to a $K_{Ic}$ slow-bend test specimen (Figure 17.4). Since the $\delta_c$ test is regarded as an extension of $K_{Ic}$ testing, the British test method is very similar to the ASTM E-399 test method for $K_{Ic}$ (Chapter 3). Similar specimen preparation, fatigue-cracking procedures, instrumentation, and test procedures are followed. The displacement gauge is similar to the one used in $K_{Ic}$ testing, and a continuous-load-displacement record is obtained during the test.

### 17.2.2.  Calculation of CTOD Values

Depending upon the material tested and the test temperature, different load-crack-mouth opening displacement ($P$-$\Delta$) records can be obtained as shown in Figure 17.5. The clip-gauge displacement shown in Figure 17.4 measures the displacement at the edge or "mouth" of the crack (CMOD). The CTOD value at the tip of the crack is calculated from the clip-gauge displacement measured at the edge or mouth of the specimen. The different type $P$-$\Delta$ records can be described as follows:

Type a: Unstable (brittle) fracture with no prior crack growth.

Type b: Brittle crack initiation or pop-in ($v_{ic}$) which is arrested followed by subsequent increase in load. Final failure is by cleavage, $v_c$.

Type c: Stable crack extension (ductile fibrous thumbnail, $v_{id}$) followed by unstable cleavage fracture at $v_u$. Stable crack extension starts at $P_{id}$. To determine $P_{id}$, special techniques are required such as multiple specimens or unloading compliance. Thus, use of $P_{id}$ to determine $v_{id}$ is not widely used.

Type d: Stable crack extension starting at $P_{id}$, followed by subsequent increase in load and cleavage failure at $v_u$. Cleavage pop-in ($v_{ic}$) may also occur, but not necessarily.

Type e: Generally stable ductile fibrous tearing resulting in a roundhouse

SCHEMATIC SHOWING INTERPRETATION OF TYPES OF P–CMOD RECORDS

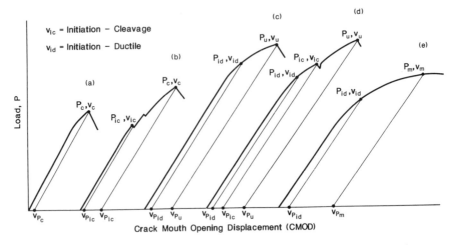

**Figure 17.5** Schematic CTOD load-clip-gage displacement (CMOD) records showing different types of behavior.

curve. Failure is by stable ductile tearing at $P_m$. The value of $v_m$ nominally is selected at the first attainment of maximum load.

Five different clip-gauge (crack-mouth) displacement values can be obtained from $P$-$\Delta$ records and used to calculate CTOD ($\delta$) values as follows:

$\delta_c$ (from $v_c$) = CTOD value at unstable cleavage fracture. This value applies only when there is no evidence of slow crack growth (ductile tearing).

$\delta_u$ (from $v_u$) = CTOD value at unstable cleavage fracture after slow crack growth (ductile tearing, $v_{id}$).

$\delta_m$ (from $v_m$) = CTOD value at first attainment of maximum load.

$\delta_{ic}$ (from $v_{ic}$) = CTOD value at first pop-in (or brittle crack extension) which is followed by subsequent increase in load.

$\delta_{id}$ (from $v_{id}$) = CTOD value at onset of stable crack extension (ductile fibrous thumbnail). Special test methods and instrumentation are required to determine this value.

After one of the displacement values has been selected ($v_c$, $v_{ic}$, $v_{id}$, $v_u$, or $v_m$), the plastic portion of $v$ is determined as shown in Figure 17.6. In this illustration, the value is $v_u$. If the record were of Type a, $v_p$ would

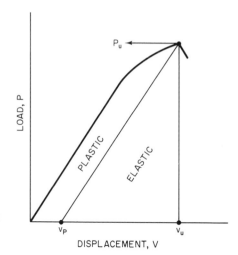

**Figure 17.6**  Schematic load-clip-gage displacement (CMOD) record showing determination of plastic component (Vp) of total displacement (Vu)

be essentially zero; if the record were of Type e, $v_p$ generally would be quite large.

The following equation from British Standard 5762 (1979) is used to analyze the load-clip gauge displacement record to obtain the critical CTOD value:

$$\delta = \frac{K^2 (1 - \nu^2)}{2\sigma_{ys}E} + \frac{0.4(W - a)v_p}{0.4W + 0.6a} \tag{17.8}$$

where  $a$  = crack length.
   $B$ = specimen thickness.
   $W$ = specimen depth.
   $E$ = modulus of elasticity.
   $\nu$ = Poisson's ratio.
   $v_p$ = plastic component of clip-gauge displacement (see Figures 17.5 and 17.6).  These values would be $v_{Pc}$, $v_{Pic}$, $v_{Pid}$, $v_{Pu}$, and $v_{Pm}$; Figure 17.5.
   $K$ = stress-intensity factor at *maximum* load (not at 5 percent secant offset).

This formula is based on an elastic and a plastic component of CTOD and implies the existence of a rotation point below the crack tip.  The first (elastic) term comes from the following relations:

$$J = \frac{(1 - \nu^2) K_{Ic}^2}{E} \tag{17.9}$$

and

$$J = m\sigma_{ys}\delta \tag{17.10}$$

where $J$ is the $J$-integral as described later in this chapter and $m$ is a constraint factor believed to be 2 for plain strain. Thus for elastic behavior,

$$\delta = \frac{K_{Ic}^2(1 - \nu^2)}{m\sigma_{ys}E} \tag{17.11}$$

For $K_{Ic} = K$ at maximum load $P_u$ in Figure 17.5, and $m = 2$,

$$\delta = \frac{K^2(1 - \nu^2)}{2\sigma_{ys}E} \tag{17.12}$$

which is the first term in the expression for $\delta$ given in Equation (17.8). Note that if $v_p = 0$, for elastic behavior, $\delta$ given in Equation (17.8) is identical to that obtained from Equation (17.4). The plastic term

$$\frac{0.4(W - a)v_p}{0.4W + 0.6a}$$

assumes an apparent center of rotation during the plastic portion of the test as shown in Figure 17.7, where $r$ is assumed to be 0.4.

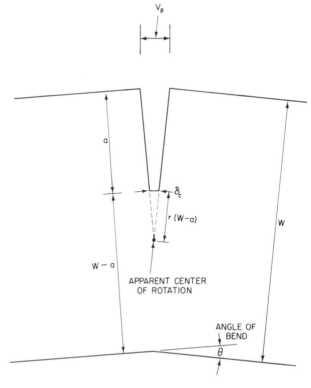

**Figure 17.7**   Assumed rotation of bend specimen during plastic portion of test $(r = 0.4)$.

### 17.2.3. Effect of Temperature, Loading Rate, and Thickness on CTOD

As would be expected, the general effects of temperature, loading rate, and specimen thickness on $\delta_c$ are similar to the effects of these same conditions on $K_{Ic}$ or CVN test results; that is, increasing the temperature increases $\delta_c$, whereas increasing the loading rate or plate thickness decreases $\delta_c$. An example of this behavior for mild steel[4] is presented in Figure 17.8. These results show that for slow-bend loading the CTOD value increases very rapidly at about $-50°F$ for this particular steel using specimens 0.394 in. thick. For thicker test specimens or for test specimens loaded under impact conditions, the transition temperature is higher and is less well defined.

Test results by Nishioka and Iwanaga[5] on HT-50, X-60, X-70, and HT-70 steels (Table 17.1) are presented in Figures 17.9 and 17.10 and demonstrate the effects of loading rate on $\delta_c$, namely, that increasing the loading rate decreases the $\delta_c$ at a given temperature in the same general manner as for $K_{Ic}$ and $K_{Id}$ results.

Wellman, Rolfe, and Dodds[6,7,8] have studied the engineering aspects of CTOD fracture-toughness testing as applied to behavior of steel structures. Three of the pressure vessel and structural steels they studied were A533, A516, and A131 (ABS-B), as described in Table 17.2. CTOD test results for A516 steels using $\frac{1}{2}$-in.- and 1-in.-thick specimens are presented in Figures 17.11 and 17.12, respectively. Although the scatter in results for the smaller specimens is greater than it is for the larger specimens, the use of a lower-

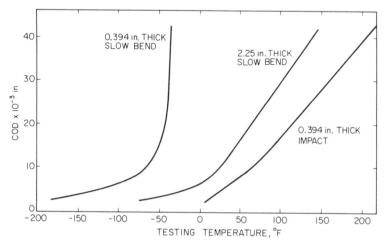

**Figure 17.8**    Effect of strain rate and thickness on the transition temperature of mild steel.

**TABLE 17.1  Materials Used to Obtain Results of Dynamic CTOD Test Values
Presented in Figures 17.9 and 17.10**
Chemical Compositions (%)

|      | Thickness (mm) | C | Si | Mn | P | S | Cu | Cr | Ni | Mo | V | Nb |
|------|------|------|------|------|------|------|------|------|------|------|------|------|
| HT50 | 22 | 0.20 | 0.33 | 1.30 | 0.018 | 0.015 | 0.04 | 0.03 | — | — | — | — |
| X-60 | 14 | 0.17 | 0.46 | 1.46 | 0.025 | 0.016 | 0.03 | — | — | — | 0.08 | — |
| X-70 | 14 | 0.14 | 0.41 | 1.30 | 0.015 | 0.005 | 0.04 | 0.09 | — | — | 0.09 | 0.029 |
| HT70 | 25 | 0.12 | 0.30 | 0.79 | 0.014 | 0.013 | 0.02 | 0.43 | 1.26 | 0.47 | 0.03 | — |

| | Mechanical properties | | |
|------|------|------|------|
| | Yield Strength MPa (ksi) | Tensile Strength MPa (ksi) | Elongation (%) |
| HT50 | 436( 63.2) | 646( 93.7) | 31.0 |
| X-60 | 417( 60.4) | 627( 90.9) | 30.0 |
| X-70 | 535( 77.5) | 620( 89.9) | 37.7 |
| HT70 | 695(100.8) | 758(109.9) | 26.7 |

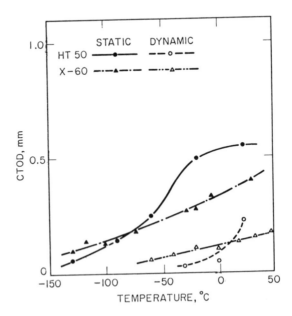

**Figure 17.9**  Static and dynamic CTOD test results. (From K. Nishioka and
H. Iwanga, "Some Results of Dynamic COD-Test," *Significance of Defects in
Welded Structures,* University of Tokyo Press, Tokyo, 1973.)

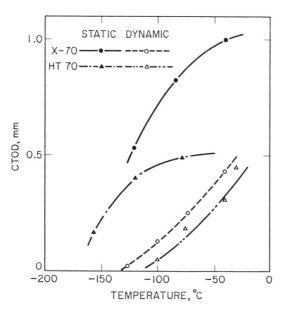

**Figure 17.10**  Static and dynamic CTOD test results. (From K. Nishioka and H. Iwanga, "Some Results of Dynamic DOD-Test," *Significance of Defects in Welded Structures*, University of Tokyo Press, Tokyo, 1973.)

bound curve gives essentially identical results in the lower transition region. This behavior has been observed for several structural steels and appears to represent a weakest-link theory; that is, there is an increase in data scatter as the specimen size decreases. Because less of the material is sampled, there is less change of the weakest link being present. This theory suggests that tests of a few large specimens or tests of many smaller specimens

**TABLE 17.2.**    **Material Properties of Steels Used to Obtain CTOD Results Presented in Figures 17.11–17.14**

| Steel | C | Mn | P | S | Si | Ni | Cr | Mo | V |
|---|---|---|---|---|---|---|---|---|---|
| A533 Grade B | 0.20 | 1.23 | 0.015 | 0.017 | 0.26 | 0.49 | — | 0.52 | 0.05 |
| A516 | 0.23 | 1.15 | 0.009 | 0.22 | 0.25 | — | — | — | — |
| A131 Grade B | 0.16 | 1.01 | 0.009 | 0.010 | 0.070 | — | — | — | — |

| Steel | Yield Strength MPa (ksi) | Tensile Strength MPa (ksi) | Elongation in 2 in., percent | Reduction in Area, percent | Ultimate Strain |
|---|---|---|---|---|---|
| A533 | 427 (62) | 586 (85) | 28 | 77 | 0.213 |
| A516 | 266 (38.5) | 459 (66.5) | 38 | 72 | 0.228 |
| A131 | 261 (37.8) | 471 (68.3) | 52 | 73 | 0.218 |

result in the same lower-bound value. The results presented in Figures 17.11 and 17.12 suggest that this observation may be true.

Note that the upper-shelf values increase with specimen depth ($W$) as would be expected because the rotation, and thus the CTOD increases with increasing depth of specimen.

Similar CTOD temperature results for A533 and A131 (ABS-B) steels are presented in Figures 17.13 and 17.14, respectively. Note that for both of these steels, there is considerable scatter in the transition region but that a lower-bound curve can easily be drawn.

### 17.2.4. Correlations between CTOD and $K_c$

The need for correlations between various fracture-toughness tests was described in Chapter 5 along with the details of specific correlations. The results of various of these $K_c$–CVN–CTOD and $J$-integral correlations are presented in Figures 17.15 through 17.17 for the three steels described in Table 17.2.

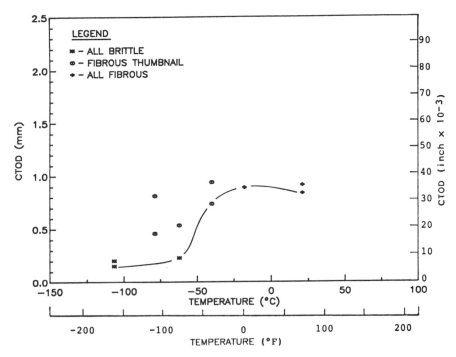

**Figure 17.11** CTOD-temperature transition curve for an A516 steel. [Specimen size (12.5 × 25 × 100 mm)]

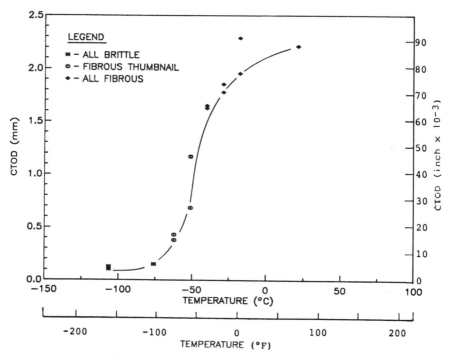

**Figure 17.12** CTOD-temperature transition curve for an A516 steel. [Specimen size (25 × 50 × 200 mm)]

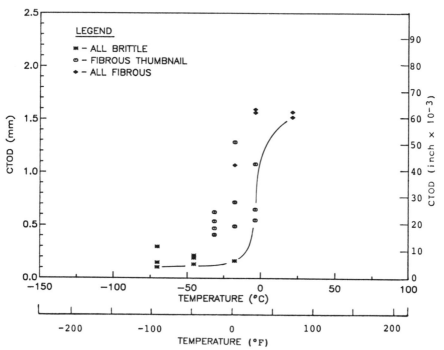

**Figure 17.13** CTOD-temperature transition curve for an A533 steel. [Specimen size (25 × 50 × 200 mm)]

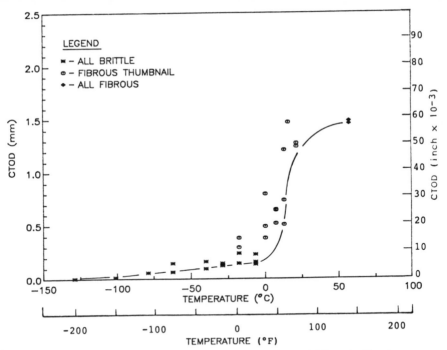

**Figure 17.14** CTOD-temperature transition curve for an A131 steel. [Specimen size (25 × 50 × 200 mm)]

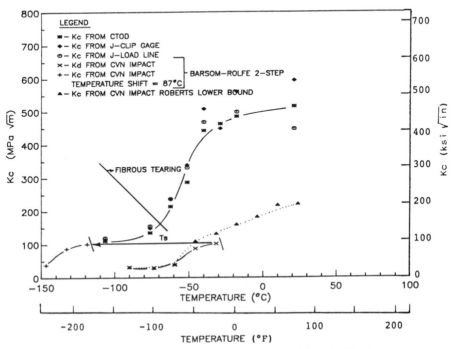

**Figure 17.15** $K_c$-CVN-CTOD-J correlations for an A516 steel. [Specimen size (25 × 50 × 200 mm)]

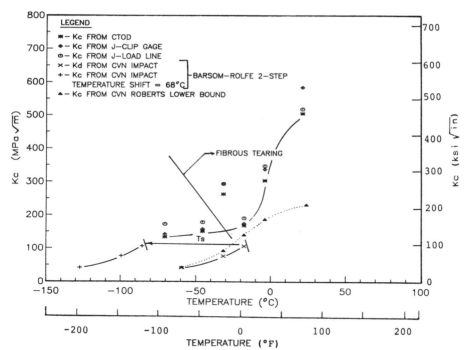

**Figure 17.16** $K_c$-CVN-CTOD-J correlations for an A533 steel. [Specimen size (25 × 50 × 200 mm)]

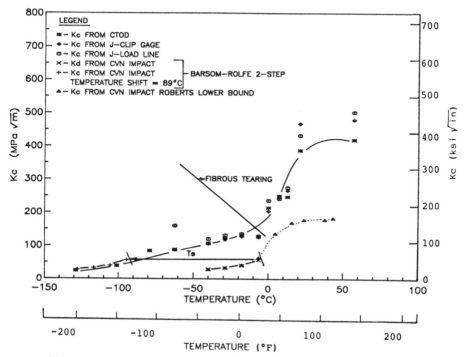

**Figure 17.17** $K_c$-CVN-CTOD-J correlations for an A131 steel. [Specimen size (25 × 50 × 200 mm)]

## $K_c$–CTOD

$$K_c = \sqrt{1.6\,E\sigma_{flow}\delta} \tag{17.13}$$

$E$ = modulus of elasticity, ksi

$$\sigma_{flow} = \frac{\sigma_{ys} + \sigma_{ult}}{2}$$

$\delta$ = CTOD (in.)

$m$ = 1.6 for plain strain and 1.2 for plane stress

**$K_c$–CVN two-stage correlation.**   Estimation of $K_{Id}$ versus temperature from CVN impact test results:

$$K_{Id} = \sqrt{5E(CVN)} \qquad (psi\sqrt{in.}, psi, ft\text{-}lb) \tag{17.14}$$

$$K_{Id} = \sqrt{0.64E(CVN)} \qquad (MPa\sqrt{m}, MPa, J) \tag{17.15}$$

Obtain the $K_{Ic}$ versus temperature by shifting the $K_{Id}$ versus temperature curve to lower temperatures by using the temperature shift, $T_{shift}$, relationship:

$$T_{shift} = 215 - 1.5\sigma_{ys} \qquad (°F, ksi) \tag{17.16}$$

$$T_{shift} = 119 - 0.12\sigma_{ys} \qquad (°C, MPa) \tag{17.17}$$

## $K$–$J$-integral

$$K = \sqrt{JE} \tag{17.18}$$

where $J$ is the $J$-integral as described later in this chapter.

Low- to moderate-yield-strength steels (for example, the steels described in Table 17.2) exhibit a typical plane-strain rise in toughness with increasing temperature from the lower shelf through the lower transition region. These steels then undergo a rapid toughness increase (the upper transition region) which is characterized by the development of a ductile fibrous thumbnail on the fracture surface. In the lower transition region, the Barsom-Rolfe two-stage CVN–$K_c$ correlation yields essentially the same results as do the CTOD–$K_c$ and the $J$–$K_c$ correlations. In the upper transition region and upper shelf, specimen size effects dominate the material behavior. Thus, after the initiation of stable cracking, the correlations may not be meaningful even though the trend appears to be reasonable.

### 17.2.5. Application of CTOD Test Results

Three general approaches to the application of CTOD test results in fracture control or analysis are as follows:

1. Use CTOD–$K_c$ correlations to estimate $K_c$ values at the service tem-

perature as described in the previous section. Use these $K_c$ values directly in design as described in Chapter 6.

2. Use the Burdekin and Dawes CTOD design curve as described in the next section.

3. Use finite element analysis techniques to model the specific structural geometry being analyzed. Develop stress, load, pressure, and so on versus CTOD curves for the specific flaw geometry assumed to be present in the structure. Conduct CTOD tests on the material of interest to determine the critical CTOD and the critical load directly from the load-CTOD curve. Although probably the most accurate method, this is certainly the most expensive and, in the case of three-dimensional analyses, is extremely complex. Wellman has used this technique to predict the behavior of actual pressure-vessel tests within 7 percent, as described in Reference 8.

A design curve, proposed by Burdekin and Dawes,[9] for $\delta_c$, strain, and crack-size relationships is shown in Figure 17.18(a). It is based on the strip-yielding model discussed earlier and on considerable experimental work done at the Welding Institute in the United Kingdom, Figure 17.18(b). This design curve was developed primarily for pressure vessels, but it can be used for other structural applications as well. It should be emphasized that the design curve, Figure 17.18(a), is based primarily on the results of actual experimental results rather than on theoretical predictions, Figure 17.18(b).

The design curve relations in Figure 17.18(a) are

$$\phi = \left(\frac{\varepsilon}{\varepsilon_{ys}}\right)^2 \qquad \text{for } \frac{\varepsilon}{\varepsilon_{ys}} \leq 0.5 \tag{17.19}$$

$$\phi = \frac{\varepsilon}{\varepsilon_{ys}} - 0.25 \qquad \text{for } \frac{\varepsilon}{\varepsilon_{ys}} > 0.5 \tag{17.20}$$

where

$$\phi = \frac{\delta}{2\pi\varepsilon_{ys}a} \tag{17.21}$$

This design curve has a "built-in" factor of safety ranging from 1 to about 5, with an average value of about 2.5.

The value of the parameter used in Equations (17.19–17.21) for CTOD results depends on the appropriate design stress level and stress-concentration factors.

The particular parameter $\phi$ is determined from the CTOD design curve, Figure 17.18(a), as follows:

1. Determine $\varepsilon/\varepsilon_{ys}$, where $\varepsilon$ is the actual strain at the location being

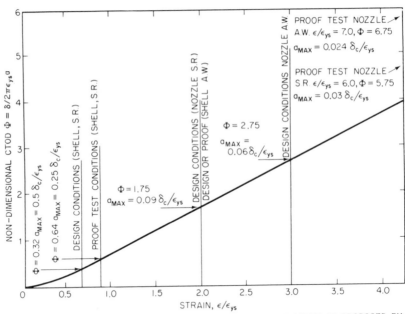

(a) DESIGN CURVE FOR CTOD, STRAIN, AND CRACK SIZE RELATIONSHIP AS PROPOSED BY BURDEKIN AND DAWES

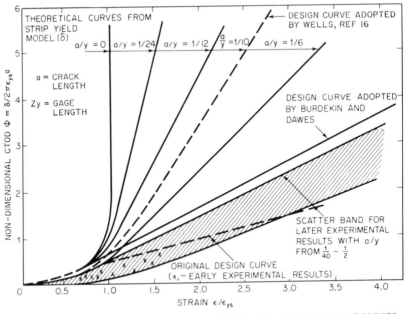

(b) THEORETICAL AND EXPERIMENTAL CTOD, STRAIN, AND CRACK SIZE RELATIONSHIPS

**Figure 17.18** Theoretical, experimental, and design curve for CTOD, strain and crack-size relationship. (a) Decision curve for CTOD, strain, and crack-size relationship as proposed by Burdekin and Dawes; (b) theoretical and experimental CTOD, strain, and crack-size relationships.

analyzed, and $\varepsilon_{ys}$ is the yield strain. That is, for general design (stress concentration equal to 1.0),

$$\varepsilon = \frac{\sigma_{des}}{E} \tag{17.22}$$

For the conditions at a stress-concentration factor, multiply the nominal design strain ($\varepsilon_{des} = \sigma_{des}/E$) by the stress-concentration factor to find the maximum strain $\varepsilon$.

2. Knowing $\varepsilon/\varepsilon_{ys}$, determine the nondimensional CTOD, $\phi$, from Figure 17.18(a).

3. Knowing $\phi$, use the relation between $a_{cr}$ and $\delta_c$ to determine $a_{cr}$:

$$a_{cr} = \frac{\delta_c}{2\pi\varepsilon_{ys}\phi} \tag{17.23}$$

where $a_{cr} = \frac{1}{2}$ through-thickness crack length, $2a$—hence the factor 2.

$\delta_c =$ critical CTOD for the actual material being used at the location being evaluated, measured at the service temperature and loading rate (note, however, that most $\delta_c$ values are measured under conditions of slow or "static" loading).

$\varepsilon_{ys} =$ yield strain of actual material being used.

For example, for a nominal design stress of $\sigma = \sigma_{ys}/2$ and a stress-concentration factor of 4, $\varepsilon/\varepsilon_{ys} = 4 \cdot \frac{1}{2}\varepsilon_{ys}/\varepsilon_{ys} = 2$. For this case, $\phi = 1.75$ and

$$a_{cr} = \frac{1}{2\pi(1.75)} \frac{\delta_c}{\varepsilon_{ys}} = 0.09 \frac{\delta_c}{\varepsilon_{ys}}$$

For a more severe condition of $\varepsilon/\varepsilon_{ys} = 3.0$, $\phi = 2.75$ and

$$a_{cr} = \frac{1}{2\pi(2.75)} \frac{\delta_c}{\varepsilon_{ys}} = 0.06 \frac{\delta_c}{\varepsilon_{ys}}$$

The CTOD design curve described in Figure 17.18(a) takes into consideration the effects of geometric concentrations and residual stresses and has been widely used to analyze the fracture behavior of welded structures.[6-10] However, it is based on a semiempirical analysis with several simplifying assumptions and is a *design* curve with an actual factor of safety varying between about 1 and 5.

Figure 17.19 compares the allowable crack sizes predicted by actual CTOD test results and three critical crack sizes at fracture in wide-plate tests. Note that all results fall above the $S = 1$ line (factor of safety $= 1$) and that $S = 2.5$ is a reasonable average representation of the test results, although there is considerable scatter.

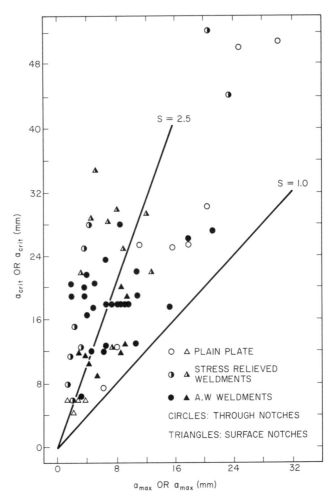

**Figure 17.19** Comparison of critical and maximum allowable crack sizes showing safety factors (Ref. 32).

As an example of the use of the CTOD design procedure, assume that the behavior of a structural steel with a yield strength of 40,000 psi is being analyzed for the more severe condition of $\varepsilon/\varepsilon_{ys} = 3.0$. In this case,

$$\varepsilon_{ys} = \frac{\sigma_{ys}}{E} = \frac{40,000}{(30 \times 10^6)} = 0.00133$$

$$a_{cr} = \frac{0.06}{0.00133}\delta_c = 45\delta_c$$

If test results show that $\delta_c$ at the service temperature is 0.02 in. (20 mils),

$$a_{cr} = 0.9 \text{ in.} \quad \text{and} \quad 2a_{cr} = 1.8 \text{ in.}$$

A more typical case might be

$$\sigma = \tfrac{2}{3}\,\sigma_{ys}$$

with no stress-concentration factor. In this case, $\phi = 0.32$, Figure 17.18(a), and

$$a_{cr} = \frac{1}{2\pi(0.32)}\,\frac{\delta_c}{0.00133}$$

$$a_{cr} = 3758\delta_c$$

For a material with the same $\delta_c = 0.020$ in.,

$$a_{cr} = 7.5 \text{ in.} \quad \text{and} \quad 2a = 15 \text{ in.}$$

Remember that this value, $2a = 15$ in., is based on a *design* curve with an average factor of safety of 2.5 (Figure 17.19). Thus, the actual critical crack size would be about $2.5 \times 15 = 37.5$ in. From a practical viewpoint, having a critical crack size, $2a$, of 45 in. essentially means that some other mode of failure, for example, yielding or excessive deflection, probably controls and that the probability of brittle fracture is extremely low.

As a comparison between the Burdekin-Dawes design curve approach, Figure 17.18(a), and the $K_c$ approach using CTOD–$K_c$ correlations, the critical crack size will now be determined using the latter approach.

For the same conditions, that is, $\sigma_{ys} = 40$ ksi and CTOD $\approx 20$ mils, see Figures 17.14 and 17.17. Figure 17.14 gives the CTOD result for an A131 steel with a yield strength of about 40 ksi. The 20-mil temperature is about $+10°C$. From the CTOD–$K_c$ correlations shown in Figure 17.17 for this steel, the $K_c$ value at $+10°C$ is about 220 ksi$\sqrt{\text{in.}}$.

Thus for the same through-thickness crack, for $K_c = 220$ and $\sigma_{des} = \tfrac{2}{3}\sigma_{ys}$,

$$K_c = \sigma\sqrt{\pi a} \tag{17.4}$$

$$a_c = \frac{K_c^2}{\sigma_{des}^2 \pi}$$

$$2a_c \cong \frac{2(220)^2}{(27)^2\pi} = 42 \text{ in.}$$

(compared with 37.5 in. from the Burdekin and Dawes design curve). Thus both approaches are indeed consistent.

In summary, the CTOD fracture parameter provides a method to extend fracture-mechanics concepts throughout the entire elastic-plastic transition-temperature region unencumbered by the problems of "valid" results as is the case for $K_{Ic}$ testing, or by the restriction of stable crack growth as is the case for $J_{Ic}$ (discussed later in this chapter).

## 17.3. *R*-Curve Analysis

As described in Chapters 2, 3, and 4, $K_{Ic}$ is the critical stress-intensity factor under conditions of plane strain ($\varepsilon_z = 0$) with attendant small-scale crack-tip plasticity. Conversely, $K_c$ is the critical stress-intensity factor under plane-stress conditions ($\sigma_z = 0$) with attendant large-scale crack-tip plasticity. Thus the behavior represented by $K_c$ is the opposite of that represented by $K_{Ic}$, that is, negligible rather than complete through-thickness elastic constraint (stress) at fracture. $K_c$ values are generally 2–10 times larger than $K_{Ic}$ and vary not only with temperature ($T$) and strain rate ($\dot{\varepsilon}$), as does $K_{Ic}$, but with plate thickness ($B$) as well. Furthermore, for fixed conditions of temperature, strain rate, and plate thickness ($T$, $\dot{\varepsilon}$, and $B$), the $K_c$ values may also vary with initial crack length, $a_0$.

The operating temperatures, rates of loading, and thicknesses of most structural materials used in actual structures are generally such that plane-stress rather than plane-strain conditions actually exist in service. Accordingly, considerable effort has been devoted toward plane-stress fracture-toughness evaluations using an *R*-curve or resistance curve analysis as one of several extensions of linear-elastic fracture mechanics into elastic-plastic fracture mechanics.

*R*-curves characterize the resistance to fracture of materials during incremental slow-stable crack extension. They provide a record of the toughness development as a crack is driven stably under increasing crack-driving forces, that is, loads. They are dependent upon specimen thickness, as well as on test temperature and loading rate. An *R*-curve is a plot of crack-growth resistance in a material as a function of actual or effective crack extension. $K_R$ is the crack-growth resistance expressed in units corresponding to $K$, namely, ksi$\sqrt{\text{in}}$. $K_c$ is the plane-stress fracture toughness and is equal to the value of $K_R$ at a particular instability condition determined during an *R*-curve test.

The *R*-curve describes the variation in $K_R$ with crack length, $a_0$. It consists of a plot of $K_R$ versus $\Delta a$, where $K_R$ represents the driving force required to produce stable crack extension ($\Delta a$) prior to unstable crack growth at $K_c$. The $K_c$ value that results for a given crack length, $a_0$, is the value associated with the point of tangency between the line representing the applied load and the *R*-curve itself (Figure 17.20).

The dashed lines in Figure 17.20 represent the variation in $K_I$ with crack length, $a$, for constant load $P_1$, $P_2$, or $P_3$, where $P_3 > P_2 > P_1$. That is, for a given load level and increasing crack length, $a$, $K_I$ will increase because $K_I = f(P, \sqrt{a})$.

The solid lines represent the increase in $K_R$ with increasing load and increasing crack length for two different initial crack lengths. The two points of tangency where $K_R = K_c$ for $a_0 = a_1$ and $a_0 = a_2$ represent points of instability, or the critical-plane–stress-intensity factor, $K_c$, at the particular

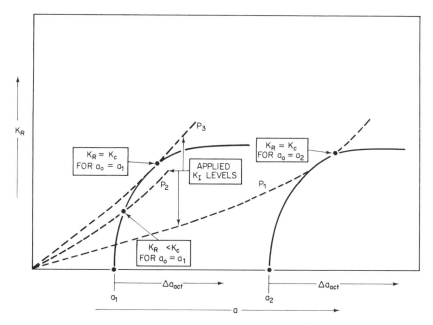

**Figure 17.20**   Basic principle of *R*-curves for use in determining $K_c$ under different conditions of initial crack length $a_0$ (Ref. 15).

crack length and, of course, for the given conditions of temperature, loading rate, and plate thickness. The $K_R$ value is always calculated by using the effective crack length, $a_{eff}$, and is plotted against the actual crack extension, $\Delta a_{act}$, that takes place physically in the material during the test.

Under plane-strain conditions, the fracture toughness of a material depends on only two variables—$K_{Ic} = f(T$ and $\dot{\varepsilon})$—because, by definition, conditions of maximum constraint exist. Under plane-stress conditions, the fracture toughness depends on four variables: $K_c = f(T, \dot{\varepsilon}, B,$ and $a_0)$. For given test and material conditions ($T$, $\dot{\varepsilon}$, and $B$), a $K_c$ value merely represents a singular point on an *R*-curve. On the other hand, an *R*-curve describes the complete variation of $K_c$ with changes in initial crack length, $a_0$. Also, an *R*-curve can be used to determine the changing resistance to fracture with increasing crack length. This increase in resistance to brittle fracture with stable crack growth can be significant. The slope of the *R*-curve (referred to as tearing modulus) is also an important indicator of the increase in notch toughness (resistance to fracture) with increasing crack length prior to final fracture. As such, a single *R*-curve is a highly efficient method of fracture characterization since it is equivalent to a large number (10 to 15) of direct $K_c$ tests conducted with various (initial) crack lengths, $a_0$. Thus, the *R*-curve is the most general characterization of plane-stress-fracture behavior and depends on only three variables: $R$-curve $= f(T, \dot{\varepsilon},$ and $B$).

An ASTM special technical publication[11] describing the state of the art of $R$-curve testing has been published, and numerous studies of $R$-curve methods exist. Novak[12] has published an extensive study of the $R$-curve behavior, and some of his results are described in this chapter as an example of the use of $R$-curves. ASTM also has developed a "Standard Practice for $R$-Curve Determination": E-561.[13]

$R$-curves can be determined by using either of two experimental methods: *load control* or *displacement control*. The load-control technique can be used to obtain only that portion of the $R$-curve up to the $K_c$ value (where complete unstable fracture occurs), whereas the displacement-control technique can be used to obtain the entire $R$-curve and therefore offers a fundamental advantage. The equivalence of the two techniques for determining $K_c$ has been demonstrated by the work of Heyer and McCabe,[14,15] the originators of the displacement-control technique. However, their demonstration of equivalence for the two test techniques was restricted to very high-strength steels and aluminum alloys, where the principles of linear-elastic fracture mechanics (LEMF) are directly applicable as a result of limited crack-tip plasticity.

The evaluation of $R$-curves for relatively low-strength, high-toughness alloys is more complex. Because such materials exhibit large-scale crack-tip plasticity ($r_y$) at fracture, relative to the test-specimen in-plane dimensions ($W$ and $a$), LEFM principles cannot be applied directly. As a consequence, a nonlinear, elastic-plastic approach is required. In this elastic-plastic approach, the crack-opening displacement ($\delta$) at the physical crack tip is measured and used in calculating the equivalent elastic $K$ value. The equivalent elastic $K$ value is the analog $K$ value that would be measured under elastic conditions for which LEFM principles can be used directly when specimens of the same thickness, $B$, but much larger planar dimensions, $W$ and $a$, are tested. This nonlinear approach is based on theoretical considerations advanced earlier by Wells[16] and reviewed more recently by Wells[17] and Irwin.[18] This elastic-plastic crack model is designated the crack-opening-stretch (COS) method, where $\delta$ and COS are equivalent terms. The application of the COS analysis method to $R$-curve testing has been developed to an advanced degree by McCabe and Heyer.[19] Furthermore, this method can be used with either the load-control or displacement-control test procedures.

The procedure for conducting $R$-curve tests is described briefly in this chapter. Generally the thickness of the test specimens, $B$, is equal to the plate thickness being considered for actual service usage. The other dimensions, $W$ and $H$, are made considerably larger, Figure 17.21. This is done so that the plastic-zone size ahead of the crack tip, $r_y$, is small with

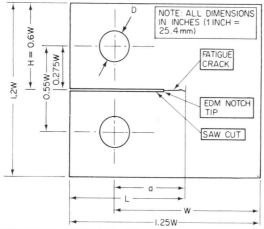

(a) CT SPECIMEN (H/W = 0.600) USED FOR <u>LOAD-CONTROL</u> TESTS. TYPES 2T AND 4T

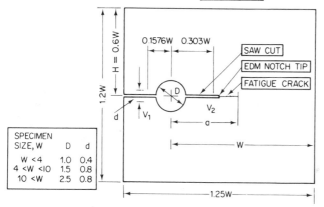

(b) CT SPECIMEN (H/W = 0.600) USED FOR <u>DISPLACEMENT-CONTROL</u> TESTS. TYPES 4C AND 7C

**Figure 17.21**  Compact-tension specimens used for load-control and displacement-control tests (Ref. 12). (a) CT specimen ($H/W = 0.600$) used for load-control tests, types 2T and 4T; (b) CT specimen ($H/W = 0.600$) used for displacement-control tests, types 4C and 7C.

respect to the remaining ligament length, $W - a$, Figure 17.22. In load control, the specimen is loaded in a large-tension machine with loading pins, Figure 17.21(a). For displacement control, special wedges are pushed into a hole on the crack line to wedge the crack surfaces apart, Figure 17.22. The overall testing setup for displacement control is shown in Figure 17.21(b), where this particular test specimen is about 2 × 2 ft. The displacement-control method offers a primary advantage in being able to obtain the entire *R*-curve for many values of crack length, whereas the load-control technique

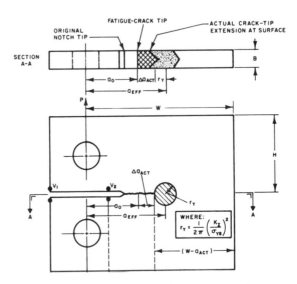

**Figure 17.22**   R-curve specimen showing $a_{\text{eff}} = a_0 + \Delta a_{\text{actual}} + r_y$.

can be used only to obtain that portion of the $R$-curve up to the $K_c$ value, where complete unstable fracture occurs. However, the advantage of the displacement-control technique is partially offset by the necessity for new or unique loading facilities and sophisticated instrumentation, whereas with

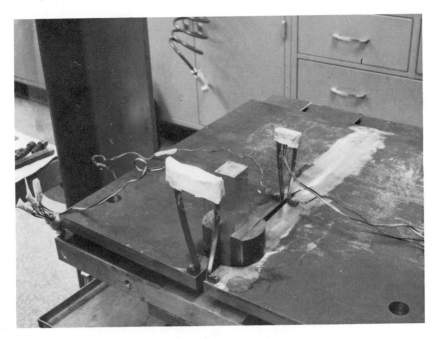

**Figure 17.23**   Displacement-control $R$-curve test setup.

the load-control method a conventional tension machine may be used in conjunction with relatively simple measuring devices (clip gauge and vernier micrometer).

After fabrication of the test specimens, they are fatigue-cracked and then tested at various testing temperatures.

For linear-elastic behavior, that is, no back surface yielding,

$$K_R = \frac{P \cdot f(a/W)}{B \cdot W^{1/2}} \qquad (17.24)$$

where $f(a/W)$ is a specific function of the relative crack length as given in ASTM Standard E-561[13] and $a$ is the effective crack length corresponding to

$$a_{\text{effective}} = (a_0 + \Delta a + r_y) \qquad (17.25)$$

$a_0$ = original crack length
$\Delta a$ = physical crack growth at tip of original crack
$r_y$ = plastic-zone size

$$r_y = \frac{1}{2\pi}\left(\frac{K_R}{\sigma_{ys}}\right)^2 \qquad (17.26)$$

The expression for $r_y$ is most accurate for high-strength low-toughness materials. For lower-strength, high-toughness materials, that is, $K_R/\sigma_{ys} > 1$, $r_y$ becomes large, and the expression for $K_R$, Equation (17.24), becomes increasingly invalid. In these cases, either plot $J$-resistance or CTOD-resistance curves directly (i.e., measure $J$ or CTOD as a function of $\Delta a$), or use correlations to estimate $K_R$ as described previously.

That is,

$$K_c = \sqrt{m\sigma_{\text{flow}}E\delta} \qquad (17.27)$$

or

$$K_c = \sqrt{J_c E} \qquad (17.28)$$

In either case, an elastic-plastic fracture parameter, $J$ or $\delta$, must be determined for low-strength, high-toughness materials, inasmuch as the elastic parameter, $K$, is not valid.

To illustrate the types of results that might be expected in terms of $K_c$, results of an $R$-curve investigation for a 50- and 62-ksi yield-strength structure steel (A572) by Novak[12] are described. Novak used two types of compact-tension (CT) test specimens (Figure 17.24), namely, load-control and displacement-control specimens, in his investigation.

Typical $R$-curves for various-sized specimens of A572 steel tested at +72°F are presented in Figure 17.24. Test results at +40°F and −40°F for this same steel are presented in Figures 17.25 and 17.26, respectively, and show that the effect of temperature on $R$-curve behavior is similar to the effect obtained using other types of fracture-toughness tests such as $K_{Ic}$ or CVN impact test specimens; that is, the values decrease significantly at lower temperatures.

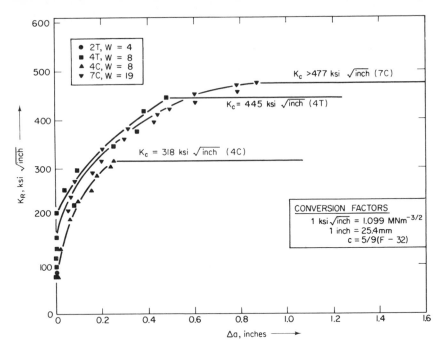

**Figure 17.24** *R*-curve and $K_c$ results for full-thickness ($B = 1.5$ in.) specimens of A572 Grade 50 steel tested at $+72°F$ (Ref. 12).

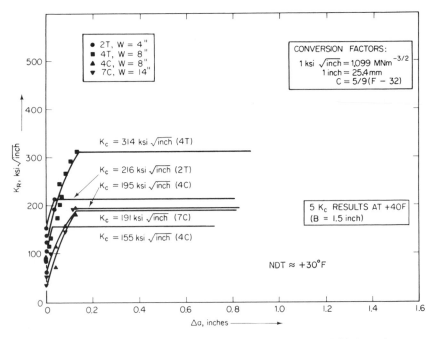

**Figure 17.25** *R*-curve and $K_c$ results for full-thickness ($B = 1.5$ in.) specimens of A572 Grade 50 steel tested at $+40°F$ (Ref. 12).

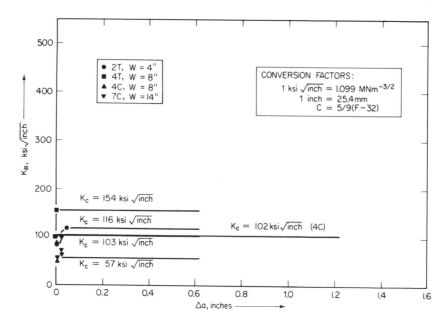

**Figure 17.26**   *R*-curve and $K_c$ results for full-thickness ($B = 1.5$ in.) specimens of A572 Grade 50 steel tested at $-40°F$ (Ref. 12).

The results in Figures 17.24 through 17.27 show, collectively, that a rapid increase in plane-stress crack tolerance occurs with increasing temperature. Evidence for this can be seen in both the increasing amounts of stable crack extension preceding fracture, $\Delta a_c$, and the increasing $K_c$ values that occur with increasing temperature. The specific variation in $K_c$ values with temperature for all the 1.5-in.-thick specimens tested is presented in a summary plot (Figure 17.27). This figure shows that the $K_c$ transition is quite steep at temperatures above $0°F$ ($-18°C$).

Static $K_{Ic}$ and $K_c$ test results for this same steel are compared in Figure 17.28 and show that at the higher temperatures, the plane-stress $K_c$ transition behavior is extremely large, much larger than the plane-strain transition in $K_{Ic}$ described in Chapter 4. This behavior would be expected as $K_c$ values extend well into the elastic-plastic region, whereas $K_{Ic}$ values are limited to plane-strain behavior. The significance of the $K_c$ values obtained from an *R*-curve analysis is in the calculation of the critical flaw size ($a_{cr}$) required to cause fracture instability under the same material and test conditions ($T$, $\dot{\varepsilon}$, and $B$) used to measure the specific $K_c$ value. The specific $a_{cr}$ value is further related to the level of the design stress, $\sigma_{des}$, relative to $\sigma_{ys}$ for a given specimen or structural geometry. A normalized plot showing the general relationship of $a_{cr}$ to such design parameters for a large center-cracked tension (CCT) specimen subjected to uniform tension is presented in Figure 17.29. Because of the normalized basis of the plot,

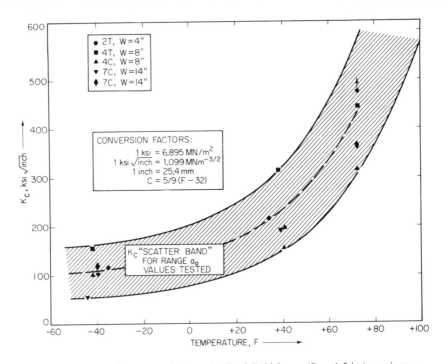

**Figure 17.27** Summary of $K_c$ results for full-thickness ($B = 1.5$ in.) specimens of A572 Grade 50 steel and A572 steel processed to 62-ksi strength level (Ref. 12).

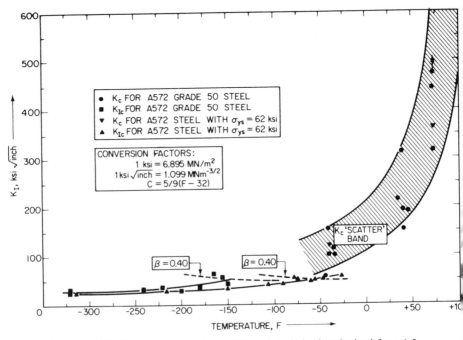

**Figure 17.28** Summary comparison of $K_c$ and $K_{Ic}$ behavior obtained from 1.5-in.-thick plates of A572 steel (Ref. 12).

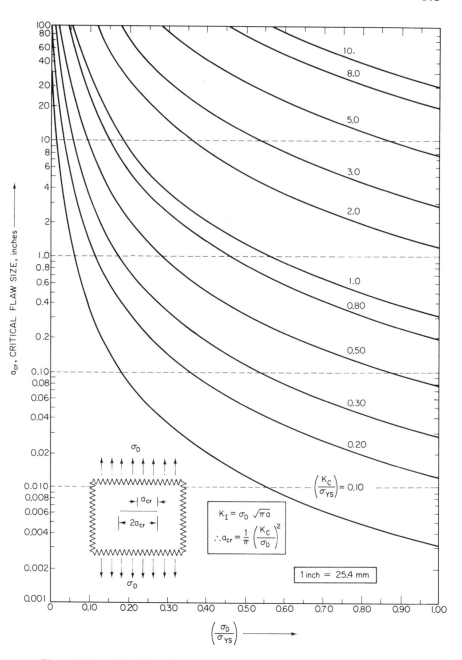

**Figure 17.29** Critical-flaw-size ($a_{cr}$) requirements for the initiation of a critical event ($K_{Icr}$) for cracks contained in an infinite center-cracked tension specimen and subjected to a uniform tension stress ($\sigma_D$) applied remotely (Ref. 12).

Figure 17.29 can be used to calculate $a_{cr}$ values for a CCT specimen of any material (i.e., with any yield strength $\sigma_{ys}$) for which fracture-mechanics results ($K_{Ic}$, $K_{Id}$, $K_c$, $K_{Iscc}$) are available under the loading rate, temperature, and state of stress of interest.

An example of the applicability of the $a_{cr}$ values can be given in terms of a typical structural member, such as a large H-beam (girder) with typical thicknesses for both the flange and the web on the order of $\frac{1}{2}$–$1\frac{1}{2}$ in. Specifically, the $a_{cr}$ values cited would have application for through-thickness cracks located in the web of such a beam, where plane-stress conditions would exist and where the crack is surrounded by a thin plate having large in-plane dimensions. However, the same $a_{cr}$ values would have no application for partial-thickness cracks (PTC) emanating from the top surface of the tension flange of the beam (such as would occur at the base of a cover plate due to fatigue).

These same stress-state (plane-stress) and structural (large planar dimensions) requirements are necessary for the interpretation of essentially *all* R-curve measurements, since such measurements intrinsically deal with materials exhibiting high levels of crack tolerance. In turn, high levels of crack tolerance ($K_c/\sigma_{ys} > 1$) under plane-stress conditions imply the existence of either very large critical flaws ($a_{cr}$) under low levels of elastic stress ($\sigma_D \leq \frac{1}{2}\sigma_{ys}$) (Figure 17.29) or high $K_c$ levels that translate, for short cracks ($a$), into large values of the corresponding critical crack-tip plastic zone, $r_y$ under the action of high elastic stress ($\frac{1}{2}\sigma_{ys} \leq \sigma_D \leq \sigma_{ys}$). In either case, *containment* of such values within a large elastic-stress field is *necessary* before $a_{cr}$ calculations can be valid. Accordingly, to accomplish this containment for plane-stress conditions, large planar dimensions relative to the thickness, $B$, are necessary for either a specimen or a structural element.

## 17.4. J-Integral

The path-independent $J$-integral proposed by Rice[20] is a method of characterizing the stress-strain field at the tip of a crack by an integration path taken sufficiently far from the crack tip to be analysed and then substituted for a path close to the crack-tip region. Thus, even though considerable yielding may occur in the vicinity of the crack tip, if the region away from the crack tip can be analyzed, behavior of the crack-tip region can be inferred. Thus this technique can be used to estimate the fracture characteristics of materials exhibiting elastic-plastic behavior and is a means of extending fracture-mechanics concepts from linear-elastic ($K_{Ic}$) behavior to elastic-plastic behavior.

For linear-elastic behavior, the $J$-integral is identical to $G$, the energy release rate per unit crack extension (Chapter 2). Therefore a $J$-failure criterion for the linear-elastic case is identical to the $K_{\text{Ic}}$ failure criterion. For linear-elastic plane-strain conditions,

$$J_{\text{Ic}} = G_{\text{Ic}} = \frac{(1 - \nu^2)K_{\text{Ic}}^2}{E} \qquad (17.29)$$

The CTOD parameter, $\delta$, also is related to $J$ (and $K$) as follows:

$$\delta = \frac{G}{m\sigma_{ys}} = \frac{K^2}{m\sigma_{ys}E} \qquad (17.30)$$

because $J = G$,

$$J \simeq m\sigma_{ys}\delta \qquad (17.31)$$

where $1 \leq m \leq 2$

Recent studies[7] of the relation between $J$ and $\delta$ indicate that $\sigma_{\text{flow}} \cong (\sigma_{ys} + \sigma_{\text{ult}})/2$ should be used in Equation (17.27). Thus,

$$J = m\sigma_{\text{flow}}\delta \qquad (17.32)$$

is a preferred relation between $J$ and $\delta$.

The $J$-integral, a mathematical expression, is a line or surface integral that encloses the crack front from one crack surface to the other as shown in Figure 17.30. It is used to characterize the local stress-strain field around the crack front for either elastic or elastic-plastic behavior. Figure 17.30 also shows the three expressions used to determine values of the line integral, namely, evaluation of the integral, determination of the potential energy rate, and use of the pure-bend-area method.

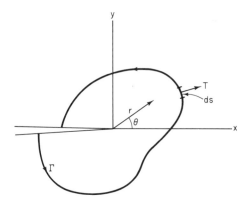

$$J = \int_r W dy - T\left(\frac{\delta\bar{u}}{\delta x}\right) ds \, (\text{LINE INTEGRAL})$$

$$J = -1/B \frac{\delta u}{\delta a} (\text{POTENTIAL ENERGY RATE})$$

$$J = 2A/Bb \, (\text{PURE BEND AREA METHOD})$$

**Figure 17.30**  Crack-tip coordinate system and typical line integral contour.

The line integral is defined as follows:

$$J = \int_\Gamma W \, dy - T\left(\frac{\partial \overline{u}}{\partial x}\right) ds \qquad (17.33)$$

where     $\Gamma$ = any contour surrounding the crack tip as shown in
              Figure 17.30 (note that the integral is evaluated in a
              counterclockwise manner starting from the lower flat
              notch surface and continuing along an arbitrary path $\Gamma$
              to the upper flat surface).

$\quad\quad\quad W$ = loading work per unit volume or, for elastic bodies,
              the strain energy density = $\int_0^\varepsilon \sigma d\varepsilon$.

$\quad\quad\quad\, T$ = the traction vector at $ds$ defined according to the
              outward normal $n$ along $\Gamma$, $T_i = \sigma_{ij}n_j$.

$\quad\quad\quad \overline{u}$ = displacement vector at $ds$.

$\quad\quad\, ds$ = arc length along contour $\Gamma$.

$T\left(\dfrac{\partial \overline{u}}{\partial x}\right) ds$ = the rate of work input from the stress field into the
                        area enclosed by $\Gamma$.

For any linear-elastic or elastic-plastic material treated by deformation theory of plasticity, Rice[20] has proven path independence of the $J$-integral. Read[21] and Dodds, et al.[22,23] have described experimental and analytical methods for direct evaluation of the $J$-integral.

The second method of determining $J$ is based on a more physical viewpoint. In this method $J$ is interpreted as the potential energy difference between two identically loaded bodies having neighboring crack sizes, or

$$J = -\frac{1}{B}\frac{\partial \overline{u}}{\partial a} \qquad (17.34)$$

This definition is shown schematically in Figure 17.31, where the shaded area is $\partial \overline{u} = JB \, da$. Note that $\Delta$ is measured in the load line.

Considerable work on developing the $J$-integral as an analytical tool for elastic-plastic crack-tip field analysis has been performed by Begley and Landes[24,25] using a compliance technique. In this method, several specimens of varying crack length are used to obtain load versus displacement curves, as shown schematically in Figure 17.32. Values of energy per unit thickness (area under the $P$–$\Delta$ curve) are obtained for different initial crack lengths at various values of deflection. These energy values are plotted against crack length as shown in Figure 17.33 for an 1196NiCrMoV steel having a yield strength of 135 ksi. The slopes of the curves in Figure 17.33 are the changes in potential energy per unit thickness per unit change in crack length and thus are equal to values of $J$. These values of $J$ are plotted versus deflection in Figure 17.34. In their studies of this particular material, Begley and Landes[24] found that the average deflection at failure was about 0.024 in. (Figure 17.32), and thus the critical $J$, or $J_{\mathrm{Ic}}$, is about 950 in.-

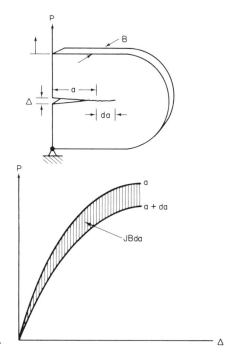

**Figure 17.31**  Interpretation of *J*-integral.

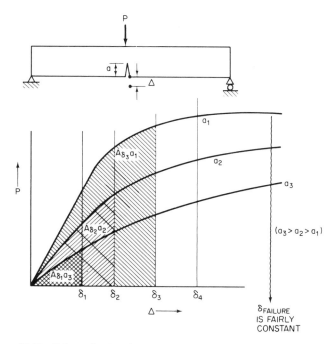

**Figure 17.32**  Schematic showing load-displacement curves for various crack depths.

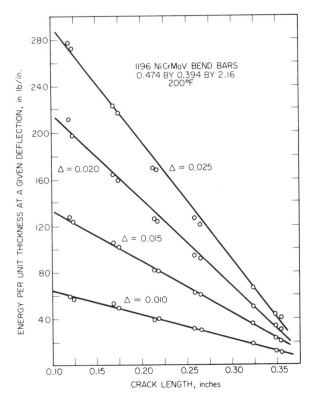

**Figure 17.33**  Energy absorbed at a given deflection versus crack length for NiCrMoV steel bars.

lb/in.$^2$ (Figure 17.34) or an equivalent $K_{Ic}$ of about 180 ksi$\sqrt{in}$. This value agreed reasonably well with $G_{Ic}$ values obtained from essentially elastic failures of 8-in.-thick compact-tension specimens of the same material.

Bucci et al.[26] summarize this procedure to compute the $J$ integral as a function of $\delta$ from a family of load-displacement records as follows:

Given a typical test specimen configuration [Figure 17.35(a)], load-displacement ($P$–$\Delta$) records are obtained for several test specimens, each having different crack lengths [Figure 17.35(b)]. For given values of deflection, $\Delta$, the area under each load-displacement record may be interpreted as pseudopotential energy of the body at that displacement. This energy can then be plotted [Figure 17.35(c)] as pseudopotential energy normalized per unit thickness, $U/B$, versus crack length, $a$, for constant $\Delta$. Following $J = -(1/B)(\delta \bar{u}/\partial a)$, $J$ may be interpreted as the area between load-displacement curves of neighboring crack sizes, or more simply as the negative slope of the $U/B$ versus crack-length curves, for given constant $\Delta$. This permits evaluation of a $J$ versus $\Delta$ relationship which is also a function of crack size, $a$, [Figure 17.35(d)]. Given

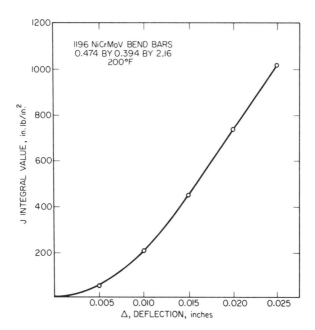

**Figure 17.34**  *J*-value as a function of deflection for NiCrMoV bend bars.

the *J* versus $\Delta$ relationship for a given crack size, an experimentally determined critical fracture displacement, which characterizes onset of unstable fracture in a particular test specimen, may be used to determine a critical *J*, $J_c$.[27,28]

**Standard test method for $J_{Ic}$.**    The third method of determining *J*, the pure-bend-area method, is widely used and is described in ASTM Standard E-813 for $J_{Ic}$—"A Measure of Fracture Toughness."[29] This Standard Test Method is used to determine the value of *J* at the initiation of crack growth. Either bend or edge-notched specimens are used, as shown in Figure 17.36. For either specimen type, the load-line displacement must be measured, not the crack-mouth opening displacement as in $K_{Ic}$ or CTOD testing. The total area under the load–load-line displacement curve should be measured as shown in Figure 17.37. For each of several tests to different $\Delta a$ (crack-growth values), determine *J* as follows:

$$J = \frac{A}{Bb} \cdot f\left(\frac{a_0}{W}\right) \qquad (17.35)$$

where  *A* = area under load–load-point displacement record, energy.
   *B* = specimen thickness.
   *b* = initial uncracked ligament, $W - a_0$.
   *W* = specimen width.
   $a_0$ = original crack size, including fatigue crack.

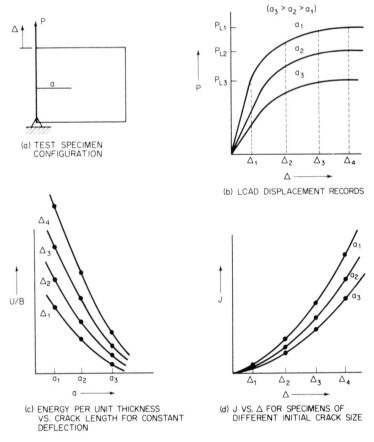

(a) TEST SPECIMEN
CONFIGURATION

(b) LCAD DISPLACEMENT RECORDS

(c) ENERGY PER UNIT THICKNESS
VS. CRACK LENGTH FOR CONSTANT
DEFLECTION

(d) J VS. Δ FOR SPECIMENS OF
DIFFERENT INITIAL CRACK SIZE

**Figure 17.35**  Schematic diagram of $J$ versus $\delta$ evaluation employing a family of load-displacement records (Ref. 26).

For the three-point bend specimen,

$$f\left(\frac{a_0}{W}\right) = 2.0$$

For the compact specimen,

$$f\left(\frac{a_0}{W}\right) \simeq 2.2$$

as given in ASTM Standard E-813.[29]

Because the $J$-value at onset of crack growth cannot be known ahead of time, several identical test specimens must be tested to different values of $\Delta a$ (crack growth). These specimens are then unloaded, heat tinted to

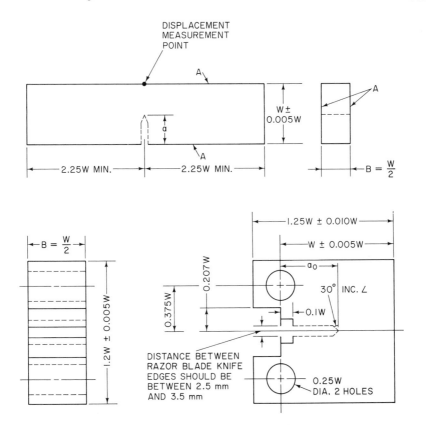

**Figure 17.36**   Typical compact specimen modified for *J*-integral testing.

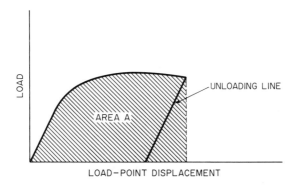

**Figure 17.37**   Illustration of work to a given displacement.

mark the crack growth, $\Delta a$, broken open (usually at very low temperatures), and values of $J$ versus $\Delta a$ plotted as shown in Figure 17.38.

A regression curve is fitted to the data points, and the intersection of this line with the blunting line ($J = 2\sigma_{ys}\Delta a_B$) is defined as $J_{Ic}$, the value of $J$ at the onset of crack growth. For additional details of testing, validity of data, and so on, the reader is referred to the current ASTM E-813 Standard.[29]

**Fracture control using the $J$-integral approach.** The objective in elastic-plastic fracture control using the $J$-integral method is similar to that using the linear-elastic $K_{Ic}$ approach, namely, keep

$$J_I < J_{Ic}$$

or

$$K_I < K_{Ic}$$

for all service loadings. $J_I$ and $K_I$ are the driving forces, which are a function of crack size and applied load, whereas $J_{Ic}$ and $K_{Ic}$ are the resistance forces, which are essentially material properties at the appropriate service temperature and loading rate.

One of the more widely used applications of $J$-integral test results is merely to relate $J_{Ic}$ to $K_{Ic}$ as follows:

$$K_{Ic}^2 = \frac{J_{Ic}E}{(1 - \nu^2)} \tag{17.36}$$

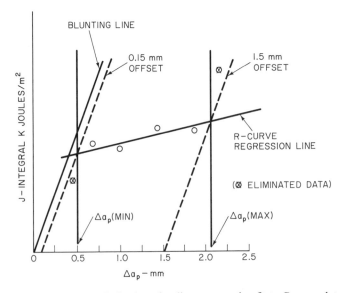

**Figure 17.38**  Data elimination after linear regression fit to $R$-curve data.

This assumes linear-elastic behavior, and thus the $K$ value is presumed to be that for a material thick enough to be in plane strain. Obviously, the error involved in the estimate of $K_{Ic}$ or $K_c$ increases with increasing values of $J$, particularly because measurements of $J_{Ic}$ require stable crack growth to occur. However, this approach does give reasonable estimates of $K_{Ic}$ or $K_c$ as shown in Figures 17.15 to 17.17. Knowing $K_{Ic}$ or $K_c$ in this manner, use linear-elastic calculations of $K_I$ described in Chapter 2 and compare $K_I$ to $K_{Ic}$.

Another application of *J*-integral in fracture control is the use of a *J*-analysis curve, similar to the CTOD design curve described previously. Because it has been shown that the *J* and CTOD values are linearly related[7] as shown in Figure 17.39, such an analysis curve seems to be a logical tool

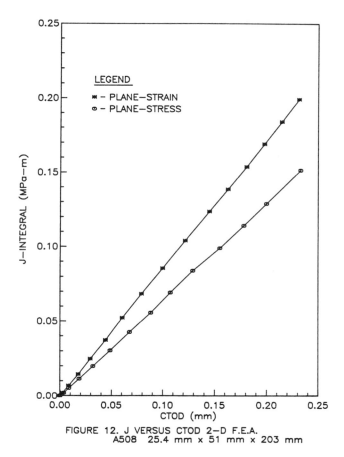

FIGURE 12. J VERSUS CTOD 2–D F.E.A.
A508   25.4 mm × 51 mm × 203 mm

**Figure 17.39**   *J*-integral versus CTOD for two-dimensional finite element analysis of A508 steel in 25.4-mm × 51-mm × 203-mm specimen.

to use $J$-values. That is,

$$J = m\sigma_{flow}\delta \tag{17.32}$$

The previously described CTOD design curve (Figure 17.18) initially was developed on the basis of the Dugdale model.[2] However, the current design curve is based primarily on experimental evidence. Previously, it was shown to be a reasonable design curve with actual test results having factors of safety between about 1 and 5 (Figure 17.19).

In contrast, the $J$-analysis curve is based on analytical, or computational, studies. Turner[30,31] has noted that inasmuch as the two curves are similar, each supports the other.

The $J$-analysis curve is represented by the following expressions:

$$\frac{J}{E\pi\bar{a}\varepsilon_{ys}^2} = \left(\frac{\varepsilon}{\varepsilon_{ys}}\right)^2 \qquad \text{for } \left(\frac{\varepsilon}{\varepsilon_{ys}}\right) < 1.0 \tag{17.37}$$

and

$$\frac{J}{E\pi\bar{a}\varepsilon_{ys}^2} = \frac{2\varepsilon}{\varepsilon_{ys}} - 1 \qquad \text{for } \left(\frac{\varepsilon}{\varepsilon_{ys}}\right) > 1.0 \tag{17.38}$$

Substituting

$$J = m\sigma_{ys}\delta \tag{17.32}$$

into the foregoing expressions and remembering from the CTOD design curve that

$$\phi = \frac{\delta}{2\pi\varepsilon_{ys}a} \tag{17.21}$$

results in the following relation:

$$2m\phi = \left(\frac{\varepsilon}{\varepsilon_{ys}}\right)^2 \qquad \text{for } \frac{\varepsilon}{\varepsilon_{ys}} \leq 1.0 \tag{17.39}$$

and

$$2m\phi = \frac{2\varepsilon}{\varepsilon_{ys}} - 1 \qquad \text{for } \frac{\varepsilon}{\varepsilon_{ys}} \geq 1.0 \tag{17.40}$$

The factor, $m$, is a plastic-stress-intensification factor that ranges from about 1.0 to 2.0. For plane-strain conditions, $m \cong 2$,

$$\phi = \tfrac{1}{4}\left(\frac{\varepsilon}{\varepsilon_{ys}}\right)^2 \qquad \text{for } \frac{\varepsilon}{\varepsilon_{ys}} \leq 1.0$$

and

$$\phi = \tfrac{1}{2}\left(\frac{\varepsilon}{\varepsilon_{ys}} - 0.5\right) \qquad \text{for } \frac{\varepsilon}{\varepsilon_{ys}} \geq 1.0$$

For plane-stress conditions, $m \simeq 1.0$,

$$\phi = \tfrac{1}{2}\left(\frac{\varepsilon}{\varepsilon_{ys}}\right)^{2} \qquad \text{for } \frac{\varepsilon}{\varepsilon_{ys}} \leq 1.0$$

and

$$\phi = \frac{\varepsilon}{\varepsilon_{ys}} - 0.5 \qquad \text{for } \frac{\varepsilon}{\varepsilon_{ys}} \geq 1.0$$

Both *J*-design curve relations ($m = 2$ and $m = 1$) are plotted in Figure 17.40, along with the CTOD design curve presented earlier in Figure 17.18(a).

The CTOD curve[32] is a *design* curve with a factor of safety of about 2.5 (Figure 17.19), whereas the *J*-analysis curve is an analytical curve used to predict critical conditions and does not have a built-in factor of safety. Thus, the *J*-analysis curves for $m = 1$ and $m = 2$ should fall below the CTOD design curve as shown in Figure 17.40.

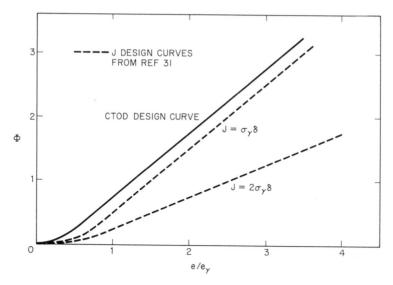

**Figure 17.40**  Comparison between CTOD design curve and curves based on Ref. 31.

In summary, one way to use the *J*-integral analysis curve in fracture control parallels the procedure followed using the CTOD design curve, namely,

1. Determine $\varepsilon/\varepsilon_{ys}$ for the structure under consideration.
2. Enter the analysis curve (Figure 17.40) to determine $\phi$.
3. Because

$$\phi = \frac{J}{2\pi m E \varepsilon_{ys}^2 \bar{a}} \qquad (17.41)$$

determine $J = J_{applied}$ for the specific strain level and crack size in the structure being analyzed.
4. Compare $J_{applied}$ with $J_{Ic}$ from material testing as described previously. Alternately, if $J_{Ic}$ is known from testing, and the failure strain level or critical crack size is desired, use the same analysis curve but start with $J = J_{Ic}$ and either (a) determine $\varepsilon_{failure}$ for a given crack size, $a$, or (b) determine $a_{cr}$ for a given strain level, $\varepsilon$.

**Finite element approach for J-fracture analysis.** A more accurate method of analyzing the fracture behavior of flawed structures is the use of a finite-element analysis (FEA) to determine the $J$-integral as a function of applied

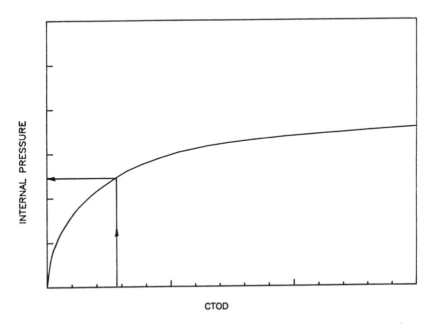

**Figure 17.41** Schematic showing relationship between internal pressure and CTOD for a notched pressure vessel.

loading, that is, an applied $J$-integral analogous to a driving force. Using this method, a curve of load, pressure, or stress versus applied $J$ at the site of a flaw or crack is developed for the structure of interest. A schematic curve of internal pressure in a notched pressure vessel versus applied $J$ at a long surface flaw is shown in Figure 17.41. Using either a 2-D or a 3-D finite element analysis, similar curves can be developed for the value of applied $J$ that characterizes the driving force as a function of a specific load input in any type of structural member. It should be noted that at present, this is an extremely complex and expensive analytical procedure, particularly for 3-D analyses.

Next, $J_{Ic}$ tests of material either from the actual structure or from material similar to that from the structure are conducted. The curve of pressure versus applied $J$ is entered at the $J_{Ic}$ value for the material to predict the failure pressure for the notched pressure vessel, as is shown schematically in Figure 17.41.

Wellman[8] has used this approach to predict the behavior of five notched pressure vessels. Both 2-D and 3-D analyses were used. The 3-D FEA mesh for a surface-notched pressure vessel is shown in Figure 17.42. The internal pressure versus CTOD ($J$ could also have been used, except that it is not yet well defined in 3-D) for this particular pressure vessel is shown in Figures 17.43(a) and (b). It should be noted that although his results are presented in terms of CTOD, similar curves of $J$ versus pressure could have been developed inasmuch as there is a direct relation between $J$ and CTOD.

Finally, the critical $J_{Ic}$ value (or, in this example, CTOD value of 0.325 mm) from material tests is used to enter the analysis curve, Figure 17.43(b). Using this technique, the predicted burst pressure (190 MPa) was 1 percent less than the actual burst pressure. This procedure offers tremendous potential,[8] but it is expensive and fairly complex.

**Engineering approach for elastic-plastic fracture analysis using EPRI curves.**    Kumar, German, and Shih[33] have developed an engineering approach using handbook solutions, graphical methods, and estimation procedures to construct crack-driving force ($J_1$) solutions for various specimen geometries and pressure-vessel elements. These estimations of $J_1$ then can be compared with actual $J_{Ic}$ values in the same manner as described previously. Elastic solutions for $J$ are combined with plastic solutions for $J$ to determine $J_{applied}$ values. To apply this technique, the stress-strain curve for the material being analyzed must be known so that it can be characterized analytically. Values of $J_{elastic}$ and $J_{plastic}$ are tabulated for different geometries, specimen aspect ratios, crack length, and so on in Reference 33.

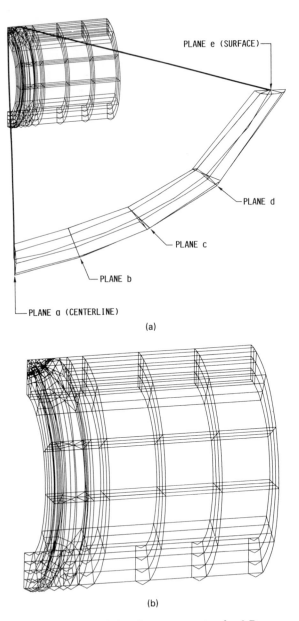

**Figure 17.42** (a) Crack-tip finite element geometry for 3-D pressure vessel, V-2; (b) 3-D finite element mesh for HSST pressure vessel, V-2.

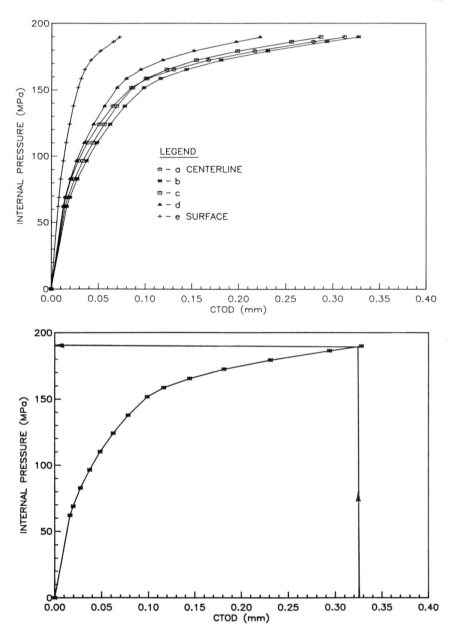

**Figure 17.43** (a) Internal pressure versus CTOD for vessel HSST, V-2, at five planes along crack front; (b) internal pressure versus CTOD for vessel HSST, V-2, using maximum CTOD curve, plane *b*.

# References

1. A. A. WELLS, "Unstable Crack Propagation in Metals—Cleavage and Fast Fracture," *Cranfield Crack Propagation Symposium, 1*, Sept. 1961, p. 210.
2. D. S. DUGDALE, "Yielding of Steel Sheets Containing Slits," *Journal of the Mechanics and Physics of Solids, 8*, 1960, p. 100.
3. "Methods for Crack-Tip Opening Displacement (CTOD) Testing," *British Standards Institution*, BS5762, 1979.
4. J. E. HOOD, "Fracture Initiation in Tough Materials," Conference of Metallurgists, CIMM, Montreal, Aug. 31, 1971.
5. K. NISHIOKA and H. IWANAGA, "Some Results of Dynamic COD-Test," *Significance of Defects in Welded Structures*, University of Tokyo Press, Tokyo, 1973.
6. G. W. WELLMAN and S. T. ROLFE, "Engineering Aspects of CTOD Fracture Toughness Testing," *Welding Research Council Bulletin 299*, Nov. 1984, pp. 3–14; and "Elastic-Plastic Fracture Test Methods: The User's Experience," *ASTM STP 856*, American Society for Testing and Materials, Philadelphia, 1985, pp. 230–262.
7. G. W. WELLMAN, S. T. ROLFE, and R. H. DODDS, "Three-Dimensional Elastic-Plastic Finite Element Analysis of Three-Point Bend Specimens," *Welding Research Council Bulletin 299*, Nov. 1984, pp. 15–25; and "Fracture Mechanics: Sixteenth Symposium," *ASTM STP 868*, American Society for Testing and Materials, Philadelphia, 1985, pp. 214–237.
8. G. W. WELLMAN, S. T. ROLFE, and R. H. DODDS, "Failure Prediction of Notched Pressure Vessels Using the CTOD Approach," *Welding Research Council Bulletin 299*, Nov. 1984, pp. 26–35.
9. F. M. BURDEKIN and M. G. DAWES, "Practical Use of Linear Elastic and Yielding Fracture Mechanics with Particular Reference to Pressure Vessels," *Practical Application of Fracture Mechanics to Pressure-Vessel Technology*, Institute of Mechanical Engineering, London, May 1971.
10. G. D. FEARNEHOUGH, G. M. LEES, J. M. LOWES, and R. T. WEINER, "The Role of Stable Ductile Crack Growth in the Failure of Structures," *Practical Application of Fracture Mechanics to Pressure-Vessel Technology*, Institute of Mechanical Engineers, London, May 3–5, 1971.
11. "Fracture Toughness Evaluation by R-Curve Methods," *ASTM STP 527*, American Society for Testing and Materials, Philadelphia, 1973.
12. S. R. NOVAK, "Resistance to Plane-Stress Fracture (*R*-Curve Behavior) of A572 Structural Steel," *ASTM STP 591*, American Society for Testing and Materials, Philadelphia, 1976.
13. "Standard Practice for R-Curve Determination," ASTM Standard E-561—81, Volume 03.01, *Metals—Mechanical Testing*.
14. R. H. HEYER and D. E. MCCABE, "Plane-Stress Fracture Toughness Testing Using a Crack-Line-Loaded Specimen," *Engineering Fracture Mechanics, 4*, 1972, p. 393.
15. R. H. HEYER and D. E. MCCABE, "Crack Growth Resistance in Plane-Stress Fracture Testing," *Engineering Fracture Mechanics, 4*, 1972, pp. 413–430.

16. A. A. WELLS, "Notched Bar Tests, Fracture Mechanics, and the Brittle Strengths of Welded Structure," *British Welding Journal, 12*, No. 2, 1962.

17. A. A. WELLS, "Crack Opening Displacements from Elastic-Plastic Analyses of Externally Notched Tension Bars," *Engineering Fracture Mechanics, 1*, No. 3, April 1969, pp. 399–410.

18. G. R. IRWIN, "Linear Fracture Mechanics, Fracture Transition, and Fracture Control," *Engineering Fracture Mechanics, 1*, No. 2, Aug. 1968, pp. 241–257.

19. D. E. MCCABE and R. H. HEYER, "*R*-Curve Determination Using a Crack-Line-Wedge-Loaded (CLWL) Specimen," *ASTM STP 527*, American Society for Testing and Materials, Philadelphia, 1973, pp. 17–35.

20. J. R. RICE, "A Path Independent Integral and the Approximate Analysis of Strain Concentration by Notches and Cracks," *Journal of Applied Mechanics, Transactions ASME, 35*, June 1968.

21. D. T. READ, "Experimental Method for Direct Evaluation of the *J*-Contour Integral," Fracture Mechanics: Fourteenth Symposium, Vol. II: *Testing and Applications, ASTM STP 791*, American Society for Testing and Materials, Philadelphia, 1983, pp. II 199–213.

22. R. H. DODDS and D. T. READ, "Elastic-Plastic Response of Highly Deformed Tensile Panels Containing Short Cracks," *ASME Special Publication, Computational Fracture Mechanics—Nonlinear and 3-D Problems*, Vol. 85, PVPD, June 1984, pp. 25–34.

23. R. H. DODDS, D. T. READ, and G. W. WELLMAN, "Finite-Element and Experimental Evaluation of the *J*-Integral for Short Cracks," Fracture Mechanics, Vol I: *Theory and Analysis, ASTM STP 791*, American Society for Testing and Materials, Philadelphia, 1983, pp. I 520–542.

24. J. A. BEGLEY and J. D. LANDES, "The *J* Integral as a Fracture Criterion," *ASTM STP 514*, American Society for Testing and Materials, Philadelphia, 1972, pp. 1–20.

25. J. D. LANDES and J. A. BEGLEY, "The Effect of Specimen Geometry on $J_{Ic}$," *ASTM STP 514*, American Society for Testing and Materials, Philadelphia, 1972, pp. 24–39.

26. R. J. BUCCI, P. C. PARIS, J. C. LANDES, and J. D. RICE, "*J* Integral Estimation Procedures," *ASTM STP 514*, American Society for Testing and Materials, Philadelphia, 1972, pp. 40–69.

27. J. R. RICE, P. C. PARIS, and J. G. MERKLE, "Some Further Results on *J*-Integral Analysis and Estimates," *ASTM STP 536*, American Society for Testing and Materials, Philadelphia, 1973, pp. 231–245.

28. J. D. LANDES and J. A. BEGLEY, "Test Results from *J*-Integral Studies: An Attempt to Establish a $J_{Ic}$ Testing Procedure," *ASTM STP 560*, American Society for Testing and Materials, Philadelphia, 1974, pp. 170–186.

29. "Standard Test Method for $J_{Ic}$, a Measure of Fracture Toughness," ASTM E-813—81, Vol. 03.01, *Metals—Mechanical Testing*, 1985.

30. C. E. TURNER, "The *J*-Estimation Curve, *R*-Curve, and Tearing Resistance Concepts Leading to a Proposal for a *J*-Based Design Curve Against Fracture," Proceedings of the Conference Fitness for Purpose, *Validation of Welded Constructions*, The Welding Institute, Paper 17, 1982.

31. C. E. TURNER, "A *J*-Based Design Curve," *Advances in Elasto-Plastic Fracture Mechanics*, edited by L. H. Larsson, Applied Science Publishers, New York, 1979.

32. M. G. DAWES, "The COD Design Curve," *Advances in Elasto-Plastic Fracture Mechanics*, edited by L. H. Larsson, Applied Science Publishers, New York, 1979.

33. V. KUMAR, M. D. GERMAN, and C. F. SHIH, "An Engineering Approach for Elastic Plastic Fracture Analysis," EPRI Report NP 1931, July 1981.

# Problems

The following problems can be divided into two broad categories:

INDIVIDUAL PROBLEMS FOR CHAPTERS 1–9

These problems pertain specifically to individual chapters and should be assigned as each chapter is finished.

ADVANCED DESIGN PROBLEMS

These problems are more complex in nature and require the student to synthesize portions of Chapters 1–9.

### Chapter 1

**1.1.** Define brittle fracture.

**1.2.** Define ductile fracture.

**1.3.** Which type of fracture is preferable? Why?

**1.4.** Define toughness.

**1.5.** Define notch toughness.

**1.6.** Describe how you would measure the toughness and the notch toughness of a structural steel at $+70°F$ using a tension test and a CVN impact test, respectively. Be specific as to units, method of measurements, and so on.

**1.7.** What are the three levels of structural performance? Briefly describe the physical significance of each.

**1.8.** What is the purpose of notch-toughness testing?

**1.9.** Define fracture mechanics.

**1.10.** What are the *primary* factors that affect the susceptibility of a structure to brittle fracture?

**1.11.** Define $K_{Ic}$ and $K_I$, and describe the difference between the two. Be specific.

**1.12.** Describe the difference between $\sigma$ and $\sigma_{ys}$.

**1.13.** Define subcritical crack growth.

**1.14.** Briefly describe the regions of subcritical crack growth.

**1.15.** Define fracture criterion.

**1.16.** Define fracture-control plan.

### Chapter 2

**2.1.** Given a structural material with a yield strength of 80 ksi that is loaded to a $K_I$ value of 60 ksi$\sqrt{\text{in.}}$, plot the stress distribution, $\sigma_y$, directly ahead of a crack ($\theta = 0$) between $r = 0$ and $r = 1.0$ in.

**2.2.** If the crack in Problem 2-1 is an edge crack 1.0 in. long in a 2-inch thick infinite plate, what is the nominal stress at failure if $K_{Ic} = 60$ ksi$\sqrt{\text{in.}}$?

**2.3.** Calculate $a_{cr}$ for $\sigma_{des} = \sigma_{ys}$, $\sigma_{ys}/2$, and $\sigma_{ys}/4$ for an edge crack in a 4-inch thick infinite plate in a material that has $\sigma_{ys} = 100$ ksi and $K_{Ic} = 120$ ksi$\sqrt{\text{in.}}$

**2.4.** For an edge crack in a 30-in.-wide plate with $\sigma_{ys} = 40$ ksi and $K_{Ic} = 100$ ksi$\sqrt{\text{in.}}$, calculate $a_{cr}$ for $\sigma_{des} = \sigma_{ys}$, $\sigma_{ys}/2$, and $\sigma_{ys}/4$.

**2.5.** A steel is being considered for the design of a pedestrian bridge over a busy street. The steel has a $K_{Ic}$ value of 80 ksi$\sqrt{\text{in.}}$ at 0°F and a yield strength of 100 ksi. Prepare a curve of allowable design stress versus crack depth for an edge notch in a 40-in.-wide plate loaded in tension.

**2.6.** Plot a stress–flaw-size, $K_{Ic}$ curve for an edge crack in a plate of infinite width for the following three steels. Plot all graphs on the same sheet of graph paper.

| Material | $K_{Ic}$ (ksi$\sqrt{\text{in.}}$) | $\sigma_{ys}$ (ksi) |
|----------|------------------|-----------|
| A | 80 | 260 |
| B | 110 | 220 |
| C | 140 | 180 |

**2.7.** To study the effect of crack geometry on $a_{cr}$, calculate and *draw* $a_{cr}$ for the following conditions.

(a) Edge crack—$K_I = 1.12\ \sigma\sqrt{\pi a}$.

(b) Through-thickness crack—$K_I = \sigma\sqrt{\pi a}$.

(c) Surface crack—$K_I = 1.12\ \sigma\sqrt{\pi a/Q} \cdot M_k$—for three conditions $a/2c = 0.1$, and 0.25, and 0.5.

Given that $K_{Ic} = 55$ ksi$\sqrt{\text{in.}}$, $\sigma_{ys} = 100$ ksi, and $\sigma_{des} = \sigma_{ys}/4$, calculate $a_{cr}$ and compare the results using *neat half-scale* drawings of the actual crack geometries. Assume a plate thickness of 2 in. and an "infinite" width.

**2.8.** Given the following data, calculate the critical stress, $\sigma_{cr}$, for the crack geometries specified.

(a) The surface area of the crack can be 6 in.$^2$, in any combination of width and depth.

(b) $K_{Ic} = 120$ ksi$\sqrt{\text{in.}}$

(c) $\sigma_{ys} = 100$ ksi.
(d) $\sigma_{des} = 40$ ksi.
(e) Infinite width.
(f) Plate thickness = 2 in.
   (i) A through-thickness crack.
   (ii) A single-edge crack.
   (iii) A circular crack.
   (iv) A surface crack.

**2.9.** A long 1-in.-thick steel plate loaded in tension is 8 in. wide and has an edge crack 2 in. deep. If the steel has a yield strength of 60 ksi and a $K_c$ of 200 ksi$\sqrt{\text{in.}}$, what load can the plate withstand before failure. What is the mode of failure? Explain your answer.

**2.10.** Using an edge-crack analysis, compare four steels that could be used in bridges in Alaska where service temperatures as low as $-60°$F will occur. The properties of the steels are as follows:

| Steel | Minimum Yield Strength (ksi) | Yield Strength at $-60°$F (ksi) | $K_{Ic}$ at $-60°$F (ksi$\sqrt{\text{in.}}$) |
|---|---|---|---|
| A36 | 36 | 44 | 60 |
| A441 | 50 | 69 | 53 |
| A572 | 50 | 57 | 100 |
| A514 | 100 | 110 | 60 |

Prepare graphs of stress versus flaw size for each of these steels. If the design stress of each steel is $0.6 \times$ the minimum yield strength, compare the critical crack sizes of each steel at the design stress level.

**2.11.** Plot neatly on the same graph the stress–flaw-size relation for a through-thickness crack in an infinite plate for the following three materials.

| Material | Modulus of Elasticity (psi) | $K_{Ic}$ (ksi$\sqrt{\text{in.}}$) | $\sigma_{ys}$ (ksi) |
|---|---|---|---|
| Steel | $30 \times 10^6$ | 100 | 120 |
| Titanium | $15 \times 10^6$ | 80 | 100 |
| Aluminum | $10 \times 10^6$ | 40 | 60 |

**2.12.** Given a longitudinal surface crack in a pressure vessel with the following dimensions and properties, what is the factor of safety against (1) yielding and (2) fracture, if the internal pressure is 3000 psi?

Length = 16 ft      $2c$ = 4.0 in.
Diameter = 4 ft      $\sigma_{ys}$ = 100 ksi
Thickness = 2 in.      $E$ = $30 \times 10^6$ psi
    $a_i$ = 1.0 in.      $K_{Ic}$ = 120 ksi$\sqrt{\text{in.}}$

**2.13.** An infinite plate of A36 steel with the properties shown has a 0.4-in.-long-crack propagating from a 2-in.-diameter hole. What stress level will cause failure?

$$K_{Ic} = 50 \text{ ksi}\sqrt{\text{in.}}$$
$$\sigma_{ys} = 40 \text{ ksi}$$
$$E = 30 \times 10^6 \text{ psi}$$

**2.14.** In Problem 2-13, assume that cracks grow from both sides of the hole. Determine the total crack length that will cause failure at a stress level of 20 ksi and show this crack with a sketch.

**2.15.** Given a steel with the following properties,

$$E = 30,000,000 \text{ psi}$$
$$RA = 60\%$$
$$\sigma_{ys} = 100 \text{ ksi}$$
$$\sigma_T = 120 \text{ ksi}$$
$$\sigma_{des} = 40 \text{ ksi}$$
$$K_{Ic} = 120 \text{ ksi}\sqrt{\text{in.}}$$

Assume that this steel is used in an infinite plate that has a 2-in. diameter hole in the center. Furthermore, assume that cracks are growing uniformly from both sides of the hole. What is the total defect length (crack plus hole) at failure?

**2.16.** Assume that the steel described in the previous problem is used in a plate 7.2 in. wide and that this plate has a 2-in.-diameter hole in the center. Also, assume that cracks are growing uniformly from both sides of the hole. What is the total defect length (cracks plus hole) at failure?

**2.17.** For the flaw shown in the figure below, what is the $K_I$ value in terms of $\sigma$? Label any factors you use.

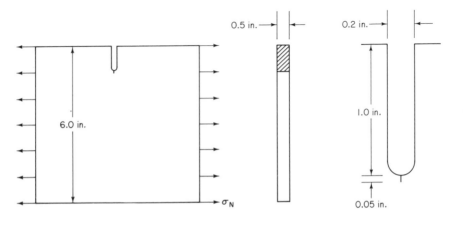

STRUCTURE                    SECTION                    NOTCH DETAIL

**2.18.** If $K_{Ic}$ is 80 ksi$\sqrt{in.}$ and $\sigma_{ys}$ = 200 ksi, what load can the structure shown in the figure below withstand before failure? Define failure specifically.

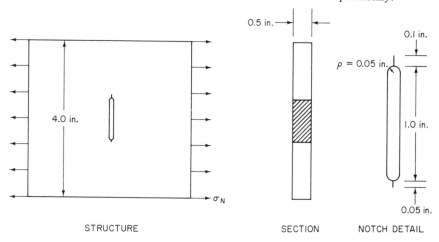

| STRUCTURE | SECTION | NOTCH DETAIL |

*Chapter 3*

**3.1.** What are the minimum specimen size requirements to ensure elastic plane-strain behavior? What is the relationship of the minimum thickness to the radius of the plane-strain plastic zones?

**3.2.** Briefly describe the significant steps involved in the ASTM Standard Test Method for $K_{Ic}$, ASTM Standard E-399.

**3.3.** Given a bend-test specimen 12 in. long, the following information was found:

$P_Q$ = 3700 lb
$B$ = 1.457 in.
$W$ = 2.0 in.
$a$ = 0.940 in.

Following the procedure used in the standard $K_{Ic}$ test method, determine $K_Q$.

**3.4.** A $K_{Ic}$ test was conducted on an A36 steel ($\sigma_{ys}$ = 36 ksi) according to the ASTM E-399 test procedure. The three-point bend specimen was tested at a temperature of $-45°F$ and a loading rate of 12.5 kips/min. At $-45°F$, A36 steel has a yield strength of 42 ksi. Fatigue cycling was conducted for 33,000 cycles at $P_{max}$ = 4.50 kips and $P_{min}$ ≈ 0.0 kips. The terminal stage for the fatigue crack was conducted for 100,000 cycles at $P_{max}$ = 3.5 kips and $P_{min}$ ≈ 0.0 kips. The finished specimen dimensions are as follows:

$S$ = 10 in.
$B$ = 1.25 in.
$W$ = 2.50 in.
$a$ (center plane) = 1.261 in.
$a$ (quarter plane) = 1.262 in.
$a$ (quarter plane) = 1.261 in.
$a$ (surface) = 1.215 in.
$a$ (surface) = 1.220 in.

The specimen exhibited linear-elastic behavior up to a load of 5.0 kips and a Crack Mouth Opening Displacement (CMOD) of 0.00450 in. Remaining load-CMOD values are given in the following table. Plot the load-CMOD data and determine the $K_{Ic}$ value for the material according to ASTM Standard E-399.

| Load (kips) | CMOD (in.) |
|:-----------:|:----------:|
| 5.0 | 0.00450 |
| 5.5 | 0.00495 |
| 6.0 | 0.00545 |
| 6.5 | 0.00595 |
| 7.0 | 0.00650 |
| 7.5 | 0.00710 |
| 8.0 | 0.00770 |
| 8.5 | 0.00840 |
| 9.0 | 0.00920 |
| 9.5 | 0.01010 |
| 10.0 | 0.01120 |
| 10.5 | 0.01260 |
| 10.9 | 0.01440 |

**3.5.** A similar test was performed on the same steel as described in Problem 3-4. The three-point bend specimen was tested at a temperature of $-320°F$ and a loading rate of 12.5 kips/min. At $-320°F$, assume that this A36 steel has a yield strength of 115 ksi. Fatigue cycling was conducted for 32,000 cycles at $P_{max} = 4.5$ kips and $P_{min} \approx 0$ kips. The terminal stage for the fatigue crack was conducted for 82,000 cycles at $P_{max} = 3.5$ kips and $P_{min} \approx 0$ kips. The finished specimen dimensions are as follows:

$S = 10.0$ in.
$B = 1.245$ in., 1.248 in., 1.249 in.
$W = 2.498$ in., 2.500 in.
$a$ (center plane) $= 1.308$ in.
$a$ (quarter plane) $= 1.301$ in.
$a$ (quarter plane) $= 1.300$ in.
$a$ (surface) $= 1.244$ in.
$a$ (surface) $= 1.240$ in.

The specimen showed linear-elastic behavior up to a load of 4.6 kips and a CMOD of 0.0045 in. Remaining load-CMOD values are as follows. Plot the load-CMOD data and determine the $K_{Ic}$ value for the material according to ASTM Standard E-399.

| Load (kips) | CMOD (in.) |
|:-----------:|:----------:|
| 4.60 | 0.00450 |
| 5.00 | 0.00495 |
| 5.50 | 0.00550 |
| 5.90 | 0.00595 |

**3.6.** A $K_{Ic}$ test was conducted on steel from the Carquinez Straits Bridge ($\sigma_{ys}$ = 52 ksi) according to the ASTM E-399 test procedure. The three-point bend specimen was tested at a temperature of $-20°F$ and a loading rate of 12.5 kips/min. Fatigue cycling was conducted for 100,000 cycles at $P_{max}$ = 8.25 kips and $P_{min} \approx 0$ kips. The terminal stage for the fatigue crack was conducted for 20,000 cycles at $P_{max}$ = 6.0 kips and $P_{min} \approx 0$ kips. The finished specimen dimensions are as follows:

$S$ = 12.0 in.
$B$ = 1.492 in., 1.491 in., 1.490 in.
$W$ = 3.006 in., 3.004 in.
$a$ (center plane) = 1.585 in.
$a$ (quarter plane) = 1.568 in.
$a$ (quarter plane) = 1.579 in.
$a$ (surface) = 1.487 in.
$a$ (surface) = 1.519 in.

The specimen showed linear elastic behavior up to a load of 7.9 kips and a CMOD of 0.0066 in. Remaining load-CMOD values are as follows. Plot the load-CMOD data and determine the $K_{Ic}$ value for the material according to ASTM Standard E-399.

| Load (kips) | CMOD (in.) |
|---|---|
| 7.9 | 0.00660 |
| 9.1 | 0.00765 |
| 10.4 | 0.00880 |

**3.7.** A similar test was performed on the same steel as described in Problem 3.6 ($\sigma_{ys}$ = 52 ksi). The three-point bend specimen was tested at a temperature of 66°F and a loading rate of 12.5 kips/min. Fatigue cycling was conducted for 142,000 cycles at $P_{max}$ = 8 kips and $P_{min} \approx 0$ kips. The terminal stage for the fatigue crack was conducted for 12,000 cycles at $P_{max}$ = 6.0 kips and $P_{min} \approx 0$ kips. The finished specimen dimensions are as follows:

$S$ = 12.0 in.
$B$ = 1.480 in., 1.483 in., 1.481 in.
$W$ = 3.002 in., 3.004 in.
$a$ (center plane) = 1.630 in.
$a$ (quarter plane) = 1.623 in.
$a$ (quarter plane) = 1.604 in.
$a$ (surface) = 1.506 in.
$a$ (surface) = 1.504 in.

The specimen showed linear-elastic behavior up to a load of 9.0 kips and a CMOD of 0.008 in. Remaining load-CMOD values are as follows. Plot the load-CMOD data and determine the $K_{Ic}$ value for the material according to ASTM Standard E-399.

| Load (kips) | CMOD (in.) |
| --- | --- |
| 9.0 | 0.0080 |
| 10.5 | 0.0094 |
| 12.0 | 0.0111 |
| 13.5 | 0.0127 |
| 15.0 | 0.0146 |
| 16.3 | 0.0166 |

**3.8.** A $K_{Ic}$ test was conducted on a structural aluminum alloy ($\sigma_{ys}$ = 49.8 ksi) according to the ASTM E-399 test procedure. The three-point bend specimen was tested at a temperature of 71°F and a loading rate of 250 lb/sec. Fatigue cycling was conducted for 87,000 cycles with $P_{max}$ = 4000 lb and $P_{min}$ = 340 lb. The finished specimen dimensions are as follows:

$S$ = 8 in.
$B$ = 1.001 in.
$W$ = 1.998 in.
$a$ (center plane) = 1.050 in.
$a$ (quarter plane) = 1.030 in.
$a$ (quarter plane) = 1.026 in.
$a$ (surface) = 0.989 in.
$a$ (surface) = 0.990 in.

The specimen showed linear-elastic behavior up to a load of 8.5 kips and a CMOD of 0.013 in. Remaining load-CMOD values are as follows. Plot the load-CMOD data and determine the $K_{Ic}$ value for the material according to ASTM Standard E-399.

| Load (kips) | CMOD (in.) |
| --- | --- |
| 8.5 | 0.013 |
| 9.50 | 0.0145 |
| 10.38 | 0.016 |
| 11.25 | 0.0175 |
| 12.00 | 0.019 |
| 13.00 | 0.0213 |
| 13.75 | 0.0235 |
| 14.08 | 0.0255 |
| 14.18 | 0.0275 |
| 14.20 | 0.029 |

**3.9.** In the ASTM "Standard Test Method for Plane-Strain Fracture Toughness of Metallic Materials" (E-399), $R_{sb}$ is defined as the strength ratio for bend specimens. This is the ratio of bending moment at $P_{max}$ to bending moment at yielding. Derive the expression given in E-399—A3, as well as a similar expression for the ratio of the fully plastic moment to the yield moment, assuming an elastic-plastic stress-strain curve with no strain hardening.

## Chapter 4

**4.1.** A thick rotor forging made of 124K406CrMoV steel (Chapter 4) has a surface flaw with $a/2c = 0.1$. Operating temperature is $+100°F$ and $\sigma_{des} = \sigma_{ys}$. How safe is the rotor?

**4.2.** For an infinite plate with an edge crack being loaded at $+70°F$, use the results in Figure 4.12 and calculate $a_{cr}$ for both a 2-in.-thick and a 0.5-in.-thick plate, $\sigma_{ys} = 180$ ksi. Use a design stress based on $\frac{2}{3}\sigma_{ys}$.

**4.3.** Prepare *neat* curves of $\sigma$ versus $a$ for the following situations:

(a) A36 steel at $-100°F$ tested slowly, Figure 4.28.
(b) A36 steel at $-100°F$ tested dynamically, Figure 4.28.
(c) A36 steel at $-100°F$ tested at an intermediate loading rate, Figure 4.28.

Using a scale of 1 in. vertical equal to 10 ksi and 1 in. horizontal equal to 0.5 in., plot three curves on the same sheet of paper for each of two crack geometries specified as follows:

(1) An edge crack in an infinite plate.
(2) A surface crack in an infinite plate with $a/2c = 0.25$.

On a third sheet of clean paper, compare the *actual* crack sizes (to scale) for each condition at a stress level of 20 ksi.

**4.4.** Work Problem 4.3 for an A517 steel having a yield strength of 100 ksi (Chapter 4). Prepare curves for a service temperature of $-100°F$.

## Chapter 5

**5.1.** The following Charpy V-notch impact specimen results were previously obtained for an ABS-B steel with a yield strength of 40 ksi:

| Temperature (°F) | Absorbed Energy (ft-lb) | Lateral Expansion (mils) | Percent Shear (%) |
|---|---|---|---|
| −90 | 1.5 | 1.0 | 0 |
| −90 | 1.5 | 2.0 | 0 |
| −90 | 2.0 | 1.0 | 0 |
| 0 | 6.5 | 10.0 | 5 |
| 0 | 7.5 | 6.0 | 5 |
| 32 | 8.5 | 13.0 | 10 |
| 32 | 12.5 | 11.0 | 15 |
| 32 | 13.0 | 10.0 | 20 |
| 32 | 15.5 | 19.0 | 10 |
| 70 | 24.0 | 23.0 | 30 |
| 70 | 28.0 | 26.0 | 30 |
| 70 | 52.0 | 55.0 | 50 |
| 70 | 60.0 | 65.0 | 60 |
| 70 | 68.0 | 66.0 | 50 |
| 120 | 67.0 | 76.0 | 80 |
| 120 | 73.0 | 67.0 | 85 |
| 120 | 87.0 | 80.0 | 90 |

Plot the resultant curves on separate pages and sketch the levels of performance on each of the three curves:

(a) Absorbed energy versus temperature.
(b) Lateral expansion versus temperature.
(c) Percent shear versus temperature.

**5.2.** The following Charpy V-notch impact specimen results were previously obtained for an A572 steel with a yield strength of 53 ksi:

| Temperature (°F) | Absorbed Energy (ft-lb) | Lateral Expansion (mils) | Percent Shear (%) |
|---|---|---|---|
| −90 | 4.0 | 2 | 0 |
| −90 | 2.0 | 2 | 0 |
| −40 | 6.5 | 5 | 5 |
| −40 | 6.0 | 3 | 5 |
| 0 | 9.0 | 9 | 10 |
| 0 | 12.5 | 11 | 10 |
| 32 | 38.0 | 34 | 35 |
| 32 | 21.5 | 19 | 20 |
| 50 | 17.0 | 19 | 25 |
| 50 | 23.0 | 24 | 30 |
| 80 | 49.5 | 44 | 80 |
| 80 | 30.0 | 31 | 50 |
| 100 | 35.0 | 36 | 65 |
| 100 | 48.5 | 44 | 75 |
| 165 | 56.0 | 53 | 90 |
| 165 | 62.0 | 59 | 100 |

Plot the resultant curves on separate pages and sketch the levels of performance on each of the three curves:

(a) Absorbed energy versus temperature.
(b) Lateral expansion versus temperature.
(c) Percent shear versus temperature.

**5.3.** The following CVN impact results were obtained from aluminum specimens:

| Temperature (°F) | Absorbed Energy (ft-lb) |
|---|---|
| −90 | 7 |
| −90 | 7 |
| 10 | 8 |
| 10 | 7 |
| 10 | 7 |
| 32 | 9 |
| 70 | 10 |
| 70 | 7 |

Plot the absorbed energy versus temperature for the aluminum specimens and also for the ABS-B steel from Problem 5.1. Why does the steel exhibit transition behavior and why doesn't the aluminum exhibit this type of behavior?

**5.4.** Using the CVN impact results of ABS-B steel analyzed in Problem 5.1, plot an appropriate curve of $K_{Id}$ and $K_{Ic}$ versus temperature.

**5.5.** Using the CVN impact results of A572 steel analyzed in Problem 5.2, plot an appropriate curve of $K_{Id}$ and $K_{Ic}$ versus temperature.

**5.6.** Determine the $K_{Id}-K_{Ic}$ temperature shift for

| | |
|---|---|
| (a) ABS-B steel | $\sigma_{ys} = 40$ ksi |
| (b) A514 steel | $\sigma_{ys} = 120$ ksi |
| (c) Grade 200 maraging steel | $\sigma_{ys} = 220$ ksi |
| (d) 7076 aluminum | $\sigma_{ys} = 40$ ksi |

**5.7.** Given the CVN results of Problems 5.1 and 5.2, determine both $K_{Id}$ and $K_{Ic}$ at $-100°F$. If the aluminum referred to in Problem 5.3 has a $K_{Ic}$ of 40 ksi$\sqrt{in.}$ at $-50°F$ and 50 ksi$\sqrt{in.}$ at $50°F$, estimate the $K_{Ic}$ at $-100°F$.

**5.8.** An A514 steel ($\sigma_{ys} = 120$ ksi) was tested in the laboratory, and the following CVN results were obtained:

| Temperature<br>(°F) | Absorbed Energy<br>(ft-lb) |
|---|---|
| −250 | 8 |
| −150 | 15 |
| −150 | 18 |
| −100 | 23 |
| −100 | 26 |
| − 50 | 45 |
| − 50 | 48 |
| − 50 | 52 |
| 0 | 60 |
| 0 | 62 |
| 0 | 70 |
| 70 | 65 |
| 70 | 75 |
| 70 | 78 |
| 120 | 65 |
| 120 | 75 |

(a) Estimate the $K_{Ic}$ and $K_{Id}$ values at a temperature of 80°F.

(b) Estimate the $K_{Ic}$ and $K_{Id}$ values at a temperature of $-100°F$.

**5.9.** The absorbed energy test results obtained in the laboratory for an A36 steel ($\sigma_{ys} = 36$ ksi) have been misplaced. Only three specimens remain unbroken and no more material is available. The broken specimens are still available and their measurable values are as follows:

| Temperature (°F) | Lateral Expansion (mils) | Percent Shear (%) |
|---|---|---|
| −90 | 3 | 0 |
| −90 | 2 | 0 |
| −30 | 5 | 5 |
| −30 | 3 | 0 |
| 10 | 10 | 10 |
| 10 | 8 | 5 |
| 40 | 15 | 12 |
| 40 | 18 | 15 |
| 70 | 38 | 45 |
| 70 | 45 | 55 |
| 110 | 56 | 70 |
| 110 | 60 | 70 |
| 150 | 80 | 85 |
| 150 | 85 | 100 |

One of the unbroken specimens was tested at −30°F, one at 70°F, and the last specimen at 150°F. Their results are as follows:

| Temperature (°F) | Absorbed Energy (ft-lb) | Lateral Expansion (mils) | Percent Shear (%) |
|---|---|---|---|
| −30 | 8 | 4 | 5 |
| 70 | 54 | 40 | 45 |
| 150 | 110 | 80 | 90 |

Plot an estimated curve of absorbed energy versus temperature using the data given for this particular steel. Estimate the absorbed energy values at temperatures of 40°F and 110°F.

## Chapter 6

**6.1.** A 1.0-in.-thick deck plate in a ship hull is fabricated from a steel with the following properties:

$$E = 30 \times 10^6 \text{ psi}$$
$$\sigma_{ys} = 120 \text{ ksi}$$
$$K_{Ic} = 60 \text{ ksi}\sqrt{\text{in.}}$$
$$K_{Id} = 40 \text{ ksi}\sqrt{\text{in.}}$$
$$RA = 60\%$$
$$\sigma_T = 150 \text{ ksi}$$

(a) For a design stress of 30 ksi for both static and dynamic loading, what is the maximum crack size, $2a$, that the structure can tolerate before fracture?

(b) If a crack with a total length of 4 in. is discovered at sea, and the maximum stress at that time is 10 ksi, what is the factor of safety against fracture?

**6.2.** Determine the critical crack depth for design stress levels of 80, 60, 40, and 20 ksi for a semi-infinite plate with an edge crack. The steel plate has a $K_{Ic}$ of 90 ksi$\sqrt{in}$. and a yield strength of 120 ksi. Tension test results indicate a percentage reduction in area at fracture of 60 percent. Compare your answers graphically with the values for a steel with $K_{Ic} = 180$ ksi$\sqrt{in}$.

**6.3.** An aluminum used in the aerospace industry has a $K_{Ic}$ of 27 ksi$\sqrt{in}$. and a $\sigma_{ys}$ of 70 ksi. Assume that $\sigma_{ys}$ in tension and compression are equal. This material is to be used in two design situations. For each of these situations, determine the factor of safety against the possible modes of failure for each condition.

CONDITION I:  Long internally pressurized cylindrical vessel with hemispherical ends

Pressure: 400 psi
Diameter of vessel: 6 ft
Thickness of vessel: 0.5 in.
Length of vessel: 30 ft
Surface flaw with depth of 0.25 in. and length of 10 in.
Assume crack is in longitudinal direction.

CONDITION II:  Solid column loaded in compression as shown in the figure below

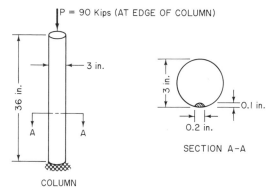

SECTION A–A

COLUMN

**6.4.** Two steels, A and B, are being considered for use in a 0.4-in.-thick pressure vessel that may have a surface flaw such that $K_I = 1.12\sigma\sqrt{\pi a/Q} \cdot M_k$ (assume $Q = 1.0$). The factor of safety against both fracture and yielding must be 2.0. Assume that fabrication quality and cost are the same for both steels.

Compare the behavior of both steels on a $\sigma$–$a$–$K$ plot. Select a steel for this application and a recommended design stress level for that steel. Justify your answer.

| Steel | $K_{Ic}$ (ksi$\sqrt{in}$.) | Yield Strength (ksi) |
|-------|--------------|----------------|
| A | 200 | 150 |
| B | 300 | 200 |

**6.5.** You are expected to design an offshore structure out of a high-strength steel that has the following properties at 30°F, the service temperature.

$\sigma_{ys}$ = 100 ksi
$\sigma_T$ = 120 ksi
$E$ = 30 × $10^6$ psi
RA = 65%
$K_{Ic}$ = 120 ksi$\sqrt{\text{in.}}$
$K_{Id}$ = 80 ksi$\sqrt{\text{in.}}$

A tentative tensile design stress of 60 ksi for both static and dynamic loading has been proposed, but no one knows what size flaws might be present as a result of fabrication.

Assume that various possible *shapes* of flaws might be present, and *determine and show to scale* the critical crack sizes for the design conditions given. Analyze the possibility of both static and dynamic loading. The various possible shapes of flaws in the tensile members (which will be 3 in. thick and 24 in. wide in cross section) are as follows:

(a) Single-edge crack.
(b) Through-thickness center crack.
(c) Surface crack along the 24-in. side
  (i) $a/2c$ = 0.1.
  (ii) $a/2c$ = 0.25.
  (iii) a/2c = 0.50.
(d) Embedded circular crack.

**6.6.** A major structure will be built using a high-strength aluminum that has a $K_{Ic}$ of 100 ksi$\sqrt{\text{in.}}$ and a $\sigma_{ys}$ of 70 ksi. Two design situations are contemplated as follows:

(a) $K_I$ = 50 ksi$\sqrt{\text{in.}}$ and $\sigma_{design}$ = 50 ksi.
(b) $K_I$ = 50 ksi$\sqrt{\text{in.}}$ and $\sigma_{design}$ = 25 ksi.

For these two design conditions, determine the actual flaw size that can be tolerated if the following possible flaw geometries can exist in very large plates.

(a) Center crack.
(b) Edge crack.
(c) Circular embedded crack.
(d) Surface flaw
  (i) $a/2c$ = 0.1.
  (ii) $a/2c$ = 0.25.
  (iii) $a/2c$ = 0.50.

A full-scale comparison of these flaw sizes should be made.

**6.7.** The following information on two steels and two aluminums has been obtained.

| Material | $\sigma_{ys}$ (ksi) | $K_{Ic}$ (ksi$\sqrt{\text{in.}}$) | Fabricated Cost ($/lb) |
|---|---|---|---|
| Steel A | 180 | 120 | 2.00 |
| Steel B | 120 | 160 | 1.50 |
| Aluminum C | 60 | 50 | 2.50 |
| Aluminum D | 30 | 80 | 2.00 |

You are responsible for the analysis of a 10-ft-long section of a pressure vessel (neglect any end effects—consider only the 10-ft-long section) with the following general conditions:

GENERAL CONDITIONS

(a) Internally pressurized vessel (10 ft long). Cost given is fabricated (cost/lb). Neglect any other costs of fabrication.
(b) Vessel diameter = 40 in. (O.D.).
(c) Internal pressure, $p$, is $p = 2000$ psi.
(d) Yielding occurs when the hoop stress exceeds $\sigma_{ys}$.

THREE SPECIFIC CONDITIONS

(I) Flaw-free vessels—perfect fabrication—factor of safety of 2.0 against yielding.
(II) Flawed vessel with poor quality control—0.5-in.-deep longitudinal crack $(a/2c = 0.25)$—factor of safety of 2.0 against fracture and yielding.
(III) Flawed vessel with improved quality of fabrication—0.1-in.-deep longitudinal crack $(a/2c = 0.25)$. Increase cost per pound by 50 percent because of improved quality of fabrication. Factor of safety of 2 against fracture and yielding.

REQUIRED SOLUTION

Determine the design stress levels, wall thicknesses, weight of 10-ft section and cost of a 10-ft section for each condition for your material.

## Chapter 8

**8.1.** A 6-in.-wide-steel plate with a yield strength of 60 ksi has a 0.5-in.-deep edge notch. Determine the fatigue-crack-initiation threshold. Plot $\Delta\sigma$ versus root radius ($\rho$) for the tension specimen.

**8.2.** An edge-notched initiation tension specimen of steel with a yield strength of 60 ksi has an applied $\Delta\sigma = 30$ ksi. If the width of the specimen is 6 in., plot the notch depth versus root radius ($\rho$) required for fatigue crack initiation.

**8.3.** A surface notch has been found in an infinitely wide steel plate with a yield strength of 80 ksi and a thickness of 1 in. The plate is loaded in tension with $\sigma_{min} = 0$ ksi and $\sigma_{max} = 30$ ksi. If the notch has an $a/2c$ ratio of 0.25, plot the notch depth versus root radius ($\rho$) required for fatigue crack initiation.

If the notch has an $a/2c$ ratio of 0.25 and a root radius ($\rho$) of 0.05 in., calculate the notch depth required for fatigue-crack initiation to begin.

**8.4.** Plot the fatigue-crack-propagation threshold versus $\sigma_{min}/\sigma_{max}$ for a ferrite-pearlite steel with a yield strength of 40 ksi.

**8.5.** A cyclic load is applied to a tension member of HY-130 steel. If a stress fluctuation of 20 ksi is applied, plot a graph showing the effect of root radius ($\rho$) on fatigue-crack initiation cycles.

**8.6.** A through-thickness notch is found in a tension member of A-517 steel. If the yield strength of the steel is 100 ksi, determine the fatigue-crack-initiation threshold. If the root radius of the notch is 0.02 in., $\Delta\sigma$ is 25 ksi, and $\sigma_{max}$ is 50 ksi, calculate the initiation notch length. If a 0.2-in. notch is present, determine the number of cycles required for fatigue-crack initiation.

### Chapter 9

**9.1.** Plot a curve of crack depth versus number of cycles of fatigue loading to failure for the following structural case:

(a) A36 structural steel (ferrite-pearlite).
(b) $K_{Ic} \cong 50$ ksi$\sqrt{\text{in.}}$ at service temperature.
(c) Minimum yield strength is 36 ksi.
(d) $\sigma_{des} = 23$ ksi (live load + dead load).
(e) Live-load stress = 10 ksi.
(f) Dead-load stress = 13 ksi.
(g) Surface flaw with $a_i = 0.1$ in.
(h) $a/2c = 0.25$ and remains that way throughout the life of the structure.
(i) Plate thickness is 4 in. and width is semi-infinite.
(j) Use $\Delta a = 0.1$ in.

Tabulate your results, and summarize the initial and final conditions including sketches. Also, determine $N$, neglecting the $R$-ratio term $(1 - R)^{1/2}$, to establish the effect of neglecting $R$-ratio.

**9.2.** Plot a curve of crack depth versus number of cycles of fatigue loading to failure for the following structural case:

(a) A514 steel (martensitic).
(b) $K_{Ic} = 120$ ksi$\sqrt{\text{in.}}$ at service temperature.
(c) Minimum yield strength is 100 ksi.
(d) $\sigma_{des} = 45$ ksi (live load + dead load).
(e) Live-load stress is 12 ksi.
(f) Dead-load stress is 33 ksi.
(g) Surface flaw with $a_i = 0.2$ in.
(h) $a/2c = 0.25$ and remains that way throughout the life of the structure.
(i) Plate thickness is 6 in. and width is semi-infinite.
(j) Use $\Delta a = 0.1$ in.

Tabulate your incremental crack-growth results and summarize the initial and final conditions, including a sketch. Also, determine $N$, neglecting the $R$-ratio term $(1 - R)^{1/2}$, to establish the effect of neglecting $R$-ratio.

**9.3.** Which of the following three steels would have the longest design fatigue life for the conditions given here? Document your answer.

| Steel | $\sigma_{ys}$ (ksi) | $K_{Ic}$ (ksi$\sqrt{\text{in.}}$) |
|-------|---------------------|-----------------------------------|
| Austenitic stainless | 50 | 80 |
| Ferrite-pearlite | 70 | 80 |
| Martensitic | 90 | 80 |

STRUCTURAL CONDITIONS

(a) Edge crack in an "infinite" plate.
(b) $a_0 = 0.3$ in.
(c) $\Delta\sigma = 15$ ksi.
(d) $R = 0$.
(e) $K_{I\ des} = K_{Ic}/2$.

Sketch to scale the conditions at failure for each of the three steels.

**9.4.** Given the following conditions, calculate the number of cycles required to grow an edge crack to failure in an infinitely wide plate:

$$\sigma_{ys} = 100\ \text{ksi}$$
$$K_{Ic} = 100\ \text{ksi}\sqrt{\text{in.}}$$
$$E = 30,000,000\ \text{psi}$$
$$\Delta\sigma = 30\ \text{ksi}$$
$$\sigma_{des} = 50\ \text{ksi}$$
$$a_i = 0.10\ \text{in.}$$
$$da/dN = \frac{0.66 \times 10^{-8}\ (\Delta K_I)^{2.25}}{(1 - R)^{1/2}}$$

**9.5.** Assume that you are responsible for the design of a wide-plate tension member built from a structural aluminum. The member has a center-crack defect ($2a$) and can be subjected to various service conditions. Information on the aluminum is as follows:

$$E = 10 \times 10^6\ \text{psi}$$
$$\sigma_{ys} = 40\ \text{ksi}$$
$$da/dN = \frac{6 \times 10^{-8}\ (\Delta K)^{2.5}}{(1 - R)^{1/2}}$$

Determine the propagation fatigue life for the following design conditions:

(a) Stress range = 10 ksi (0 to tension)
   $2a_0 = 0.5$ in.
   $K_{Ic} = 50$ ksi$\sqrt{\text{in.}}$
(b) Stress range = 20 ksi (0 to tension)
   $2a_0 = 0.5$ in.
   $K_{Ic} = 50$ ksi$\sqrt{\text{in.}}$

(c) Stress range = 20 ksi (0 to tension)
   $2a_0$ = 2.0 in.
   $K_{Ic}$ = 40 ksi$\sqrt{in.}$
(d) Stress range = 20 ksi (0 to tension)
   $2a_0$ = 0.1 in.
   $K_{Ic}$ = 80 ksi$\sqrt{in.}$
(e) Stress range = 10 ksi (0 to tension)
   $2a_0$ = 0.5 in.
   $K_{Ic}$ = 160 ksi$\sqrt{in.}$

Discuss briefly the relative importance on the fatigue-crack-propagation life for this aluminum of

(1) Lowering the stress range.
(2) Decreasing the initial flaw size.
(3) Using a tougher material.

**9.6.** Develop a series of "fatigue–crack-propagation design curves" for a *ferrite-pearlite* structural-grade steel that can be heat treated to the following conditions:

| Condition | $\sigma_{ys}$ (ksi) | $K_{Ic}$ (ksi$\sqrt{in.}$) |
|-----------|---------------------|----------------------------|
| A         | 60                  | 120                        |
| B         | 70                  | 100                        |
| C         | 80                  | 80                         |
| D         | 90                  | 60                         |

This steel is to be used in a structure that will be subjected to a dead-load stress of $0.2\sigma_{ys}$ and a live-load stress of $0.3\sigma_{ys}$.

Assuming that the design curves for each condition are for an infinitely wide plate with an edge crack, *carefully* plot a design curve of *initial crack size* (vertical axis) versus *propagation life* (horizontal axis). A semilog plot may be desirable. Note that this is not a typical crack-growth curve as plotted before. Rather, it is a series of four design curves showing the relation between initial crack size and propagation life.

**9.7.** Given the following conditions, determine the fatigue life and plot a curve of the crack depth versus number of cycles of fatigue loading for each of the given wall thicknesses. The structure is an internally pressurized long cylindrical vessel with a 40-in. diameter and a crack oriented in the longitudinal direction.

WALL THICKNESS

(1) 0.5 in.
(2) 1.0 in.
(3) 2.0 in.

STRUCTURAL CONDITIONS

(a) Ferrite-pearlite steel.
(b) $K_{Ic} \cong$ 50 ksi$\sqrt{in.}$ at service temperature.
(c) Minimum yield strength is 36 ksi.

(d) Maximum hoop stress = 23 ksi.
(e) Minimum hoop stress = 13 ksi.
(f) Surface flaw with $a_i$ = 0.1 in.
(g) $a/2c$ = 0.25 and remains that way throughout the life of the structure.

Also, summarize the initial and final conditions including sketches.

**9.8.** An infinitely wide tension member with a 0.5-in. thickness is subjected to a cyclic stress. A surface crack occurs on the face of the member at fabrication. Given the following conditions, determine the fatigue life and plot a curve of the crack depth versus number of cycles of fatigue loading. Assume that once the surface crack grows through the member, it becomes a through-thickness crack. Sketch this through-thickness crack to scale and explain your selection of $2a_i$.

STRUCTURAL CONDITIONS

(a) Ferrite-pearlite steel.
(b) $K_{\mathrm{Ic}} \cong 50$ ksi$\sqrt{\mathrm{in.}}$ at service temperature.
(c) Minimum yield strength is 36 ksi.
(d) $\sigma_{\max}$ = 23 ksi.
(e) $\sigma_{\min}$ = 13 ksi.
(f) Surface flaw with $a_i$ = 0.1 in.
(g) $a/2c$ = 0.25 and remains that way until the crack grows through the member.

**9.9.** To emphasize the importance of stress range on fatigue-crack-growth behavior, plot curves of initial surface flaw depth ($a_i$) versus number of cycles to failure ($N$). A large nuclear pressure vessel is fabricated from 10-in.-thick plates of A533 steel with a yield strength of 100 ksi. Fatigue-crack behavior is

$$K_{\mathrm{I}}^2 = \frac{1.21\pi\sigma^2 a}{Q}$$

$$\frac{da}{dN} = \frac{0.5 \times 10^{-8}\,(\Delta K)^{2.0}}{(1 - R)^{1/2}}$$

$$K_{\mathrm{Ic}} = 100 \text{ ksi}\sqrt{\mathrm{in.}} \text{ at the service temperature}$$

$$\frac{a}{2c} = 0.25$$

Determine the required curves for the following conditions:

(a) $\sigma_{\max}$ = 60 ksi    $\sigma_{\min}$ = 0 ksi
(b) $\sigma_{\max}$ = 60 ksi    $\sigma_{\min}$ = 30 ksi
(c) $\sigma_{\max}$ = 30 ksi    $\sigma_{\min}$ = 0 ksi
(d) $\sigma_{\max}$ = 30 ksi    $\sigma_{\min}$ = $-30$ ksi

Neglect any effect of $M_K$.

**9.10.** A tension member of 1.5-in. thickness and infinite width has cracks emanating from both sides of a 2-in. hole. The cracks are initially 0.5-in. each and grow at equal rates. How far can the cracks grow before instability occurs? And how many loading cycles occur before failure if the following conditions apply?

(a) Ferrite-pearlite steel.
(b) Yield strength = 50 ksi.
(c) $\sigma_{des}$ = 30 ksi.
(d) $\Delta\sigma$ = 15 ksi.
(e) $K_{Ic}$ = 100 ksi$\sqrt{in.}$

## Advanced Design Problem 1

A 48-in.-outside-diameter, 1-in.-thick pressure vessel is to be fabricated from steel A, B, or C. The vessel is 200 in. long with hemispherical ends and will be subjected to an internal pressure of 4000 psi. Assume "perfect" welding with all reinforcement ground smooth. Steels A, B, and C have the following properties:

| Steel | Yield Strength (ksi) | $K_{Ic}$ (ksi$\sqrt{in.}$) |
|-------|----------------------|----------------------------|
| A     | 200                  | 100                        |
| B     | 200                  | 150                        |
| C     | 200                  | 200                        |

(1) Carefully sketch the worst possible location of an external surface flaw that could exist on this vessel.
(2) If a surface crack is 1.0 in. long and 0.3 in. deep, determine the factor of safety against fracture for each steel with the flaw located as sketched in item (1).
(3) Assuming that steel A is used, how *deep* can a surface flaw grow by fatigue or stress corrosion before failure occurs, assuming that the $a/2c$ ratio of the crack is 0.3.
(4) If steel C is used, and a 0.3-in. deep, 1.0-in.-long surface flaw grows by fatigue with a constant aspect ratio ($a/2c$ constant), describe the failure condition. (Hint: Do not forget to account for $M_k$.)
(5) If steel C is a martensitic steel and is pressurized from 0 to the maximum pressure with $a_i$ = 0.3 in. and $2c_i$ = 1.0, as described in (4), determine $N_p$.

## Advanced Design Problem 2

Two-in.-thick plates of A517 martensitic steel with $\sigma_{ys}$ = 100 ksi and having the CVN impact test results shown in the following table will be used to fabricate cylindrical pressure vessels having a nominal diameter of 6 ft. and an overall length of about 30 ft. This steel has a 0.2 percent offset yield strength of 100 ksi.

Surface cracks with $a/2c$ = 0.3 and a depth of 0.4 in. may go undetected. Note that the plates may be oriented in either the longitudinal or transverse direction and that the surface flaws may be oriented in either direction. For a factor of safety of 2.0 against both yielding and fracture, determine the maximum *allowable* pressure to which this vessel should be subjected. Service temperature will be +75°F.

Assume that the ends of the vessels and the connections to these ends are to be analyzed by another division and your concern is only for the longitudinal section of the vessels.

Explain your answer clearly and include clear sketches of the vessel, plate orientation, and crack orientation that controls the design.

| Temperature (°F) | $CVN_{transverse}$ (ft-lb) | $CVN_{longitudinal}$ (ft-lb) |
|---|---|---|
| −100 | 3 | 6 |
| −75 | 5 | 15 |
| −50 | 7 | 28 |
| −25 | 10 | 37 |
| 0 | 12 | 43 |
| 25 | 13 | 47 |
| 50 | 14 | 49 |
| 75 | 14 | 50 |
| 100 | 15 | 50 |

### Advanced Design Problem 3

Assume that you are responsible for determining the overall safety of a long-span steel bridge being built in St. Louis. Your inspection people have found a 6-in.-long initial surface flaw having a depth of 0.10 in. oriented perpendicular to the direction of primary stress in the tension flange of a main structural member. This member is a welded wide-flange section with flange dimensions 15.0 in. by 2.50 in. and web dimensions 55.0 in. by 1.15 in. The member has a design moment of 90,400 kip in., which corresponds to a maximum bending stress of 35 ksi. The stress consists of a live-load design stress of 25 ksi plus a dead-load design stress of 10 ksi. The fatigue-crack-growth behavior of A514 steel can be characterized by the following expression:

$$\frac{da}{dN} = \frac{0.45 \times 10^{-8}(\Delta K)^{2.25}}{(1 - R)^{1/2}}$$

(1) Because the member is located such that it cannot be removed or repaired, how many cycles of load will it take to develop a 1.0-in.-deep crack, assuming the $a/2c$ ratio remains constant?

(2) If this steel has a yield strength of 100 ksi, and the following CVN impact values,

| Temperature (°F) | CVN Absorbed Energy (ft-lb) | Percent Shear (%) |
|---|---|---|
| −50 | 5 | 20 |
| −40 | 10 | 40 |
| −20 | 20 | 60 |
| 0 | 35 | 80 |
| +20 | 36 | 100 |
| +40 | 38 | 100 |
| +60 | 36 | 100 |

approximate the factor of safety against fracture after the crack has grown to a depth of 1.0 in. *Be very specific* as to what you mean.

(3) As the crack continues to grow, how does the safety of the bridge change?

(4) Compare the toughness of this steel with that required by AASHTO as shown in Table 16.1. Prepare a typed two-page memo commenting on the safety of this bridge steel in light of the AASHTO specification and the initial flaw depth of 0.1 in. The bridge is being built near St. Louis, where the minimum service temperature is 20°F.

### Advanced Design Problem 4

Compare the cycles to failure, $N$, versus the initial flaw size, $a_i$, for five A517 martensitic steels having different toughness levels, $K_{Ic}$, for a specific stress range, $\Delta\sigma$, equal to 6 ksi.

Construct a semilog plot of $a_i$ (vertical arithmetic axis, 0–1.5 in.) versus $N$ (horizontal log axis $10^4$ to $10^7$) for the following conditions:

Steels: A517 martensitic with $K_{Ic}$ values of 55, 70, 100, 150, 190 ksi$\sqrt{\text{in.}}$

Stress range: 33 ksi to 39 ksi.

Crack geometry: edge crack in an infinite plate.

Because the stress range is low, and $\sigma_{max}$ is only about one-third of $\sigma_{ys}$, the $R$-ratio should not have that much of an effect. Therefore, neglect $R$-ratio in this problem.

### Advanced Design Problem 5

The structure shown is built from an A517 quenched-and-tempered martensitic steel with the following properties:

$\sigma_{ys} = 120\,\text{ksi}$

$\sigma_{ult} = 140\,\text{ksi}$

CVN IMPACT PROPERTIES

| Temperature (°F) | CVN Impact (ft-lb) |
|---|---|
| −150 | 5 |
| −100 | 10 |
| − 50 | 25 |
| 0 | 35 |
| + 50 | 40 |
| +100 | 40 |

The structure is loaded in fatigue from a minimum stress of 40 ksi to a maximum stress of 60 ksi. To perform its design function, the U-shaped notch

shown below was carefully machined into one edge. The structure must operate
at 60°F.

(1) How many cycles of loading can it take before total failure? Define failure
very specifically.

(2) Estimate the fatigue life if the maximum stress were decreased to 50 ksi.

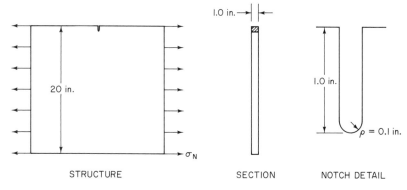

STRUCTURE                      SECTION          NOTCH DETAIL

*Advanced Design Problem 6*

The attached test results are for a newly developed steel with a yield strength
of 100 ksi. A 1-in.-thick (1T) Type B WOL specimen was used (Chapter 2). Using
the raw data given in the accompanying table,

(1) Determine a $da/dN$ expression for this specific steel for constant amplitude
loading of $\Delta P = 3$ kips.

(2) Using this $da/dN$ expression, calculate the fatigue lives for the following
conditions (neglect the effect of $R$-ratio):

(a) $K_{Ic} = 150$ ksi$\sqrt{\text{in.}}$
    $\sigma_{max} = 55$ ksi
    $\sigma_{min} = 40$ ksi
    $a_0 = 0.75$-in. edge crack in an infinite plate

(b) $K_{Ic} = 100$ ksi$\sqrt{\text{in.}}$
    $\sigma_{max} = 55$ ksi
    $\sigma_{min} = 40$ ksi
    $a_0 = 0.75$-in. edge crack in an infinite plate

(c) $K_{Ic} = 150$ ksi$\sqrt{\text{in.}}$
    $\sigma_{max} = 45$ ksi
    $\sigma_{min} = 40$ ksi
    $a_0 = 0.75$-in. edge crack in an infinite plate

(3) Compare your results using the $da/dN$ expression given in Chapter 9 for
martensitic steels.

(4) Why is there a difference in the two expressions?

| Data Point No. | $N$ ($\times 10^3$) | Measured Crack Length (in.) |
|---|---|---|
| 0 | 0 | 1.000 |
| 1 | 35.12 | 1.033 |
| 2 | 76.55 | 1.077 |
| 3 | 101.49 | 1.114 |
| 4 | 129.82 | 1.154 |
| 5 | 141.95 | 1.172 |
| 6 | 151.16 | 1.189 |
| 7 | 160.19 | 1.208 |
| 8 | 179.40 | 1.237 |
| 9 | 189.29 | 1.262 |
| 10 | 200.84 | 1.294 |
| 11 | 210.45 | 1.319 |
| 12 | 218.40 | 1.341 |
| 13 | 227.23 | 1.360 |
| 14 | 236.64 | 1.396 |
| 15 | 244.02 | 1.426 |
| 16 | 249.74 | 1.453 |
| 17 | 255.27 | 1.478 |
| 18 | 261.58 | 1.511 |
| 19 | 264.88 | 1.535 |
| 20 | 271.86 | 1.579 |
| 22 | 280.50 | 1.653 |
| 24 | 286.90 | 1.703 |
| 26 | 290.59 | 1.758 |
| 28 | 293.21 | 1.804 |
| 30 | 295.63 | 1.860 |
| 32 | 298.06 | 1.912 |
| 34 | 300.00 | 2.006 |

## Advanced Design Problem 7

Select a steel for use in high-pressure cylindrical containment vessels for the next generation of nuclear submarines.

Two steels are being considered for this application, HY-130 and HY-180, which are both martensitic steels. The material properties for these two steels are as follows:

|  | HY-130 | HY-180 |
|---|---|---|
| Yield strength, ksi | 130 | 180 |
| Tensile strength, ksi | 150 | 190 |
| $K_{Ic}$, ksi$\sqrt{\text{in.}}$ | 280 | 300 |
| $K_{Iscc}$ in sea water, ksi$\sqrt{\text{in.}}$ | 260 | 180 |
| Cost, $/lb* | 0.50 | 1.00 |

*Hypothetical costs.

Design parameters for the containment vessels are as follows:

(a) Internal pressure = 5000 psi.

(b) Internal diameter of cylindrical portion = 30 in.

(c) Overall length of each vessel is 20 ft. Ends are to be hemispherical.

(d) Welded fabrication will be used. Assume that weld metal properties are the same as base metal properties.

(e) The vessels must have a factor of safety of at least 2.0 against both yielding and fracture of a 5-in.-deep surface flaw. Assume that $a/2c$ is 0.4.

(f) The vessels will be cycled from 0 to full design pressure (5000 psi).

(g) Inspection is such that all flaws greater than 0.05 in. can be found during fabrication. In service, the vessels will be in the forward-flooding zone of the submarine and cannot be inspected although they can be protected by painting.

On the basis of *performance, weight,* and *cost,* recommend which steel you would use. Justify your answer.

### Advanced Design Problem 8

The accompanying material property data are for a pressure-vessel steel to be used in a nuclear reactor. Note that properties are given for both the "as-received" condition and the "after-an-irradiation-treatment equivalent" to that which the vessel will receive after ten years of service.

In addition, the following information will be useful:

$\sigma_{ys}$ = 60 ksi

$\sigma_{ult}$ = 90 ksi

$E = 30 \times 10^6$ psi

NDT = 60°F

After irradiation, the tensile properties are

$\sigma_{ys}$ = 90 ksi

$\sigma_{ult}$ = 120 ksi

$E = 32 \times 10^6$ psi

Service temperature of the vessel is 240°F, and wall thickness is 12 in.

Analyze the safety of the vessel for each of the following conditions:

(a) Initial operation during first year (before any irradiation damage) at 240°F. Initial stress due to pressure is 25 ksi, and there is no thermal gradient.

(b) Operation after ten years of service. (Other conditions are the same as in condition a.)

(c) During shutdown at 180°F, where the pressure stress is 15 ksi and the temperature difference through the wall is 25°F.

(d) Final stages of shutdown at 70°F where the stress due to pressure is 5 ksi and there is no thermal gradient.

If you can *quantify* the factor of safety for any of these four cases, do so. Comment specifically on the validity or reasonableness of your answers.

During shutdown, what special precautions would you recommend? Explain your answer.

**Toughness Values for Pressure-Vessel Steel**

| Temperature (°F) | CVN Impact Values (ft-lb) | | Lateral Expansion (mils) | |
|---|---|---|---|---|
| | Unirradiated | Irradiated | Unirradiated | Irradiated |
| 0 | 5 | 5 | 5 | 2 |
| 20 | 5 | 5 | 10 | 2 |
| 40 | 10 | 5 | 15 | 2 |
| 60 | 15 | 5 | 20 | 5 |
| 80 | 20 | 5 | 25 | 5 |
| 100 | 30 | 10 | 30 | 5 |
| 120 | 40 | 15 | 35 | 10 |
| 140 | 50 | 20 | 40 | 15 |
| 160 | 80 | 25 | 50 | 15 |
| 180 | 100 | 30 | 60 | 20 |
| 200 | 100 | 30 | 60 | 20 |
| 240 | 100 | 30 | 60 | 20 |
| 280 | 100 | 30 | 60 | 20 |

# Index

## A

AASHTO fatigue design curves, 438
AASHTO material toughness requirements, 220
Acceleration in fatigue-crack growth, 281
$a_{cr}$, critical flaw size, 15, 19, 192, 201, 502, 561, 573
Advanced design problems, 612
$a_{eff}$, 568
Allowable $K_{IR}$ values, 517
Alternate wet and dry conditions, effect of on corrosion fatigue, 414
Alternating stress, 225, 321
Aluminum, $K_{Ic}$ values, 89
Aluminum alloys
  corrosion-fatigue crack propagation, 406, 408
  fatigue crack propagation, 294
Analysis of failure—260-in diameter motor case, 202
Analysis of $P$-$\Delta$ records. 84
Applied potential, effects of on $K_{Iscc}$, 366
Arc-shaped specimen, 76, 77, 88
Arc welding, 425
Arrest fracture toughness, 95, 97, 132, 516
$a/2c$ ratio, 47
Austenitic stainless steels, fatigue crack propagation, 290

## B

Bend specimen, 68, 75, 82, 86
$\beta_{Ic}$, 113, 132, 142, 476, 480
Blunt notches, 165, 254

## A

Bolt-loaded $K_{Iscc}$ WOL specimen, 351, 354
Brittle:
  -ductile transition, 10, 132, 471
  fracture, 1, 8, 12
    analysis, 193
    surface, 121, 130, 131, 461
Burdekin and Dawes CTOD design curve, 559
Butt welds, 428

## C

Cantilever-beam $K_{Iscc}$ test specimen, 348
Cathodic potentials, 366
Charpy V-notch (CVN) impact specimens.
  *See* CVN
Chevron notch, 78
Circular crack, 45
Column instability, 21, 199
Compact tension specimen, 49, 69, 76, 81, 87, 374
  center of rotation, 550
Complex cyclic behavior, 234, 236
Composite behavior, effect of on fatigue-crack growth, 306
Composition, effect of on $K_{Iscc}$, 366
Conditional $K_{Ic}$ ($K_Q$), 86
Constant:
  displacement, 348, 352
  load, 348, 352
Constant-amplitude, 341, 403
  cyclic-load fluctuation, 225, 278
  fatigue-crack propagation, 278
Constrained flow, 121
Constraint, 119, 124